Dryland Horticulture

Dryland Horticulture

Edited by
M.K. Jatav
Principal Scientist
P.L. Saroj
Director
B.D. Sharma
Principal Scientist

ICAR-Central Institute for Arid Horticulture
Bikaner-334 006, Rajasthan

CRC Press
Taylor & Francis Group
Boca Raton London New York

CRC Press is an imprint of the
Taylor & Francis Group, an **informa** business

NEW INDIA PUBLISHING AGENCY
New Delhi – 110 034

First published 2022
by CRC Press
2 Park Square, Milton Park, Abingdon, Oxon, OX14 4RN

and by CRC Press
6000 Broken Sound Parkway NW, Suite 300, Boca Raton, FL 33487-2742

CRC Press is an imprint of Informa UK Limited

British Library Cataloguing-in-Publication Data
A catalogue record for this book is available from the British Library

Library of Congress Cataloging-in-Publication Data
A catalog record has been requested

ISBN: 978-1-032-15837-2 (hbk)
ISBN: 978-1-003-24590-2 (ebk)

DOI: 10.1201/9781003245902

Preface

Agriculture is a primitive science as well as art being practiced over the globe since unmemorable time not only for livelihood security of the people but supplying of raw materials to the industries. Agriculture sector is facing intolerable population pressure due to its exponential growth and additive growth in agriculture production. Consequently, it resulted continuous increase in inputs to achieve more and more yield due to which great ecological imbalance and diminishing return have been realized after green revolution. Moreover, sustainable livelihood of resource poor farmers is the top priority for the nation today. The productivity of various horticulture commodities are far below than the potential productivity in arid and semi arid regions of the country. Hence, sustained and steady growth in rural income is critical for positive impact on living standard of various stakeholders. Recently, our Hon'ble PM has proposed an agenda of doubling the farmers' income by 2022. In this context, the farm income of resource poor farmers in the arid regions could be improved substantially through advance approaches/technologies in the field of fruits and vegetables production and their post harvest management. Major constraints in fruit and vegetable production of arid ecosystems are; (i) low crop productivity, (ii) poor soil conditions and ground water resource, (iii) yield loss due to abiotic stresses mainly low and erratic rainfall (iv) less availability of quality planting material (v) inadequate crop diversification, (vi) post harvest losses (vii) poor adoption of recommended package of practices due to resource constraints *etc*. However, to double the income of farmers by 2022, in numerical terms would imply restructuring agriculture processes and policy interventions. The net income of farmers can surely be doubled well within the stipulated period by efficient management of water and agri-input, integrated horticulture based farming system, better market price realization and thereby improving productivity of arid horticultural crops. This includes (1) horticulture- a sunrise sector, (2) better resource use efficiency or saving in cost of production, (3) focused initiatives for additional income, by integrating off farm enterprises.

Horticulture has emerged as a core sector of agriculture over the past decade, growing steadily in annual area coverage and output. In the year 2017-18, horticulture stood at nearly 25 million ha and production was at a record 306.80 million tonnes of which 270 million tonnes was in fruits and vegetables alone.

By 2022, 6.5 lakh ha of additional area will be brought under horticultural crops through area expansion and 1.75 lakh ha of additional area will be brought under protected cultivation. Besides, horticulture is seen as optimal option for crop diversification in agriculture. The per unit earning capacity of farmers is much higher than in case of food grains and it also addresses environmental concerns. Nevertheless, most horticultural output requires specialized market linkages and connectivity, post harvest infrastructure and to make this sector more remunerative, there is a need to address the constraints involved. As reported one fourth of horticultural produce are being spoiled due to their perishability. Major thrust areas for horticulture development are cluster approach linked with post-harvest management and marketing, market intelligence to promote market led production, quality seeds and planting material, adoption of improved varieties and rejuvenation with improved cultivars, hi-tech horticulture and precision farming, high density plantations, *etc*. Diversification of present farming system with horticulture crops, availability of micro irrigation, resource use efficiency, increase in cropping density, better agronomic practices, incorporation of improved varieties and technologies, area expansion may be taken under consideration as sources of income growth in horticulture sector. Arid and semi-arid areas can be fruitfully utilized for commercial cultivation of horticultural crops because majority of horticultural crops are perennial in nature, widely spaced, low water requirement in comparison to field crops, deep and extensive root system capable of extracting water from deeper layers, large canopy to harvest optimum natural resources and high yielding. The information compiled in form of a book will address the major issues of farmers income through advance approaches and dissemination of technologies developed in arid and semi-arid regions, so that these can be translated on the farmer's field for optimum utilization of these scarce resources.

We sincerely thank all the contributors who have made it possible to bring out first edition of this book. Probably the book may not be as complete, thus, suggestions are welcomed for future editions. We deeply appreciate the help received from the scientists and staff of ICAR-CIAH, Bikaner. We place on record our sincere gratitude to "New India Publishing Agency" and their production staff who have strived hard for speedy publication of this book without compromising the quality of the work. Lastly, but not in the least we shall consider our efforts successful and gratifying if this book serves the purpose for which it is intended.

September, 2018 **Editors**
Bikaner, Rajasthan

Contents

Section 01: Production, Protection and Physiology of Arid Horticultural crops

Section 02: Economics, Management and Extension of Arid Horticultural Crops

Contributors

A.K. Singh
ICAR-Central Institute for Arid Horticulture
Central Horticultural Experiment Station
Godhra
Panchmahals, Gujarat-389 340

A.K. Verma
ICAR-Central Institute for Arid Horticulture
Bikaner-334 006
Rajasthan

Chet Ram
ICAR-Central Institute for Arid Horticulture
Bikaner-334 006
Rajasthan

Alemwati Pongener
ICAR-NRC on Litchi
Muzaffarpur-842 002, Bihar

Anita Meena
ICAR-Central Institute for Arid Horticulture
Bikaner-334 006
Rajasthan

Anshuman Singh
ICAR-Central Soil Salinity Research Institute, Karnal-132 001

B.L. Nagar
Agriculture University, Kota-324001
Rajasthan

B.D. Sharma
ICAR-Central Institute for Arid Horticulture
Bikaner-334 006
Rajasthan

B.M. Yadav
ICAR-Central Arid Zone Research Institute
Regional Research Station
Bikaner-334 004, Rajasthan

B.R. Choudhary
ICAR-Central Institute for Arid Horticulture
Bikaner-334 006, Rajasthan

Birbal
ICAR-Central Arid Zone Research Institute
Regional Research Station
Bikaner-334 004, Rajasthan

Chavlesh Kumar
ICAR-Indian Agricultural Research Institute
New Delhi-110 012

D.D. Nangare
ICAR-National Institute of Abiotic Stress Management
Baramati-413 115 Maharashtra

D.K. Sarolia
ICAR-Central Institute for Arid Horticulture
Bikaner-334 006
Rajasthan

D.K. Samadia
ICAR-Central Institute for Arid Horticulture
Bikaner-334 006, Rajasthan

D.S. Mishra
ICAR-Central Institute for Arid Horticulture
Central Horticultural Experiment Station
Godhra
Panchmahal -389 340, Gujarat

Dhurendra Singh
ICAR-Central Institute for Arid Horticulture
Bikaner-334 006, Rajasthan

G.L. Bagdi
ICAR-Central Arid Zone Research Institute
Regional Research Station
Bikaner-334 004, Rajasthan

Gopal Lal
ICAR-National Research Centre on Seed Spices, Tabiji, Ajmer -305 206, Rajasthan

Hare Krishna
ICAR-Indian Institute of Vegetable Science
Varanasi-221 305
Uttar Pradesh

J.S. Gora
ICAR-Central Institute for Arid Horticulture
Bikaner-334 006, Rajasthan

K.R. Sheetal
ICAR-Central Arid Zone Research Institute
Regional Research Station
Bikaner-334 004, Rajasthan

Kamlesh Kumar
ICAR-Central Institute for Arid Horticulture
Bikaner-334 006
Rajasthan

Kanak Lata
ICAR-Krishi Vigyan Kendra
Panchmahal, Vejalpur
Godhra 389 340, Gujarat

Lokesh Kumar
ICAR-Central Institute for Arid Horticulture
Bikaner-334 006
Rajasthan

M.K. Verma
ICAR-Indian Agricultural Research Institute
New Delhi-110 012

M.L. Soni
ICAR-Central Arid Zone Research Institute
Regional Research Station
Bikaner-334 004, Rajasthan

M.K. Jatav
ICAR-Central Institute for Arid Horticulture
Bikaner-334 006, Rajasthan

Mukesh K. Berwal
ICAR-Central Institute for Arid Horticulture
Bikaner-334 006
Rajasthan

N.D. Yadava
ICAR-Central Arid Zone Research Institute
Regional Research Station
Bikaner-334 004, Rajasthan

N.S. Nathawat
ICAR-Central Arid Zone Research Institute
Regional Research Station
Bikaner-334 004
Rajasthan

Narendra Pratap Singh
ICAR-National Institute of Abiotic Stress Management
Baramati- 413 115

N.N. Gaikwad
ICAR-National Research Centre on Pomegranate
Solapur-413 255

N.V. Singh
ICAR-National Research Centre on Pomegranate, Solapur-413 255

O.P. Awasthi
ICAR-Indian Agricultural Research Institute
New Delhi-110 012

O.P. Aishwath
ICAR-National Research Centre on Seed Spices, Tabiji, Ajmer -305 206
Rajasthan

P. L. Saroj
ICAR-Central Institute for Arid Horticulture
Bikaner-334 006, Rajasthan

P.P. Singh
ICAR-Central Institute for Arid Horticulture
Bikaner-334 006, Rajasthan

P.S. Renjith
ICAR-Central Arid Zone Research Institute
Regional Research Station
Bikaner-334 004, Rajasthan

Pankaj Kumar
MGS University
Bikaner-334004, Rajasthan

Prabodh Chander Sharma
ICAR-Central Soil Salinity Research Institute
Karnal-132 001, Haryana

R.K. Meena
ICAR-Central Institute for Arid Horticulture
Bikaner-334 006, Rajasthan

R.C. Balai
ICAR-Central Institute for Arid Horticulture
Bikaner-334 006, Rajasthan

R. Kumar
ICAR-Central Institute for Arid Horticulture
Bikaner-334 006, Rajasthan

R.S. Singh
ICAR-Central Institute for Arid Horticulture
Bikaner-334 006
Rajasthan

Radhakrishanan
ICAR-Central Arid Zone Research Institute
Regional Research Station
Bikaner-334 004, Rajasthan

Raj Kumar
ICAR-Krishi Vigyan Kendra
Panchmahal, Godhra 389 340, Gujarat

Rajesh Jalwania
KVK
Chittorgarh-312001, Rajasthan

Ramesh Kumar
ICAR-Central Institute for Arid Horticulture
Bikaner-334 006, Rajasthan

Ramkesh Meena
ICAR-Central Institute for Arid Horticulture
Bikaner-334 006, Rajasthan

R.K. Pal
ICAR-National Research Centre on
Pomegranate, Solapur-413 255

S.K. Maheshwari
ICAR-Central Institute for Arid Horticulture
Bikaner-334 006, Rajasthan

S.R. Meena
ICAR-Central Institute for Arid Horticulture
Bikaner-334 006, Rajasthan

S.V.R. Reddy
ICAR-Central Institute for Arid Horticulture
Bikaner-334 006, Rajasthan

Sanjay Kumar Singh
ICAR-NRC on Litchi
Muzaffarpur-842 002, Bihar

Sanjay Singh
ICAR-Central Institute for Arid Horticulture
Central Horticultural Experiment Station
Godhra
Panchmahals, Gujarat-389 340

Shakti Khajuria
ICAR-Krishi Vigyan Kendra
Panchmahal Godhra 389 340
Gujarat

S.M. Haldhar
ICAR-Central Institute for Arid Horticulture
Bikaner-334 006
Rajasthan

Sunil Kumar
ICAR-Indian Agricultural Research Institute
New Delhi-110 012

Tanushree Sahoo
ICAR-Indian Agricultural Research Institute
New Delhi-110 012

V.S. Rathore
ICAR-Central Arid Zone Research Institute
Regional Research Station Bikaner-334 004
Rajasthan

V. Subbulakshmi
ICAR-Central Arid Zone Research Institute
Regional Research Station
Bikaner-334 004, Rajasthan

V. V. Appa Rao
ICAR-Central Institute for Arid Horticulture
Central Horticultural Experiment Station
Godhra
Panchmahals, Gujarat-389 340

Vikas Yadav
ICAR-Central Institute for Arid Horticulture
Central Horticultural Experiment Station
Godhra, Panchmahals
Gujarat-389 340

Vishal Nath
ICAR-NRC on Litchi
Muzaffarpur-842 002
Bihar, India

Yogeshwar Singh
ICAR-National Institute of Abiotic Stress Management
Baramati-413 115
Maharashtra

Section 01: Production, Protection and Physiology of Arid Horticultural Crops

1

Approaches for Improving Productivity and Profitability in Arid Horticulture

P.L. Saroj and M.K. Jatav

Introduction

In Indian agriculture, horticulture sector has made considerable progress after independence and as a result, horticulture has emerged as one of the major drivers of growth in the agricultural and allied sector. The horticulture sector includes large number of crops *viz*., fruits, vegetables, ornamental crops, spices, tuber crops, plantation crops, medicinal and aromatic crops, mushroom and their post harvest management. On the production and productivity front, the horticulture sector outperformed conventional food crops. Between 2004-05 and 2014-15, horticultural output achieved an annual growth of about 7 per cent as compared to around 3 per cent growth in food grain production. This increase in production has come from an increase in acreage and even larger increase in productivity. Over the last decade, the area under horticulture grew by about 3% per annum and annual production increased by 5.4%. Thereby, India has maintained its second rank in the global production of fruits and vegetables, next only to China. India's fruit productivity is better than that of China, though China is the largest fruit producing country.

The country is currently producing 295.2 million tonnes of horticulture produce from an area of 24.9 million hectares, which has surpassed the estimated food grain production of 275.68 million tonnes (2016-17). Though the production of food grains and horticultural produce are not meaningfully comparable due to fundamental differences in resource requirements, their farming practices, characteristics of produce and most importantly, their nutritional and medicinal properties as well as energy value. The special attention was given to the this sector, especially after the introduction of the Horticulture Mission for North East and Himalayan States (HMNEH) and the National Horticulture Mission

(NHM) during 11th Plan period, owing to bumper fruit production. Given the increasing pressure on land, growth strategies have been focusing on raising productivity through high density plantations, protected cultivation, micro irrigation, quality planting material, rejuvenation of senile orchards and an emphasis on post harvest management and marketing of produce for better price realization. A comparative picture of horticulture and food grain production over the last few years is given in figure 1. The trend in horticultural production growth over the last 10 years is depicted in figure 2 and the share of different commodities in horticulture production is given in figure 3.

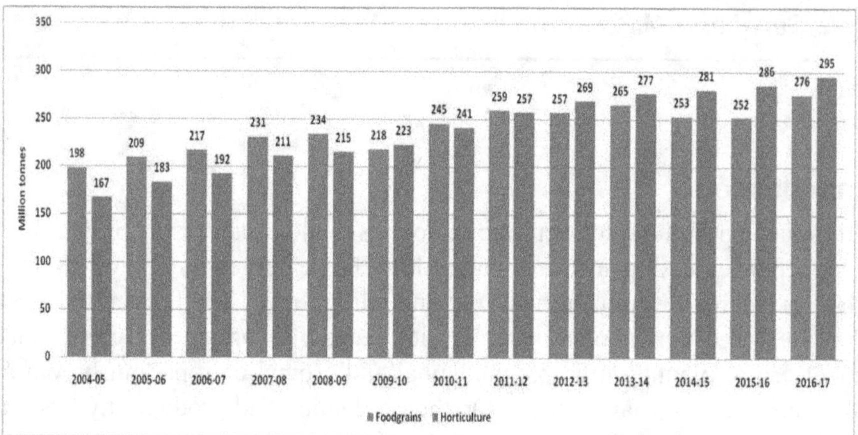

Fig. 1: Trend in food grain and horticulture production

A large variety of fruits, such as banana, mango, citrus, papaya, guava, grape, sapota, pomegranate, pineapple, aonla, litchi, pear, plum and walnut are grown in India. India accounts for about 13 per cent of the total world production of fruits and leads in the production of mango, banana, papaya, sapota, pomegranate, acid lime and aonla. Special efforts are being made to improve the productivity of fruit crops by enhancing the supply of quality planting materials from accredited nurseries and improved package of practices. The per capita availability of fruit to the Indian population is 189 g/person/day and has been helping in supplementing nourishment. The vegetable production has also increased significantly and India continues to be the second largest producer (next to China) of vegetables in the world and occupies first position in the production of cauliflower, second in onion and third in cabbage in the world. In the past three decade, India made a quantum jump in vegetable production, but below the requirement for the balanced diet of @ 300 g/day/person. The major vegetable crops grown in the country are potato, tomato, onion, brinjal, cabbage, cauliflower, peas, okra, chillies, beans, melons, *etc*. Due emphasis is being given for production of vegetables under protected cultivation under Mission for Integrated development of Horticulture (MIDH). India is a leader in the production of vegetables like peas and okra. Besides, India occupies the second position in terms of production of brinjal, cabbage, cauliflower and onion and the third position in the production potato and tomato in the world.

A wide variety of spices, like black pepper, chillies, ginger, turmeric, garlic, cardamom and a variety of tree and seed spices are being produced in India and India is the largest producer and exporter of spices in the world. The major spice producing states are Gujarat (18%), Andhra Pradesh (14%), Rajasthan (11%), Madhya Pradesh (8%) and Karnataka (6%). Spice production in India is currently estimated at 5.7 million tonnes from an area of about 3.2 million hectares. The production of spices in the country has registered a substantial increase over the last decade. Chilli is the major spice crop, accounting for about 30 per cent of total spice production in the country. Garlic accounts for a 23 per cent share in production, while turmeric accounts for a 15 per cent share in production. Besides these, various plantation crops like coconut, areca nut, cashew nut, oil palm *etc*; tuber crops like sweet potato, colocasia, yam, tapioca etc; medicinal and aromatic plants like kalmegh, safed musali, asparagus, tulasi, stevia, palmarosa, mentha *etc*; ornamental crops like rose, gladiolus, marigold gervera, jusmine, hibiscus, bougainvillea *etc*; musrooms are being grown in the country on large scale with different nutrional, nutraceutical, socio-economic and ecological dimentions.

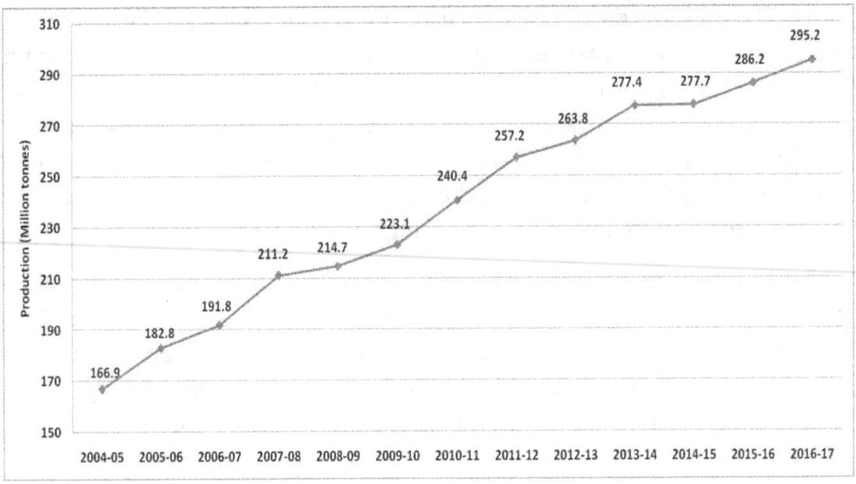

Fig. 2: Growth trend in horticulture production

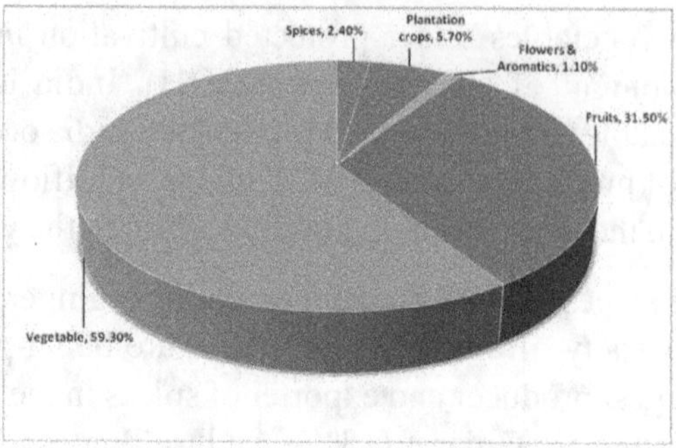

Fig. 3: Share of different commodity groups in horticuture production

Arid Horticulture

In India, arid region is spread over 39.54 million ha including 31.71 million ha as hot arid region mainly in the states of Rajasthan, Gujarat, Andhra Pradesh, Punjab and Haryana and 7.83 million ha as cold arid region in the states of Jammu and Kashmir and Himachal Pradesh (Saroj and Dhandar, 2004). Out of total hot arid region of 31.71 million ha, 49.61% area is alone in the state of Rajasthan. The hot arid region having 20 million population with population density of 61 persons per square km making up a population of nearly people. The Indian arid zone is characterized by high temperature and low and variable precipitation which limit the scope for high horticultural productivity. However, these conditions greatly favour development of high quality production in number of fruits such as date palm, ber, pomegranate, citrus, aonla, bael, grapes, guava and in vegetables such as cucurbitaceous crops, spices and some medicinal plants. The optimized technologies and inputs could increase the existing low productivity. It is now realized that there is a limited scope for quantum jump in fruit and vegetable production in the traditional production areas. The recent aware-ness regarding the potential of these ecologically fragile lands for production of quality horticultural produce has not only opened up scope for providing economic sustenance for the people of this region, but also for bringing new areas to increase production through horticulture.

The soils of arid region are very poor in fertility. The proper soil depth, hard layer/pan are also a problem in some parts for better plantations. The soils of the north-western arid region described as 'desert soils' and 'grey brown soils' of the Order Aridisols are light textured. The water holding capacity of soil is also poor. Soils are generally rich in total potassium and boron but are low in nitrogen, phosphorus and micronutrients such as copper, zinc and iron. The soils often have high salinity. The ground water resource is not only limited owing to poor surface and sub-surface drainage but is also saline in quality. The irrigation water resources in the region are seasonal rivers and rivulets, surface wells and some runoff water storage devices (*e.g.,nadi, tanka, khadins*) and canal irrigation in arid region. Thus, the water resources in arid region are limited and can irrigate hardly 4% of the area.

The annual average rainfall in arid region is very low and varies from 100 mm in north-western sector of Jaisalmer to 450 mm in the eastern boundary or arid zone of Rajasthan. Most of the precipitation in north western arid region occurs during July-September in about 19-21 rain spells. Due to low and erratic rainfall pattern in arid region, appropriate technology is needed to increase productivity. Water is precious input in hot arid region of the country therefore, adoption of micro-irrigation system is desirable to save water and enhance productivity. For arid environment, the variety is needed which are resistant to biotic and abiotic stresses for sustainable production. In western parts of arid region, occurrence of frost is also common features during winter season which affects vegetative growth of plants as well as productivity and quality of fruits especially in ber, lasoda and aonla.

One of the major problems in development of horticulture scenario in arid region is lack of sufficient quality seed and planting materials. Seed is a precious input to increase quality production of vegetables, spices and flowers as well as some fruit crops. Thus, production of quality seed material and their supply to farmers will boost up the arid horticultural crops production. The post harvest management is essential to overcome the losses at different stages of grading, packing, storage, transport and finally marketing of both fresh and processed products. The weak processing infrastructure, as it exists today, has been one of the limiting factors for ineffective utilization of the raw materials resulting in huge post harvest losses. Lack of proper storage facility and knowledge and equipment for grading and packaging of fruits and vegetables is also constraint for the growers of hot arid region.

Marketing of horticultural produce is a major constraint in the production and disposal system and has a major role to play in making the industry viable. The high capital cost involved in establishing orchards, or rejuvenation of existing old unproductive plantation poses serious constraint in area expansion. The situation becomes further difficult in view of the large number of small holdings devoted to these crops which are essentially owned by weaker section, who have limited resource to invest, also cannot afford burden of credit, even if available. High cost of inputs and lack of enough incentives for production of quality varieties /species, product diversification, value addition, etc. also hinder crops development.

As far as vegetation is concerned, the arid horticultural crops have developed and/or modified their organs to perform certain vital physiological functions such as strong deep root system (ber, bael, aonla, wood apple, jamun, *etc.*), synchronize their flowering and fruit development with the season of moisture availability (ker, lasora, aonla, pilu, *etc.*) and other xerophytic characters *i.e.* leaf shedding in summer (ber), scanty foliage (ker), spiny cladode (cactus pear), mucilaginous sap in plant part (ker, gonda, pilu, bael, *etc.*), sunken stomata and fur/ hairiness and waxy coating on the leaf surface (phalsa, ber, lasoda, fig, etc.) thorny nature, and selective or reduced absorption of cation (Na^+) and anions (Cl^-, $SO4^-$) for survival under adverse arid conditions. Therefore, cultivation of suitable species of fruit and vegetables in the arid areas will increase the sustenance of the inhabitants and provide the alternate sources of income through development of nurseries, small scale industries and exports.

Extent of Area and Production

The vast land resources offer great opportunity for horti-cultural development in the Indian arid zone which is spread over nearly 31.71 million ha area of which 41.5 per cent is arable and 19 per cent is culturable wastelands (Table 1). In arid region, considerable area has come up under fruits like aonla, ber, pomegranate, tamarind, annona, fig and phalsa in different parts of the country. Ber has spread from northern states to the western and southern India from a mere 12000 ha in 1978 to nearly 86,000 ha in 2004-2005 with a production of about 0.90 million tonnes. Similarly, the area under pomegranate has also leaped to over 1.25 lakh ha. Likewise aonla, presently cultivated on 50,000 ha with the production of 2,50,000 tonnes (Table 2). This has become possible as a result of the research and developmental efforts. Besides, All India Coordinated Research Project on Arid Zone Fruits initiated since the IVth Five Year Plan lead to the development of new varieties and agro-techniques for fruit crops based on ecological regions. Under-utilized horticultural crops are another area which needs attention. There is need to develop large orchards of under-utilized fruits like jamun, mahua, chironji, khirni, wood apple, fig, karonda, lasoda, manila tamarind *etc.* for commercial exploitation both for internal consumption and export as these fruits are of high nutritive value and are rich in anti-oxidants for health conscious populace of modern world. Majority of Indian vegetable farmers are small scale cultivators and in arid region, it confined to irrigated resource situation. Some traditional vegetables like guar, moth mateera, kachari, tinda, snap melon *etc.* are grown under rainfed conditions also. The vegetable seed industry is very much developed in the past decade, but the focus is primarily on a few vegetables like onion, tomato, chilli, bell pepper and okra. Public sector should come forward for diversification to specialized indigenous vegetables in each state for domestic market as well as for export.

Table 1: Area under arid region in India

States	Total area (m ha)	Arable area (m ha)	Culturable wasteland (m ha)
Rajasthan	19.61	9.20	4.20
Gujarat	6.20	0.02	1.66
Punjab	1.50	1.05	0.03
Haryana	1.30	1.15	0.02
Peninsular India	3.10	1.63	9.11
Total	31.71	13.15	6.03

Table 2: Area and production of various arid horticultural crops.

Crops	1993-94		2016-17		2030	
	Area (ha)	Production (MT)	Area (ha)	Production (MT)	Area (ha)	Production (MT)
Ber	41,256	3,30,048	85,500	8,95,000	1,10,000	12,10,000
Pomegranate	4,500	45,000	2,02,050	26,68,950	3,50,000	50,00,000
Aonla	26,000	2,86,000	75,000	8,25,000	1,00,000	13,00,000
Date palm	5,000	41,000	25,000	2,00,000	40,000	3,75,000
Sapota	49,000	6,44,000	80,000	9,13,000	1,10,000	12,80,000
Tamarind	10,500	52,500	15,280	82,300	20,000	1,20,000
Others (Custard apple, Fig & Phalsa)	6,600	30,450	15,700	1,31,700	35,000	3,50,000
Total	1,42,856	14,28,998	4,98,530	57,15,950	7,65,000	96,35,000

Prospects of Arid Horticulture

There is limited possibility of horticulture expansion on good lands and assured irrigation areas but there is a tremendous scope in expansion of horticultural crops in arid and semi- arid regions. Vast land resource, surplus family labours, increasing canal irrigated area, developing infrastructural facilities, plenty of solar and wind energy, *etc.* are the strength in arid region for research and development of arid horticulture. Further, minimum pressure of diseases and insects in the region is another dimension for production of different horticultural produce. Ber is commercially grown in about 85,000 ha. with production of 0.89 million tonnes under semi arid and arid regions of the country. It requires more attention for value addition otherwise it would not lure to the peoples. Pomegranate area (0.202 million ha) and production (2.66 million tonnes) is increasing very fast in dry part of the country, since, it has vast scope of export from semi arid and arid regions of the country. At present, pomegranate is being exported of worth Rs.92 million. The crops like aonla, custard apple, tamarind are also coming very well under dry land conditions and having potential for export because of its demand in the International market. Similarly, fig is cultivated in more than 3,000 ha area in Maharashtra, Karnataka. there is good demand for dehydrated fig. Likewise, custard apple is grown in the state of Maharashtra, Andhra Pradesh, Karnataka, Rajasthan and Tamil Nadu. In foot hills of Aravali, custard apple is growing naturally and its potential should be exploited as few plantations have been established well and fruit production have started in Chittorgarh district. Aonla is a medicinal fruit plant and cultivated in more than 75,000 ha area and produced 0.82 million tonnes fruits/year. Date palm is most suitable fruit tree of hot arid region and it is grown in Rajasthan, Gujarat, Punjab and Haryana in more than 25, 000 ha and producing 0.2 million tonnes fresh fruit. However, date is imported to India from Gulf countries due to its meager production. Now, imported tissue culture plants have been planted in districts of western Rajasthan and Gujarat. These plantations are under vegetative growth stage. Bael is also an important fruit crop of semi arid and arid region. Earlier there were no systematic orchards of bael. Now, looking to its nutritional and medicinal value, attention is being given on its commercial production.

With increasing irrigation facilities, vegetable cultivation is also increasing in arid and semi-arid regions of the country. However, there is a need for development of heat, salinity and frost tolerant varieties of different vegetables. Also, improvement in production technologies, many seed spices likes coriander, cumin, fenugreek, ajwain, fennel, dill and nigella, are being cultivated on large scale and also exported to earn foreign exchange. There is a vast potential of floriculture in some parts of Rajasthan because of low infection of disease/insects good market demand for decorating and other uses. The prospect of floriculture under hot arid condition is also important with the increasing demand of cut flowers. At present, Roses, Marigold, Chrysanthemum and other flowers are being cultivated nearby cities of Udaipur, Ajmer, Jaipur, Kota and Sri Ganganagar districts in Rajasthan. There is a good scope of Tourism development with development in gardening, landscaping and floriculture in the state besides associated industries.

The region receives high solar radiation. Multistory cropping system models based on fruit trees can be developed to uti-lize this valuable resource for production of quality fruits, vegetables and medicinal and aromatic plants. In high radiation zone, which also has low humidity, the attack of insect pests and diseases are very low. However, conditions of high radiation and temperature coupled with low humidity cause water stress in plants owing great water losses from soil and plant surface but by following proper soil and water management techniques, optimum produc-tivity can be maintained. The high summation of heat units especially during the long summer is a valuable resource for development of high total soluble solids in fruit crops such as date palm, pomegranate, ber, grapes *etc*. Besides, the conditions of high temperature and low humidity help in solar drying of fruits and vegetables. The practice is already common for drying pods of khejri *(sangri)* and fruits of kachri *etc*. The sharp fluctuations in day and night temperatures during autumn, spring and summer help in development of sweetness in sweet orange, kinnow, ber and date palm, flesh colour and sweetness in pomegranate arils and mateera pulp.

The use of chemicals in horticulture is much less in arid region than intensive cultivation of various horticultural crops under assured irrigated conditions of humid region. Also, problem of biotic pressure is less in arid region. The farming in arid region largely supported by animals and the ratio of man and animal population is about 1:1. This is the biggest advantage for adoption of organic farming in arid region. The farmers can fetch better income through organic farming of horticultural crops in the arid and semi-arid regions.

Technological Advancement

Adoption of Potential Varieties

While selection of fruit crops for arid and semi-arid regions, one of the basic requirements is that those crops who complete their vegetative growth and reproductive phase during the period of maximum moisture availability should be given preference. The fruit such as ber, guava, pomegranate, custard apple, aonla and sour lime, conform to this prerequisite. Among vegetable crops, those who can survive in these regions belong to the cucurbitaceous and solanaceous groups. Cluster bean and moth bean are also very drought hardy. In these crops also, there is need to select the most drought hardy cultivars for rainfed production. Several varieties were also identified in fruits, vegetables, flowers and medicinal and aromatic plants through researches conducted in different parts of the country. The important fruit and vegetable crops and cultivars which can be grown successfully in hot arid region are presented in table 3.

Table 3: Promising crops and their varieties.

Crops	Cultivar
Fruits	
Aonla	Kanchan, Krishna, Balwant, NA-6, NA-7, NA-10, Goma Aiswrya
Bael	NB-5, NB-9, Pant Aparna, Pant Suwarna, Goma Yashi
Ber	Gola, Mundia, Kaithali, Banarasi Karaka, Umran, Goma Kirti
Custard apple	Balanagar, Mammoth, Island Gem, Arka Sahan
Date palm	Halawy, Barhee, Medzool, Khuneji
Fig	Poona Fig, Conadria, Diana, Excel, Black Ischia
Guava	Sardar Guava, Allahabad Safeda, Kohir Safed, Safed Jam, Chittidar, Arka Amulya

Pomegranate	G-137, Mridula, Bhagawa, Phule Arakta, Ruby, Jalore Seedless
Sapota	Kalipatti, Cricket Ball
Tamarind	PKM -1, Pratisthan, Yogeshwari, Goma Prateek
Sweet orange	Malta Blood Red, Mosambi, Valencia
Minor Fruits	Lasoda-Thar Bold, Karonda-Thar Kamal, Khirni-Thar Rituraj, Chirounji-Thar Priya
Vegetables	
Cluster bean	Pusa Navbahar, Durga Bahar, HBG 4, ARG 80, Thar Bhadevi
Cowpea	Pusa Dofasali, Pusa Phalguni, Pusa Rituraj, Pusa Barsati
Chilli	Pusa Jwala, Mathania
Bottle gourd	Pusa Summer Prolific Round, Pusa Summer Prolific Long, Pusa Meghdoot, Pusa Manjari, Thar Samridhi
Mateera	Thar Manak, AHW-19, AHW-65
Ridge gourd	Thar Karni
Sponge gourd	Thar Tapish
Kachri	AHK-119, AHK-200

The ICAR- Central Institute for Arid Horticulture, Bikaner (Rajasthan) has released 32 varieties of different arid fruits and vegetable crops which includes, Thar Bhubharaj, Thar Sevika, Goma Kirti, Thar Malti of ber; Thar Shobha of khejri; Goma Aishwarya of aonla; Thar Bold of lasoda; Thar Samridhi of bottle guard; AHW-19, AHW-65, Thar Manak of mateera; AHK-119, AHK-200 of Kachari; AHS-10, AHS-82 of snap melon; AHC-2, AHC-13 of kakri and Goma Manjri the bhadevi of cluster bean. At Institute level, Goma khatta, a pomegranate genotype for anardana purpose has released. Goma Pratik of Tamarind and Goma Yashi of bael and Goma Priyanka of Jamun has released by the institute for cultivation in semi-arid parts of the country. Thar Mahi of Sward bean; Thar Kartiki and Thar Maghi of Indian bean have also been released by the Institute. A promising line of chironji CHES C-7 (Thar Priya) has been identified during 2013 for cultivation in semi-arid region. Further, a number of varieties of arid fruit and vegetable crops have recommended for cultivation in different parts of the country after evaluation of germplasm for high yield and quality. Varietal variation in endurance to drought has also been observed in horticultural crops. Early ripening cultivars seem to escape stress conditions caused by the receding soil moisture stored in the soil profile during the monsoon. Ber cultivars Gola, Seb and Mundia for extremely dry areas, Banarasi Kadaka, Kaithli, Umran and Maharwali for dry regions, and Sanaur-2, Umran and Mehrun for comparatively humid regions have been recommended. Apart from morphological parameters, plants should also have physiological parameters for endurance to drought for commercial cultivation in this region. Some physiological parameters identified in ber are no-mid-day depression in photosynthetic rate, low rate of transpiration, maintenance of leaf water balance, growth, canopy development, dry matter allocation, high water use efficiency, *etc.* It has been demonstrated that plant having capacity for drought endurance are able to maintain turgor, dry matter allocation, leaf and fruit growth even under low soil moisture level.

Use of Quality Planting Materials

The quality planting materials must be available to the doorsteps of the growers at reasonable price. The planting materials must be true-to-type, propagated by appropriate technique, healthy and disease free and as per standard. The appropriate technique of propagation is very important for mass multiplication of various fruit plants. Techniques for propagation of ber, pomegranate, fig, custard apple, date palm and other fruit crops have been standardized and details are presented in table 4.

Table 4: Standardized methods of propagation

Name of fruits	Methods of propagation	
	Standardization	Commercial
Aonla	Budding (Patch, Shield)	Patch budding
Bael	Root Cutting, Budding, (patch and shield), Air Layering	Patch budding
Ber	Cutting, Air Layering, Budding, (T, I, ring and forkert)	Patch budding
Custard apple	Soft wood grafting	T-budding, Inarching,
Date palm	Seed, Sucker	Sucker and Tissue culture
Fig	Cutting, Budding and Air Layering	Hard wood cuttings
Grape	Hard wood stem cutting, chip budding, layering and micro propagation	Hard wood stem cuttings
Guava	Cutting, Air Layering, Budding, Stooling and Soft Wood Grafting	layering, Tissue culture grafting
Jamun	Seed and Budding	Shield and Patch budding
Karonda	Seeds, Soft Wood cuttings in mist	Seeds, Air layering
Phalsa	Seeds and Cutting	Seeds
Pomegranate	Cutting, Budding (chip, patch and forkert),Air Layering, Tissue Culture	Hard-wood cuttings and Air layering, Tissue culture

For propagation of ber, seedling rootstocks are raised in the nursery in 300 gauge polythene tubes (25 cm length and 10 cm diameter, open at both ends), filled with a mixture of farm yard manure (FYM), sand, and clay in 1:1:1 ratio. The seedlings can be budded when about 90-100 days old. These plants become ready for transplanting, 1-2 months after budding. This technique helps to retain the straight growth of the tap root as the tubes are open at the bottom. Thus, the tubes neither restrict root growth nor induce coiling. Budded plants raised by this technique are also suitable for transportation to distant place (Pareek, 1978). Similarly, propagation of aonla by raising seedlings in poly-containers and patch budding just after 5-6 months with >90 per cent success was standardized by Saroj *et al.* (2000). Planting of stem cuttings of pomegranate, phalsa, fig and mulberry in such polythene tubes would also induce straight roots. For various biotic and abiotic stresses, suitable rootstocks in different fruit crops have also been standardized to achieve the optimum productivity. The details for each crop are given in table 5.

Table 5: Rootstocks of arid fruit crops

Fruit crops	Rootstocks	Purpose
Ber	*Zizyphus nummularia, Zizyphus rotundifolia*	Dwarfingvigorous
Fig	*Ficus glomerata*	Nematode resistant
Guava	Inter-specific rootstock (*P. molle* x *P. guajava*) *Psidium pumilum, Psidium mole*	Wilt resistant semi-dwarfing
Jamun	*Syzigium fruticosum*	Termite resistant
Grapes	Dogridge, Salt Creek, St. George	Tolerant to moisture stress and salinity

Looking to vast potential and demand of quality planting materials, nursery could be a very good enterprise to provide opportunity of employment in rural areas. There is a lot of scope of nursery raising in arid region for production of quality plating materials. Due to low pests/disease infestation, different kinds of plants can be multiplied in the nursery. Plants multiplied through vegetative methods are true to type and produce fruits early in comparison to orchards developed by seedling plantations. The fruit plants propagated in the nursery are generally used to establish orchards. Such plants invariably lose their tap roots as a result of repeated transplanting. Therefore, proper nursery management and timely disposal of quality planting materials with appropriate plant standard are important considerations. Proper establishment of mother block, raising of rootstocks, use of poly-containers, rooting media and after care are also important considerations (Saroj, 2004).

Orchard Establishment

Orchard establishment in arid region needs special consideration as the soil conditions are very poor, water availability is limited, extremes of temperature and high wind velocity. Therefore, proper land preparation, proper pit digging and appropriate filling mixture, establishment of wind breaks, use of hardy rootstocks, *in-situ* system of orchard establishment *etc.* are very important.

Plants raised in containers, if not sold timely, develop coiled roots. For success in dry lands, plants must have root architecture with a strong tendency to penetrate deep into the soil. *In situ* technique of orchard establishment is found suitable under arid conditions (Vishalnath *et al.*, 2000). The pit size 2 x 2 x 2 ft and filling mixtures (FYM + pond silt + soil in 1:1:1 ratio) has standardized for planting of ber and pomegranate in arid region.

The plant density mainly depend upon the plant type, soil fertility status and management practices, while, planting system to be adopted in arid region depends largely upon the topography of the land, fruit species and soil type. In the plains, planting, is generally done is square or rectangular system. On sloppy lands, fruit trees are planted on contour terraces, half moon terraces, trenches and bunds, and micro-catchments. On marshy and wet areas mounding and ridge-ditch method of planting have been suggested. The trenches and bunds made across the slope are staggered (Saroj *et al.*, 1994). In a micro-catchment, which may be triangular or rectangular, trees are planted at the lowest point where runoff accumulates. The planting distance 6 x 6 m or 8 x 8 m for ber cultivation is optimum while for date palm, bael and aonla; it should be 8 x 8m or 10 x 10 m under arid region.

Water Management

Infact, major constraint in commercial cultivation of arid horticultural crops is water resources under hot arid region. Hence, the need of the hour is to develop technologies, which not only requires low water input but also have high water use efficiency. Water being a rare commodity in arid eco-system, the first and foremost requirement is to conserve the available soil or rain water. Also, there is a need to commercially exploit the existing stress hardy genetic resources for development of horticulture in this region. Studies have shown that following strategies could be adopted to enhance the water productivity in arid region.

- Use of water conservation techniques such as mulching, pitcher irrigation, double walled pot, use of soil amendments, water absorbing polymer, *etc.*
- Use of pressurized irrigation, tank irrigation, *ex-situ* and *in-situ* water harvesting methods can be used.
- Application of water at critical times in relation to growth and development stages of fruit and vegetable crops and varying agro climatic conditions.

For conservation of rain water both *in situ* and *ex-situ* technologies have been developed. It has been reported that micro-catchment slopes greater than 5 per cent did not significantly affect run off at Jodhpur and that the highest ber yield was obtained when 0.5 per cent and 5 per cent slopes had 8.5m and 7m length of run, and 72 m^2 and 54 m^2 catchment area per tree, respectively. Mulching with organic materials (e.g., hay, straw, dry leaves, and local weeds) has been found highly beneficial in reducing evaporation loss. The practice also suppresses weed growth, prevents erosion, and adds organic matter to the soil. Black polythene mulch is very effective in ber orchards in western India, although, local organic mulch materials are cheaper than polythene mulches but these require proper care to maintain effective cover thickness. Leaf mulch has been used to conserve soil moisture in sapota orchards in Karnataka, Tamil Nadu, and Andhra Pradesh. At ICAR-CIAH, Bikaner, the work on *in-situ* water harvesting has been undertaken in Pomegranate, aonla and vegetable. It has been demonstrated that application of black polythene mulch and local weeds helps in conserving soil moisture status. It has been demonstrated that plant growth and development remains optimum with use of above mulching materials. Mulching studies with respects to soil hydro thermal regimes in brinjal revealed that organic mulches curtailed soil temperature during warm months, while an increase was recorded during the winter month. Significant increase in fruit yield by 66 and 58% could be obtained through lasoda *(Cordia* sp.*)* and kheep *(Leptodenia pyrotechnica*) mulching (Awasthi *et al.,* 2006).

Among the *ex-situ* water conservation methods, in arid ecosystem, emphasis has been given mostly on pressurized irrigation system. It has been demonstrated that fruits and vegetables can be grown economically by use of drip or sprinkler irrigation system. At CIAH, Bikaner and its regional station it has been demonstrated that crops such as pomegranate and ber can be grown successfully under drip irrigation system with the water saving of 25 per cent as compared to conventional flooding or bubbler system of irrigation. The use of drip alone or in combination with mulching has been demonstrated as a successful technology for cultivation of pomegranate at Anantapur (Anon., 2006). The studies have shown that highest number of 'B' grade pomegranate can be harvested under drip + mulch. Application of pitcher irrigation was attempted in cactus pear at CIAH, Bikaner and it was recorded that growth of cactus pear was better under this treatment as compared to control. The use of double ring system to conserve the moisture applied for production of fruit crops was attempted in aonla. It was observed that by this method the water is applied in zone having functional roots and hence, water use efficiency is enhanced.

Water loss due to transpiration can be reduced by use of radiation reflectants, stomata closing chemicals, and plastic films. Spraying of 4-6 per cent Kaolin, 0.5-1.0 per cent liquid paraffin, and 1.5 per cent power oil, after occasional rains in low rainfall areas, considerably reduce plant water losses (Pareek and Sharma, 1991). Chemicals such as Decinyl succinic acid (DSA), Abscisic acid (ABA), and Cetylalcohal cause stomata clousure and thereby reduce transpiration. Anti transpirants like 10% Kaolin has been found effective to check moisture losses from plant surface resulting in increased fruit production in pomegranate. Shelterbelt and windbreak plantations can reduce evapo-transpiration by reducing the wind speed and stabilizing microclimate.

Rainfed Region

Most part of the vast arid region depends on rain water from the little rainfall. Horticulture production under these conditions is only possible through proper management of rain water and utilization hardy plant species. Watershed manage-ment has been used by mankind since antiquity. For this, special techniques have been used for growing trees and crops in desert areas. The *khadins* or *kharins* in Jaisalmer which are low laying basins collecting run-off from sur-rounding rocky catchments for cultivation are believed to be about 300 500 years old. In the year 1908, 377 *khadins* have been reported to exist around Jaisalmer. These, however, remained abandoned for about 100 years, but are now being revived again. Even under the extremely arid agro climate of Jaisalmer, fruit trees like mango, Citrus, *Cordia, etc.* can be seen growing in the *khadins*. The *in situ* run-off concentration technique is most suited particularly for the fruit crops because: i) being deep rooted they are able to utilize the moisture stored in lower soil layers and can produce optimum crops even in low rainfall year, and ii) once estab-lished they continue to produce for many years. The watersheds are divided into micro catchments considered optimum either for a tree or a group of trees, arranged in a row or other patterns suited to the location. In nearly flat areas, catchments could be provided on two sides of a tree row or around each tree. The optimum catchments are worked out considering water requirement and root architecture of the fruit species, on one hand, and the expected water input through rainfall and run-off, on the other (Table 6). Study led to estimate the catchments area required per tree and tree per hectare of different fruits for growing in sandy catchments. Using the formula, A= TS/RC, where A is catchments area per tree in m^2, T is canopy are of the tree in m^2, S is run-off supplement required in mm per tree to meet the rainfall deficit, R is average annual rainfall in mm and C is run-off coefficient of the soil.

Table 6: Catchments area per tree and plant population per hectare with different slopes of sandy catchments

Fruit	Canopy area (m^2)	Runoff supplement (mm)	Catchment's Slope (%)	Catchment's area (m^2)	Area per tree (m^2)	Population per ha
Ber	36	300	0.5	100	136	74
			5.0	75	111	90
			10.0	60	96	104
Pomegranate	20	800	0.5	148	168	60
			5.0	111	131	76
			10.0	89	109	92
Guava	56	800	0.5	415	471	21
			5.0	311	367	27
			10.0	249	305	33
Fig	56	800	0.5	415	471	21
			5.0	311	367	27
			10.0	249	305	33
Aonla	56	500	0.5	259	375	32
			5.0	194	250	40
			10.0	156	212	47
Sour Lime	36	1000	0.5	333	369	27
			5.0	250	286	35
			10.0	200	236	42
Custard apple	20	300	0.5	56	76	132
			5.0	42	62	161
			10.0	33	53	189

Ber plantation has been raised with *in-situ* runoff concentration system under hot arid climatic conditions of North West India. The fruit yield in Gola and Seb cultivars of ber increased with increase in degree of slope and decreased with length of run as a result of increase in runoff and moisture storage in a 3 m soil profile. The highest fruit yields were obtained when 0.5 and 5 per cent respectively had 8.5 and 7 m length of run and 72 and 54 m^2 catchments area per tree giving 2 and 1.5 contributing per planted area ratios. To conserve the moisture losses from the surface of plant and soil, use of mulches, cover crops and anti transpirants have been used. Black polythene (400 gauge) has been found effective mulch material in ber, pomegranate, date palm and aonla orchards. Some local weeds like *bui (Aerva tomentosa), sarkanda (Saccharum munja),etc.* are effective live mulch materials.

Irrigated Arid Region

The main sources of irrigation in arid region are wells and tube wells as in Jodhpur, Barmer, Churu, Sikar, Jhunjhunu, Pali, Jaipur, Nagaur and Jaisalmer districts of Rajasthan and in a considerable part of Gujarat and Haryana; canals as in Ganganagar, Hanumangarh, Bikaner and Jaisalmer districts of Rajasthan, northern Gujarat and in a considerable part of Punjab and Haryana. About 36 per cent area of the arid region is irrigated by wells or tube wells having brackish water and 60 per cent area is under canal command (Pareek, 1982). At present most of the horticultural production in arid areas comes from the small pockets having assured irrigation covering less than 4 per cent of these lands. Efficient water use technique such as mulching for soil moisture conservation, use of barriers in the soil to reduce percolation, irrigation by drip and sprinklers have been developed for fruit and vegetable crops.

Drip irrigation or more broadly known as micro-irrigation is mainly suited for orchard and plantation crops where it saves 30-70 per cent irrigation water and increase yield by 25-80 per cent (Table 7). Evidences from drip irrigation trials have clearly indicated the advantages like water saving, higher productivity, limited weed growth, better management of assets, off-season maturity, better fruit quality and reduced incidence of insects, pests and diseases. Salient findings of micro-irrigation research at various institutions in India are summarized by Sivanappan, (1998) as detailed in table 7.

Magar and Firke (1994) reported the results in respect of the quantity of water applied in different fruit crops on the basis of age of the plant and open pan evaporation (Table 8).

Table 7: Relative performance of crops with drip irrigation in comparison with that of traditional irrigation methods

Crop	Location	Yield (q ha^{-1})		Irrigation water (cm)		WUE (Q ha^{-1}cm^{-1})		Advantages of drip irrigation	
		Surface	Drip	Surface	Drip	Surface	Drip	Saving of water (%)	Increase in yield (%)
Bottle gourd	Jodhpur	380	558	84	74	4.5	7.5	12	31.9
Bitter gourd	Chalakudy	32	43	76	33	0.42	1.3	56.6	25.6
Brinjal	Akola	91	148.0	168.0	64.0	0.55	2.3	62.0	38.5
	Delhi	280	33.8	45.0	35.0	6.2	9.7	22.2	17.2
	NCPA	280	320.0	90.0	42.0	3.11	7.6	53.3	12.5
	Pune	225	245.0	78.0	51.0	2.9	4.8	34.6	8.2
	Rahuri	280	280.0	90.0	42.0	3.11	6.7	53.3	0.0
Cauliflower	Akola	83	116.0	39.0	26.0	2.1.0	4.5	33.3	28.4
Chilly	Pantnagar	171	274.0	27.0	18.0	6.3.0	15.2	33.3	37.6
Cucumber	NCPA	42.3	60.9	109.0	41.7	0.30	1.5	61.7	30.5
Okra	Pune	155	225.0	54.0	24.0	2.9	9.4	55.6	31.1
	Coimbatore	100	113.1	53.5	8.6	1.87	13.2	84.0	11.6
	Delhi	360	480.0	42.0	26.0	8.6	18.5	38.1	25.0
	Rahuri	189	203.0	219.0	113.0	0.86	1.8	48.4	7.0
Onion	Delhi	284	342.0	52.0	26.0	5.5	13.2	50.0	17.0
	Hisar	93	112.0	50.0	45.0	1.6	2.5	25.0	17.0
Potato	Delhi	172	291.0	60.0	27.5	2.9	10.6	54.2	41.0
	Hisar	235.7	344.2	20.0	20.0	11.8	17.2	0.0	31.5
	Parbhani	334	480.0	30.0	22.0	11.1	21.8	26.7	30.4
Raddish	Coimbatore	10.5	11.9	46.0	11.0	0.23	1.1	76.1	11.8
Sugarbeet	Hisar	418	489	50.0	37.0	8.4	13.2	26.0	14.5
Sweet potato	Coimbatore	42.4	58.9	63.0	23.0	0.67	2.4	60.3	28.0
Tomato	Akola	45	58.0	102.0	77.0	0.44	0.75	24.5	22.4

	Coimbatore	6108	88.7	49.8	10.7	1.24	8.28	78.5	30.3
	Delhi	257	396.0	47.0	25.0	5.5	15.8	46.8	35.1
	Ncpa	320	480.0	30.0	19.0	10.7	25.3	36.7	33.3
	Pantnagar	104.0	137.0	22.0	14.0	4.7	9.8	36.4	24.1
	Parbhani	320/0	480.0	32.4	22.2	9.9	21.6	31.5	33.3
	Pune	292.0	413.o	31.0	20.0	9.4	20.7	35.5	20.3
	Rahuri	16.4	17.2	29.7	20.8	0.6	0.82	30.0	4.7
	Udaipur	144.0	175.0	41.0	28.0	3.5	6.3	31.7	17.7
Ber	Belvatgi	13.7	18.0	15.4	12.5	0.9	1.4	18.8	23.9
Grapes	Dharwad	101.0	101.0	53.0	28.0	1.91	3.6	47.2	0.0
Guava	Allahabad	0.16/plant	0.22/plant	6.4	5.21	0.03	0.04	18.6	27.3
Kinnow	Delhi	68.0	98.0	22.1	17.3	3.1	5.7	21.7	30.6
Lemon	Delhi	15.0	27.0	23.0	17.5	0.65	1.54	23.9	44.4
Pomegranate	Belvatgi	7.4	14.4	10.7	8.7	0.7	1.7	18.7	48.6
	Hyderabad	15.0	37.0	183.0	178.0	0.08	0.21	2.7	59.5
Watermelon	Jodhpur	294.6	882.0	80.0	80.0	3.7	11.0	0.0	66.6
	Pune	82.1	504.0	72.0	25.0	5.9	20.2	65.3	16.3

Table 8: Average water requirement of fruit crop by drip method.

Name of fruit crops	Age (Y)	Water requirement (Litres/tree/day) Jan.	Feb.	Mar.	Apr.	May	Jun.	Jul.	Aug.	Sep.	Oct.	Nov.	Dec.
Sapota, Tamarind	1	-	-	-	-	-	22	23	27	24	22	25	21
	2	27	38	56	77	82	60	41	34	34	41	31	26
	3	45	63	93	128	137	94	68	56	56	62	51	43
	4	65	92	135	186	199	137	98	81	81	97	74	62
	5	76	107	158	217	232	159	195	95	101	114	86	73
Aonla	1	-	-	-	-	-	-	24	20	20	24	18	16
	2	18	25	36	50	54	37	27	22	22	26	20	17
	3	22	31	46	66	67	47	33	27	27	33	25	21
	4	32	45	66	91	97	67	48	40	40	48	36	31
	5	37	53	77	106	114	78	56	47	47	56	42	36
Orange, Lime, Guava	1	-	-	-	-	-	-	18	15	15	18	13	11
	2	13	18	28	37	40	27	20	16	16	19	15	15
	3	3	16	23	34	46	49	34	24	20	24	18	16
	4	23	33	49	6	72	49	35	29	29	35	27	23
	5	29	41	61	84	89	61	41	37	47	47	32	28
Ber, Pomegranate, Fig	1	-	-	-	-	-	-	13	10	10	12	9	8
	2	12	17	25	34	36	25	18	15	15	18	13	11
	3	16	23	34	46	52	34	18	15	15	18	13	11

The study carried out at MPKV, Rahuri (Maharashtra) revealed that although tree growth was not affected, the fruit yield and water use efficiency were considerably increased by drip irrigation in *cv*. Ganesh of pomegranate. (Table 9)

Table 9: Effect of methods of irrigation on growth and yield of Ganesh pomegranate

Method	Water used (litres)	Trea hight (m)	Stem diameter (m)	Fruit yield yield (q/ha)	WUE (q/ha cm)
Check basin					
0.6 IW/CPE	4800	2.35	3.78	40.6	0.82
0.8 IW/CPE	6600	2.56	4.02	47.5	0.69
Trickle Daily					
20% WA	3580	2.16	3.79	52.5	1.25
30% WA	5322	2.44	4.02	55.9	0.72
Alternate day					
20% WA	3580	2.30	3.84	48.6	1.16
30% WA	5322	2.24	4.01	57.3	0.92

Integrated Nutrient Management

The balanced nutrition in plants is required at appropriate age of plants. The application methods also play important role for availability of nutrients to the plants. In ber orchards, besides 10-15 kg organic manure, annual application of 100 g N, 50 g P_2O_5 and 50 g K_2O per tree is recommended. Fertilizer doses should be raised according to the age of plants and soil fertility of the region. Application of 15-20 kg FYM per tree has been found beneficial in aonla, custard apple, and tamarind. At MPKV, Rahuri, in addition to 50 kg FYM, 625 g N, 225 g P_2O_5 and 225 g K_2O has been recommended for application to 5-year-old pomegranate trees At Bangalore, application of 500 g N + 250 g P + 125 g K produced six times higher yield than in the control. In 6 to 7 years old fig trees planted at 5m x 5m spacing, fertilization with 900 g N + 250 g K improved fruit production. The studies conducted on Date palm at AICRP-AZF, Abohar showed that application of 300-400 g N/tree/year gave maximum number and weight of bunch. Similarly in pomegranate it has been demonstrated that application of 50 per cent recommended dose of nitrogen at monthly interval gave best performance. In order to conserve the costly input such as fertilizer, attempts were made to supply this along with water under pressurized irrigation system. The studies conducted in pomegranate and ber has demonstrated that fertilizer saving to the tune of 25 per cent can be achieved if plants are fertigated through drip.

For substitution of inorganic with organic fertilizers attempt was made in pomegranate. The results have demonstrated that a good crop of pomegranate can be harvested by giving 50 per cent RD of NPK through Vermicompost and 50 per cent through inorganic fertilizers. Thus, the use of inorganic fertilizers can be reduced to half through this technology. In general, micronutrients are deficient in semi-arid and arid region soils. Foliar feeding of nutrients such as nitrogen (0.5-2.0% urea), zinc (0.05-1.0% zinc sulphate), and boron (0.05- 1.0% borax) has given beneficial results in these areas (Pareek and Sharma, 1991). In the medium rainfall region of eastern Uttar Pradesh, application of FYM, pond soil, gypsum, and pyrite in sodic soils resulted in better establishment and growth of aonla and bael plants. Foliar spray of micronutrients (Fe 0.50% + Zn 0.50% + Cu 0.25%) improved the yield and fruit quality in kinnow mandarin in arid region. Foliar spray of zinc sulphate 0.5-1.0 per cent improved fruit quality in ber cv. Seb in semi arid region.

Plant Canopy Architecture Management

The canopy of plant plays a vital role to increase quality production of fruit trees. Plant architecture can be maintained through proper training and pruning of fruit plants. Training at initial stages of growth gives proper shape and strong frame to the trees. The bushy pomegranate should be trained keeping 3-5 stems from the ground level while in other fruits; single stem training keeping 3-4 main branches is adopted. However, pruning is essential to regulate reproductive phase of plants. Ber is pruned during January in Tamil Nadu, by the end of April in Maharashtra, and by the end of May in North India. The main shoots of the previous season are cut back retaining 15-25 nodes, depending upon location, cultivar, and age and vigour of tree. All the secondary shoots are completely removed. As a result of light pruning for several years, long non-flowering shoots were developed. To eliminate this, half the number of shoots on the tree should be pruned keeping normal length and remaining half should be pruned keeping one to two nodes to induce new growth for fruiting in the following year. In phalsa, the time of pruning should be regulated according to the flowering period and should result in maximum number of new shoots on which bearing takes place. Established phalsa bushes should be pruned at 20-30 cm height once a year during January in north India and twice a year (December and June) in south India. Pruning from ground level is done either to rejuvenate old bushes or to train young plants into bush from. Defoliation of leaves in lasoda trees in the month of December-January produces early flowering and fruiting in arid region. High density planting in aonla fruit trees was found remunerative to achieve high yield and better returns under semi arid conditions (Singh *et al.*, 2009).

Fruit Based Cropping Systems

Monoculture in arid zone is highly risk prone due to crop failures, hence a suitable tree crop combinations is essential for alleviating the risk, generation of income, improving productivity per unit area/volume as a result of efficient use of natural resources and inputs. This will also moderate microclimate and improve site conditions in long run. In areas with large livestock population, horti-pastoral system would be beneficial. In the arid areas, the system could have combinations such as khejri *(Prosopis cineraria)*+ber+ dhaman *(Cenchrus ciliaris, C. setigerus) or* sewan *(Laisurus sindicus).* Fruit trees can also be planted in association with forest trees and they yield wood for packaging and fuel. Multi-storey combinations incorporating large trees, small trees and ground crops can be used. Agri-horticultural combinations with legume intercrops such as moong bean, moth bean, cluster bean, and cowpea are beneficial. In the rainfed orchards of guava and ber, cluster bean, okra, and cowpea in kharif (rainy season) proved good in the medium rainfall region of Gujarat. Under South Indian conditions of Hyderabad; cowpea, green gram, cluster bean and horse gram in ber orchards and bitter gourd, tomato and okra in acid lime orchards have been grown as intercrops. In low rainfall (300-500 mm) zone, combinations such as khejri or ber + ber or drumstick + vegetables (legumes and cucurbits) and in 500-700 mm rainfall zone, combination of mango or ber or aonla or guava + pomegranate or sour lime or lemon or drumstick + solanaceous or leguminous or cucurbitaceous vegetables can be adopted.

In arid ecosystem, attempts have been made to develop models for crop diversification. Keeping in view the traditional over storey crops as ber and new introduction aonla, the cropping models have been developed. It has been demonstrated that in ber (*Ziziphus mauritiana*) based cropping system cultivation of Indian aloe (*Aloe barbedensis*) can be taken up as a remunerative model (Dhandar *et al.*, 2004) while in bearing stage Ber+ cluster bean (*Cyamopsis tetragonoloba)*-mustard (*Brassica juncia*) is a viable option (Saroj *et al.* 2003). Similarly, in aonla (*Emblica officinalis*) based cropping system, it has been demonstrated that model consisting of aonla + ber along with moth bean or fenugreek can be adopted as a sustainable model for nutritional and income security of the inhabitants (Awasthi *et al.*, 2007).

Plant Health Management

Besides wild animals, rodents and birds there are many insects and diseases causing loss to horticultural crops. Termites cause considerable damage particularly in low rainfall areas. Methyl parathion dust (5%) should be applied in the pits (50 g pit^{-1}) dug for planting fruit trees. Subsequently, water soluble insecticides (Chloropyriphos) should be applied with irrigation water. Fruit fly

(*Carpomyia vesuviana*) causes serious damage in ber fruits. To keep the infestation under check, the chemical spray schedule should consist of spray at pea stage with 0.03 per cent monocrotophos, second spray after 15 days with 0.1 per cent carbaryl. During maturity of the fruits, if necessary, sprays should be done with 0.5 per cent malathion mixed with 0.5 per cent *gur* or sugar solution. This schedule has also been found effective against fruit borer (*Meridarchis scyrodes*) which causes serious damage in southern and western India. Pomegranate butterfly (*Virachola isocrates*) causes considerable damage to pomegranate fruits. Bagging of fruits with butter paper gives good protection. For control, 0.02 per cent Deltamethrin and 0.2 per cent Carbaryl 50 WP sprayed in rotation at 21days interval starting from fruit set is the most cost effective.

For the control of ber powdery mildew, fungicides such as 0.1 per cent Dinocap or Carbendazim or Triademorph or Thiophenate methyl and 0.2 per cent wettable sulphur have been found most effective when sprayed 2-4 times at 15 to 20 day interval starting from initiation of the disease. One spray of the fungicide at initiation of new growth after pruning is an effective prophylactic measure. Black leaf spot (*Isariopsis indica*), found under more humid conditions, can be controlled by 2-3 sprays of 0.2 per cent Captafol or Copper oxychloride or Mancozeb and 0.1 per cent Carbendazim at 15 day intervals. For the control of leaf and fruit spot in pomegranate, four sprays with 0.25 per cent Ziram and 1 per cent Bordeaux mixture at 15 day intervals are most effective. Since, the intensity of the disease is more under humid conditions during mrig bahar as many as 10 sprays at 10 day intervals may be necessary. Fungicides such as Captafol, Mancozeb, Carbendazim, Copper oxychloride, and Thiophenate methyl could also be used. For the control of rust in aonla, 4 sprays of 0.2 per cent Chlorothalonil at 15 day intervals soon after initiation of symptoms give the best control.

Apart from chemical control, attempts have also been made to use bio pesticides for control of pests in arid fruit crops. It has been demonstrated that application of Neem Seed Kernal Extract (NSKE @ 2.5-5%) on various crops effective in controlling pests in pomegranate, aonla, chilli and brinjal. Similarly use of bio-control measures to control ber powdery mildew was also attempted. An isolate, CIAH-196 of *Trichoderma* has potential to be used as bio-control of ber powdery mildew. The isolate thus obtained is resistant even to fungicides and hence can be used in combination with pesticides.

Post Harvest Management

In order to dough-tail the amount of spoilage caused due to gluts and inappropriate post harvest handling methods like losses at different stages of grading, packing, storage, transport and finally marketing. To obtain a quality dehydrated product, factors like composition of commodity, its maturity, pre-drying treatments and methods of dehydration have to be taken into consideration. It is mentioned that the physico-chemical composition with regard to moisture content, TSS, sugars, acidity, ascorbic acid, tannins, alkaloides, *etc.* determine the dehydration quality of the commodity and also the dehydrated product. Therefore, it must be harvested at appropriate stage of maturity. In arid region, horticulture produce suffers heavy post-harvest losses in the absence of adequate post-harvest and marketing infrastructure *viz*; pre-cooling units, packing and grading sheds, short and long term cold storage facilities, refrigerated containers, storage and phyto-sanitary facilities at airports. Value addition of perishable commodities is needed to achieve better price of produce in the market, which signifies the steps and series of operations like delineation of criteria for maturity, pre harvest treatments to reduce post harvest losses, techniques of harvesting to minimize on farm losses, standards for grading and packing for distance transportation, post harvest treatments and conditions of storage to improve shelf life, processing techniques to develop more useful product and utilization of waste to develop byproducts. Besides, dehydration brining, pickling, beverage making, preserve making, *etc.* are the other methods of value addition being adapted to various arid commodities. Proper packaging is an essential prerequisite to reduce losses of produce.

Tropical region in general and arid region in particular has greater significance of dehydration technologies of fruits and vegetables. In arid region, intense solar radiation is available in major part of the year, therefore, sun drying is very common and easy method. Various fruits like ber, date palm, aonla, fig and ker and vegetables such as kachari, snap melon (*khelra*), roundmelon peel (*phophalia*), khejri (*sangri*), kumat, cluster bean (*guar*), leafy vegetables, *etc.* are being dried in sun. Presently, the technologies of mechanical dehydration like solar, osmotic, freeze, fludized bed drying, *etc.* have been developed and attained sophistication.For making dry dates (*chhuhara*), the *doka* fruits of date palm are dehydrated; whereas full mature sour fruits of wild pomegranate are dried for *anardana*. Pre-drying treatments like blanching or brining or sulphitation improves the quality of dehydrated products. It has been found that blanching of aonla for 4 minutes in boiling water improved the quality of dried product. Boiling of *doka* fruits for 4-6 minutes in Medjool, Khadrawy and Shamran cultivars gave better quality dry dates. Curing of dates at low temperature had blistered skin, greater percentage of moisture and reducing sugars and lesser percentage of non- reducing sugars but by curing in boiling water, the product was of better quality without blisters. The cooking of *doka* fruits Khadrawy and Shamran date palm cultivars for 10-25 minutes was required for making dry dates (*chhuhara*). For making soft dates, curing for much shorter period i.e. for 20-25 seconds followed by dehydration at 30-40°C was required.

Dipping of ker fruits in 4-6 per cent salt solution or 0.2 per cent NaOH solution for about a week improves the quality of dehydrated product. Blanching of immature green pods of khejri (*sangri*) in 2 per cent salt solution for 5 minutes retains light green colour of pods after dehydration. Further, it was observed that blanching of pods in 2 per cent salt solution along with 0.6-0.8 percent potassium meta-bisulphite (KMS) improves the quality of dehydrated pods of khejri. Among the various methods of dehydration, sun drying and solar dehydration is most common in arid fruits and vegetables. Date biscuits prepared by adding 10% and 20% date pulp powder were found nutritive in terms of protein and sugars (Singh *et al,* 2013). For sun drying in open, produce is spread in single layer on *pucca* floor or bamboo structure, or wooden crates or metal trays depending upon availability, convenience in handling, kind of fruits and the product type. It is the cheapest method but at the same time the process remains slow and product develops deep brown colour due to the lack of sufficient sunshine. In arid region, open sun drying of product may contain deposited dust, which reduces their market value. Under such circumstances, it is observed that covering of product with news paper (at small scale) or by thin clothes can improve the quality. Large scale sun drying requires a very large area to collect solar heat. The rate of drying depends on the temperature during the drying season. It is slower during winter (20-25°C) than summer (35-40°C). The air circulating type electric dehydrators have précised temperature control system inside the chamber, which yields desired quality product without spoilage. Such dryer gave good quality product by drying dates. Dehydration of vegetables *viz*. chillies, sangri, methi, coriander leaves, cluster bean pods, kachari, *etc*. is a common practice in arid region.

Future Challenges

Traditional farming in arid region is risky proposition and production cost is also rising than productivity gain and thus eroding profit margins. Therefore, it is essentially to adopt new technologies for lowering production costs, improving crop yields, enhancing the quality of produce and reducing the post harvest losses. Integration of horticulture and animal husbandry besides other secondary off farm enterprises are better option for improving productivity and profitability. Market driven diversification in a global perspective has become the new paradigm driving future horticultural growth of India. Specialty commodities such as off-season varieties and production systems, new crops and novel varieties should be identified to capture new opportunities. Hort-business centres equipped with facilities like cold storage, processing unit, packaging units, quality control facilities, financial linkage points, insurance facilities, *etc*. should act as a price determination centres at the producer level. Value addition of horticultural produces offers great scope in domestic as well as export markets. Strict adherence of international standards of quality is essential in this area through good management practices and health hazards and critical control points. Farmer's organizations, prospective entrepreneurs, cooperatives, self help groups and better marketing infrastructure at the farm level need to be encouraged.

References

Anonymous 2013-14. Annual Report, CIAH, Bikaner, p.132.

Awasthi, O. P., Singh, I. S. and Sharma, B. D. 2006. Effect of mulch on soil hydro thermal regimes, growth and fruit yield of brinjal under arid conditions. *Indian Journal of Horticulture*, 63(2): 192-194.

Dhandar, D.G., Saroj, P.L., Awasthi, O.P. and Sharma, B.D. 2004. Crop diversification for sustainable production in irrigated hot arid eco-system of Rajasthan. *Journal of Arid Land Studies*, 148: 37-40.

Magar, S. S and Firke, D. 1994 Water management of fruit crops under limited water supply. *Shetkari*, Pp. 17-21.

More, T.A. and Singh, R.S. 2008. Conserving biodiversity in different areas, *The Hindu Survey of Indian Agriculture*,Chennai, Pp. 50- 54.

More, T.A., Samadia, D.K., Awasthi, O.P. and Hiwale, S.S. 2008. Varieties and Hybrids of CIAH. Tech. Bull. No. 30, Bikaner, 11p.

Pareek, O.P. 1978. Qicker way for raising ber orchards. *Indian Horticulture*, 23:5-6.

Pareek, O. P. 1982 Proceedings of workshop on Problems of deserts in India, jaipur, 16-18 September 1975. Geological Society of India, Calcutta, Misc. Publ. No. 49: 313-320.

Pareek, O. P. 1993 Water management in fruit crops. In: Advances in Horticulture, Vol. 2. MPH, New Delhi (Eds: Chadha, K. L. and Pareek, O. P.) pp. 726-761,

Pareek, O.P. and Sharma, S. 1991. Fruit trees for arid and semi-arid lands. *Indian Farming*, 41:25-30.

Saroj, P.L. 2004. Nursery management in arid ecosystem. In: Advances in Arid Horticulture, Vol.1. (Editors: P.L. Saroj *et al.*) Pub. International Book Distributing Company, Lucknow (UP). Pp-233-251.

Saroj, P.L. and Dhandar, D.G. 2004. Changing scenario in arid horticulture. In: Advances in Arid Horticulture, Vol.1. (Editors: P.L. Saroj *et al.*) Pub. International Book Distributing Company, Lucknow (UP). Pp-1-27.

Saroj, P.L., Dubey, K.C. and Tewari, R.K. 1994. Utilization of degraded lands for fruit production, *Indian Journal of Soil Conservation*, 22:162-176.

Saroj, P.L., Vishal Nath and Vashishtha, B.B. 2000. Effect of polycontainers on germination, seedling vigour, root characters and budding success in aonla. *Indian Journal of Horticulture*, 57(4):300-304.

Saroj, P.L., Dhandar, D.G., Sharma, B.D., Bhargava, R. and Purohit, C.K. 2003. Ber (*Ziziphus mauritiana* L.) based agrihorti system: A sustainable landuse for arid ecosystem. *India J. Agroforestry*, 5(1&2): 30-35.

Singh, A.K.; S. Singh; V.V. Apparao; Meshram, D.T; Bagle, B.G. and More, T. A. 2009. High density planting systems in Aonla, Tech.Bull. No. 34, CIAH, Bikaner,15p.

Singh, R.S.,Bhargava, R, Garima Pal and Sharma, B.D. 2013. Utilization of date palm pulp for making delicious biscuits, *Annals of Arid Zone*, 52: 113-115.

Sivanappan, R. K. 1998. Status and perspective of micro-irrigation research in India. In: proceeding of the National Seminar on Micro-Irrigation Research in India: Status and Perspectives for the 21st century, Bhubaneswar, July 27-28, 1998.

Vishal Nath, Saroj, P. L.; Singh, R. S., Bhargava, R; and Pareek, O. P. 2000. *In-situ* establishment of ber orchards under hot arid eco-system in Rajasthan. *Indian Journal of Horticulture*, 57(1):21-26.

2

Integrated Nutrient Management for Horticultural Crops in Arid Region

B.D. Sharma, M.K. Jatav, R.C. Balai and Anita Meena

Introduction

The Indian hot arid zone occupies an area of 0.32 million km to forming a continuous stretch in the north western states of Rajasthan, Gujarat, Punjab, Haryana and scattered land masses in the states of Maharashtra, Karnataka and Andhra Pradesh, 70 per cent of it falls in western Rajasthan. Low and erratic rainfall, extreme temperatures, long sunshine duration (6.6-10 hours), low relative humidity (30-80 per cent), high wind velocity (9-13 km/h) and high evapo transpiration (1600-1800 mm) and characteristic features of the region. Moreover the soils are poor in nutrients and scarcity of water and recurring droughts are perennial constraints. Despite hostile conditions, the western arid Rajasthan supports a large human and livestock population and a variety of flora and fauna. However, the ever increasing human and livestock population and development activities exert enormous pressure on the natural resource in the region. These areas experience an annual rainfall between 100 and 500 mm with a coefficient of variation varying from 40 to 70%. The region is characterized by low and erratic rainfall with extremes of temperature (1–48 °C), high wind velocity and sandy soils. In dry and arid region, wind erosion affects 13.5% in India (Sehgal *et al.*, 1994, Kar *et al.*,1959). Wind erosion is very active in the Indian *Thar* Desert and poses severe multifaceted problems (Dey *et al.*, 2016). Loss of nutrient-rich particles from agricultural fields has been occurring due to suspension of fine particles in air and deposition of eroded soil. Among terrain properties, soil aggregate distribution, surface roughness, soil moisture and vegetation cover are important factors influencing wind erosion. Indiscriminate grazing in the region also further destroys vegetation and exposes the land surface, thus making it more vulnerable to wind erosion.

High summer temperature (often reaching 50°C) and very low winter temperature (sometimes below 0°C), with large diurnal and spatial variability, as well as high wind speed between March and July with speed gusts of >50 km/h during dust storms, are the other major climatic characteristics. The mean annual potential evapo transpiration exceeds precipitation by a wide margin (1400-2000 mm). Broadly, the soils in large parts of the western sandy plains and in the dune-covered areas are deep, excessively drained, calcareous or non-calcareous sandy.

High population pressure has forced us to plough 46% of our country area for food production. The present productivity is 12.36 tones/ha of fruits needs to be raised to 40.0 tones/ha by using N; P_2O_5: K_2O with other inputs, since recommended ratio being 0.75:0.50:1.00. To achieve this target, the present day requirement of 0.67 million tones N, 0.46 million tones P_2O_5 and 0.91 million tones K_2O has been raised to 1.07, 0.58 and 1.14 million tones, respectively in order to supply the recommended ratio of fertilizers to an area of 3.65 million ha growing fruits (Yadav, 2009). Similarly in vegetables, the present production level of 13.41 tones/ha to be raised to 40 tones/ha requiring 1.25 million tones of N, P_2O_5 and K_2O combined at the ratio of 1.33: 0.88 and 1.00.

The soils of India and especially of arid and semi-arid regions are impoverished and hungry to plant nutrients. The soils are very poor in fertility and single grained structure and needs proper nutrient management for optimum production of the horticultural crops. Considering economy, energy and environment, it is imperative that nutrients are to be used effectively by adopting the appropriate doses of nutrients to be applied, their placement and correct timings to ensure higher yields and to sustain the available nutrients in soil at the optimum level.

Since cost of the inputs like fertilizers and manures are very high, thus, efficient nutrient management not only help in increasing the present fruit and vegetable production level but also sustained the fruit production and protect the environment from different types of hazards occurring due to misuse of costly fertilizers. Therefore, sustainable nutrient management is the passport to enter into the 21st century. In view of these situations, the application of nutrients from the combinations of inorganic, organic and biological sources not only fulfills the nutrient requirement of the crop but also improves the soil health. Hence integrated nutrient management practices will help to increase the productivity of the crop and also enrich the biota of soil. So first one should know the integrated nutrient management, objectives of INM and their need in the agricultural system so that practice can be implemented in the field effectively.

Integrated Nutrient Management

The maintenance or adjustment of soil fertility and plant nutrient supply to an optimum level for sustaining the desired productivity through optimization of the benefits from all possible sources of plant nutrients in an integrated manner (R. N. Roy, FAO, Rome ,1995). Integrated nutrient management, developed on the principles of eco-friendly and efficient balanced fertilization and based on optimization of nutrient supplies from all the available sources, inorganic and organic, for pre-defined yield targets of the crop through an efficient combination of soil, water, organic matter etc. INM is a system that helps to restore and sustain crop productivity and also assist in checking the emerging micro-nutrient deficiencies.

Objectives of INM

The objectives of integrated plant nutrient management are:

- To increase the availability of nutrients from all sources in the soil during growing season
- To reduce the inorganic fertilizer requirement
- To match the demand of nutrients by the crop and supply of the nutrients from all sources
- To optimize the functioning of the soil biosphere with respect to specified function
- To minimize the losses of nutrients to the environment through volatilization, denitrification, surface runoff and leaching beyond the rooting zone.

Need of INM

- Decline in productivity of the crop due to decrease in effective nutrient supply
- Poor utilization of the nutrients by the crop
- Accelerated appearances of P, S and Zn deficiencies associated with more N fertilizer use
- On acute P-deficient soils, N application alone depresses
- Decline in Soil Organic Matter (SOM) associated with continuous application of nitrogen fertilizer
- Accentuation of soil acidity by continuous application of acid forming fertilizers.
- Changing land use pattern from the forest ecosystem to agro ecosystem is responsible for depletion in SOM, impoverishment of soil fertility

In India, effective nutrient management has played a major role in accomplishing the enormous increase in food production. However, it is not possible to supply all the nutrient requirements of crops through inorganic fertilizers. So by taking into consideration the above facts, integrated nutrient management (INM) has been developed. INM has multifaceted potential for the improvement of plant performance and resource efficiency while also enabling the protection of the environment and resource quality. Integrated nutrient management (INM) involves the use of manures, chemical fertilizers and biological agent achieve sustainable crop production and improved soil health. INM is the best approach for better utilization of available resources and to produce crops with less expenditure. In soils of India, Nitrogen, Phosphorus and sulfur deficiencies are principal yield-limiting factors in crop production. INM, which entails the maintenance of soil fertility to an optimum level for crop productivity to obtain the maximum benefit from all possible sources of plant nutrients organics as well as inorganic in an integrated manner (Aulakh *et al.* 2005 & 2010) is essential to address the twin concerns of nutrient excess and nutrient depletion. INM is also beneficial for marginal farmers who cannot afford to supply all crop nutrients through costly chemical fertilizations (Meena *et al*., 2013, Singh *et al.*, 2014).

For making the INM practice effectively fruitful, status of soil health should also be known; therefore soil health can be defined as given below:

Soil Health Management

- Soil Health Capacity of a soil to function within ecosystem boundaries to sustain biological productivity, maintain environmental quality and promote plant and animal health.
- In the context of agriculture, it may refer to its ability to sustain productivity.
- A healthy soil would ensure proper retention and release of water and nutrients, promote and sustain root growth, maintain soil biotic habitat, respond to management and resist degradation.

Components of Nutrients

- Labile soil nutrient pool
- Inorganic or mineral fertilizers
- Organic manures /composts / vermicompost
- Biofertilizers

Nutrient Management of Arid and Semi-arid Fruit Crops

Plant nutrient refers to those compounds, which are required by the plant as a source of body building material and for the energy, without which, it will not be able to complete its life cycle. The fruit tree nutrition is concerned with the provision of plant with nutrients as well as nutrient uptake and their distribution in their plant system. Integrated nutrient management refers to maintain the soil fertility and plant nutrient supply to an optimum level for sustaining the desired crop productivity through optimization of benefits from all possible sources of plant nutrients in an integrated manner. Therefore, it is holistic approach, where we first know what exactly required by the plant for optimum level of production, in what different forms these nutrients should be applied in soil and at what different timings with the best possible method, and how best these forms should be integrated to obtain the highest productive efficiency on the economically acceptable limits in an environmental friendly manner.

Sixteen elements are essential for the satisfactory development of any plant. On the basis of requirement by the plant, the nutrients have been classified into two groups: macronutrient and micronutrient (Table 1). Generally carbon, hydrogen, oxygen, nitrogen, phosphorus and potassium are absorbed in large quantities and therefore, referred to the macronutrients nutrient. The nutrients are required in small quantities are zinc, boron, iron, manganese, molybdenum, copper and chlorine and fall in the micronutrients category.

Table 1: List of essential macro and micro plant nutrient

Macronutrients		Micronutrients	
Carbon	Potassium	Zinc	Boron
Hydrogen	Calcium	Iron	Molybdenum
Oxygen	Magnesium	Copper	Chlorine
Nitrogen	Sulphur	Manganese	
Phosphorus			

At harvest of one tone fruits of some species absorbs different quantities of the nutrients are presented in table 2.

Table 2: Nutrient harvested by fruit crops

Fruit	Nutrient absorbed in kg/t of the fruits		
	N	P	K
Aonla	4.3	1.7	7.3
Citrus	1.1	0.6	6.3
Date palm	4.2	2.0	9.0
Pomegranate	1.6	0.5	5.0
Ber	1.4	0.4	4.8

(Iyengar, 1993)

The data presented in the table 2 revealed that the fruit plants need higher amount of nitrogen and potassium and a small quantity of phosphorus. The nutrient requirement of fruit and vegetable crops may varies from species to species, therefore, precise estimate of the nutrient requirement in fruit trees, it is necessary to determine not only the quantities removed annually by the fruits and new shoots but also those accumulating over the years in other plant parts and the proportion of the currently fertilizer nutrients moving into different parts.

The requirement of nutrients depends upon soil fertility and age of the plant. The studies carried out at AICRP centre Faizabad revealed that in aonla orchard, 100 g nitrogen, 50 g phosphorus and 50 g potash along with 10 kg FYM has been recommended for one year old plant. This dose should be increased every year in the same proportion upto the age of 10 yeares and there after the doses should be stabilized. Hence according to this recommendation the manuring and fertilization should be done as per detailed here:

Table 3: Manure and fertilizer recommendations for aonla fruit crop

Age of plant	FYM (kg)	Nitrogen (g)	Phosphorus (g)	Potash (K)
1	10	100	50	50
3	30	300	150	150
5	50	500	250	250
10 and onwards	100	1000	500	500

Studies carried out at Faizabad revealed that in sodic soils breaking of hard pan is essential before filling the pits. Pit should be dug in May and June and rainwater should be allowed to collect in the pit and collected water should be thrown outside. After 2-3 washing the pit must be filled as give below for successful cultivation of aonla in sodic soils.

Table 4: Pit filling mixtures in sodic soils for aonla plantation

Pit size (m^3)	Soil pH	FYM (kg)	Sand (kg)	Soil amendments (kg)		
				Gypsum	Pyrite	Usar tor masala
1.0	8.0	40-60	10-15	-	-	-
1.0	8.5	40-60	10-15	5	4	3
1.0	9.0	40-60	15-20	10	8	6
1.0	9.5	40-60	15-20	15	12	9

Considering the organic matter is poor in soil, one third of nutrients should be given through FYM and rest through inorganic sources. Site preparation is an integrated practice aimed at achieving higher percent of survival and rapid initial growth of tree plants. A study was carried out at ICAR-CIAH, Bikaner for micro site development for faster establishment of pomegranate plants. In this study different pit size and various filling mixtures were tried and results revealed that in 60 cm^3 pit size filled with pond silt, sand and FYM gave the maximum survival and rapid plant growth of pomegranate var. Jalore Seedless as depicted in the following table.

Table 5: Response of pit size and filling mixture on the establishment of pomegranate

Treatment	Survival (%)	Plant height (cm)	Plant spre N-Sx E-W (cm)	RWC of leafad
No pit	59.5	60.50	62 x 65	82.50
45 cm^3 pit size				
TS +M	71.1	88.20	95 x 99	89.50
TS+M+PS	72.0	90.10	100 x 96	87.60
TS+M+PS+F	72.5	95.60	98 x 102	86.50
60 cm^3 pit size				
TS +M	81.5	112.50	108 x 115	88.50
TS+M+PS	82.0	112.10	110 x 111	89.50
TS+M+PS+F	88.0	115.00	115 x 116	90.50

TS= Top soil, M= Manure, PS= Pond silt F= Fertilizers

To ensure high economic productivity to sustain the available soil nutrients status at the desired level, correct dose of fertilizers and manures must be applied by use of suitable diagnostic tools designated to avoid nutrients shortage and excesses. Several diagnostic methods have been developed to arrive at need based manuring schedule.

Nutrient management is extremely important for fruit and vegetable crops where many time quality of the produce assumes greater consideration than productivity per se. Application of too much nitrogen (N), in general, results in oversized, poorly coloured fruit with poor keeping quality. Too little N also cause problems poor fruit set, small fruit, pale foliage and stunted growth. Excessive or deficient of other nutrients also can result in serious problem in fruits. Many fruit crops are heavy removers of nutrients and high yield can only be sustained through the application of optimal doses of nutrients in balanced proportion. Diagnosing nutrient deficiency symptoms in fruit and vegetable crops can help in proper nutrient management. Use of soil and or tissue analysis may help to confirm whether symptoms are nutritional or due to some other factors.

The productivity of fruits is comparable to China and other leading fruit producing countries of the world but India still lags behind USA and Indonesia in terms of productivity. This might be because of improper orchard management particularly nutrient management. The continuous use of chemical fertilizers particularly N,

P and K has impaired the soil fertility and decreased the factor productivity. The increasing cost of fertilizers and poor purchasing capacity and their negative effect on soil health has led to intensified attempts to the use of biofertilizers and organic matter with inorganic fertilizers. The Indian soils are mostly poor in nitrogen and phosphorus and 50% soils are low in potassium.

The main objective behind INM system is to manage and sustain the agricultural productivity and improve the farmers profitability through the judicious and efficient use of chemical fertilizer, organic farming , green manure and compost including vermicompost , crop residues and biofertilizers. However, this does not mean adding everything everywhere, rather, a well considered practical and efficient blend of diverse nutrient source is required which can produce desired yield and maintain soil health on long term basis. INM system helps to restore and sustain crop productivity, and also assists in checking the emerging micronutrient deficiencies. Further, it brings economy and efficiency in use of fertilizer.

Major components of integrated nutrient management are (i) integration of soil fertility restoring crops like green manures, legumes, etc.; (ii) recycling of crop residues; (iii) use of organic manures like FYM, compost, vermicompost, biogas, slurry, poultry manure, Biological composts, Press mud cakes, Phospho-compost (iv) utilization of biological agent; (v) efficient genotypes; (vi) balanced use of fertilizer nutrients as per the requirement and target yields. Strong and convincing evidence indicates that INM practice could be an innovative and environment friendly practice for sustainable horticulture in arid region. In general following are the common sources of nutrients and type of material which can provided nutrients are given here:

Inorganic sources	Straight & Mixed fertilizers
Organic sources	Manure, Compost, Green manure
Biofertilizers	Nitrogen fixer, PSB, VAM
Crop management	Legumes, crop rotation, *etc.*

Integrated approach of nutrient application

- Soil management
- Soil moisture management
- Fertilizer management
- Crop management

Before adopting the integrated nutrient management approach in horticultural crops, farmer should analyze their field soil and evaluate the soil conditions on the basis of the criteria given in the table 3.

Soil and Nutrient Management

Assessment of soil nutrient status is based on the assumption that root extract nutrients from the soil in a manner comparable to chemical soil extractants and that there is a direct relationship between levels of extractants and nutrients in soil and their uptake by plants. Chopra and Kanwar (1986) stated that the soil chosen for plantation of any fruit and vegetable crops, initially, should be brought to an optimum fertility level after its chemical analysis. The fertility parameters are presented in table 3 for the soil categorization.

Table 6: Soil categorized for the fertility parameters

Soil Parameters		Category	
	Low (<)	Medium	High (>)
Soil pH (1.2.5)	6 to 7		
Organic carbon (%)	0.50	0.50 to 0.75	0.75
Available N (kg /ha)	280	280 to 560	560
Available P_2O_5 (kg /ha)	25	28 to 56	56
Available K_2O (kg /ha	140	140 to 280	280
Sulphur (ppm)	10	10 to 20	20
Zinc (ppm)	0.5	0.5 to 1.0	1.0
Iron (ppm)	5	5 to 10	10
Copper (ppm)	0.2	0.2 to 0.4	0.4
Manganese (ppm)	5	5 to 10	10
Molybdenum (ppm)	0.05	0.05 to 1.0	1.0
Boron (ppm)	0.1	0.1 to 0.5	0.5

Chopra and Kanwar (1986)

An assessment of nutrient additions, removals, and balances in the agricultural production system generates useful, practical information on whether the nutrient status of a soil (or area) is being maintained, built up or depleted. Estimates of nutrient input and output allow the calculation of nutrient balance sheets both for individual fields and for geographical regions. This is primarily because nutrient removals by crops far exceed the nutrient additions through manures and fertilizers. For the past 50 years the gap between removals and additions has been estimated at 8 to 10 Mt $N+P_2O_5+K_2O$ per year (Tandon, 2004). For example, nutrient losses through soil erosion are alarmingly large, but are rarely taken into account. Nutrient loss through soil erosion is second only to nutrient removal as a result of crop production. An annual loss of 8 Mt plant nutrients has been mentioned through 5.3 billion t of soil lost by water erosion (Prasad and Biswas, 2000). Estimates of removals through leaching and gaseous losses are not available in Cropping based system (Tondon, 2004).

Table 6. Nutrient balance sheet in dry land agriculture

	$N+P_2O_5+K_2O$ (Mt)
Estimated additions(fertilizers)	1.0 mt
Estimated removals (crops)	7.4mt
Balance	-6.4mt

(Tandon 2004)

Leaf Analysis

Leaf is the center of physiological activity in plants where translocated minerals are known to involved in such vital reactions such as photophosphorylation, synthesis of chlorophyll, amino acids, nucleic acid, proteins etc., and consequently any deficiency of a nutrient would drastically effect the related activity. Leaf analysis methodology consists of leaf sampling technique, sample preparation, analysis and making diagnosis and nutrient recommendations using leaf nutrient guide. A large number of fruit plants, environmental and procedural factor vitiate leaf nutrient concentration, however, a careful worked out sampling technique for index tissue will make a sound foundation of leaf analysis programme. Modern methods of interpretation of leaf analysis data such as DRIS and Boundary Line Concept can be adopted to diagnose the growth/yield limiting nutrients and to recommend the optimum use of nutrients. The merits of leaf analysis are: (1) understand the internal function of the nutrient in the plant, (2) confirm toxicity/deficiency (3) identify mineral imbalance (4) ascertain whether applied nutrients have entered in the plant system and (5) identify hidden hunger

Method of Leaf Sampling

There are various steps in carrying out tissue analysis (a) Leaf sampling (b) Sample handling and processing (c) Leaf analysis methodology (d) Leaf nutrient norms (e) Diagnosis of nutrient disorders (f) Interpretation of data and nutrient recommendation.

Leaf sampling: The general principle in leaf sampling is to collect recently youngest mature leaf or petiole considering the following factors.

a) Factors associated with crops

(i) Plant part (ii) Leaf part (iii) Leaf age (vi) Leaf portion (v) Flush (vi) Type of shoot (vii) Leaf size (viii) Leaf health (ix) Alternate bearing (x) Cultivars/Root stock (xi) Crop load (xii) Plant size (xiii) Sample size

Table 7: Leaf sampling technique and nutrient norms for fruit crops

Crop	Plant part, age, stage and position	Sample size	Optimum leaf nutrient (%)		
			N	P	K
Mango	Collect 4-5 month-old leaf from current season's growth from middle part of the shoot	30	0.84-1.53	64-147	0.52-1.10
Mandarin	Basal 6 month-old leaf from current growth in march	50	3.00-3.50	150-250	0.90-1.10
Acid lime	Basal leaf at 5- month-age from current Season's growth	50	1.96-2.30	120-290	1.60-1.90
Sweet Orange	Basal leaf 6- month-age from current season's growth emerged in march to September	50	2.00-2.20	100-110	0.40-1.20
Lemon	Basal 6- month-old leaf from current Season's growth	50	2.20-2.70	150-300	1.00-2.00
Guava	Third pair of leaf from apex in August/December	30	1.60-2.40	150-300	1.30-1.70
Pomegranate	Eight leaf pair from tip in 15 April flush for February crop and from August flush for June crop	50	1.20-1.40	100-200	1.00-1.40
Ber	Fifth leaf from tip of secondary and tertiary shoot in June	50	1.50-2.20	140-450	1.60-2.00
Phalsa	Fourth leaf from growing tip one month after pruning	30	1.50-1.60	150-200	1.60-2.00
Custard apple	Fifth leaf from growing point in May/June	50	1.40-1.80	70-100	0.80-1.20
Fig	Basal leaf from mid-summer growth	40	2.00-2.50	90-100	0.70-0.90

(*Source*: B.S.Bhargava, 1999. Leaf analysis for diagnosing nutrients need in fruit crops)

b) Environmental factors

(i)Temperature (ii) Leaf exposure (iii) Season (vi) Geographical direction (v) Moisture (vi) Sampling height (vii) Spray of chemicals (viii) Dent (ix) Soil fertility variations

Sampling Handling

(a)Time taking during transit (b) Working technique (c) Temperature of even (d) Processing equipment (e) Analytical technique .

Common guidelines for leaf and petiole sampling

1. Select a vegetative terminal unless otherwise specified
2. Sample at chest height (>2 m)
3. Collect composite sample from north, south, east and west
4. Select leaves which are fully exposed to light
5. Collect sample prior to irrigation and fertilizer application
6. Collect 50 for small size leaves, 30 leaves in medium sized and 20 leaves having large leaves
7. Avoid sampling from soiled, diseased/insect
8. Avoid sample contamination
9. Index tissue sampling technique in fruit crops for yield and quality

The leaf sampling technique has been standardized for most of the arid zone fruits using the leaf sampling norms and the details are given in table 7

Interpretation of Leaf Analysis Data

Critical Nutrient Concept

This technique has been used by Subramanian *et al.* (1974), Samra *et al.* (1978), Chaddha *et al.* (1980) and Bhargava (1992) to divide orchards expected to give a relatively large response to a particular nutrient from those expected to give little or no response, assuming that other nutrients are present in adequate amounts. A dividing line between two categories (high and low) might be determined approximately by a graphical technique in which vertical and horizontal lines are superimposed on a scatter diagrams so as to maximize the number of points in the positive quadrants. The horizontal lines are 90% probability line and the vertical line is so drawn that maximum points of the scatter diagram are on two positive quadrants. The vertical line, which is determined by eye judgment, is known the critical level.

Yield Groups and Nutrient Content

Samra *et al.* (1978) adopted grouping of yield level in mango e.g. less than 150, 150-300 and more than 300 fruits/tree to work out leaf nutrient standards (Table 6). There were variations between the critical limits reported for young trees in sand culture and those actually observed in fully-grown trees in orchards.

Table 8: Leaf nutrient guides for the mango cv. Dashehari

Nutrient	Critical limits
Nitrogen (%)	1.23
Phosphorus (%)	0.06
Potassium (%)	0.54
Calcium (%)	1.71
Magnesium (%)	0.91
Sulphur (%)	0.12
Iron (ppm)	171
Manganese (ppm)	66
Zinc (ppm)	25
Copper (ppm)	12

Diagnosis and Recommendation Integrated System (DRIS)

The DRIS was developed by Beaufils (1957, 1971 and 1973) as an objective means of coping with the difficulties inherent in diagnostic procedures. It has long been recognized that the nutrient concentrations change markedly as plant age. In general the foliar concentrations of N, P, K and S tend to decrease during aging. In contrast Ca and Mg concentrations tend to increase. The dynamic nature of plant tissue nutrient concentration, which has been illustrated by the examples cited, imposes severe limitations on the use of foliar analysis of diagnostic purposes (Bates, 1971& Bouma, 1983). Critical nutrient value or sufficiency range system generally depend on diagnostic norms derived from plant tissue of a specific age and categorize plant health based solely on nutrient concentration. Thus, the age of growth of the sampled plant is a prime concern. If the concentrations of the nutrients are decreasing during aging, then the ratios N/P, N/K and P/K or their reciprocals should remain fairly constant. Similarly, because Ca and Mg concentrations usually decrease during aging, a quotient formed from these nutrients (Ca/Mg or Mg/Ca) should also produce a content value. Furthermore, products of two nutrients, one of whose content decreases with time should also be fairly constant nevertheless, these forms of expression are often less affected by aging processes and so present an opportunity to expand the usefulness and accuracy of foliar diagnoses.

$$N\%/P\% = (100N/DM)/(100P/DM) = (100N/DM) * (DM/100P)$$

Where DM is dry matter.

Even though a ratio such as N/P can be easily calculated from N% and P%, commonly reported values, interpretation may be more difficult (Sumner, 1978). For example, if the optimum leaf N/P ratio is 10.04 and an individual sample has a value of 14.00, it is impossible to determine whether N is to high, P is too low or N is too low and P is too low. Taking a holistic approach and considering a single ratio for a single nutrient (N/P, N/K, *etc.*) is even more confusing than considering a single ratio and is probably one reason that dry matter ratio have been preferred than nutrient ratios.

DRIS Norms

In contract to the classical field experiment approach to leaf analysis the DRIS employ survey technique where a large number of randomly distributed sites throughout the area are selected. At each site samples of leaf and soil are taken for analysis and the details of applied manures and fertilizers are recorded. Experiment conducted at one place employing various levels of 1, 2 and 3 plant nutrients can also be used to develop DRIS taking the following steps:

1. Define the parameters to ber improved and the factors likely to affect them.
2. Collect the reliable data available from fields and experimental plots.
3. Study the relationship between yield and environmental and external factors such as soil composition, weather conditions, *etc.* so that the most favourable condition can be determined in a particular case.
4. Study the relationship between yield and internal parameters such as petioles compositions using following steps
    a. Each internal plant parameter is expressed in as many forms as possible i.e. N/DM, DM/N, N/P, P/N, N*P, etc.
    b. The whole population is divided into a number of sub-groups based on yield into two groups of high and low yield.
    c. The means of such population are calculated for various forms of expression.
    d. If necessary, class interval limits between average and outstanding yields are readjusted to so that means of below average and average yielding population remains comparable.
    e. Chi-square test is used to know that sub population confirm to a normal distribution.

f. The variance rations between yields of sub-populations for all norms of expressions are calculated together with the coefficient of variation.

g. The forms of expressions, for which significant variance ratio are obtained and the mean values for the population are the essentially the same are selected and related to one another in expression with common nutrient.

DRIS Index Interpretation

The nutrient indices should some to zero. The more negative an index, the more lacking is the nutrient it represent relative to other nutrients used in the diagnosis. Alternatively, a large positive nutrient index indicates that the corresponding nutrient is present in relatively excess quantity. For example indices for N, P and K are as follows:

N index= [f (N/P) + f (N/K)]/2 = 45.96 + 40.27)/2 = 43

P index = [-f (N/P) – f (K/P)/2 = (-45.96 + 5.61)/2 = -20

K index = [-f(N/K) + f(K/P)/2 = (-40.27 - 5.61)/2 = -23

The N index is (+43)> > P index (-20) > K index (-23).

This is interpreted to mean that in terms of relative importance to yield, K is more required than P, which is more required than N. In plant sample with optimal nutrient balance, all nutrient indices would equal zero. However, it is important to recognize that an individual nutrient is not necessarily present in optimum concentration if its index equal zero.

Optimum Leaf Nutrient Status of the Important Fruits

Leaf is the principle site of plant metabolism and changes in the nutrient supply are reflected in composition of leaf. The concentrations of the nutrient in leaf at a specific growth stage are related to the performance of the crop. The leaf nutrient levels required for optimum production in some important fruit crops have been presented in table 7. Efforts in the nutritional management programme should be directed towards the maintenance of these levels in tree leaves before flowering.

Table 9: Optimum leaf nutrient status of some important fruit crops

Fruit	Nutrient											
	N (%)	P (%)	K (%)	Ca (%)	Mg (%)	S (%)	B (ppm)	Fe (ppm)	Mn (ppm)	Zn (ppm)	Cu (ppm)	References
Mango	1.23	0.06	0.54	1.71	0.91	0.12	-	171.0	66.0	25.0	12.0	Samra *et al.* (1978)
Citrus	2.50	0.14	0.90	4.20	0.43	0.25	-	90.0	112.5	62.5	10.5	Reuther *et al.* (1962)
Guava	1.70	0.50	1.30	2.10	0.46	-	10.0	-	-	77.5	75.0	Singh (1981)
Grape	1.75	0.56	1.66	0.94	-	0.20	-	-	125.0	92.0	-	Bhargava and Chadha (1993)
Litchi	1.47	1.03	-	-	-	-	-	-	-	-	-	Ghosh *et al.* (1986)
Sapota	1.66	0.08	0.80	0.83	0.48	0.07	-	100.0	39.3	15.7	6.7	Annapurna *et al.* (1988)
Pomegranate	2.50	0.19	1.47	-	-	-	96.0	199.0	196.0	55.0	-	Shende (1977) and Bambal (1987)
Ber	2.50	0.25	1.82	1.21	0.45	-	-	-	-	-	-	Khanduja and Garg (1984)

Manure and Fertilizer Recommendations for Some Tropical and Subtropical Fruits

To optimize the soil fertility level and to keep the plant nutrient status above the critical levels, a number of recommendations for the fruit crop are available region wise (Tandon, 1987). Based on the averages of these recommendations and results available under the All India Coordinated Research project on fruit crops, the recommendations are being made for different fruit crops are presented in Table 10. Besides these recommendations application of zinc and iron has become common. Spray of boron has helped in aonla and mango. Therefore, regular application of micronutrients is necessary to get an optimum production per unit area.

Table 10: Manure and fertilizer recommendations for some tropical and subtropical fruits

Fruit	FYM (Kg/plant)	Nutrient recommendation (g/plant /annum)		
		N	P_2O_5	K_2O
Mango	75	775	500	700
Citrus (Acid lime)	50	900	250	500
Guava	60	360	180	180
Pomegranate	30	600-700	200-250	200-250
Ber	50	500	200	300
Sapota	50	1,000	500	500
Aonla	50	1,500	750	1,000
Date palm	30	1,000	500	500

To meet the requirement of nutrients recommendation for different fruit crops, it could either be inorganic source or organic source, but each one has its limitations. General consensus is that, it is desirable to meet fifty per cent of nitrogen through organic sources and fifty per cent through fertilizers for good performance; otherwise continuous use of inorganic fertilizers is detrimental to the soil health. Addition of the organic matter would not only provide the needed nutrient including micro-nutrient at a slow pace for continuous growth, but it would also improve the physical condition of soil. Furthermore, it would also improve the aeration of soil, providing a good scope for the better root growth and increased production. Fortunately, in fruit crops from the very beginning, application of organic manures has been practice and all our recommendations have organic component. It should be pursued. The evidence that the organic manures and biofertilizers reduce the need of chemicals supplying specific micronutrient and to minimize the risk of inducing deficiencies of many other micro-nutrients also commands itself a sound practice for a sustainable horticulture based on the low external chemical input.

Time and Method of Nutrient Application

To improve the efficiency of applied nutrients, they should be made available at an appropriate time, like new vegetable growth phase, fruit growth phase and they should be placed near the active roots. These timings and the zones of activity vary from fruit to fruit and are influenced by cultural practices in different regions and some other factors. Some of the standards developed for different fruit crops as reviewed by Chundawat (1997) are presented below.

Mango

In two doses, i.e. soon after the harvest, full does of FYM, phosphorus and potash and 50 per cent of nitrogen and rest of the 50 per cent nitrogen at pea stage of the fruit development should be applied.

(b) Unirrigated

On the onset of monsoon, full does of FYM, phosphorus and potash and 50 per cent nitrogen should be given and 50 per cent of nitrogen should be applied after the end of the monsoon.

Citrus

In arid conditions citrus crop can be grown successfully with assured irrigated conditions. In such situations, organic manure should be applied in the month of January and inorganic fertilizers in two or three equal split doses, i.e. first does 15 days before the expected flowering, second does after fruit set and third does soon after the rainy season is over (September-October).

Guava

In guava, the time of nutrient application is governed by the choice of crop to be taken and generally winter season crop, which is based on June-July flowering gives a good remuneration. Therefore, total quantity of the organic manure, phosphorus and potash and half does of the nitrogen should be applied with onset of monsoon (June-July). The balance nitrogen should be applied in October-November.

Grape

North India

Total quantity of the organic manure, phosphatic fertilizer and half does of the nitrogen and potash should be applied immediately after winter pruning and half of nitrogen and potash should be applied after berry set.

(b) Tropical Condition

Thirty per cent of the nitrogen and 50 per cent of the phosphorus in April; 30 per cent nitrogen and 40 per cent potash in May; 20 per cent nitrogen and 50 per cent phosphorus in October; 20 per cent nitrogen and 30 per cent potash in November and rest 30 per cent potash in December should be applied.

Pomegranate

Choice of crop determines the timings of nutrient application. In arid conditions, generally *mrig bahar* crop is being taken. Full does of the organic manure, phosphorus and potassium and half does of the nitrogen should be applied in circular trenches at 20-30cm deep and 1m away from main shoots in the month of June just after first irrigation of *bahar* treatment and second half dose of the nitrogen should be applied after three weeks.

Ber

The total quantity of the organic manure, phosphorus and potassium and half does of the nitrogen should be applied after pruning in June- July coinciding with monsoon in circular trenches at 15-20cm deep and 1-1.5m away from the main trunk and half does of the nitrogen should be applied in post monsoon period, i.e. September-October.

Sapota

(a) **Rainfed:** Under rainfed condition, the fertilizers should be applied once in a year, i.e. before monsoon.

(b) **Irrigated**: The total quantity of the organic manure and half does of the chemical fertilizer should be applied in beginning of the monsoon and second half does in the month of October.

Aonla

The total quantity of the organic manure, phosphatic, nitrogen and potash should be applied in equal two splits doses to mature bearing tree once after fruit set and again during fruit growth and development. The plants need to be irrigated after fertilizer application. All fertilizer and FYM should be placed in trenches under full canopy covered are.

Active Root zone of Fruit Crops for Nutrient Application

For the efficient utilization of applied nutrient, it is essential to know the active root zone, so that the nutrients may be placed around this zone, to be made available to the plant. The active root zones as well as place of the application of fertilizers as summarized by Chundawat (1997) have been presented in Table 11.

Table 11: Active root zone of some important fruit plants and the place of fertilizer application

Fruit	Area of the active root zone	Place of the application
Citrus	Within a redial distance of 120 cm at the depth of 15 to 30 cm.	30 to 45 cm away from the trunk under the drip followed by light working.
Guava	At 30 cm distance from the trunk at depth of 30-60 cm.	Within the drip at a depth of 15-20 cm.
Ber	50 to 75 cm deep under the drip.	
Sapota	Under the canopy within a depth of 0 to 30 cm.	Under the drip and mix upto the depth of 30 cm depth.
Date palm	Upto a radius of 1.7 m from the bole.	Under the canopy and mix in the soil Upto 15 cm depth.Around the tree upto a radius of 2.0 m.

Foliar Application of Nutrients

Foliar application is based on the principle that the nutrients are quickly absorbed by leaves and transported to different parts of the plant to fulfill the functional requirement of nutrition. Foliar application of the nutrients is obviously an ideal way of evading the problems of nutrient availability. This method is highly helpful for the correction of trace element deficiencies to restore disrupted nutrient supply and to overcome stress factors limiting their availability. This method has been commercialized in a number of fruits, like citrus, pineapple, guava *etc.*

Application of Biofertilizers in Horticultural Crops

Increased fertilizer cost and the awareness of environmental pollution due to fertilizer runoff necessitated the use of biofetilizers for the development of more efficient fertility management programme. However, very little work has been done on the use of biofertilizers in fruits. In case of pomegranate, nitrogen use efficiency has been increased through the use of 250 g *Azotobactor* culture with hundred-gram nitrogen per plant (Wavhal *et al.*,1985).

Similarly, VAM (Vescular Arbuscular Mycorrhiza) fungi are of the great significance for fruits. It forms a symbiotic association with the roots of plants and helps them in transfer of the nutrients from the soil to the root system. In India, studies on the VAM fungi have so far remained confined to some inoculation trails in some fruits, like citrus, banana, lichi, mango and papaya indicating an increased efficiency of the water and nutrient uptake by host plant. However, more studies are required to adjudge the role of biofertilizer in production of the fruits.

Management of the Farm Waste for INM

Biomass is available in the form of leaf fall or plant residue and it may become an integral part of the soil fertility management. The organic farmers, who combine earthworm culture in plant basin with the degradation of such biomass and have come out with some excellent results in sapota, mango, completely eliminating inorganic fertilizer application, are intelligently managing this. A study carried out at Central Institute for Arid Horticulture, Bikaner using the different types of manure either as a sole source or in combinations with inorganic fertilizers in pomegranate crop and results depicted in Table 5 revealed that application of vermicompost alongwith inorganic fertilizers in 50: 50 ratio gave the best results in plant growth, fruit yield and fruit quality. Thus integration of inorganic and organic sources of nutrients may be proved better in sandy soils of arid region for supplying required nutrients to the crop as well as improve the physical properties of the soil.

Nutrient Imbalance Status in Indian Soils

Nutritional stress in plants results from nutrient imbalances in the soil. Balanced plant nutrient supply ensures high yields and quality food. It refers to application of different nutrient elements in the proportion required for optimum growth of plants. Balanced nutrition eliminates all deficiencies.

When nutrition becomes imbalanced, the plant becomes stressed and does not develop normally and result is low yield and poor quality produce. Fertilizers are meant to correct nutrient deficiencies and improve soil fertility so that higher productivity is sustained. Due to intensive cultivation and imbalance of nutrients, the deficiency of secondary and micronutrients have developed in addition to N, P and K (Muralidharudu, *et al.,* 2011**).** Soil analysis data has shown the soils of 93%, 91% and 51% were low in available N, P, and K, respectively. Incidences of micronutrient deficiencies are also increasing. Among the secondary and micronutrients, 41% of soils have been found deficient in S, 44% in Zn, 15% in Fe, 8% in Cu, 6% in Mn, 33% in B and 13% in Mo (Shukla *et al.,* 2012). The consumption ratios are more towards N in vegetables than in the fruits. Out of total NPK usage, the share of N was 50% in vegetables and 44% in fruits. In contrast to the overall fertilizer consumption, where in the share of N, P and K is 64%, 26% and 10% respectively, the general nutrient uptake and removal pattern in horticultural crops shows that the highest uptake is that of potash, much above that of N. Although, potash consumption by horticultural crops, itself and in relation to N, is higher than in food grains, it is still well below the quantities removed by the fruits and vegetables (Srivastava *et al.,*2008).

Low level of nutrient input is one of the reasons for relatively low yields, much below the realizable potential. There are number of reasons why better nutrient management in horticultural crops deservers serious attention:

- Obtain higher yield and higher returns,
- Increase the returns from establishment expenses of orchard,
- Help to improve the nutritional standards of the people by providing adequate supply of fruits and vegetables at affordable prices,
- Provide greater and better quality raw material for fruit and vegetable processing industries To help in earning foreign exchange through the export of high quality produce in fresh or processed form.

Table 12. Soil nutrient balance sheet in India

	Gross balance sheet,000t			Net Balance sheet ,000t		
Nutrient	Addition	Removal	Balance	Addition	Removal	Balance
N	10923	9613	1310	5461	7690	-2229
P_2O_5	4188	3702	486	1466	2961	-1496
K_2O	1454	11657	-10202	1018	6994	-5976
NPK total	16565	24971	-8406	7945	17645	-9701

Table 13. Nutrient removal by different horticulture crops

Crop	Yield (t/ha)	Nutrient removal (kg/ha)		
		N	P_2O_5	K_2O
Fruits				
Mango	15	100	25	110
Banana	57	322	73	1180
Citrus	20	22	12	57
Apple	29	18	2	40
Pineapple	84	150	45	530
Papaya	80	225.5	60	180
Grape	20	169	40	180
Vegetables				
Potatao	28	202	50	225
Brinjal	60	175	40	300
Tomato	37	104	22	141
Cauliflower	50	250	100	350
Cabbage	37	112	28	112
Beans	35	130	40	160
Green Peas	25	55	20	40
Lettuce	30	107	30	234
Spinach	25	120	45	200
Celery	20	140	55	220
Onion	30	73	36	68

Source: Ganeshamurthy *et al.*, 2011

Fertilizer is one of the costliest inputs in agriculture and the use of right amount of fertilizer is fundamental for farm profitability and environmental protection Dumping of fertilizers by the farmers in the fields without information on soil fertility status and nutrient requirement by crop causes adverse effects on soil and crop regarding both nutrient toxicity and deficiency either by overuse or inadequate use (Ray *et al.,* 2000). Managing the location specific variability in nutrient supply is a key strategy to overcome the current mismatch of fertilizer rates and crop nutrient demand in environments (Dobermann and Cassman, 2002). To enhance farm profitability under different soil-climate conditions, it is necessary to have information on optimum doses for crops. Traditionally, to determine the optimum nutrient doses of most appropriate method is to apply fertilizer on the basis of soil test and crop response studies. Improved crop management needs to be envisaged with adequate emphasis on balanced plant nutrition for stability in production, appropriate soil nutrient resilience. Most often, prior estimation of the actual nutrient requirement of a particular crop, native soil fertility status, has been ignored. Soil test based fertilizer use is must for sustainable agriculture (Rao and Srivastava, 2000) .Thus, need based estimation of N, P, and K requirements may call for Soil Test Crop Response (STCR) based nutrient management, which can be represented in a linear relationship, correlating their requirement with a specified target yield depending on their native soil status. The fertilizer application by the farmers in the field without knowledge of soil fertility status and nutrient requirement of different crops usually leads to adverse effect on soil as well as crops by way of nutrient deficiency or toxicity due to over use or inadequate use of fertilizers.

In this regard, targeted yield approach has been found to be beneficial which recommends balanced fertilization considering available nutrient status in the soil and the crop needs. Targeted yield approach was first developed by Ramamoorthy *et al.*(1967) established theoretical basis and experimental technique to suit it to Indian conditions. This target yield equation (TYE) is considered as a soil-and fertilizer-based precision farming strategy to meet nutrient demands for a specified yield (Balasubramanian *et al.,* 1999). Location specific fertilizer recommendations are possible for soils of varying fertility, resource conditions of farmers and levels of targeted yield for similar soil classes and environment

Soil testing procedures have been developed to relate extractable nutrient levels to crop growth and yield. Nitrogen, phosphorus, potassium and micro -nutrients are most likely to be limiting crop growth. The nitrogen status in the soil is quite dynamic, and predicting its availability over time is difficult. The availability of phosphorus and potassium in the soil is fairly stable over time unless major additions are made. Additions of these two elements and micronutrient over time in manures and commercial fertilizers have caused significant increases in the available levels in the soil. Balance nutrition of crop should ensure that the nutrients in available forms are in adequate quantities and right proportions as per the requirement of the crops. Therefore, the addition of fertilizers and manures should be such that in combination with indigenous nutrient supply, they provide the nutrients in appropriate quantities and ratios matching the plant requirement for desired yield goals. Different approaches of fertilizer recommendations in vogue are briefly discussed here.

Tissue Testing for Nutrient Recommendation in Horticultural Crops

Soil tests, although useful in predicting fertilizer and soil amendment needs, are not the final measure of what nutrients a plant will absorb. Because temperature, moisture regimes, soil acidity and other soil conditions may modify the uptake of different nutrients by plants, it is sometimes necessary to determine nutrient content in the plants to evaluate the actual soil nutrient availability status. Total plant analysis and green tissue analysis are the most important diagnostic techniques for determining deficient, sufficient or excessive amounts of essential elements in plant tissue. Tissue tests are of immense help to farmers in overcoming many problems and are the better approaches for nutrient management of horticultural crops. The success of this approach depends mainly on selecting the right tissue and deciding on the right stage of sampling. The sampling guide and the optimum nutrient levels developed for different horticultural crops are given (Table 14).

Table 14: Plant tissue sampling guidelines for horticultural crops

Crop	Plant part	Growth stage/ Time
Fruit Crops		
Custard Apple	5th leaf from	Apex 2 months after new growth
Fig	Fully expanded leaves, mid-shoot current growth	July-August
Citrus	3 to 5 month-old leaf from new flush. 1st leaf of the shoot	June
Guava	3rd pair of recently matured leaves	Bloom stage (August or December)
Pomegranate	8th leaf from apex Bud differentiation.	In April for FebruaryCrop and August for June crop.
Sapota	10th leaf from apex	September
Phalsa	4th leaf from apex	One month after pruning
Ber	6th leaf from apex from secondary or tertiary shoot.	Two months after pruning

Vegetable Crops		
Cluster bean	1st fully developed leaf	
Cucumber	5th leaf from tip	Flower bud start to small fruit
Brinjal	Leaf blades with midribs minus petioles from most recent fully developed leaf	
Garlic	Most recent fully matured leaf	Pre-bulb
Peas	Most recent fully developed leaflet	First bloom
Tomato	Leaves adjacent to inflorescence	Mid bloom

Integrated nutrient management is the combined application of chemical fertilizers along with organic resource materials like, organic manures, green manures, bi-fertilizers and other organic decomposable materials for crop production. Integrated nutrient management is a practice where all source of nutrient namely organic, inorganic (chemical fertilizer), biofertilizer can be combined and applied to soils so that crop growth is enhanced and can get good yield with quality product.

Integrated nutrient management (INM) has to be considered an integral part of any sustainable agricultural system.

Soil fertility in systems under arid and semi-arid conditions referred to as dry areas or dry lands, is constrained by environmental extremes of hot and cold temperatures, as well as by low water availability. With some exceptions, these soils have inherently low fertility, low availability of nitrogen and phosphorus, low water-holding capacity, high pH, low soil organic matter shallowness, stoniness, and other speciûc problems (Matar *et al.*, 1992). These areas are given the vulnerability of these lands to degradation and the arid ecology of Rajasthan suffers from inherent soil or land degradation, caused by overgrazing, lack of adoption of modern farming technologies, limitation of the farmers in ability to replenish nutrients lost in the continuous cultivation, annual bush burring, soil and wind erosion among others. Hence, these lands are of great global signiûcance even if their agricultural production potential is relatively low. There is considerable information on the use of fertilizers and legumes to enhance soil fertility in dry areas, including timing and amount of fertilizers, application methods, crop responses, and effects of crop rotations and also on dryland fallowing and water use (Farahani *et al.,* 1998), however, despite evidence that some nutrient inputs are required to sustain higher agricultural production on these lands. It is recommended that the application of INM is needed, not only to replenish soil nutrients but also to improve the physical, chemical and biological properties of soil and increase the agriculture production with farmer income.

Influence of INM Practices on Biological, Phyco-chemical Properties of Soil and Horticultural Crops

The soil physico-chemical properties of the soil under different INM treatments were measured periodically.(Table 15) reported the changes in the different properties over the year and revealed that pH of the soil did not change much when only chemical fertilizers were applied but on the application of FYM, pH of the soil lower down. On the application of biofertilizers pH of the soil did not change much. Data regarding the organic carbon status revealed that application of FYM increased the level of OC while inorganic fertilizers and biofertilizers have not changed the OC status of the soil. Available P and K_2O also have been affected by the application of INM treatments and recommended dose of N, P and K increased the availability of P and K_2O in the soil and their maximum status were recorded on the application of application of inorganic fertilizers along with FYM. Likewise Availability of zinc and iron content in the soil has also been increased over the application of FYM (Anonymous, 2015-2016).

In general, soil mulched with organic mulches showed beneficial effect in suppressing the fluctuation of soil temperature at 20 cm depth throughout the experiment at Godhara and revealed the significant differences with the organic mulches tried, soil temperature lowered significantly with paddy straw followed by grasses. and soil moisture content was recorded maximum with paddy straw mulch at both the depths of soil (0-15 cm and 15-30 cm). Amongst the organic mulches evaluated, soil moisture ranged 19.80-14.80, 20.60-16.70% in paddy straw and it was recorded 14.80-12.10, 16.40-13.50% in control at both the depths from soil surface after mulching. (Anonymous 2015-2016).

Table 15: Effect of different INM treatments on physico-chemical properties of the soil

Treatment	pH	Organic carbon (%)	Available P (kg/ha)	Available K_2O (kg/ha)	Available Zn (ppm)	Available Iron (ppm)
Control	8.10	0.12	08.50	175.00	0.50	3.50
RDF	8.20	0.15	14.50	210.00	0.50	3.50
RDF + FYM	7.50	0.25	14.50	215.00	0.60	4.25
RDF +Azotobactor	8.00	0.15	14.00	200.00	0.60	3.80
RDF + PSB	8.00	0.15	16.00	200.00	0.60	3.80
RDF + VAM	8.00	0.15	16.50	200.00	0.60	3.80
RDF+FYM + AZB	7.60	0.28	14.50	210.00	0.65	4.80
RDF + FYM + PSB	7.60	0.28	17.50	210.00	0.65	4.80
RDF + FYM + VAM	7.60	0.28	18.50	220.00	0.65	4.80
RDF +FYM + PSB + AZB	7.60	0.28	18.50	220.00	0.65	5.00
RDF + FYM + PSB + AZB + VAM	7.50	0.28	18.50	220.00	0.65	5.00
Initial level	8.20	0.08	08.00	180.50	0.50	3.50

The soil moisture status of the soil under different INM treatment was monitored and revealed that application of FYM alone or in combination with inorganic and bio fertilizers increased the soil moisture status at both the strata. Monitoring of soil status at two depths revealed the more moisture has been accumulated at lower depths (Table 16). Application of bio fertilizers alone did not improve the soil moisture status of the soil (Anonymous, 2015-2016).

Table 16. Effect of different INM treatments on soil moisture of the soil

Treatment	Soil moisture (%) after 24 hrs of irrigation	
	0-0.30m	0.30-0.60m
Control	2.80	3.50
RDF	3.50	3.50
RDF + FYM	5.50	6.00
RDF +Azotobactor	3.40	3.50
RDF + PSB	3.50	3.50
RDF + VAM	3.80	4.50
RDF+FYM + AZB	5.50	6.50
RDF + FYM + PSB	500	6.00
RDF + FYM + VAM	6.00	6.00
RDF +FYM + PSB + AZB	6.50	6.50
RDF + FYM + PSB + AZB + VAM	6.00	6.00

Organic carbon of the soil improved by the application of organic manure with recommended dose of fertilizer, this combination significantly influenced crop growth, development and productivity (Jat, *et al.,* 2013). Most of the research results clearly demonstrated that INM enhances the yield potential of crops over and above achievable yield with recommended fertilizers (Table 16) and results in better synchrony of crop N needs due to (a) slower mineralization of organics (b) reduced N losses via denitrification and nitrate leaching; (c) enhanced nutrient use efficiency and recovery by crops, and (d) improvements in soil health and productivity and sustain high crop yields in various cropping systems ensuring long term sustainability of the system (Sharma *et al.,* 2013). Judicious application of mineral fertilizers and organic manure along with biofertilizers and micronutrients gave highest available NPK in soil as compared to other treatment combination. (Meena *et al.,* 2013, Aulakh *et al*., 2 010). Incorporation of FYM, Green Manure and BGA, through an inorganic source in the treatment increased organic carbon, mineralisable N and reduced the bulk density of soil . (Kumar, *et al.,* 2014 & Rahman, *et al.* 2014). Pintu *et al* (2010) studied the adoption of integrated nutrient management practices helped to build up soil nutrient status with respect to N, P, K, Fe, Mn, Cu and Zn contents. However, the highest organic carbon content (0.88%) was observed in the treatment where 4tonnes /ha organic manure was applied along with recommended levels of NPK and zinc at 0.5 kg/ha.

Bio-fertilizer has been identified as an alternative to chemical fertilizer to increase soil fertility and crop production in sustainable farming. The use of bio-fertilizers effectively enrich the soil and cost less than chemical fertilizers, which harm the environment and deplete non-renewable energy sources. Bio-fertilizers have definite advantage over chemical fertilizers. Chemical fertilizers supply over nitrogen where as bio-fertilizers provide in addition to nitrogen certain growth promoting substances like hormones, vitamins, amino acids, etc, crops have to be provided with chemical fertilizers repeatedly to replenish the loss of nitrogen utilized for crop growth. On the other hand bio-fertilizers supply the nitrogen continuously throughout the entire period of crop growth in the field under favorable conditions. Bio-fertilizers improve the soil structure. Bio-fertilizers, however, have no toxic effects. Bio-fertilizers are commonly called as microbial inoculants which are capable of mobilizing important nutritional elements in the soil from non-usable to usable form by the crop plants through their biological processes.

Biofertilizer is a substance which contains living microorganism which, when applied to seed, plant surfaces, or soil, colonizes the rhizosphere or the interior of the plant and promotes growth by increasing the supply or availability of primary nutrients to the host plant (Table 3). Very often microorganisms are not as efficient in natural surroundings as one would expect them to be and therefore artificially multiplied cultures of efficient selected microorganisms play a vital role in accelerating the microbial processes in soil. Use of biofertilizers is one of the important components of INM, as they are cost effective and renewable source of plant nutrients to supplement the chemical fertilizers for sustainable agriculture. Several microorganisms and their association with crop plants are being exploited in the production of biofertilizers.

Different INM practices were with different growth parameter were studied in Kinnow and bael crops and significant result were observed with application of different treatment with RDF. The fruit weight, fruit yield, TSS, acidity and juice recovery were measured in different INM treatment. It was revealed that maximum fruit weight (235 g) was recorded in RDF of N, P, K + FYM + PSB + Azotobactor + VAM which was significantly at par with RDF of N, P, K +FYM + Azotobactor treatment. The TSS was measured in mature fruits from all treatment and recorded in the range of 12.50 to 15.00 ° Brix and data revealed that addition of FYM, inorganic fertilizers increased the TSS content. The acidity content was maximum in control and inorganically fertilized treatments while FYM reduced the juice acidity. The juice recovery was ranged from 40 to 55 percent and maximum juice (55%) was recorded in those treatments where FYM was the component of the treatment. (Table 20). The Morphological parameter were also observed with the INM practices. The data presented in table 10 revealed that significantly maximum plant height (4.50 m) was recorded in RDF of N, P, K + FYM + PSB + Azotobactor + VAM treatment and minimum was in control (2.80 m). The pattern in plant height revealed that addition of RDF along with FYM and consortium of biofertilizers has the highest increment in plant growth. Likewise plant spread in both the directions was also more in the same INM treatment. The data on stem diameter was also significantly differed among INM treatments and maximum stem diameter was recorded in RDF + FYM + PSB + AZB + VAM and RDF +FYM + PSB + AZB treatments. (Table 21).

Table 20: Effect of different INM treatments on yield and fruit quality parameters of kinnow orchard

Treatment	Fruit weight (g)	Fruit yield (t/ha)	TSS (°Brix)	Acidity (%)	Juice (%)
Control	125	10.00	12.50	0.85	40.00
RDF	165	12.00	12.00	0.70	50.00
RDF + FYM	225	16.50	12.50	0.60	55.00
RDF +Azotobactor	175	12.00	12.50	0.60	50.00
RDF + PSB	165	12.00	12.50	0.70	55.00
RDF + VAM	165	12.00	12.50	0.70	50.00
RDF+FYM + AZB	190	19.00	15.00	0.70	55.00
RDF + FYM + PSB	200	19.50	15.00	0.70	55.00
RDF + FYM + VAM	200	18.50	15.00	0.70	55.00
RDF +FYM + PSB + AZB	225	20.25	15.00	0.65	55.00
RDF + FYM + PSB + AZB + VAM	235	22.50	15.00	0.65	55.00
SE±	15.25	1.95	0.65	0.26	1.20
CD 5%	42.90	5.25	1.80	NS	3.55

Table 21: Effect of different INM treatments on morphological parameter of kinnow orchard

Treatment	Tree height (m)	Tree Spread N-S (m)	Tree Spread E-W (m)	Stem diameter (cm)
Control	2.80	2.65	2.65	80
RDF	3.00	2.70	2.65	80
RDF + FYM	3.65	3.00	3.00	80
RDF +Azotobactor	3.00	2.70	3.00	80
RDF + PSB	3.00	2.90	2.95	80
RDF + VAM	3.00	2.90	2.80	80
RDF+FYM + AZB	4.00	3.00	2.95	80
RDF + FYM + PSB	4.00	3.00	3.00	80
RDF + FYM + VAM	4.00	3.00	3.00	85
RDF +FYM + PSB + AZB	3.90	3.25	3.00	85
RDF + FYM + PSB + AZB + VAM	4.50	3.45	3.00	85
SE±	0.22	0.15	0.18	6.20
CD 5%	0.54	0.42	0.40	16.00

Meanwhile we should Continue to Adopt the Following Pratices of INM.

- Before planting an orchard, soil should be brought to an optimum level by green manuring or FYM application, creation of drainage, pit making and filling mixtures and required quanity of nutrients and soil amendments
- Timing of nutrient application should be need and system based
- Placement of nutrients should be in the absorption zone.
- Use of biofertilizers like Azotobactor, Azospirillum and VAM should be standardized for various horticultural crops.

Conclusion

India has a wide variety of climate and soil on which a large range of horticultural crops such as fruits, nuts, vegetables, root tubers, palms, spices, cashew, cocoa, ornamental crops, medicinal and aromatic plants are grown. The policymakers have realized that the horticultural crops need promotion for achieving sustainability of small holdings, crop diversification, increasing employment, providing an enormous export potential, improving environment, and more than anything for achieving nutritional security. A due importance to nutrients is essential as these affect the productivity, quality and profitability of horticultural crops. To obtain desired yields in horticultural crops sufficient amounts of readily available essential plant nutrients need to be supplied to the plant. Soil nutrients, like all agricultural inputs, need to be managed properly to meet the fertility requirements of horticultural crops. Appropriate nutrient management practices for horticultural crops vary widely due to cropping, topographical, environmental, and economic conditions. With the variety of factors complicating the fertility management in horticultural crops, it is nearly impossible to recommend single set of best management practices uniformly to all the horticultural farms. Nutrient management practices for maintaining or improving farm profitability must be tailored to the unique conditions of individual farms. A number of options for improved nutrient management are available for the horticultural crops and are discussed in this bulletin.

References

Annapurna, K., Bojappa, K.M. and Bhargava B.S. 1988. Leaf sampling guide for sapota (*Manilkara achras* M. Fosberg) cv. Cricket Ball. *Crop Res*., 1: 69-75.

Anonymous 2015-16. Annual Report, ICAR-Central Institute for arid horticulture, Bikaner,334006

Aulakh, M.S. Khurana, M.P.S. and Singh, D. 2010. Water pollution related to agricultural, industrial and urban activities, and its effects on food chain: case studies from Punjab. *j. new seeds*,112-137.

Aulakh, M.S..and Adhya, T.K 2005. Impact of agricultural activities on emission of greenhouse gases - indian perspective. in 'international conference on soil, water and environmental quality - issues and strategies" 319-335

Balasubramanian V. Farmer adoption of improved nitrogen management technolo- gies in rice farming 1999: technical constraints and opportunities for improvement. *Nut. Cycling in Agro–Ecosystem.* 53:93-101.

Bates, T. F. 1971. Factors affecting nutrient concentrations in plants and their evaluation. A Review, *Soil Sci*. 112: 116-130.

Beaufils, E. R. 1971. Physiological diagnosis- A guide for improving maize production based on principles developed for rubber trees. *Fertil. Soc. S. Afr. J*. 1: 1-28.

Beaufils, E. R. 1973. Diagnosis and recommendation integrated system (DRIS). *Soil Sci. Bull. No. 1*, University of Natal, S. Africa.

Beaufils, E.R. 1957. Research for rational exploitation of hevea using a physiological diagnosis based on the mineral analysis of various parts of the plants. *Fertilite*. 3: 27-38.

Bhargava, B.S. and B.M.C. Reddy 1992. Selection of index tissue in banana for nutritional diagnosis. *Indian J. Hort*,. 49(2): 120-126.

Bhargava, B.S. and Chadha, K.L. 1993. Leaf nutrient guide for fruit crops. *Advances in Hort.*, Vol. 2. pp.973-1029.

Bouma, D. 1983. Diagnosis of mineral deficiencies using plant tests. In: A. Lauchli and Bieleski, R. L. (eds) Encyclopedia of plant physiology. Vol. 15A. *Inorganic plant nutrition.* Springer- Verlag, Berlin, pp. 120-148.

Chadha, K. L., J. S. Samra and R. S. Thakur 1980. Appraisal of recommended manurial schedule in mango. *Punjab Hort. J.* 20: 13-23.

Chopra, S.L. and Kanwar, J.S. 1986. *Analytical Agricultural Chemistry*. pp.250-252. Kalyani Publishers, Ludhiana.

Chundawat, B.S. 1997. *Nutritional Management in Fruit Crops*. Agrotech Publishing Academy, Udiapur.

Dey, B.N. 2016. Reclamation of Sand Dunes and Other Shifting Sand Dunes. Available online: http://www

Dobermann A, Cassman K.G. 2002. Plant nutrient management for enhanced productivity in intensive grain production systems of the united states and Aasia. *Plant and Soil.* 247(1):153–175.

Ganeshamurthy, A. N. Satisha, G. C. and Prakash Patil. 2011. Potassium nutrition on yield and quality of fruit crops with special emphasis on banana and grapes", *Karnataka Journal of Agricultural Sciences*, 24.1 .

Ghose, B., Bose, T.K. and Mitra, S.K. 1986. Chemical induction of flowering and control of fruit drop in litchi (*Litchi chinesis* Sonn.) *Proc*., 22nd *Int. Hort. Cong*., California. Abst., 1189.

Iyenger, B. R. V. 1993. Fertilizer management. In: *Advances in Horticulture* Vol. 2 pp. 1055-1072. Chaddha, K. L. and Pareek, O. P. (eds). Malhotra Publishing House, New Delhi.

Jat G, Majumdar SP, Jat NK, Majundar SP. 2013. Potassium and zinc fertilization of wheat (Triticum aestvum) in Western arid zone of India. *Indian J Agron*, 58(1):67-71.

Kar, A.; Moharana, P.C.; Raina, P.; Kumar, M.; Soni, M.L.; Santra, P.; Ajai; Arya, A.S.; Dhinwa, P.S.. Kaul, R.N. 1959. Shelterbelt to stop creep of the desert. *Indian For*., 85, 191–195.

Kumar M.. Dwivedi R, Anand A.K. and Kumar. A 2014. Effect of nutrient on vegetative growth, fruit maturity and yield attributes of phalsa (grewia subinaequalis d.c.), g.j.b.b., vol.3 (3) , 264-268.

Matar, A., Torrent, J., and Ryan, J. 1992. Soil and fertilizer phosphorus and crop responses in the dryland Mediterranean zone, *Adv. Soil Sci*.,18, 81– 146 .

Meena, R.S., Ramawatar, Kamalesh, Meena, V.S. and Ram, Kalu. 2013. Effect of organic and inorganic source of nutrients on yield, nutrient uptake and nutrient status of soil after harvest of greengram. A*n asian j soil sci*.. 8, 80-83.

Muralidharudu, Y., Reddy, K.S.,Mandal, B.N., Subba Rao, A., Singh,K.N. and Sonekar, S., 2011. GIS based soil fertility maps of different states of India. IISS, Bhopal

Pintu Sur Mandal,M. and Das , D. K. 2010. Effect of integrated nutrient management on soil fertility and organic carbon in cabbage (Brassica oleracea var. capitata) growing soil, *Indian Journal of Agricultural Sciences* 80 (8): 38–00, August 2010.

Prasad, R.N. and P.P. Biswas. 2000. In Singh, G.B. and Sharma, B.R. (eds). 50 years of natural resource management research, ICAR. New Delhi, pp 644.

R. N. Roy (1995). FAO, Rome

Rahman, M.H., Islam, M.R. Jahiruddin, M., Rafii, M.Y., Hanafi, M.M., Malek, M.A. 2014. Integrated nutrient management in maize-legume-rice cropping pattern and its impact on soil fertility. *J. Food Agric. Environ*., 11: 648-652

Ramamoorthy B, Narasimham R.L, Dinesh R.S. 1967. Fertilizer application for specific yield target of sonara-64 wheat. Indian Farming. 17:43–45.

Rao S, Srivastava S. 2000. Soil test based fertilizer use a must for sustainable agriculture. *Fertilizer News*.45:25–38.

Ray P.K. Jana A.K Maitra DN, Saha M.N., Chaudhaury J, Saha S, Saha R. 2000. Fertilizer prescriptions on soil test basis for jute, rice and wheat in typic ustochrept. *Journal of Indian Society of Soil Science*. 48: 79–84.

Samra, J.S., R.S. Thakur and K.L. Chadha. 1978. Evaluation of existing critical limits of leaf nutrient standards in mango. *Scientia Hort*. 8: 349-355.

Sehgal, J.; Abrol, I.P.(1994). Soil Degradation in India: Status and Impact; Oxford and IBH: New Delhi, India,1994; p. 80.

Sharma, R. 2013. Effect of long term integrated nutrient management system on soil and crop productivity in Rice -Wheat crop sequence. Indian ETD Repository @ inflibnet, Chaudhary Sarwan Kumar Himachal Pradesh Krishi Vishvavidyalaya, Department of Agronomy Forages and Grassland Management.

Shukla, A.K., Behera, S.K., Subba Rao, A. and Singh, A.K. 2012. State-wise micro and secondary nutrients recommendations for different crops and cropping systems. *Research Bulletin No*. 1/2012, IISS, Bhopal

Singh, N.P. 1981. Studies on guava nutriation with special referance to leaf analysis. *National Symp.Trop.Subtrop. Fruit Crop.* Bangalore, pp. 75

Singh, S.K.., Shahi, S.K. Singh, D.K. and Meena, V.S. 2014: Soil properties and productivity of rice (oryza sativa) as influenced by urea, fym and biofertilizers. *ann. of boil*, 30 (2) : 335-339.

Srivastava, A.K. and Shyam Singh 2008. DRIS norms and their field validation in Nagpur mandarin (Citrus reticulata Blanco). *Journal of Plant Nutrition* 31, 91-107.

Subramanian, T.R., Chadha, K.L, Melanta, K.R. and Srinivasa Murthy, H.K. 1974. Studies on nutrition of pineapple. III. Effect of varying levels of NPK on leaf nutrient status in variety 'Kew' in high fertility soil. *Indian J. Hort*. 31: 219-23.

Sumner, M. E. 1978. Diagnosing the sulphur requirements of corn and wheat using foliar analysis. *Soil Science Society of American Journal* 45, 87-90.

Tandon, H.L.S. 2004. Fertilizers in Indian Agriculture – from 20th to 21st century. FDCO, New Delhi, India. pp. 240.

Yadav, R.L. and Meena, M.C. 2009. Available micronutrients status and relationship with soil properties of Degana soil series of Rajasthan. *Journal of the Indian Society of Soil Science* 57(1) 90-92.

3

Minimization of Pesticide Residues in Horticulture Produce for Ensuring Higher Returns in Export

O.P. Awasthi and Sunil Kumar

Introduction

Horticultural crops comprising of fruits, vegetables, flowers, medicinal and aromatic plants, and mushrooms play an important role in meeting the livelihood, nutritional security and socio-economic development of farmers. India's horticulture production (287 mt) surpassed food grain output (265 mt) fifth year in a row in 2016-17 (First Advance Estimates, 2016-17). Horticultural crops together contribute around 30.4% to the agricultural GDP from nearly 14% of the total cropped area, and contribute nearly 37% to the export of the agricultural commodity from the country (Anonymous, 2015). Fruits and vegetables are an important part of the human diet and a major source of biologically active substances such as vitamins and secondary metabolites. India is the second largest producer of fruits and vegetables in the world, with a combined production of 260.7 million metric tons (91.7 million metric tons of fruits and 169 million metric tons of vegetables) during 2016-17.

Rapid growth in horticultural production has been accompanied by heavy use of pesticides thus causing major concern over health effects. Heavy pesticide use occurs, in part, because numerous pests attack horticultural crops, including the fruit itself, reducing market value and yield on high-value crops. Pesticide use raises safety concerns for agricultural workers who apply pesticides. Concern is particularly high in flower production because of heavy spraying in enclosed conditions. Potential food safety risks from pesticide residues are also a significant issue for importers of fresh fruits and vegetables and a market-risk factor for exporters who may have shipments detained or rejected if residues exceed allowable limits (George *et al*., 2003).

The increased use of chemical pesticides on horticultural crops has raised a number of economic, ecological and health concern. Economic concerns arise from the over reliance and use of chemical pesticides which increase the costs of production. Indiscriminate use of pesticides has resulted in ecological problems of common pests developing resistance, elimination of natural enemies and other beneficial arthropods, and environmental pollution. Human health concerns focus on risks from shortcomings in protective clothing, large deviations from recommended doses/situations, and excessive run-off into the soil and water sources. These concerns are exacerbated by poorly regulated internal markets for pesticides that have fostered usage of banned or outdated products; creating a situation that if not stopped will negatively impact on horticultural exports to countries with more stringent regulatory requirements for fresh crop produce. Meeting these food safety requirements has become a major challenge for the fresh produce export sector of many countries. To ensure and maintain export compliance, grower and consumer safety, and environmental integrity; farmers, government and development partners are developing programmes designed to improve pesticides usage, regulation and management on horticultural crops (Karungi *et al.*, 2011). In this chapter, major important horticultural crops grown in India have been selected to illustrate trends in pesticide usage and regulation, and the development and application of alternative pest management practices through farmer participation and training.

Pesticides

The term "pesticides" can be defined as "a substance or association of substances".

- That is intended to repel, destroy or combat:
- undesirable species of plants or animals causing damage or otherwise harming the production, processing, storage, transport or marketing of foodstuffs, agricultural products, wood and wood products, or animal feed.
- That is used to:
- limit losses in crop yields
- protect stored foodstuffs
- increase the comfort of the farmer
- limit the development of pathogens for humans and animals
- destroy undesirable plants (herbicides, algaecides, moss killers)
- destroy parts of plants, slow down or prevent undesirable growth of plants (haulm destroyers, anti-sprouting agents, etc.).

Pesticide Names

Pesticides have a trade name, which is written in a large letter, on the pesticide label. The active ingredient (poison part) has a common name, which appears on much smaller print on the label. There are several products with different trade names, which all contain the same pesticide.

Pesticide Hazard Classification and Uses

All pesticides can be dangerous, some very dangerous while others are less dangerous. The level of danger is usually marked on the pesticide label as hazard classification in the World Health Organization (WHO) systems. The Categories are Ia, IIb, II, III, or unclassified. Some countries indicate these hazards by a colour code (Table 1).

Table 1: Classification of pesticide based on toxicity

WHO Class	Color of the toxicological band	Danger symbol	LD 50 (mg/kg body weight			
			ORAL		DERMAL	
			Solid	Liquid	Solid	Liquid
Ia. Extremely hazardous EXTREMELY TOXIC	Red		<5	<20	< 10	<40
IIb. Very hazardous TOXIC	Red		5–50	5–50	20–200	10–100
II. Moderately hazardous HARMFUL	Yellow	X	50–500	50–500	200–2 000	100–1 000
III. Slightly hazardous CAUTION	Blue		>500	>2 000	>1 000	>4 000

Source: http://www.who.int/ipcs/publications/pesticides_hazard_2009.pdf?ua=1

Pesticide Formulations

There are different types of formulation such as:

- Emulsifiable concentrate (EC)
- Wettable powders or granules (WG) and Dust

Pesticide Mode of Action

- Insecticides may affect insects by:
- Direct contact
- Secondary contact (insect touching the sprayed leaves or any part of the plant)

- Ingestion (insects eat or suck the liquid of sprayed plant)
- Repelled (insects run away from the smell or taste)
- Fumigant action (the insects breathe in the pesticide vapour).

Herbicides may be

- Non-selective herbicide that kill all the plants
- Selective herbicides, that kill some plants
- Contact herbicide – kill only the leaves they touch
- Translocated herbicides – travel down the roots and kill them too
- Foliar herbicides – applied on the leaves
- Soil applied herbicides –applied to the soil to kill the weeds.

Fungicide may be

- Non-systemic fungicides, only stay on the surface of the leaves and kill fungus on contact. These are preventive fungicides.
- Systemic fungicides – move inside the plant system and kill fungus inside the plant. These are curative fungicides.
- Bio-pesticides: Bio-pesticides are new age crop protection products manufactured from natural substances like plants, animals, bacteria and certain minerals. They are eco-friendly, easy to use; require lower dosage amounts for same performance as compared to chemical based pesticides. The bio-pesticides category currently is a small proportion of the market but has a huge growth potential considering its non-toxic nature.

The Indian crop protection market is dominated by Insecticides, which form almost 60% of domestic crop protection chemicals market. Fungicides and herbicides are the largest growing segments accounting for 18% and 16% respectively of total crop protection chemicals market, respectively. The key growth drivers for fungicides include a shift in agriculture from cash crops to fruits and vegetables and government support for exports of fruits and vegetables. Bio-pesticides include all biological materials organisms, which can be used to control pests. Currently bio-pesticides constitute only 3% of Indian crop protection market; however there are significant growth opportunities for this product segment due to increasing concerns of safety and toxicity of pesticides, stringent regulations and government support. Andhra Pradesh (including Telangana & Seemandhra), Maharashtra and Punjab are top three states contributing to 45% of pesticide consumption in India. Andhra Pradesh is the leading consumer with24% share.

Factors to be considered while choosing pesticide

- Efficacy of active ingredient
- Safer active ingredients
- Safer formulations
- Safer packing
- Mode of action
- Resistance – change the type of pesticide active ingredient regularly to prevent pest resistance building up

Maximum Residue Limits (MRL)

MRL are defined as the Maximum Concentration of Pesticide Residue (expressed as milligrams of residue per kilogram of food/animal feeding stuff) likely to occur in/or on food and feeding stuffs after the use of pesticides according to GAP, i.e. when the pesticide has been applied in line with the product label recommendations and in keeping with local environmental and other conditions.

What is the Purpose of MRL?

MRL are intended primarily as a check that the pesticide is being used correctly (i.e. that the GAP for pesticide usage is being observed) and to assist international trade in treated produce.

How are MRL set by Codex?

The activity of the Codex Alimentarius related to Pesticide Residues is carried out under the umbrella of the Joint FAO/WHO Meetings on Pesticide Residues (JMPR). It was established in 1963 following a decision by FAO Conference that the CAC should recommend MRL for pesticide and environmental contaminants in specific food products to ensure the safety of foods containing residues.

Pesticide Residue in Fruits and Vegetables

Most farmers and commercial producers treat fruit and vegetable with pesticides. This is because pesticide decrease toxins produced by food infecting organisms, increase productivity and are less labour intensive. The presence of pesticide residues is a concern for consumers because pesticides are known to have potential harmful effects on human being. Methomyl is a carbamate insecticide with restricted use because of its high toxicity to humans. The MRLs for methomyl and thiodicarb (sum of methomyl and thiodicarb expressed as methomyl) for the most of fruits and vegetables is set at the 0.02 mg/kg in the

Regulation (EC) 396/2005. The other pesticide residues found in cucumber samples were boscalid, chlorpyrifos, cyprodinil, fenhexamid, imidacloprid, metalaxyl and tebuconazole. Many commodities contained more than one residue per product, up to 9 residues in grapes and tea were determined, up to 5–9 residues in citrus fruits like orange, mandarins, lemons, peaches, pears and up 3–5 residues in pomegranates, plums, cucumbers, tomatoes, strawberries. The most frequently detected pesticide residues were imazalil, thiabendazole, chlorpyrifos, maneb group, procymidone, methidathion, lambda-cyhalothrin, carbendazim, iprodione, orthophenylphenol, vinclozolin, endosulfan, pyrimethanil, fenhexamid, prochloraz, cyprodinil, boscalid. The reason of high detection frequency of thiabendazole, imazalil residues is that these pesticides are widely used for post-harvest treatment of citrus fruits and bananas in order to preserve fruit during the transport process which may take several weeks. Residues of these pesticides were found in 70% of the imported citrus fruits and bananas at or below the MRL. High level of pesticide residues was determined in imported blackberries, strawberries and other berries as well as mangoes, papaya, pepinos, bitter melon, peas, beans, eggplant, spinach and other vegetables.

Table 2: Major products across crop protection chemicals segments

Molecule	Type	Application	Crop
Acephate	Insecticide	Control of severe infestations of sucking & chewing insects	Chillies, vegetables and fruits
Fipronil	Insecticide	Control of diamond moth	Cole crops, Chilli
Flonicamid	Insecticide	Control of all aphid species	Apples, peaches, potato, vegetables
Imidacloprid	Insecticide	Control of sucking pests -aphids, jassids, whitefly	Vegetable crops
Glyphosate	Herbicide	Control of weeds and grasses	Variety of crops
Quizalofop	Herbicide	Control of narrow leaf weeds	Broad leaf crops
Hexaconazole	Fungicide	Control of powdery mildews, rusts & leaf spots	Horticultural & plantation crops

Status of Pesticides in India

There are more than 234 registered pesticides in India and the Indian pesticide industry includes more than 125 large and medium scale producers of more than 500 pesticide products. Among the various pesticide formulations produced, dust formulations constitute about 85% of the total followed by water-soluble dispersible powder (12%) and dispersible powder (2%).

Methods Used to Remove Pesticide Residue

Ever since synthetic organic pesticides were first used in agriculture, investigators have been concerned with the fate of the residues during normal food processing.

Washing, peeling and cooking were soon recognized as important, but not necessarily reliable, means of reducing residue levels in fruits and vegetables. Adventitious removals by various unit processes in the food preparation industry and by home preparation of foods have been carefully studied in several recent investigations. Many of these were carried out by the National Canners Association (NCA) in the United States, as well as by various state agricultural experiment stations.

Washing and Blanching

Loosely held residues of several pesticides on various fruits and vegetables are removed with reasonable efficiency by types of washing processes normally used in home or commercial preparation. This was the characteristic fate of residues of DDT applied as wettable powder to tomatoes where cold water washing by commercial techniques removed approximately 90% and home-type washing some 78% of the total DDT residues. Residual DDT on spinach, which is more closely associated with waxes, was removed less well, even by warm water washing. This removal was considerably improved by adding detergent to the wash water. Detergents also improved the removal of parathion from spinach, carbaryl from tomatoes, DDT from potatoes, and carbaryl and DDT from apricots. However, in some cases detergent washing was decidedly less effective than plain water washing. Washing was ineffective in removing DDT residues from market samples of green beans. This was attributed to the residues being in the inner tissue of the bean and not on the surface. The deeper location may have been related to the formulation used and/or to "field weathering" of the original DDT deposit, which could greatly reduce the surface deposits.

Blanching effectively removes certain residues, although it may not remove much more than will thorough prewashing. In the NCA studies, blanching of green beans removed substantial portions of DDT, carbaryl and malathion; water blanching was generally superior to steam.

Peeling and Trimming

Compounds that penetrate the epidermis are not completely removed by washing, and peeling may completely remove the remaining residue. The recent NCA studies showed peeling to be very effective in the removal of carbaryl, malathion and DDT from tomatoes, and of DDT from potatoes. Similar observations have been made with other vegetables and fruits.

Cooking

The effects of cooking on pesticide residues are extremely varied, but may contribute greatly to residue removal. DDT partially converts to DDD (a less hazardous compound) during canning of several fruits and vegetables. Metal ions, especially iron, are involved in the conversion, which does not occur when the canning is done in glass.

Alternative Methods Used to Control Pests

To ensure food security without affecting the environment and avoiding human health problems, it is important to develop alternative methods of pest management, which include the use of bio-pesticides.

1. Integrated Pest Management

Integrated Pest Management (IPM) is a sustainable approach to pest management using a combination of techniques like biological control, habitat manipulation, modification of cultural practices, and use of resistant varieties (Flint, 2012). In this process, pesticides are used only after ensuring their necessity as per established guidelines, and treatments are made with the goal of removing only the target organism. Pest control materials are selected and applied in a manner that minimizes risks to human health, beneficial and non-target organisms, and the environment. The most effective, long-term way to manage pests is by using a combination of methods that work better together than separately. Approaches for managing pests can be grouped in the following categories:

Biological Control

Biological control is the beneficial action of parasites, pathogens, and predators in managing pests and their damage. Bio-control provided by these living organisms, collectively called natural enemies is important for reducing the numbers of pest insects and mites.

Cultural Control

Cultural controls are preventive practices that reduce pest establishment, reproduction, dispersal, and survival. There are four main strategies for cultural control of pest insects:

1. Reduce and/or disrupt pest habitat in and around crop
2. Adjust crop planting to disrupt pest habitat and nutrition requirements
3. Divert pest population away from crop
4. Reduce yield loss from insect injury

Mechanical and Physical Control

Mechanical and physical controls kill a pest directly, block pests out, or make the environment unsuitable for it. Traps for rodents are examples of mechanical control. Physical controls include mulches for weed management, steam sterilization of the soil for disease management, or barriers such as screens to keep birds or insects out.

Chemical Control

Chemical control involves use of crop protection chemicals (pesticides) to control a range of insect and vertebrate pests, diseases and weeds. In IPM, pesticides are used only when needed and in combination with other approaches for more effective, long-term control. Pesticides are selected and applied in a way that minimizes their possible harm to people, non-target organisms, and the environment.

2. Organic Farming Systems in Reducing Pesticide Residues

Extensive and highly sensitive pesticide residue testing carried out by the U.S. Department of Agriculture shows that conventional fresh fruits and vegetables are:

- Three to more than four times more likely on average to contain residues than organic produce;
- Eight to 11 times more likely to contain multiple pesticide residues than organic samples;
- Shown to contain residues at levels three to 10 times higher, on average, than corresponding residues in organic samples.

Accordingly, growing of organic fruits and vegetables offers consumers an option proven to significantly reduce dietary exposure to pesticides. For many people on most days, consumption of organic fruits and vegetables will virtually eliminate dietary exposure to pesticides and this will, in turn, reduce the frequency and magnitude of one risk factor that can contribute to a variety of diseases and health problems. The opportunity to nearly eliminate pesticide exposure via the diet by consuming organic food is borne out by extensive testing in both the United States and countries abroad.

3. Good Agricultural Practices (GAP)

GAP have been defined as "practices that address environmental, economic and social sustainability for on-farm processes, and result in safe and quality food and non-food agricultural products" by the Food and Agriculture

Organization (FAO, 2003). In India, the Bureau of Indian Standards (BIS) adopted the 'Requirements for Good Agricultural Practices' in 2010. It recommends practices for every stage of farming from land preparation to post harvest supply chain.

Conclusion

The increasing applications of pesticide in different areas of the world initiate researchers to conduct their study on pesticide residue in food commodities and set optimum levels which may be safe for health. MRLs set by different countries/ or organizations legislation are good indication for consumers in aspect of human health prospective. The ability to determine residues of toxicants in food is crucial to support efforts that are designed to protect human health. Residues of organophosphorus pesticides (Ops) were obtained in fruit and vegetables and some of them are above MRLs which may be toxic to human being. Most residues of organochlorines (OC) are also found in grain storages which are reported to cause health of the people. Consumption of organic foods than conventional was recommended to reduce pesticide residue and good for health of human being. Different processing methods like washing, peeling, cooking or boiling, frying, fermentation, grinding of food commodities were also investigated to reduce pesticide residues. Appropriate use pesticide for crops before and after harvest is crucial in management of the residues.

References

Anonymous 2015. Area and Production estimates for horticultural crops. In: Indian Horticulture Database-2014. National Horticulture Board, Ministry of Agriculture, Government of India, 85, Institutional Area, Sector-18, Gurgaon-122015 INDIA. *nhb.gov.in/area-pro/NHB-Database-2015.*

Flint, M. L. 2012. IPM in practice: principles and methods of integrated pest management (Vol. 3418). UCANR Publications.

George, W. Norton, Guillermo, E. Sanchez, Dionne Clarke-Harris, and Halimatou Koné Traoré 2003. Food safety in food security and food trade. *Reducing Pesticide Residues on Horticultural Crops*, IFPRI. Brief, 10.http://*www.who.int/ ipcs/publications/ pesticides_hazard_2009.pdf?ua=1*

Karungi, J., Kyamanywa, S., Adipala, E., and Erbaugh, M. 2011. Pesticide utilisation, regulation and future prospects in small scale horticultural crop production systems in a developing country. *In Pesticides in the modern world-pesticides use and management. In Tech.* www. fao.org/prods/GAP/index_en.htm

4

Varietal Development in Arid Vegetable Crops for Higher Productivity

B.R. Choudhary, D.K. Samadia and P.L. Saroj

Introduction

Vegetables are important constituents of Indian agriculture and play important role in nutritional security due to their short duration, high yield, nutritional richness, economic viability and ability to generate on-farm and off-farm employment. Vegetables form an integral part of staple diet in India where 42% population is vegetarian and vegetables are sources of daily requirement of nutrients, vitamins and minerals. Our country is blessed with diverse agro-climates with distinct seasons, making it possible to grow wide array of vegetables. The role of vegetable cultivation in the country's nutritional security, poverty alleviation and employment generation is becoming increasingly important. There are vast opportunities available in export sector, which need to be harnessed to their fullest ability. Vegetable crops have high yield potential and calories per unit area compared to that of cereals in the shortest time. Additionally, vegetable cultivation creates employment in rural sector, as it is a labour intensive farming practice. Vegetables are good source of ready farm income, besides being rich in health building substances, especially protective nutrients (Kalia, 2010 and Meena *et al.*, 2016).

Globally, India ranks second in vegetable production, contributing 15.8 and 14.0% of global vegetable area and production, respectively. According to National Horticulture durign 2016-17 India produced 178.17 million tonnes of vegetables from 10.24 million ha area with productivity of 17.40 t/ ha. Thus, in the past three decades, India has made a quantum jump in vegetable production as it was merely 15 million tonnes before independence. Due to advent of improved varieties/ hybrids and general awareness of nutritional security among the people, vegetable production of the country is getting momentum continuously. However, considering the shrinkage and fragmentation of land, fast urbanization and increasing population, crop diversification along with seed replacement and the cultivation of improved cultivars of major vegetables is still a challenge in improving the productivity, quality and overall production of vegetables.

The hot arid region of the country is spread over nearly 31.7 million hectare land area in different states. The major part of hot arid region of the country lies in western Rajasthan (19.62 m ha) followed by North-western Gujarat (2.16 m ha). The vegetable production in the arid region is hampered due to very high potential evapo-transpiration rate, intense solar radiations (320-619/ cm^2/day), high wind velocity (20 km/ hr), high infiltration (9 cm/ hr), extremes of temperature (0 to 48^{0}C), low relative humidity, low soil fertility, high soil salinity, very deep ground water, *etc*. These bio-physical constraints reduce the production and productivity of vegetable crops many fold. In Rajasthan, the total production of vegetables is very less *i.e.* only 1.81 million tonnes from an area of 0.17 m ha with a productivity of 10.66 t/ ha (Anonymous, 2016-17). The state has diverse agro-climatic conditions (arid, semi-arid and sub-humid tropics) and the nature has imposed some restrictions which limits the scope for diversification through vegetable crops (More and Samadia, 2008). However, these conditions favour successful cultivation of several vegetable crops such as watermelon/ *Mateera* (*Citrullus lanatus*), *Kachri* (*Cucumis callosus*), snapmelon/ *Phoot* (*Cucumis melo* var. *momordica*), long melon/ *Kakri* (*Cucumis melo* var. *utilissimus*), round melon/ *Tinda* (*Praecitrullus fistulosus*), cluster bean (*Cyamopsis tetragonoloba*), bottle gourd (*Lagenaria siceraria*), ridge gourd (*Luffa acutangula*), *etc*. Other vegetable crops like solanaceous (tomato, brinjal, chilli), bulb crops (onion, garlic), root crops (carrot, radish), cole crops (cabbage, cauliflower, knol-khol), leafy (palak, *Chenopodium*, fenugreek, *Amaranthus*), legumes (cowpea, Indian bean) and okra also have good potential under limited irrigation water facility by adopting suitable production and protection technologies.

Particularly traditional vegetable crops have great potential in arid pockets of country because they are well adapted to existing adverse environmental conditions and possess unprecedented ability to withstand against stresses like moisture stress, high soil pH and load of pests and diseases (Samadia, 2007). The family Cucurbitaceae includes about 118 genera and 825 species, many of which are economically important crops, notably those of the genera are *Cucurbita*, *Cucumis*, *Citrullus*, *Momordica*, *Sechium*, *Lagenaria*, *Luffa*, *etc*. which are best adapted to arid conditions (Pandey and Choudhary, 2012). Despite the several constraints, the hot arid region offers very good opportunity for cultivation of cucurbits like, watermelon, muskmelon, longmelon, bottle gourd, ridge gourd, *tinda*, summer squash, cucumber, *etc*. (Choudhary *et al.*, 2014 and

Choudhary *et al.*, 2015). Snapmelon is good source of carbohydrate (3.0%), protein (0.3%), fat (0.1%) and vitamin A (265 IU) and minerals (0.4%). It is much liked by the people suffering from sugar related disorders. *Kachri* contains carbohydrate (7.45%), protein (0.28%), fat (1.28%), fibre (1.21%), vitamin C (29.8 mg) and iron (0.18 mg). It is one of the components of the delicious vegetable popularly known as *Panchkuta* in the desert districts of north-western India. The mature fruits of snapmelon and *kachri* are usually cooked with various vegetable preparations. It is also used for garnishing the vegetables or used as salad. Spine gourd is good source of protein (3.1 g/ 100 g edible part) and iron (4.6 mg/ 100 g edible part). It is beneficial to those suffering from diabetes, piles, ulcer and problems related to digestion. *Mateera* fruits contain 90-95% water, 3.5-7.4% carbohydrate, 0.18-0.25% protein, 0.10-0.20% fat, 0.25-0.35% fibre, 0.2-0.3% minerals and good source of lycopene. Rind contains citrulline, an amino acid which is involved with atheletic ability and immunity. Besides the nutritional and medicinal value, *Mateera* has a significant role in quenching thirst of dessert masses and keep them cool during hot summers. Its seed kernel contains 35-40% fat and consumed by arid population.

Vegetable Crops Suitable for Arid Region

A large number of vegetable crops are being cultivated in arid region of India under limited irrigation facilities by adopting site management approach (More and Khan, 2009 and Choudhary *et al.*, 2014).

Table 1: List of vegetable crops cultivated in arid regions

Group	Vegetable crops
Cucurbits	Watermelon, muskmelon, snapmelon, *kachri*, longmelon, roundmelon, bottle gourd, ridge gourd, sponge gourd, bitter gourd, ivy gourd
Pea and beans	Pea, cowpea, cluster bean, Indian bean, sword bean, moth bean
Solanaceous	Tomato, brinjal, chilli
Root crops	Carrot, radish
Cole crops	Cauliflower, cabbage, knol-khol
Bulb crops	Onion, garlic
Greens/ leafy	Palak, spinach, amaranthus, fenugreek, coriander, chenopodium

Varietal Development at ICAR-CIAH

A lot of breeding work on vegetable crops has been done at many research institutes in the country which resulted in release of several high yielding disease resistant varieties or hybrids. But the work in respect to arid vegetable crops is still requiring attention. ICAR-Central Institute for Arid Horticulture, Bikaner is the only institute working on arid vegetables and released several improved varieties of *mateera*, snapmelon, *kakri*, *kachri*, ridge gourd, bottle gourd, pumpkin, brinjal, khejri and drumstick suitable for arid environment (More, 2010). The following varieties of different vegetables has been identified and recommended by ICAR-CIAH, Bikaner for arid zone of India (Saroj *et al.*, 2017).

Table 2: Improved varieties of arid vegetables

S.N.	Crop	Variety	Characteristics	Photograph
1	Watermelon (*Mateera*)	AHW -19	Medium-early maturing variety produces 3.0-3.5 fruits per vine, flesh dark pink, solid (firm) with good eating quality and taste having 8.0 to 8.4^{0}brix TSS. High yielder (460-500 q/ ha) and tolerate high temperature.	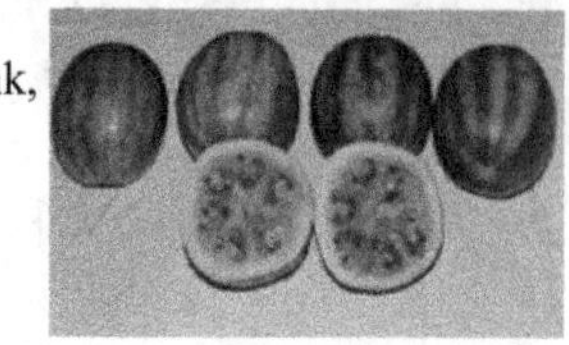
		AHW -65	Very early maturing variety produces 3-4 mature fruits per vine and gives yield of 375-400 q/ ha. The flesh is delicious, pink, solid (firm) having 8.0-8.5^{0}brix TSS.	
		Thar Manak	Developed through selection from the local land races found in arid region. It is very early. First marketable harvesting 75-80 days after sowing (DAS). The yield potential is 50-80 t/ ha under arid conditions.	
2	*Kachri*	AHK-119	Fruits are small; egg shaped weighing 50-60 g. Fruits are ready for picking in 68-70 days after sowing, 22 fruits per vine, and yields of 95-100 q/ ha.	
		AHK-200	The fruits are 100-120 g in weight, become ready for harvest in 65-67 days after sowing, bears about 20 fruits per vine, yield of 115-120 q/ ha.	
3	Snapmelon	AHS-10	Fruits can be harvested	

Contd.

			68 days after sowing. Fruits are oblong and medium in size (900), flesh whitish pink and sweet in taste having 4.5-5.0^{0}brix TSS. Bears 4.0-4.5 fruits/ vine giving yield of 225-230 q/ ha under arid conditions.	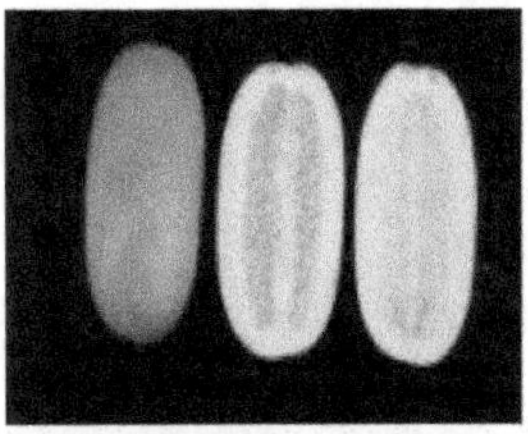
		AHS- 82	Fruit harvest starts 67-70 days after sowing, each vine bears 4.5-5.0 fruits giving an yield of 245-250 q/ ha. The flesh is light pink, sweet having 4.3-4.9^{0}brix TSS.	
4	Kakdi	AHC-2	Very early maturing and bearing uniform, medium long fruits. Fruits are light green, skin without furrows. Harvesting of tender fruits can be done 8-12 days after fertilization for salad or for garnishing vegetables. Fruits become ready for harvesting in 53-55 days after sowing and harvesting continues upto 95-110 days. Fruits weighing 275-300 g are suitable for slicing when their length is 30-35 cm and diameter is 3.0-3.5 cm. The flesh is crisp textured. Produces 12-15 tender fruit *i.e.* 4 kg/ vine and 175-202 q/ ha.	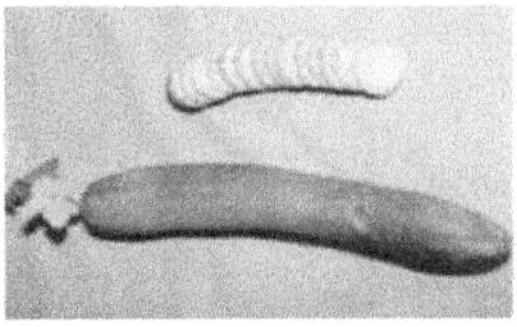
		AHC-13	Very early and highly productive variety with profuse hermaphrodite flowers. For slicing, the fruits can be harvested at very early stage (3-6 days after fertilization). First harvest can be obtained 50 days after sowing and harvesting continue upto 95-100 days. Continuous picking results in higher yield. About 20-25 fruits are borne/vine. The tender fruits weighing 75-110 g are harvested	

Contd.

			when the length is 5.5-7.0 can and diameters are 4.4-5.0 cm. The flesh is crispy and tasty. On an average 2.15 kg tender fruits/vine and 85-125 q/ ha. Can be harvested. The variety also has high heat tolerance.	
5	Longmelon	Thar Sheetal	Early in harvesting and took 45-50 days in first harvesting from sowing. The fruit length, number of marketable fruits per plant and marketable fruit yield varied from 25-30 cm, 18-22 and 132-142 q/ ha, respectively. Bear light green coloured and tender fruits at edible stage which are free from bitterness.	
6	Bottle gourd	Thar Samridhi	High yield potential (3.82-5.82 kg/ plant) under hot arid environment. Fruits weighing 450-700 g are ready for first harvesting after 50 to 55 days from sowing. The fruit yield potential is 240-300 q/ ha.	
7	Ridge gourd	Thar Karni	Profuse branching with vine length of 3.51 m. Long ovary (6.39 cm), fruit cylindrical with 10 shallow ridges and light green colour. Takes 51-55 days to first picking from sowing. Tolerent to mosaic disease and resistant melon fruit fly under field condition. Its fruit set at high temperature during April-May. Fibre content in fruits at edible stage is 1.4-1.6 g/ 100g. Fruit length (21.24 cm), fruit weight (95.74 g), fruit diameter (2.88 cm), number of marketable fruits/ plant (21.03).	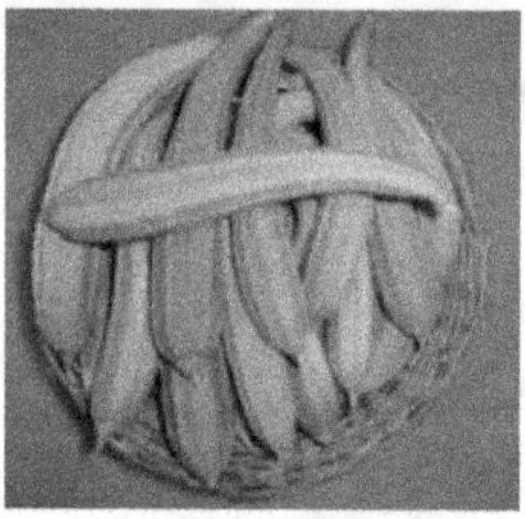

Contd.

8	Pumpkin	Thar Kavi	Fruit are dark green to yellowish, pulp cream to orange colour, number of fruits 8 per plant, yield 8 kg/ plant TSS 8.5^{0}brix	
9	Cluster bean	Thar Bhadavi	High yielding vareity. Pod yield poentital is 75 g/ plant and 9.0 cluster/ plant. Plant height is 65-70 cm. First harvest is 55-60 days after sowing.	
10	Sword bean	Thar Mahi	The pod yield potential is 56.0 q/ ha. First harvest 90-95 days after sowing. Pod yield is 1.73 kg/ plant.	
11	Indian bean	Thar Kartiki	Days to flowering, DAS is 70-75 after 5.94. First harvest is 90-95 DAS. Pod yield is 1.75 kg/ plant and yield 110 q/ ha.	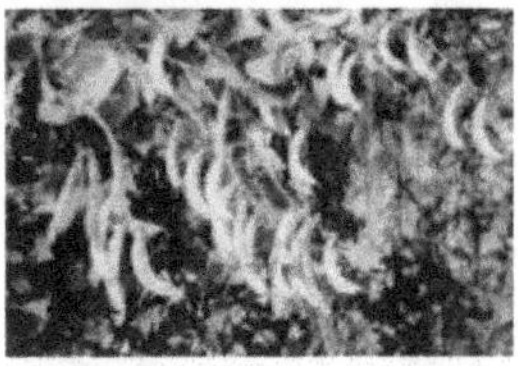
		Thar Maghi	It is high yielding, stable and found to be the most potential for commercial cultivation. For first marketable harvesting, it took 90-95 days in comparison to AHDB-3 (95-100 days). Average pod yield 1.77 kg/ plant and tender post yield is 113.25 q/ ha.	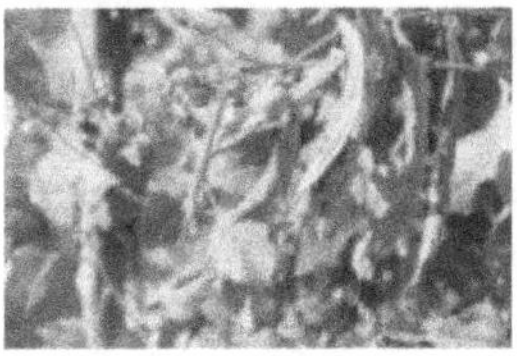
12	Drumstick	Thar Harsha	Plant height: 4.0 m, spread: 3.6 m, Late maturing type (160-180 DAS), Flowering from Dec.-Jan. and harvesting from April-June. Straight green pods (1.0 m long) pointed toward tip, drought hardy. Fruit yield 45 kg/ plant, pod weight 155.3 g. Resistant to be leaf eating caterpillar and moderately resistant to fruit fly,	

Contd.

			Tolerant to moisture and heat stress. Pod suitable for culinary purpose and leaves for both vegetable and fodder purpose.	
13	Khejri	Thar Shobha	High yielding and better quality variety. It has been recommended for uniform tender pod harvesting for vegetable use. A grafted khejri plant yield a harvest of about 4.25 kg tender pods (*Sangri*) and 6 kg dry fodder per year.	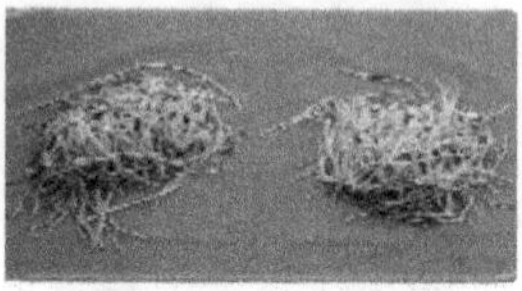
14	Brinjal	Thar Rachit	It is very early (45 days after transplanting). Suitable for cultivation under limited rains (<250 mm) with minimal irrigations adopting channel technology and better with drip irrigation. Plant height at 90 days 51.5-64.4 cm, Plants are short stature and semi-erect with purple hairy leaves. Fruit oblong-round, dark purple colour and small sized. Suitable both for rainy-winter and spring-summer crop cultivation. Fruit weight is 39.8-46.2 g, number of fruit per plant varies from 79.5-92.3 and yield 3.16-4.35 kg/ plant under high temperature and abiotic stresses of hot arid environment.	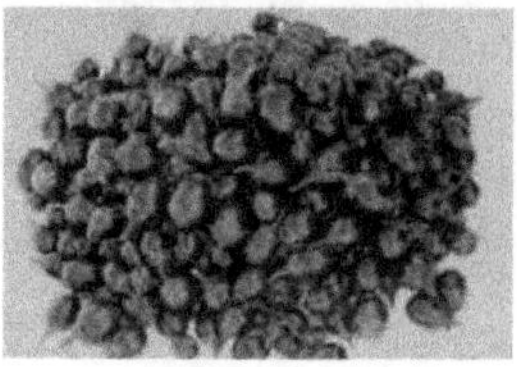

Future Scope and Research Need

The production and productivity of vegetables in arid regions is very low which is due to shortage of seed of improved varieties suitable for stress conditions and lack of suitable low cost/ pro-farmer production technologies. The other challenges which require strategic attention are sub-optimal use of resources, unorganized and scattered areas of vegetable production sites, post harvest losses, very weak agro-based industry, *etc.* which can be addressed by planning R&D at grass root level. Presently there is a great scope in expansion of vegetable crops in arid regions and it has vast potential for changing scenario of vegetable production. The arid region is endowed with surplus family labour, increasing canal irrigated area, plenty of solar and wind energy which offers high potential for successful utilization of wastelands for profitable commercial vegetable cultivation. The future programmes need to be based on the following challenges:

i. Breeding of low and high temperature tolerant lines.

ii. Development of high yielding nutrient rich improved varieties of traditional as well as underutilized vegetable crops.

iii. Standardization of advanced production technologies adapted to diverse socio-economic conditions of arid ecosystem.

iv. Exploration of heterosis for desirable traits like earliness, preferred fruit size, extended harvesting, uniform quality coupled with high marketable yield.

v. Conservation of indigenous germplasm and introduction of desired genotypes for breeding purpose.

vi. Farmers' participatory breeding and seed production approach of arid vegetable crops.

vii. Development of varieties/ hybrids suitable for processing and export purpose.

viii. Adoption of low cost protected cultivation technology under fragile climatic conditions.

References

Anonymous 2015-16. National Horticulture Board, Gurgaon. http://nhb.gov.in.

Choudhary, B.R., Haldhar, S.M., Maheshwari, S.K., Meena, S.R. and Sharma, S.K. (2014). Maximum profit from cucurbits in hot arid regions under changing conditions. *Phal Phool* (March-April). pp. 25-29.

Choudhary, B.R., Sharma, S.K. and T. Janakiram 2015. Cultivation of cucurbits in arid Rajasthan fetches more. *Indian Horticulture* (May-June, 2015) 60(3):22-24.

Kalia, P. 2010. Vegetable crops and nutritional security. *Agricultural Spectrum*, 1(10): 8-11.

Meena, S.R., Sharma, B.D., Haldhar, S.M. and Choudhary, B.R. 2016. Traditional vegetables of the hot arid region and their agro-economic importance. In: Good management practices for Horticultural Crops. M.K. Jatav *et al.* (ed.). NIPA, New Delhi. pp. 237-245.

More, T.A. 2010. Arid horticulture: Technological interventions. NAAS News, 10(1): 3-7.

More, T.A. and Khan, H. 2009. Breeding strategies for abiotic stresses with special reference to vegetable crops. In: *Proceedings of Winter School on Advances in Water and Nutrient Management in Arid Horticultural Crops* held at CIAH, Bikaner from 22 Oct.-11 Nov., 2009.

More, T.A. and Samadia, D.K. 2008. Breeding strategies for vegetable crops with special reference to arid region. In: *Proceedings of National Seminar on Opportunities and Challenges of Arid Horticulture for Nutritional and Livelihood* held at CIAH, Bikaner from 8-9 March, 2008.

Pandey, Sudhakar and Choudhary, B.R. 2012. Underutilized Cucurbits. In: Potential Perspective of Underutilized Plant Species. K.K. Behera (ed.). Narendra Publishing House, Delhi. pp. 87-128.

Samadia, D.K. 2007. Arid Vegetables. In: Underutilized and under exploited horticultural crops. Vol. 2. K.V. Peter (ed.). New India Publishing Agency, New Delhi. pp. 153-174.

Saroj, P.L., Bhargava, R., Singh, R.S. and Krishna, H. 2017. ICAR-CIAH: An overview. CIAH/Tech/Pub./No. 60. pp. 11-14.

5

Integrated Disease Management in Arid Horticulture Crops

S.K. Maheshwari, P.P. Singh and M.K. Jatav

Introduction

Arid Horticulture denotes cultivation of fruits and vegetables under dry and arid regions. Arid region in India is stretched over 39.54 million ha (31.71 million ha is under hot arid region and 7.38 million ha under cold arid region) comprising about 12.02% of the total area of the country. Most of the precipitation in north-western arid region takes place from July to September. Occasional frost has also been a grave issue for horticultural crops, especially, aonla and moringa (Dhandar and Saroj, 2004). The region is exemplified by low and erratic rainfall with extremes of temperature (1–48°C), high wind velocity and sandy soils (Bhandari *et al.* 2014). Growing of annual crops restricted only a particular season of the year is the merely source of livelihood for the inhabitant of these region. These crops often encounter abiotic stresses like drought, hot and desiccating winds during their growth period leading to poor productivity. These regions have tremendous potential for cultivation of perennial fruit trees.

Horticultural crops have great importance in these days due to their potential in terms of nutritive values. Development of suitable varieties/hybrids/technologies in horticultural crops through research and their wide spread dissemination among the farmers and developmental policies of the government culminated in tremendous increase in area (23.4 million ha) and production (283.5 million t) in India. India is the second largest producer of fruits and vegetables in the world (Anon., 2016). The success or failure of crop production relies on interaction among varieties/genotypes, inputs used, environment and management factors. Plant diseases caused by fungi, bacteria, viruses, phytoplasmas and nematodes reduce crop yields (Khoury and Makkouk, 2010). Arid horticultural crops like ber, bael, aonla, date palm, pomegranate, lasora, watermelon, kachri, snapmelon, muskmelon, bottle gourd and ridge gourd are mostly grown in Central and Western India. Horticultural crops can be infected by *Alternaria* leaf blight, *Cercospora* leaf spot, downy mildew, *Fusarium* wilt, mosaic disease, sucker rot, *Alternaria* fruit rot and post harvest diseases, etc. These diseases are major challenges in the cultivation of horticultural crops in India, which inflict heavy losses to the crops every year. Disease incidence in a crop is governed with the seed materials used and environment surrounding seed/plant. Thus, before proposing the symptoms and etiology of any disease, a complete knowledge of disease management strategies are essential. Major diseases of fruits and vegetables have been managed in hot arid system for higher returns through following advance approaches:

Present Status in India and Global Scenario

A) Integrated Disease Management in Arid Fruit Crops

1) Aonla Diseases

Aonla or Indian gooseberry (*Emblica officinalis* Gaertn.) is an ancient, indigenous fruit of India. This crop is quite hardy, prolific bearer and is remunerative for the farmers having marginal land. There has been increase in efficient utilization of natural resources, rehabilitation of wastelands and improved economic return to farmers coupled with enhanced employment. Area and production of aonla in India is 0.91 lakh ha and 1.025 million tonnes, respectively during 2016-17 (Anonymous, 2017). The edible *aonla* fruit tissue has 3 times the protein concentration and many times the ascorbic acid concentration of an apple. It is rich in polyphenols, tannins, minerals and is regarded as one of the richest source of vitamin C. The fruit also contains a high amount of salt, carbohydrates, phosphorous, calcium, iron and vitamins (Table 1).

Table 1: Nutritional value of fruit of *aonla* (per 100g edible portion).

Edible portion (%)	Moisture (%)	Protein (g)	Fat (g)	Minerals (g)	Fiber (g)
89	81.8	0.5	0.1	0.7	3.4
Carbohydrates (g)	**Calcium (%)**	**Phosphorous (%)**	**Iron (mg)**	**Vitamin C (mg)**	**Nicotinic acid (mg)**
14.1	0.05	0.02	1.2	600	0.2

Rust disease, anthracnose and post harvest diseases are major constraints in *aonla* crop production in some parts like Rajasthan and Uttar Pradesh states. In other countries like China, the brown spot, false anthracnose and powdery mildew were reported. The disease caused considerable losses in major aonla growing areas of Uttar Pradesh (Rawal, 1993). The disease was prevalent in growing areas in all popular and local cultivars in Jobner (Jaipur), Ajmer, Jodhpur, Jhunjhunu and Nagaur districts as well as Bawal and Mandi (Anon., 2010).

i) Rust Disease

The importance and its severity has first observed in Rajasthan by Tyagi (1967). It has also been observed in Lucknow (U.P.) and Rajgarh area of Pratapgarh by Misra (1988). It has been reported from Dehradun, Nagpur, Tanjore and Bihar. Most of the wild or *desi* types grown under Chomu area of Rajasthan are affected during the months of December and January.

Symptoms

Reddish spots are formed on plant leaves from the beginning of August month. Conspicuous brown rust pustules develop on the leaves and fruits and later these pustules become dark brown to black. Under severe condition, many pustules coalesce together and cover a large fruit area. Severely infected fruits may lead to premature fruit drop. The reddish brown uredo pustules can be seen on both surfaces of the leaflets. In susceptible leaves, pinkish brown pustules develop.

Casual organism- *Revenelia emblicae*

Disease Cycle

Uredia of this pathogen are typically sub-epidermal erumpent bearing smooth walled uredospores produced singly on pedicel. The uredosori rupture the epidermis and uredospores are exposed. Growth stages of uredospores and the reactions host tissue and fungal colony can be observed. Telia are sub-epidermal in origin, erumpent bearing pedicellate teleospores. The uredospores are easily blown away by air which infects the new leaflets and fruits during growing season.

Epidemiology

Under Rajasthan conditions, disease development was more when relative humidity increases, while temperature decreases (Anon., 1989). At Faizabad, the rust symptoms on fruits were observed in 2nd week of October (40th meteorological week) and reached on peak in the month of December (52nd meteorological week). Temperature showed highly significant negative correlation whereas RH (morning) and sunshine hrs/day showed significant positive correlation with per cent disease intensity (Anon., 2015).

Integrated Disease Management

Cultural Control

(a) Sanitation and clean growing of aonla.

(b) Orchard should be clean for growing of aonla plantation.

(c) Proper pruning should be done in orchard for reducing humidity.

(d) Selection of site of the field

(e) Use of disease free planting material

(f) The orchards should be free from weeds.

Host Resistance

a) Cultivar Francis and NA-6 has been found PDI with 5.50% and 15.50%, respectively against this disease under Faizabad conditions (Anon., 2015).

Chemical Control

a) Three sprays with dithane Z -78 (0.2%) at monthly interval during the months of July to September proves effective against this disease.

b) This disease can also be controlled by wettable sulphur (0.25%) during July- September.

ii) Anthracnose

It is a serious disease in aonla growing areas. This disease results heavy loss in fruit yield in susceptible cultivars in humid region.

Symptoms

This disease appears on leaflets and fruits in August-September (Mishra and Shivpuri, 1983). At the beginning, symptoms are minute, circular, brown to grey spots with yellowish margin in leaflets. Leaves dried up at advance stages. In fruits, pin lead like spots appear with dark brown to pink having yellow halo, while on fruits, depressed lesions develop which later turn dark. In severe cases, many spots coalesce with each other to form bigger lesions. The central areas of spots remain grayish raised with dot like fruiting bodies. Under severe condition, plenty spots result the fruits become shrivel and rot.

Causal organism- *Colletotrichum gloeosporioides*

Disease Cycle

The fungus perpetuates in infected crop debris and in the soil. Spores are released from last year's diseased tissue. Conidia germinate by germ tube which form appressorium after coming in contact with surface of the host. The infection peg arises from the appressorium and penetrates the host. The spores are carried out by rain splash or wind and cause secondary infection to soft and succulent tissues. The fungus obtains nutrients from plant cells and these cells are killed by leaf lesion. The lesion expands as the fungus spreads.

Epidemiology

Temperature and moisture are one of the most important factors which influence the disease development. The disease is favoured under hot and humid weather. This disease responds positively to increase in the temperature along with cloudy weather.

Integrated Disease Management Strategies

Cultural Control

(a) Discard affected fruits and leaves at initial stage from the orchard.

(b) Proper pruning should be done in orchard for air circulation.

(c) Plant trees on a wide spacing and keep the surrounding area clear of vegetation.

Chemical Control

Spraying of carbendazim (0.1%) or difolatan (0.2%) can reduce the disease.

iii) *Alternaria* Fruit Rot

It was noticed in Allahabad (Uttar Pradesh). Pandey *et al.* (1984) found *Alternaria alternata* with dropped fruits of aonla. It is also a serious problem in aonla growing areas. This disease results heavy loss in fruit yield up to 15-25% approx. in humid region.

Symptoms

Initial disease symptoms appear as small brownish to black spherical necrotic spots on fruits and increase in circular fashion with the development of the disease. In advance stage, the spots become dark brown to black and neighbouring spots coalesce. The middle part of this infected tissues become soft and pulpy.

Causal organism- *Alternaria alternata*

Disease Cycle

The fungus perpetuates as mycelium or spores in infected debris of plant parts. Conidia can remain viable up to few months at room temperature. The infection starts from the fruits, where a large number of conidia are formed. These are spread by wind, water, rain splash and cause secondary disease. The conidia germinate in the presence of moisture at temperature nearly 25-30 °C giving rise to germ tubes, which enter the host tissue.

Epidemiology

Presence of high moisture, followed by warm and dry weather conditions is most favourable for disease development. Very high temperatures along with continued dry spells check this disease. Dew and frequent rains accompanied with moderate to high temperature results in disease development.

Integrated Disease Management

Cultural control

a) Use of disease free planting material

b) Removal of infected plant debris from the aonla orchard.

c) Harvesting should be done very carefully to avoid any injury to the fruits.

Chemical Control

a) Mancozeb (0.25%) sprays at 10 days interval to reduce this disease.

b) One spray of carbendazim (0.1%) should be done 15 days prior to fruit harvest.

c) Treatment of fruits with borax (0.5%) checks the rot.

Biological Control

Use of bio-agents (*Trichoderma* spp. and *Pseudomonas* spp.) could be useful in managing the disease.

2) Ber (*Ziziphus mauritiana* Lamk.)

Ber (*Ziziphus mauritiana* Lamk) is one of the most important fruit crops grown in arid and semi-arid regions of India (Pareek, 2001) and worldwide. *Ber* also called as desert apple, jujube, *Badari* (Sanskrit), *Boroi*, Beri, Indian plum is a tropical and subtropical fruit native to the northern hemisphere. Mostly, *ber* cultivation is common in water deficient and low rainfall areas of Rajasthan, Gujarat, Maharashtra, Haryana, Uttar Pradesh, Bihar, Madhya Pradesh and Andhra Pradesh. India with an area of approximately 1 lakh ha, ranks second among jujube growing countries in the world after China. The annual production of jujube in India is estimated to be 10 lakh tons with the productivity of 10 metric tons/ ha (Bal, 2016). In view of the gaining popularity, area under this fruit is being increased gradually day by day. *Ber* fruits are natural antioxidants for people in arid regions. The fruits are quite nutritious, rich in vitamin C, second only to aonla and guava and much higher than citrus and apple. Its fruit products are getting commercial angle in different countries, particularly for their health benefiting properties (Vithlani and Patel, 2010). Ber diseases like powdery mildew, leaf spots and fruit rots in arid and semi-arid region and post harvest diseases in marketing locations are economically important. Complete loss of ber fruits due to havoc nature of powdery mildew is being experienced in semi-arid and humid regions. It affects productivity and quality of fruits. Being an obligate parasite, recurrent incidence results in heavy loss up to 80% in susceptible cultivars round in humid region every year.

i) Powdery Mildew

It is a major disease causing great loss in production and quality of *ber* fruits in semi-arid and sub-tropical region but it is not serious in Bikaner and nearby regions from last few years. This disease was first reported from Allahabad. Kumar *et al.* (1978) noted the occurrence of this disease of *ber*, maximum disease incidence (38- 60%) was observed in local cultivars Kali and Bagwari whereas Seb was least susceptible. This disease causes heavy yield losses up to 10-30% (approx.) in humid region.

Symptoms

Initial symptoms appear as white floury patches on young fruits at pea size and later cover the entire fruits. The infected fruits become misshapen, corky and finally are dropping. In severe incidence, floral parts, whole fruits, tender branches and leaves would appear with fungal conidia and less number of fruit setting and malformed fruits. *Ber* orchards are devastated completely when the powdery mildew occurs in severe form at flowering and fruit setting stages.

Causal organism: *Oidium erysiphoides f.* sp. *ziziphi*

Epidemiology

The cloudy, humid with moderate temperature favour its occurrence in ber orchards. In North India, infection appeared from October and reaching peak by December. Maximum severity could occur at max. temperature of 21.9°C, 59- 88 % relative humidity and 9.6 hours sunshine per day.

Integrated Disease Management

Cultural Control

a) Removal and destruction of infected leaves, twigs and collateral hosts from orchard.

b) Ber orchards should be free from wild species of *ber* (*Ziziphus nummularia*).

c) Summer ploughing of orchards help in desiccation of soil borne pathogen.

Biological Control

a) *Psendomonas fluorescens* CIAH- 196 (1%) was also moderately effective against this disease (Nallathambi, *et al.*, 2006).

Host Resistance

Plantation of resistant varieties is the best method for reducing the disease. Use of resistant/tolerant varieties such as Sanur-2, Chinese, Jogia and Vikas must be grown. Resistance types (Chinese, Jogia and Vikas) identified by AICRP on Arid Zone Fruits may be the potential source of resistance for varietal improvement.

Chemical Control

i) Baylaton (0.1%) recorded the highest disease control (90.5%) , followed by Karathane (0.1%) (Sharma *et al* 1983).

ii) Karathane (Dinocap) at 0.1% dose resulted the best control of powdery mildew of ber (Singh *et al.*, 1995)

ii) Black Leaf Spot

This disease was first reported from Haryana by Gupta and Madan (1977). It is common in Southern and Northern part of the country. Verma *et al* (1992) have recorded this disease from Punjab. Disease incidence was noted during survey program of Rajasthan and Uttar Pradesh from 01- 8.0 per cent and 4.75 to 38.75 per cent, respectively (Anon., 2010). It can cause heavy damage on leaves and thereby the yield is reduced up to 10-25%.

Symptoms

The disease is characterized by sooty tuft like circular to irregular black spots on lower surface of leaves. Leaves and twigs are dried rapidly under severe infection.

Causal organism- *Isariopsis indica* var. *ziziphi*

Epidemiology

Cloudy weather with moderate temperature in October- November is favourable for disease development. In

Integrated Disease Management

Cultural

a) Removal and destruction of infected leaves, weeds and twigs from orchard.

b) To keep the ber orchard free from all weeds.

c) Pruning should be done properly in the orchard.

Host Resistance

Use of resistant varieties such as ZG-3, Seb, Bahadurgarhi and Safeda Rohtak.

Chemical Control

Two fungicidal sprays of mancozeb or copper oxychloride @ 0.2% at 15 days interval were effective (Verma and Cheema, 1988).

3) Date palm (*Phoenix dactylifera* L.)

Date palm is of economic importance and represents a source of income to many farmers in large parts of Iraq, Iran, Arabian Peninsula and countries in North Africa and represents a good cash crop for many farmers. Palm diseases are among the major factors that affecting the products. Fungi and Phytoplasma are known as the most causal pathogens on date palm trees. At Bikaner, survey of date palm diseases was carried out at farmers field orchards like Bikaner, Bajju (under marshy land), Raiser, Khajuwala, Dantour, Nokha, Birmana (Sriganganagar under marshy land), Lakhuwali (Hanumangarh marshy land) and Fatehpur (Anon., 2010). Some diseases were also recorded from Saudi Arabia and Iraq. Several trees showed symptoms of wilt and dieback disease in Saudia Arabia and in particular in Al Qassim and Medina Al Monawara regions.

i) *Alternaria* Leaf Spot

It has been observed in severe form at Date Palm Research Centre, SKRAU, Bikaner and disease incidence was recorded from 14.16 to 72.50% depending upon cultivars as well as climatic conditions.

Symptoms

The spots are most common on the lower leaves (pinnae) of the plants, whereas, on upper leaves the spots are few and small size. The disease cause heavy losses to the date industries in both quality and quantity of production (Pal *et al*., 2006).

Causal organism- *Alternaria alternata*

Epidemiology

It requires hot and humid weather conditions for disease initiation. Disease development is favoured between 20 to 30° C.

Integrated Disease Management

Cultural Control

a) Removal and destruction of infected leaves, twigs and weeds from orchard.

b) Pruning should done.

c) To disinfect all tools in pruning, etc. and cut surfaces.

Host Resistant

Use of resistant varieties such as Medjool, Nagal Hilali, Sriganganagar and Medini (Pal *et al*., 2006).

Chemical Control

a) Fungicidal sprays at 15 days interval (mancozeb @ 0.2%) was given to minimize the disease.

b) Mancozeb and copper oxychloride (0.2%) are also useful for retarding the fungal growth.

ii) Graphiola Leaf Spot

The disease is also called false smut on date palm. It is the most widely spread disease and occurs wherever the date palm is cultivated under humid conditions and hyper-arid partially irrigated western plains of Bikaner, Jaisalmer, Barmer and Jodhpur. This disease was not observed at Date Palm Research Farm, SKRAU, Bikaner and demonstration field of farmers (Anon., 2010). This disease was also found in Egypt.

Symptoms

Symptoms of the disease appear as sub-epidermal spots on both sides of the pinnae (leaf flat) and on the rachis with small black sori (fruiting bodies) developing in abundance on old

fronds. Fruiting structures emerge as small-yellow/brown to black sori. These sori are abundant on three year-old leaves. The sori are 1-3 mm in diameter.

Causal organism- *Graphiola phoenicis*

Epidemiology

The disease is widely spread and occurs whenever the date palm is cultivated under humid conditions but absent in less humid regions.

Integrated Disease Management

Cultural Control

a) Removal and destruction of infected leaves, twigs and weeds from orchard.

b) Pruning should be done in orchard.

c) To disinfect all tools in pruning, etc. and cut surfaces.

d) Use of disease

Host Resistant

i) Grow genetically tolerant varieties (Barhee, Adbad, Rahman, Gizaz, Iteema, Khastawy and Tadala).

ii) Use of variety 'Hatemi' was found to be resistant (Pundir, *et al.* 2006).

iii) Variety 'Khadrawi' was found resistant with disease severity below 5% under field conditions of Haryana.

iv) Ten date palm cvs grown in Egypt were tested against this disease. The most resistant were Barakawi and Aglanee.

Chemical Control

i) Carbendazim (0.1%) is most effective followed by thiophanatemethyl (Mehta et al., 1989).

ii) Two sprays and soil drenching of copper oxychloride (0.2%) at fortnight interval in checking the infection and further disease spread.

4) Pomegranate (*Punica granatum*)

It was cultivated mostly in Asia, Northern Africa, Southern Europe, Egypt, Afganistan and Balichistan.

i) Bacterial Leaf and Fruit Spot

Bacterial leaf spot was first from Delhi in 1952. It is a major constraints for productivity and quality of fruits in Southern states particularly in Tamil Nadu, Maharashtra and Karnataka as well as in some parts of Gujarat and Rajasthan. It appeared as an epidemic form in Bangalore, Karnataka, causing 60-80% fruit yield losses. Dhandar *et al* (2004) have made in extensive survey in Solapur and Sangola districts of Maharashtra against this disease and other aspects. This disease is prevalent in all the pomegranate growing states of India. Highest severity (25.8%) was recorded in Anantapur district (Anon., 2010).

Symptoms

Small, transluscent water soaked spots appeared on the leaves and became necrotic and finally may turn dark brown. Many spots may coalesce and cover big area which are surrounded by chlorotic halo. Leaves are distorted and defoliated. Water soaked spots on fruits may turn dark brown. The spots are oily in appearance.

Causal organism: *Xanthomonas axonopodis* pv. *punicae*

Integrated Disease Management

Cultural control

a) Removal of weed hosts and other infected plant parts.

b) Pruning should be done in pomegranate orchard.

Chemical Control

Spraying of 200 ppm streptocycline can be controlled followed by Pausamycin (0.05%) + Copper oxychloride (0.2%) with 3 sprays at fortnightly intervals (Suriachandraselvan *et al*., 1993).

Host Resistant

Variety 'Jalore seedless' was resistant to this disease (Nallathambi *et al*, 2006).

5) *Bael* (*Aegle marmelos* Correa)

i) *Alternaria* Leaf Spot

Alternaria leaf spot were reported by Madan and Gupta (1985). This disease was observed in Chomu area (Jaipur district) of Rajasthan (Anon., 2010). It was also found in some districts of Uttar Pradesh.

Symptoms

The minute brown spots appear on the leaves which enlarge and become prominent with reddish brown lesions. These lesions later on coalesce into irregular patches. The affected leaves drop down finally.

Causal organism: *Alternaia* sp.

Integrated Disease Management

Cultural Control

(a) Field sanitation

(b) Pruning of the plants in orchard.

Chemical Control

Two sprays of indofil M- 45 (0.2%) at 15 days regular interval.

B) Integrated Disease Management in Arid Vegetable Crops

1) Bottle gourd (*Lagenaria siceraria)*

i) *Alternaria* Leaf Blight

This disease affects the leaves of cucurbits like bottle gourd, watermelon, muskmelon, and pumpkin. It was observed in U.P., M.P. and Bihar. Disease severity varies depending on the prevailed weather conditions. Disease incidence was recorded up to 34% in bottle gourd genotypes under hot arid conditions and causes fruit yield losses from 10 to 25%.

Symptoms

Characteristic symptoms first appear on leaves as small, circular and light brown spots of different size which later enlarge in a concentric rings and margins appear. These spots mix together to form larger necrotic areas on leaves.

Causal organism: *Alternaria cucumerina*

Integrated Disease Management

Cultural Control

a) Removal of infected crop debris, collateral hosts and weeds.

b) Use of certified seed.

c) Deep ploughing of the soil

Chemical Control

a) Sprayings of indofil M- 45 (0.2%) at regular interval was very effective for reducing this disease.

b) Integrated management measures against *Alternaria* leaf blight were developed under field conditions. Combined treatment of carbendazim @ 0.1% (Seed treatment) + mancozeb @ 0.25% (Foliar spray) + *Pseudomonas fluorescens* @ 5% (Foliar spray) + neem leaf extract @ 5% (Foliar spray) was the most effective with minimum disease incidence of 9.25%, minimum disease severity of 7.07% and maximum disease control (78.23%) (Maheshwari *et al*., 2017).

ii) *Cercospora* Leaf Spot

This disease caused by *C. citrulina* Cooke was observed on cucurbits (Sarbhoy 2006). Disease incidence was noted from 6.70 to 39.25% in different bottle gourd genotypes under arid conditions. First record of *Cercospora citrulina* leaf spot on bottle gourd in Pakistan was reported by Mukhtar *et al*. (2013).

Symptoms

Symptoms occur primarily on foliage. Small, circular to irregular circular spots with tan to light brown lesions with dark margins appear on older leaves. The number and size of lesions increases. These lesions coalesce with each other and causing entire leaves to become diseased. Infected leaves turn yellow under severe conditions and finally fall off.

Causal organism: *Cercospora* spp.

Integrated Disease Management

Cultural Control

a) Removal of infected crop debris.

b) Deep plough.

c) Growing the crop with non-host crop.

d) Destruction of weeds and collateral hosts.

Host Resistance

Use of moderately resistant varieties such as Pusa Naveen, Pusa Santushti, Pusa Samridhi and Pusa Sandesh against *Cercospora* leaf spot (Maheshwari *et al.*, 2015).

iii) Powdery Mildew

Powdery mildew affects the leaves of cucurbits like bottle gourd, cucumber, muskmelon, squash and pumpkin. The disease causes maximum damage in the warm and dry areas where moisture is present as dew. In India, the disease is prevalent in almost all the states and causes losses in terms of fruit yield and also quality (Gupta *et al.*, 2001).

Symptoms

The symptoms appear as small floury patches on leaves first on the lower surface followed by upper surface. Fungal growth on these spots results in production of powdery growth. Severely affected leaves lose their dark green colour and become pale yellow green to brown and are shriveled and premature defoliation, drying and collapse. Fruits remain undersized.

Causal organism: *Sphaerotheca fuliginea*

Integrated Disease Management

Cultural Control

a) Removal of infected crop debris.

b) Deep ploughing may burry the pathogen where it may be killed by soil microflora of the pathogen present in below soil layers, after ploughing, may be brought to the soil surface and inactivated by solar radiation.

c) Destruction of weeds and collateral hosts.

d) Crop sequence with appropriate crops may be the most effective management practices in eliminating soil borne pathogens.

e) Introduction of resistance or non-host crops in rotation helps in disease reduction.

Host Resistance

Use of moderately resistant bottle gourd varieties (Pusa Naveen, Pusa Santushti, Pusa Sandesh and Arka Bahar) against this disease (Maheshwari *et al*., 2012)

Chemical Control

a) Two to three sprays of wettable sulphur (0.2%) at 20 days interval at the time of symptoms appearance was found effective.

b) Three sprays of hexaconazole (0.05%) were found effective (Gupta and Gupta, 2001).

c) Foliar sprays of non-target chemicals (monopotassium phosphate) and neem kernel extract were effective

d) Foliar sprays of systemic fungicides like benomyl and carbendazim (0.1%) are very effective.

2) Muskmelon (*Cucumis melo* L.)

Muskmelon (*Cucumis melo* L.) popularly known as 'kharbuja' in India is one of the most important cucurbits grown as a 'Dessert crop' throughout the world. In India, its production is concentrated in the tropical and sub-tropical regions. It is grown under both riverbed and irrigated conditions for local and interstate sales and it is a high value dessert fruit compared to other fruits. Area of muskmelon crop was 45,000 ha with production of 9,35,ooo MT in India during 2015-16 (Anon., 2016).

i) Wilt

It is a threat to the cucurbitaceous crops (bottle gourd, cucumber, muskmelon and watermelon) of all over the world (Radhakrishnan and Sen, 1985). Wilt in cucurbits is found in Ukraine and Japan. The disease is the most destructive at places wherever soil temperatures are high enough for the pathogen to grow but unsuitable for the plants. Yield losses up to 80% have been reported in the worst affected areas (Gupta *et al*., 2001). Disease incidence was also observed from 5.0 to 60% at experimental field of the institute.

Symptoms

Disease appears at flowering stage of the crop. The leaves are accompanied by yellowing and marginal necrosis. The infection results in lesion formation on the collar region and infected areas appear brown and water soaked. Wilting starts sudden after fungal infection. As a result of softening of the tissue, the plants shriveled, followed by rapid mortality of whole plant. The older plants wither and die during the growing season.

Causal organism: *Fusarium oxysporum* and *Fusarium acuminatum*

Epidemiology

A large population favoured by continuous cropping of susceptible cultivars is required for serious disease outbreaks. Saprophytic activity of this disease is highest at 60% moisture holding capacity of the soil, pH 5.0-6.0 and temperature of 25 to 35° C. Most severe symptoms are observed between 18 and 22° C.

Integrated Disease Management

Cultural control

a) Crop rotation with garlic, radish, onion has been found to reduce the disease in melons and cucumber.

b) Deep ploughing.

c) Field sanitation and uprooting of the infected plants.

d) Growing the crop with non-host crop.

e) Wilt suppressiveness can be induced in soil by repeated croppings of resistant varieties of melons.

Host Resistance

a) In muskmelon, The Indian cultivars (Durgapura Madhu and Punjab Sunehri) are resistant to *Fusarium oxysporum* and *F. solani* (Radhakrishnan and Sen, 1985).

b) Mark *et al.* (2005) found Hannahs Choice F1 as resistance source against *Fusarium* race 2 as well as powdery mildew in muskmelon.

Biological Control

Wilt in cucurbits can be suppressed by *Pseudomonads* isolates.

Chemical Control

Drenching of carbendazim (0.1%) is very effective against this disease.

3) Watermelon/*mateera* (*Citrullus lanatus*)

In India, watermelon production is 23,25,000 metric tons with area of 95,000 ha during 2015-16. The fruits are rich source of β-carotene, vitamin B, C and E, minerals (K, Mg, Ca and Fe), citrulline amino acid and phenolics. Watermelon is also known as *Tarbooj* in some parts of India and *Mateera,* particularly in

Rajasthan. *Mateera* is drought hardy landraces crop. It is an indigenous type of watermelon and is extensively grown with mixed cropping on sand-dunes landscape in Thar desert.

i) Mosaic Disease

It is a viral disease and causes severe losses in Rajasthan. During 1999 crop season an epidemic of mosaic disease was prevalent on muskmelon, cucumber and squash in Punjab and in adjoining states causing huge fruit losses to the crops. Disease incidence was appeared from 14.29 to 50.0% in different genotypes of watermelon/ *mateera* at experimental field of the institute.

Symptoms

Plants showed characteristic symptoms of wilting and dry, necrotic lesions on leaves and internode shortening, mottling of leaves and development of vein clearing. Plants are stunted.

Causal organism: This disease is caused by virus and transmitted by aphid.

Integrated Disease Management

Cultural Control

a) Removal of infected plant.

b) Use of disease free seed.

Chemical Control

a) Application of insecticides like dimethote (0.1%) is recommended for the control of vectors (aphid).

b) Spraying of imidacloprid (3-5 ml/ 10 lit. of water) is also very effective for vector control.

ii) Bud Necrosis

It is also a serious problem and causes yield losses in Bikaner district and also in western Rajasthan of watermelon/*mateera* growing areas. It is also a viral disease.

Symptoms

Symptoms are leaf crinkling, mottling, yellowing, necrotic streaks on vines, shortened internodes, upright branches and necrosis and dieback of the buds.

Causal organism: This disease is caused by virus and transmitted by thrips

Disease Cycle

Diseases occur throughout the entire year but the disease incidence is generally higher during dry and hot periods when thrips populations increase rapidly. The virus is transmitted by thrips species, in a persistent (Propagative) manner (vector can acquire and transmit the virus after feeding for several minutes to hours: virus replicates inside the vector) young thrips (larval stage) acquire the virus and adult thrips spread the virus from plant to plant during feeding. The virus is not seed transmitted.

Integrated Disease Management

Cultural Control

a) Use virus free seedlings

b) Remove infected plants as early as virus symptoms are observed to prevent/minimize spread of the virus by thrips.

c) Remove crop debris, weeds and other sources of thrips after the cropping season.

d) Ploughing and harrowing may help reduce vector population in the soil.

e) Control thrips population by using plastic mulch, blue sticky traps.

Chemical Control

a. Spray monocrotophos/acephate (0.15%) at fortnight intervals after transplanting till flowering stage.

b. Chemical spray followed by neem seed kernel extract (2%) is also effective in rotation with insecticides.

c. Spraying imidacloprid at the rate of 0.5 ml/litre of water to control of sucking insects.

4) Chilli Diseases

Chilli (*Capsicum annum* L.) is the fourth major crop cultivated worldwide and is one of the most important constituent of the cuisines of tropical and subtropical countries. Several varieties of chilli are grown for vegetables, spices, condiments, sauces and pickles. Besides, it is also used in other forms like medicines and beverages. Nutrition wise these are rich in Vitamin A and C with appreciably high contents of iron, potassium, and magnesium, which have the ability to boost the immune system and lower the cholesterol levels. India has been a leading producer, consumer and exporter of chilli, especially, dried one.

i) Anthracnose Disease

Chilli production, in tropical subtropical countries, is severely impacted by occurrence of anthracnose disease, which causes substantial losses. In past, India had been the largest producer and exporter of chilli. The estimated annual loss of about 29.5%, amounting US$ 491.67 million has been reported from India alone (Garg *et al.*, 2014). In India, a calculated loss of 10–54% has been reported in yield of the crop due to the anthracnose disease to both mature fruits in the field as well as during their storage (Lakshmesha *et al.*, 2005). Significant losses have been reported from other parts of the globe as well, like a significant amount of 20–80% loss has been accounted from Vietnam (Don *et al.*, 2007). Fungi *C. capsici* Syd. Butler and Bisby are reported to cause major damage (approx. 20-45%) at the ripe fruit stage of the plant (Saxena *et al.*, 2014).

Symptoms

The small lesion is the most economically important aspects of the disease as on the fruit is enough to lower its market value thereby affecting the profitable yield of the crop. The disease is reported to affect almost all aerial parts of the plant. Chiefly, it causes fruit rot at both green and red stages primarily attacking ripe fruits, hence is also known by the name ripe fruit rot of chilli (Agrios, 2005). The disease is seed borne, soil borne, water borne and airborne and hence may lead to damage at the seedling stage or on the aerial parts of the plants.

Causal organism: *Colletotrichum capsici*

Epidemiology

Environmental factors play an important role in deciding the severity and spread of any disease. The favorable host, pathogen and weather conditions lead to establishment of disease. Hot and humid environmental conditions support the spread of the disease. Leaf surface wetness has been directly linked with disease severity owing to the better establishment of the pathogen in respect of germination, attachment and penetration into host tissues (Than *et al.*, 2008). Temperature around 27°C with relative humidity of 80% have reported to be the most optimum conditions for successful establishment of the disease.

Disease Spread

Water splashes may easily spread the conidia of the pathogen from infected to uninfected plant parts. Also, relative humidity aids successful colonization of the pathogen.

Integrated Disease Management

Cultural Control

a) Proper drainage, crop rotation and removal of any infected plant parts of the field.

b) Proper distance between the plants should be maintained so as to reduce dense canopy.

c) Transplants raised from disease free seeds of the chilli variety. The transplants should be kept weed free and away from other solanaceous crops.

Chemical Control

Recommended fungicide for the control of the disease is carbendazim.

Use of resistant varieties

Resistance in host plant against the pathogen is considered to be the one of the most vital and sustainable approaches for the management of the disease. Use of resistant varieties (BS-35, BS-20, BS-28, Punjab Lal, Bhut Jolokia, Taiwan-2, IC-383072, Pant C-1, and Lankamura Collection) of *Capsicum* spp. are recommended against this disease (Garg *et al*., 2014).

Use of Bio-agents

The bio-agents (*Psuedomonas fluorescens, Trichoderma* spp., *Bacillus subtilis* and *Pichia guilliermondi)* used against *Colletotrichum* spp. include .

Gaps in Prevailing System

Following gaps need to be addressed for effective implementation of integrated disease management practices of horticultural crops:

- Lack of disease free planting material of horticultural crops.
- Lack of awareness about uses of proper doses/measurement for fungicides with farmers as higher/lower doses may lead to resistance in fungal races.
- Lack of broad spectrum activity of fungicides in its action.
- Lack of effective forecasting systems against major diseases of horticultural crops.
- Lack of timely provision of useful information, training and instruction to workers

Advance Approaches for Fruits and Vegetables in the Arid Region

Plant biotechnology heralds a new era for plant scientists working to maintain healthy plants, optimize crop yields and minimize pesticide usage. One of the ultimate aims of agricultural biotechnology is to feed an expanding world population. These survey results describe the horticultural production challenges facing the global community of farmers and consumers in the new millennium and the dimension of the debate on the risks and benefits of developing genetically engineered crop plants to meet the increasing global food demand while preserving the environment. It is possible to introduce genes conferring disease resistance into crop plants. Such gene transfer could be accomplished by two methods, direct methods and vector-mediated methods. Besides, there are several other tools such as PCR, ELISA diagnostic kits, transgenic plant disease management and molecular markers which are increasingly used for the timely and rapid diagnosis and management of plant diseases.

Potential Benefits of Adoption of Advance Approaches for Fruits and Vegetables in the Arid Region

Plant pathology/Plant protection plays an important role in the future success programme and policies intended to enhance the horticultural production. In independent India, increasingly farmers are adopting modern methods of cultivation to get the maximum returns. Irrespective of production system (conventional and organic farming), disease management is an important consideration for realization of maximum yield per unit area. Horticultural crops give higher returns to farmers/growers due to healthy and good crop stand through adoption of good management practices like crop rotation, alteration in planting time, removal of weeds, optimal and timely uses of fungicides etc. Due to minimal uses of pesticides, such foods are considered safe for consumption and; hence, farmers/growers are likely to fetch remunerative price for their produce, which in turn will also lead to improving their socio-economic status.

Future Scope and Research Need

i) Standardization of cultural practices integrated with reduced doses of fungicides.

ii) Assessment of the efficacy of botanicals and bio-agents against major diseases of arid horticultural crops. Besides, the rearing feasibilities of the promising bio-agents should be further investigated under natural field conditions.

iii) Development of fungicides/pesticides with less toxicity.

iv) Study the residual effect of fungicides/chemicals on human beings and environment and to advise residue tolerance limit.

v) Collaboration between research organization and pesticides industries for developing new molecules with minimum persistence and their commercialization.

vi) Identification of crop rotation practices for individual horticultural crops.

Conclusion

Adoption of good agronomic practices, entailing multiple facets of disease management, not only leads to enhanced yield with optimum quality in horticultural crops but are environmentally benign too. Therefore, the need of the hour is to identify a holistic strategy integrating various approaches right from the selection of planting material, crop production and harvest to till storage. However, such strategies need enormous work out to find out viable and crop-specific recommendations for the safe and economic management of major diseases of horticultural crop.

References

Anonymous 2010. Disease management, In: Annual Report, All India Co-ordinated Research Project on Arid Zone Fruits. pp. 135- 144.

Anonymous. 2015. Epidemiological studies of aonla. Annual Report, All India Co-ordinated Research Project on Arid Zone Fruits. Pp.90.

Anonymous. 2016. Indian Horticulture Database – 2015. National Horticulture Board, Gurgaon, Haryana.

Anonymous, 2017. Indian Horticulture Database. National Horticulture Board, Gurgaon, Haryana.

Bal, J.S. 2016. Development and production of Indian jujube (ber) in India. *Acta Horticulturae,* 1116, 15-22.

Bhandari, D. C., Meghwal, P. R. and Lodha, S. 2014. Horticulture Based Production Systems in Indian Arid Regions. *In: Sustainable Horticulture System: Sustainable Development and Biodiversity* 2 (*Ed.* Nandwani, D.). Springer International Publishing, Switzerland, p. 395.

Dhandar, D. G. and Saroj, P.L. 2004. Changing scenario in arid horticulture. In: Advances in Arid Horticulture, Vol. I Saroj *et al* (ed.). International Book Distributing Co., Lucknow, pp.1-27.

Dhandar, D. G., Nallathambi, P., Rawal, R. D. and Sawant, D. M. 2004. Bacterial leaf and fruit spot: a new threat to pomegranate orchards in Maharashtra. Proc. on National Seminar of Indian Society of Mycology and Plant Pathology held at University of Goa.

Don, L. D., Van T. T., Phuong Vy T. T. and Kieu P. T. M. 2007. Colletotrichum spp. attacking on chilli pepper growing in Vietnam, Country Report, in Abstracts of the First Int. Symp. on Chilli Anthracnose, eds Oh D. G., Kim K. T. (Seoul: Seoul National University), p. 24.

Garg R., Loganathan M., Saha S., Roy B. K. 2014. Chilli Anthracnose: a review of causal organism, resistance source and mapping of gene, in Microbial Diversity and Biotechnology in Food Security, eds Kharwar R. N., Upadhyay R., Dubey N., Raguwanshi R., editors. (Springer;), 589–610.

Gupta, A. K., Sharma, R. C. and Sharma, S. 2001. Fungal diseases of cucurbits. In: Diseases of vegetable crops, V. K. Gupta and Y. S. Paul (eds.). Kalyani Publishers, New Delhi, India, pp. 71-86.

Gupta, P. C. and Madan, R. L. 1977. Diseases of fruits from Haryana- a new leaf spot disease of ber. *Curr. Sci.* 46 **:** 237- 238.

Gupta, S. K. and Gupta, A. 2001. Chemical control of powdery mildew of cucurbits. *Plant Dis. Res.* 16: 186-192.

Khoury, W. El. and Makkouk, K. 2010. Integrated plant disease management in developing countries. *J. Pl. Pathol.* 92 (4): 35-42.

Kumar, A. , Bhansali, R. P. and Arya, H. C. 1978. A note on the occurrence of powdery mildew on ber in Indian arid zone. *Ann. Arid Zone*, 17 (3): 323-325.

Lakshmesha K., Lakshmidevi K., Aradhya N., Mallikarjuna S. 2005. Changes in pectinase and cellulase activity of *Colletotrichum capsici* mutants and their effect on anthracnose disease on *Capsicum* fruit. *Arch. Phytopathol. Plant Prot.* 38, 267–279.

Madan, R. L. and Gupta, P. C. 1985. A new leaf spot of *bael* caused by *Alternaria alternata* (Fr.) Keissler. *Indian J. Plant Pathol.* 3: 239.

Maheshwari, S. K.; Choudhary, B. R.; Singh, D.; Sharma, B. D. and Sharma, S. K. 2015. Evaluation of resistance in different varieties/genotypes of bottle gourd (*Lagenaria siceraria*) against *Cercospora* leaf spot under field conditions. *Indian Journal of Agricultural Sciences* 85 (10): 1069-72.

Maheshwari, S.K., Choudhary, B. R., Sharma, B. D. and Saroj, P. L. 2017. Management of *Alternaria* Leaf Blight of Bottle Gourd in Western Rajasthan, India. *Int. J. Curr. Microbiol. App. Sci.* 6 (6): 1272-1277

Maheshwari, S. K.**;** Singh, D.; Choudhary, B. R. and Sharma, S. K. (2012)**.** Screening of bottle gourd germplasms for resistance against powdery mildew under hot arid region of Rajasthan. *Crop Improvement* 39: 1073-1074.

Mark J H, Henry M M, and Moly M J. 2005. Hannahs Choice F1: a new muskmelon hybrid with resistant to powdery mildew, Fusarium race 2 and potyviruses. *Hort. Sci.* 40 (2): 492-493.

Mehta, N., Gupta, P. C., Thareja, R. K. and Dang, J. K. 1989. Varietal behaviour and efficacy of different fungicides for the control of date palm leaf spot caused by *Graphiola phoenicis. Tropical Pest Manage.* 35 (2): 117-119.

Misra, A. and Shivpuri, A. (1983). Anthracnose, a new disease of aonla. *Indian Phytopathology* 36: 406-407.

Misra, A. K. 1988. Studies on diseases of fruit crops. Annual Report, CIHNP, Lucknow.

Mukhtar I, Mushtaq S, Khokhar I and Hannan A. 2013. First record of *Cercospora citrulina* leaf spot on *Lagenaria siceraria* in Pakistan. *The Journal of Animal and Plant Sciences* 23(6): 1756-1757.

Nallathambi, P., Umamaheswari, C., Nagaraja, A. and Dhandar, D. G. (2006). Pomegranate diseases and management. Technical Bulletin, CIAH, Bikaner, pp.30.

Pal, V., Rathore, G. S. and Godara, S. L. (2006). Screening of genotypes of date palm against *Alternaria* leaf spot. *Indian Journal of Arid Horticulture* **2 :** 62-63.

Pandey, R. S., Bhargava, S. N., Shukla, D. N. and Dwivedi, D. K. (1984). Two new fruit diseases of aonla caused by *Alternaria* species. *Int. J. Tropical Plant Dis.* 2: 79-80.

Pareek, O.P. (2001). *Ber*: International Centre for Underutilized crops, Southampton, U.K. pp. 248-266.

Pundir, J. P. S., Rathore, G. S., Naqvi, A. R. and Porwal, R. 2006. Date palm. In: *Advances in Arid Horticulture:* (Vol. II-Production technology of Arid and Semi-arid Fruits), International Book Distributing Co., Lukhnow.

Radhakrishnan, P. and Sen, B. 1985. Efficacy of different methods of inoculation of *Fusarium oxysporum* and *F. solani* for inducing wilt in muskmelon. *Indian Phytopathol.* 38: 70-73.

Rawal, R. D. 1993. Fungal diseases of tropical fruits. In: *Advances in Horticulture* Vol. 3 *Fruit Crops*. Chadha, K. L. and Pareek, O. P. (eds.), Malhotra Publ. House, New Delhi.

Sarbhoy, A K. 2006. *Text Book of Mycology*, IBH Publishing Ltd., New Delhi, pp. 242-291.

Saxena, A., Raghuvanshi R. and Singh H. B. 2014. Molecular, phenotypic and pathogenic variability in *Colletotrichum* isolates of subtropical region in North Eastern India, causing fruit rot of chillies. *J. Appl. Microbiol.* 117, 1422–1434.

Sharma, M., Majumdar, V. L. and Sharma, M. 1983. Some new post harvest diseases of ber in India. *Indian Phytopath.* 46 (4): 415.

Singh, S. B., Maheshwari, S. K. and Singh, P. N. 1995. Field evaluation of fungitoxicants against powdery mildew of ber. *Ann. Plant Protec. Sci.* 3: 168- 169.

Suriachandraslevan, M., Jayasekhar, M. and Aubu, S. 1993. Chemical control of bacterial leaf spot and fruit spot of pomegranate. *South Indian Hort.* 41: 228- 229.

Than P. P., Jeewon R., Hyde K. D., Pongsupasamit S., Mongkolporn O. and Taylor P. W. J. 2008. Characterization and pathogenicity of *Colletotrichum* species associated with anthracnose disease on chilli (*Capsicum* spp.) in Thailand. *Plant Pathol.* 57: 562–572.

Tyagi, R. N. S. 1967. Morphological and taxonomical studies on the genus *Ravenalia* occurring in Rajasthan. Ph.D. Thesis, Unicersity of Rajasthan, Jaipur.

Verma, K. S. and Cheema, S. S. 1988. Chemical control trials against mouldy leaf spot of ber caused by *Isariopsis indica* var. *ziziphi*. *Plant Dis. Res.* 3(1): 32-36.

Verma, K. S., Cheema, S. S. and Kumar, S. 1992. Some physic-pathological studies on *Isariopsis indica* var. *ziziphi*. The incitant of mouly leaf spot of ber. *Pl. Dis. Res.* 7 (2): 210-214.

Vithlani V.A. and Patel H.V. (2010). Production of functional vinegar from Indian jujube (*Zizyphus mauritiana*) and its antioxidant properties. *J. Food Tech.* 8(3):143-149.

6

Plant Bio-Regulators for Improving Stress Tolerance for Higher Crop Productivity

N.S. Nathawat, V.S. Rathore, G.L. Bagdi, M.L. Soni, N.D. Yadava

Introduction

Plant growth and productivity are adversely affected by various abiotic stresses. Plants are frequently exposed to a plethora of stress conditions such as low temperature, salt, drought, flooding, heat, and oxidative stress. Various anthropogenic activities have accentuated the existing stress factors. All these stress factors prevent them from reaching their full genetic potential and limit the crop productivity. In the event of growing concerns of uncertainties in climatic conditions, the abiotic stresses have become the major threat to agriculture production worldwide (Bray *et al*., 2000). The plant responses to abiotic stress condition are believed to be complex in nature as these are the reflections of integration of stress effects and responses at various levels of plant organization. To provide tolerance against stresses, plants are equipped with several inbuilt physiological and biochemical mechanisms occurring at cellular level. An understanding of processes linked to these mechanisms is vital in optimizing the crop growth and productivity under stress conditions.

Plant bio-regulators (PBRs), are biochemical compounds stimulates plant growth and productivity when applied, even in small quantities at appropriate plant growth stages. These are being extensively used in agriculture to enhance the productivity particularly in horticultural crops. Their central role in plant growth and development is through nutrient allocation and source-sink transitions while most of the PBRs stimulate redox signaling under abiotic stress conditions. Since climate change and degrading natural resources are projected to amplify the stresses, particularly soil moisture deficit, high temperature and soil salinity, PBRs are likely to play a crucial role in plant growth regulation. Plant bioregula-tors are powerful tools for maximizing yield and quality, and increasing net income to farmers. PBRs have been used from many years to alter the behavior of agricultural and horticultural crops for the economic benefit of the grower. Control of vegetative vigor, stimulation of flowering, regulation of crop load, reduction of fruit drop, and delay or stimulation of fruit maturity and ripening are best examples of regulation with the exogenous applications. Novel PBRs with possible benefits for fruit growers are continually being made available by industries. Every aspect of plant growth and development is controlled by plant hormones and these serve as key integrators of exogenous (envi-ronmental) and endogenous (Developmental) signals. One of the important and widely discussed aspects in abiotic stress tolerance is the regulatory roles of phytohormones. Phytohormones are chemical substances that profoundly influence the growth and differentiation of plant cells, tissues, and organs and also function as chemical messengers for intercellular communication. In general, there are five major classes of PGR, and each is grouped together into one of these classes based on their structural similarities and physiological responses. These include the auxins and gibberellins that stimulate predominantly cell elongation; cytokinins, the purine bases that stimulate cell division; ethylene, the olefinic gaseous molecule that regulates among other plant event fruit ripening; and abscisic acid (ABA), the sesquiterpene that regulates senescence and abscission of plant parts and helps in maintenance of plant water relations (Stamm *et al*., 2011). It is documented that the induction in stress tolerance is induced by the manipulations in the concentrations of endogenous phytohormones under stress conditions by helping plants in many ways. Most of the progresses made in the phytohormones research especially on plant adaptation to abiotic stresses are the outcome of advances made in the precise analysis of phytohormones employing powerful and reliable physiochemical techniques as well advances made in molecular and genetic approaches. Good progress is also achieved in demonstrating the potential of such phytohormones in the amelioration of abiotic stress responses in a number of plants.

Stressful environments interrupt the balance between generation and utilization of reactive oxygen species (ROS) leading to toxicity by enhancing production of ROS such as superoxide radicals ($O_2^{"}$), hydrogen peroxide (H_2O_2), hydroxyl radicals ($OH^{"}$) etc. in plants thereby creating a state of oxidative stress in them . This increased ROS level in plants cause oxidative da-mage to bio-molecules such as lipids, proteins and nucleic acids, thus altering the redox homeostasis. When applied exogenously at suitable concentrations PBRs enhance the efficiency of antioxi-dant system, upregulates osmolytes and enhance the expression of stress responsive gens in plants and we also proposed PBRs for sustainable agri-culture by integrating redox signaling as a possible unifying mechanism. Understanding the adaptive responses of plants is of paramount importance particularly when grown under harsh environments. A series of experiments conducted in the past have clearly established that plant growth regulators and hormones play a vital role in determining growth, development and productivity of crops. This fundamental knowledge served as basis for a number of biochemical as plant growth promoters in crop production. This tends to be the integral part of modern agriculture, which is often affected by adverse environmental factors such as drought, high temperature and salt stress. The exogenous applications of bio-regulators have been found as a good alternative to counter the toxic effects of various stresses. In the current scenario, various kinds of PBRs such as SA, Si, NO, H_2O_2, hydrogen rich water, brassinosteroids (BS), jasmonates and polyamines (PA) *etc*., have been tested for enhancing the plant stress tolerance as well as the crop yield (Jisha *et al*., 2013). Most of these PBRs treatments are applied either for seed priming or as foliar spray. Seed priming is a pre-sowing treatment that partially hydrates seeds without allowing the radicle emergence. The seed priming may be induced alone or in combination with the foliar application of PBRs at the time or early flowering or grain filling stage (Pastor *et al*., 2013). Foliar application of synthetic auxin " naphthal acetic acid (NAA) to enhance apple flower (Cline, 2006a; McArtney *et al*., 2007; Stover *et al*., 2001) and also documented as flower-promoter in several tree fruit crops (Cowgilland Autio, 2009; Schwallier, 2006). Ethephon has also enhanced flowering in bearing apple trees (Cline, 2006a). The PBR prohexadione-calcium (P-Ca), an inhibitor of gibberellic acid (GA) biosynthesis reduces shoot extension in several fruit tree species (Rademacher *et al*., 2006). Exogenous cytokinin application of benzyladenine (BA) has effectively improved lateral branching of nursery trees (Hrotko *et al*., 1997) by limit blind wood and im-proves canopy development. Flowering is another important parameter is directly related to yield and productivity of plants. It is well known that SA application induces flowering in a number of plants (Cleland and Ajami, 1974). Kumar *et al*. (2000) studied the cumulative effect of SA in combination with GA, Kinetin, NAA, ethral and chloro chloro chloride (CCC), and found a synergistic effect between SA and GA on flowering as compared to other combinations of hormones. Ethylene modulates plant growth and development under normal conditions and is also a key feature in the response of plants to a wide range of stresses. Ethylene is synthesized in plants in response to various stresses typically in response to the presence of metals, organic and inorganic chemicals, cold or heat stress, drought or flood, ultraviolet light, insect and nematode damage, phytopathogens (both fungi and bacteria), and mechanical wounding (Glick, 2014). Prolific root system is important for stress tolerance and improved water uptake particularly under abiotic stress conditions in order to harvest better crop yield (Farooq *et al*., 2009; Flowers, 2004) under abiotic stress conditions. The influence of BSs on PIN Aux efflux carriers, which control mitotic activity and cell differentiation, suggests a possible mechanism that contributes to BS-mediated root growth through regulation of Aux distribution (Choudhary *et al*., 2012). SA and SNP (Nitric oxide source) promoted Fe uptake, translocation and activation; modulated the balance of mineral elements; and protected Fe deficiency induced oxidative stress (Kong *et al*., 2014). Nitric oxide (NO) or its donor has been shown to mediate chlorophyll increase, Fe availability and antioxidant enzymes (Zhang *et al*., 2012). Several models suggest that redox signalling through NO and ROS is enhanced by SA in a self-amplifying process. However, the relationship between NO, SA, and ROS in the activation of defense genes and/or induction of host cell death is not clearly defined (Zottini *et al*., 2007). SA and SNP (Nitric oxide source) promoted Fe uptake, translocation and activation; modulated the balance of mineral elements; and protected Fe deficiency induced oxidative stress (Kong *et al*., 2014). Prospects and pathways of PBRs are thrashed out as an emerging stress alleviating technology for crop production in harsh agro-ecosystems, specifically those featured by drought, heat and salinity stress.

PBRs Induced Plant Adaptation and Mitigation Under Abiotic Stress

Drought: Exposure of plants to water stress leads to serious physiological and biochemical dysfunctions including reduction in turgor, growth, photosynthetic rate, stomatal conductance and damages of cellular components (Janda *et al*., 2007). SA has significant role in controlling abiotic stresses including drought and salinity stress. SA has high potential for improving stress tolerance in agriculturally important crops. It is a well-established fact that SA potentially generates a wide array of metabolic responses in plants and also affects the photosynthetic parameters and plant water relations. However, contrary to these observations, a reduction in chlorophyll content was observed in plants pre-treated with SA (Pancheva *et al*., 1996).

Salinity: Salinity adversely affects plant growth and development, hindering seed germination (Dash and Panda, 2001), seedling growth (Ashraf *et al*., 2002), enzyme activity (Seckin *et al*., 2009), DNA, RNA, protein synthesis (Anuradha and Rao, 2001) and mitosis (Tabur and Demir, 2010). However, plant species differ in their sensitivity or tolerance to salt stress (Ashraf and Harris, 2004). There have been numerous studies of the effects of salinity on plants (Jamil *et al*., 2007; Duan *et al*., 2008). Recently, investigations have focused more on the mechanisms of salt tolerance in plants (Dajic, 2006; Munns and Tester, 2008). Some researchers have used PGRs for reducing or eradicating the negative effects of salinity (Kabar, 1987; Mutlu and Bozcuk, 2000). Increasing salinity is associated with decreases in auxin, cytokinin, gibberellins and SA in the plant tissues and an increase in ABA and JA. Changes in hormone levels in plant tissue are thought to be an initial process controlling growth reduction due to salinity.

Heat stress: Physiological processes of plants are largely affected by the alteration of surrounded environmental temperature. The ability of plants to cope with extreme temperature is a complex process and is determined by environmental factors and also by the genetic capability of the plant. The increase of temperature up to a certain level increases plant growth, photosynthesis, respiration and enzyme activity and after that these parameters tend to decline. The average rate of enzymatic reactions increases twofold with every 10°C increase in temperature within the range. The optimal temperature for structural integrity and activity of most enzymes are within the range of 30–45°C; and enzymes are irreversibly denatured and inactivated at temperatures higher than 60°C. Heat stress significantly alters cell division and cell elongation rates which affect the leaf size and weight. Exposure of plants to severe heat stress decreased the stem growth resulting in decreased plant height (Prasad *et al* 2006). Heat stressed plants had no detectable levels of cytokinin, leading to seed abortion (i.e., cessation of dry matter accumulation before initiation of rapid starch deposition in the endosperm) or cease starch deposition prematurely (Cheikh and Jones, 1994). The decreased cytokinin pool under heat stress may be a result of decreased synthesis, conjugation to less active forms, or metabolism to inactive forms. Cheikh and Jones (1994) found that enzyme cytokinin oxidase (CKO) is relatively heat stable, and its activity increases under heat stress (Table 1).

Table 1: Chemical-based PBRs their mode of action in crop plants and orchards.

PBRs	Type of Plant	Mode of action	References
Benzyladenine (BA)	Nursery trees	Improved lateral branching	Hrotko *et al.*, 1997
Naphthal Acetic Acid (NAA)	Fruit crop	Enhance flower formation	Upreti *et al.* 2012; Ghosh *et al.* 2012
Chlormequat chloride (CC); Ethephon (ETH) and Mepiquat chloride	Cereals, cotton	Crop growth, development	Jisha *et al.*, 2013
Sulphydryl compounds (thiourea, thioglycollic acid, mercaptoethanol, Dithiothreitol	Crop plants	Improves plant stress tolerance	Nathawat *et al.* 2018; Sahu *et al.*, 1993; Srivastava *et al.*, 2016; Ramaswamy *et al.*, 2007
Gibberellins (GA_3)	Tomato, Sapota, Pomegranate	Increases seed germination, fruit size and yield	Maggio *et al.*, 2010; Bhujbal *et al.*, 2012; Reddy and Prasad, 2012
1-methyl cyclopropene (1-MCP)	Tomato	cold tolerance	Lockhart (2013)
ABA	Crop plants	Stomatal closure and enhanced plant water status	Keskin *et al.*, 2010; Thompson *et al.*, 2007

Pytohormones: Key mediators of plant responses to abiotic stresses - Plants must regulate their growth and development to respond to various internal and external stimuli. Phytohormones, adiverse group of signaling molecules found in small quantities in cells, mediate these responses. Their pivotal roles in promoting plant acclimatization to ever-changing environments by mediating growth, development, source/sink transitions, and nutrient allocation have been well established (Wolters and Jurgens, 2009). Phytohormones act either at their site of synthesis or elsewhere in plants following their transport (Fahad *et al.*, 2015 a). Phytohormones are of key importance in plant development and plastic growth. They include auxin, cytokinins, abscisic acid, ethylene, gibberellins, salicylic acid, brassinosteroids, and jasmonates.

1. Auxins: Auxins are play essential roles in diverse developmental events like root growth, vascular tissue differentiation, auxiliary bud formation, apical dominance, and flower organ development (Zhao 2010) and respond to abiotic stress. However, little information is available on the relationship between stress and auxins in plants, and the role of auxins in alleviating different stress responses needs better understanding. Indole-3-acetic acid (IAA) is one of the most multi-functional phytohormones and is vital not only for plant growth and development but also for governing and/or coordinating plant growth under stress conditions. Interestingly, there is growing evidence that IAA plays an integral part in plant adaptation to salinity stress. It increases root and shoot growth of plants growing under salinity or heavy metal stresses. Dunlop and Binzel (1996) witnessed significant reduction in the indole acetic acid (IAA) levels of tomato by salinity. Sakhabutdinova *et al.* (2003) reported a progressive decline in the IAA levels in the root system due to salinity. Afzal *et al.* (2005) documented exogenous IAA application to seeds prior to sowing alleviates the growth inhibiting effect of salt stress. High temperature affects reproductive development, leading to decline in yield. Upreti *et al.* (2012) reported that maintenance of high IAA in the reproductive organs of capsicum cultivars is responsible for lower abscission and high temperature stress tolerance. Cold stress and auxin contents are potentially linked (Wyatt *et al.* 2002), as cold stress inhibits the root gravity response and also auxin. Shibasaki *et al.* (2009) suggested that auxin concentrations are regulated by changes in auxin transport in plants under cold stress (Shibasaki *et al.* 2009). Du *et al.* (2012) documented that the increased tolerance to cold stress is due to the combined effects of IAA and ABA, as ABA or carotenoid-deficient mutants show reduced IAA content and exhibit increased cold resistance (Du *et al.* 2013). For abiotic stress response, Du *et al.* (2012) and Du *et al.* (2013) found that the change in auxin homeostasis could influence ABA synthesis and that the balance of auxin and ABA homeostasis played a crucial role in diverse stress responses.NAA is a synthetic auxin plant hormone that is routinely used for the vegetative propagation of plants from stem and cutting. The effect of NAA on plant growth is greatly dependent on the time of admission and concentration. In majority of fruit plants fruit drop is controlled by spraying of NAA in different fruit crops in different concentration. Nawaz *et al.* (2008) studied the effect of foliar sprays of NAA @ 10, 15 and 20 ppm in Kinnow mandarin and maximum Vitamin C contents (45.30 mg/100g) was found in 15 ppm NAA. Iqbal *et al.* (2009) applied with 15, 30, 45, 60, 75 and 90 ppm NAA through foliar spray and reported that 45 ppm spray reduced pre-harvest fruit drop, increased yield, pulp/acid ratio (11.31), TSS (11%), total sugar (7.45%), acidity and ascorbic acid in guava. Ghosh *et al.* (2012) application of different doses of NAA @ 15, 20, 25 and 30ppm and observed that sprayed of NAA at 15 ppm was the most effective in reducing the fruit drop at different months after fruit set which resulted in doubling of fruit production as compared to control and improved fruit size in sweet orange. 2,4-D is endogenous hormones and their balance play a modulating role in the mobilization of nutrients to the developing organs and can influence the longevity of a bud. The dependence of abscission relative to the endogenous content of auxins has been proven by exogeneous application of 2,4-D or NAA, as the transportation of auxins by the plant lasts for a long time without ethylene appearing to affect it. Reddy and Prasad (2012) reported that application of 2,4-D at 20, 30 and 40 ppm three times starting at full bloom and, subsequently, at 45 and 90 days after fruit set in pomegranate cv. Ganesh, has resulted significantly increased fruit size in length, breadth and volume and higher fruit weight (262.2 g), higher aril percent and maximum number of fruits (64.0) which resulted in highest fruit yield of 16.8 kg/plant. Ashraf *et al.* (2013) conducted an experiment to see the influence of 2,4-D in Kinnow and observed improved fruit weight, more number of fruits per plant, juice percentage, total soluble solids (TSS), ascorbic acid content, acidity, TSS/acid ratio, and reduced the fruit drop.

2. Cytokinins: Cytokinins are important growth promontory compounds involved in seed germination, morphogenesis, chloroplast biogenesis, maintenance of assimilate mobilization, fruit and leaf abscission, greater partitioning of photosynthates towards reproductive sink thereby improving the harvest index and in the regulation of stomatal functioning and root-to-shoot communication in the plant experiencing stress. These are synthesized primarily in roots, besides the shoot apex and other plant tissues. They protect cell membranes against degradation by preventing oxidation of unsaturated fatty acids and hence increase drought tolerance in plants. Further, cytokinins inhibit formation and speed up break down of free radicals such as superoxide (O_2^-) and hydroxy radical (OH^-) that otherwise oxidizes membrane lipids (Werner and Schmulling 2009). Zhou *et al.* (2004) reported that the drought stress changes in cytokinins levels in the xylem sap of apple trunk and the changes depended upon drought cycles. During the first cycle of drought and rewatering, cytokinin levels in drought-stressed plants decrease significantly, while in the second, zeatin content declines with no changes in zeatin riboside. Pillay and Beyl (1990) observed reduction in cytokinin concentration in drought susceptible cultivar of tomato. Satisha *et al.* (2005) witnessed a decline in cytokinins in the grape genotypes under soil moisture-deficit conditions. Further, the stress-induced decline in cytokinins depends upon the age of the plant organ, as the young leaves of stressed plants witness greater decline in cytokinins than the old leaves (Upreti and Murti 2004b). Stomata play an important role in the control of water transpiration and gas exchange in plant leaves, and cytokinins are important in keeping stomata open by regulating ABA action (Veselova *et al.* 2006).

3. Gibberellins: Gibberellins are tetracyclic diterpenoid carboxylic acid, which influence growth and various developmental processes, such as elongation, germination, dormancy, flowering, sex expression, enzyme induction, and leaf and fruit senescence. Maggio *et al.* (2010) reported that GA_3 treatment in tomato reduced stomatal resistance and enhanced plant water use at low salinity. Likewise, GA_3-priming increases grain yield due to the GA_3-priming-induced modulation of ion uptake and partitioning as well as hormone homeostasis under saline conditions (Iqbal and Ashraf, 2013). Bhujbal *et al.* (2012) found that application of GA_3 at 50, 100 and 150 ppm was showed that spray of 150 ppm GA_3 earliness in sprouting of new shoot, increased shoot length and maximum number of leaves per shoot in sapota. Kumar *et al.* (2012) observed that the application of GA_3 in strawberry at 80 ppm improved vegetative growth, runner production, ascorbic acid and acidity. Reddy and Prasad (2012) reported that the spray with GA3 @ 75 ppm has increased fruit size and yield in pomegranate cv. Ganesh. Khalid *et al.* (2012) working in young 'Kinnow' mandarin found that the spray of gibberellic acid 10 ppm at fruit setting stage and their effect on fruit quality was evaluated immediately after harvest.

4. Ethylene: Ethylene is a naturally occurring plant growth substance that has numerous effects on the growth, development and storage life of many fruits crops. Harvested fruits may be intentionally or unintentionally exposed to biologically active levels of ethylene and both endogenous and exogenous sources of ethylene contribute to its biological activity. Ethylene synthesis and sensitivity are enhanced during certain stages of plant development, as well as by a number of biotic and abiotic stresses. Ethylene is the simplest olefinic gaseous hormone which regulates a wide range of plant developmental processes such as pollination, seed germination, abscission and senescence, flowering, fruit ripening, root formation, and gravitropism. It is biosynthesized by the conversion of methionine to ethylene via the intermediates, S-adenosyl methionine

(SAM) and 1-amino cyclopropane-1-carboxylic acid (ACC) by involving enzymes ACC synthase and ACC oxidase. Abiotic stress conditions surge ethylene production by inducing ACC synthase and ACC-oxidase activities. The magnitude of ethylene increase under stress depends upon growth stage, stress intensity, and stress duration, and higher stress levels tend to reduce ethylene concentration. In pineapple, drought stress has no effect on flower induction and produced significantly less ethylene and had lower ACC-oxidase activity in leaf and stem tissues than the control plants (Min and Bartholomew, 2005). Zhao *et al.* (2009) in tomato suggested positive relationship between ethylene and freezing tolerance from the negative effects of 1-methyl cyclopropene, an ethylene biosynthesis inhibitor on freezing tolerance. Further support for ethylene in cold tolerance is evident from the study of Lockhart (2013) that the ethylene biosynthesis inhibitor 1-methyl cyclopropene (1-MCP) reduces cold tolerance in tomato. Ethylene is also considered important in plant's adaptation to flooding. Under flooding, the lack of oxygen in flooded roots triggers the ACC synthesis, which upon transportation upward in the plants gets oxidized to ethylene to cause nastic movements of the leaves and promote aerenchyma formation (Colmer 2003). Ramburn (2001) reported that foliar application of 0.5 gm PBZ + 0.4 gm ethephon / l promoted flowering in litchi with erratic fruiting. Onaha *et al.* (2001) found higher percentage of flower bud induction in pineapple by application of ethephon.

5. Abscisic Acid (ABA): The ABA is an important chemical signal of plant responses to a range of abiotic stresses, including drought and salinity (Keskin *et al.* 2010). High cellular ABA facilitate modifications in stomatal functioning, root hydraulic conductivity, photosynthesis, biomass allocation between roots and shoots, plant water relations, osmolyte production, and synthesis of stress-responsive proteins and genes to confer stress tolerance (Finkelstein *et al.* 2008; Kim *et al.* 2010). Drought and salinity induce ABA accumulation in the leaves of many plant species. The stress release reverses ABA increase and brings back its levels to prestressed levels. The increases in ABA enable plants to restrict their water loss through transpiration following closure of stomata and enhanced plant water status following increased root hydraulic conductivity (Thompson *et al.* 2007). ABA also participates in the communication between the root and aboveground part, either by stomatal closure or metabolic changes and gene expression (Zhang *et al.* 2006; Sobeih *et al.* 2004).

6. Salicylic acid: Salicylic acid (SA) is best known for its central role in plant defense response. SA is synthesized from chorismate via isochorismate. Infection of the plants by a broad range of pathogens results in an increase in SA levels both at the site of infection and in distant tissues. Exposure of plants to water stress leads to serious physiological and biochemical dysfunctions including reduction in turgor, growth, photosynthetic rate, stomatal conductance and damages of cellular components (Janda *et al*., 2007). SA has significant role in controlling abiotic stresses including drought and salinity stress. SA has high potential for improving stress tolerance in agriculturally im-portant crops. Its utility however depends on various factors like concentration of SA applied, mode of application and the stage of plant growth. At low concentrations SA has been found to alleviate abiotic stress and at higher concentrations it induces oxidative stress (Miura and Tada, 2014). It has been suggested that SA-induced drought tolerance is associated with an enhanced antioxidant system (Horváth *et al*., 2007; Zhou *et al*., 2009). SA is known to improve germination, plant growth, rate of transpiration, stomatal regulation, photosynthesis, ion uptake and trans-port in plants (He *et al*., 2010; Khan *et al*., 2014, 2015). Low doses of SA have potential to improve seed germination and seedling establishment under diffe-rent abiotic stress conditions (Rajjou *et al*., 2006; Alonso-RamÍrez *et al*., 2009). Lee *et al*. (2010) reported the role of SA in seed germi-nation under salinity. They found that SA promotes germination un-der high salinity by modulating antioxidant activity in Arabidopsis. High salinity induces serious metabolic perturbations in plants, as it generates ROS which disturb the cellular redox system in favor of oxidized forms thereby creating an oxidative stress that may damage DNA, inactivate enzymes and cause lipid peroxidation (Tari *et al*., 2004; Szepesi *et al*., 2005).

7. **Brassinosteroids:** Brassinosteroids (BRs) comprise a relatively new group of polyhydroxy steroidal plant hormones with strong growth and development-promoting potential. They were first isolated and characterized in pollen of the rape plant (Brassica napus). More than 70 BRs have been isolated from plants. However, brassinolide, 28-homobrassinolide, and 24-epibrassinolide constitute the three most bioactive BRs and are widely used in physiological and experimental studies. They are present in almost every part of plants including pollen, flower buds, fruits, seeds, vascular cambium, leaves, shoots, and roots. They play a critical role in numerous developmental processes such as stem and root growth, floral initiation, and development of flowers and fruits. However, recent findings suggest stress-impact mitigating roles of BRs and associated compounds in various plants subjected to various abiotic stresses. Numerous studies indicate brassinosteroid potential in enhancing ability of plant to cope with drought stress, salt stress, and high and low temperature stresses (Fariduddin *et al*. 2014). The growth regulatory and stress protection capability of brassinosteroids are linked to their action on metabolic processes associated with photosynthesis and nucleic acid and protein biosynthesis (Fariduddin *et al*. 2014). Information on mechanism by which brassinosteroids exhibit stress tolerance is lacking. Most of the research on the response action of brassinosteroids to stress factors is made employing their exogenous application.

Zhang *et al.* (2008c) also witnessed improvement in photosynthesis by regulating ribulose-1,5-bisphosphate carboxylase/ oxygenase activity and sugar accumulation following brassinosteroid treatment in stressed plants. Brassinosteroids are also found effective in modulating salinity stress as evident from improvements in plant tolerance by epibrassinolide treatment. The effect is due to protective action against stress-induced oxidative damage of membrane lipids and induction in antioxidant enzymes (Hayat *et al.* 2010). Similarly, Ding *et al.* (2012) witnessed improvement in salt tolerance by epibrassinolide in eggplant. Chilling tolerance in plants is influenced by brassinosteroids. The report of Huang *et al.* (2006) showed that epibrassinolide application upregulated 17 proteins which were downregulated by chilling to confer chilling tolerance. Chilling tolerance is also attributed to increase in membrane permeability (Janeczko *et al.* 2007), high pigment accumulation, and upregulation in cold-related genes (Kagale *et al.* 2007) besides high activation of rubisco and expression of photosynthetic genes (Xia *et al.* 2009). Thus, the physiological responses of brassinosteroids though are variable due to complexities in the molecular mechanism of their action; their potential in improving abiotic stress tolerance has immense utility in managing plant responses to stress.

8. Jasmonic acid: The cyclopentanone phytohormones derived from the metabolism of membrane fatty acids including primarily methyl jasmonate (MeJA) and its free acid jasmonic acid (JA) are collectively called jasmonates (JAs) and are widespread in the plant kingdom. These multifunctional compounds are involved in crucial processes associated with plant development and survival including reproductive processes, flowering, fruiting, senescence, secondary metabolism, and direct and indirect defense responses. In addition to developmental functions of plants, JA activates plant defense responses to pathogenic attack as well as environmental stresses including drought, salinity, and low temperature (Fahad *et al.*, 2015 b).

9. Sulphydryl Compounds: Exogenous application of stress alleviating agents is an important means to improve stress tolerance in crop plants (Wahid and Shabbir 2005, Farooq *et al.* 2009). Thiourea (TU) and thioglycollic acid (TGA) with thiol (-SH) functional group are important among the stress alleviating compounds for plants. These are highly water -soluble and easily absorbed by living tissues. It has been reported that –SH compounds prevent denaturation of membrane protein and helps to maintain membrane integrity (Rikin *et al.* 1979, Giaquinta 1979), improve translocation of sucrose from source to sink via phloem loading (Giaquinta 1979), enhances photosynthetic pigments (D'Souza *et al.* 2009), net photosynthetic rate (Burman *et al.* 2004, Nathawat *et al.*, 2018), antioxidant enzyme activities (Nathawat *et al.* 2007) and N - metabolism (Garg *et al.* 2006); and decreases stress induced drop in PS I and PS II

(Ramaswamy *et al.* 2007) in plants under stress conditions. Exogenous application of –SH compounds has been shown to improve growth, yield and water use efficiency of *Triticum aestivum* L. (Sahu and Singh 1995), *Vigna radiata* (Misra *et al.* 1997), *Cyamopsis tetragonoloba* L. Taub (Garg *et al.* 2006, Burman *et al.* 2008, Nathawat *et al.*, 2016), *Pennisetum glaucum* (Parihar *et al.* 1998, D'souza *et al.* 2009) in arid and semi-arid regions (Table 2 & 3).

Table 2. The effect of exogenous –SH compounds applications on net photosynthetic rate, stomatal conductance, transpiration rate and water use efficiency of moth bean

Treatment	Net photosynthetic rate (μmol m^{-2} s^{-1})		Stomatal conductance (μmol m^{-2} s^{-1})		Transpiration rate (μmol m^{-2} s^{-1})	
	2010	2011	2010	2011	2010	2011
Control	9.3^{d}	9.9^{d}	379.3^{c}	407.9^{d}	5.3^{d}	8.1^{c}
TGA 200 mg L^{-1}	11.2^{bc}	11.6^{bc}	430.3^{b}	457.9^{c}	5.8^{bc}	8.9^{ab}
TGA 300 mg L^{-1}	12.3^{ab}	12.1^{abc}	461.1^{ab}	515.4^{ab}	6.3^{ab}	9.1^{ab}
TGA 400 mg L^{-1}	13.6^{a}	12.7^{ab}	516.9^{a}	534.2^{a}	6.9^{a}	9.3^{ab}
TU 500 mg L^{-1}	10.5^{c}	11.4^{c}	451.7^{ab}	477.4^{bc}	5.6^{c}	8.8^{b}
TU 750 mg L^{-1}	12.1^{abc}	11.9abc	476.3^{ab}	498.2^{abc}	6.2^{abc}	9.0^{ab}
TU 1000 mg L^{-1}	12.9^{ab}	12.8^{a}	489.2^{a}	519.9^{ab}	6.6^{a}	9.3^{a}
LSD ($P < 0.05$)	1.5	0.9	43.6	40.9	0.4	0.4

Nathawat et al.,2016, Proc.Natl. Acad. Sci., India, Sect. B Biol. Sci. 88:875-885.

Table 3. The effect of exogenous –SH compounds applications on pods per plant, seed per pod, seed yield of cluster bean

Treatment	Pods per plant		Seed yield(Kg ha^{-1})	
	2010	2011	2010	2011
Control	9±4.1^{d}	87±3.6^{c}	1006±45^{c}	968±35^{c}
TGA 200 mg L^{-1}	107±4.7^{c}	100±3.4^{b}	1163±43^{b}	1094±43^{b}
TGA 300 mg L^{-1}	111±5.5^{abc}	104±3.1^{ab}	1211±41^{ab}	1139±36^{ab}
TGA 400 mg L^{-1}	116±5.6^{a}	106±2.7^{a}	1247±43^{a}	1151±33^{a}
TU 500 mg L^{-1}	109±4.8^{bc}	101 ±3.5^{b}	1170±47^{b}	1113±21^{b}
TU 750 mg L^{-1}	110± 6.0^{bc}	102±4.3^{ab}	1223±53^{ab}	1129±26^{ab}
TU 1000 mg L^{-1}	114±5.3^{a}	104 ±2.5^{ab}	1233±46^{ab}	1146 ±32^{a}
LSD ($P < 0.05$)	7	5	73	63

Nathawat et al., 2016, Experimental Agriculture, **52 (3)**: 418-433.

10. Polyamines: Polyamines are small ubiquitous nitrogenous compounds. Increased polyamine levels in stressed plants are of adaptive significance because of their involvement in the regulation of cellular ionic environment, maintenance of membrane integrity, prevention of chlorophyll loss and stimulation of protein synthesis, nucleic acids and protective alkaloids (Kusano *et al.* 2008). Polyamines are also found to play an important role in conferring tolerance against drought, salinity, flooding, heat stress, and chilling stress in plants, as evident from polyamine changes under stress alleviation responses of exogenous applied polyamines (Gill and Tuteja 2010). In general, plants experiencing abiotic stresses tend to increase polyamine levels, and increase in polyamines helps in the regulation of plant tolerance to stress. The stress-induced increases in polyamine provide tolerance by stabilizing membrane integrity and functionality, altering hormonal balances and inducing antioxidant enzymes (Goyal and Asthir 2010). Generally, a polyamine response due to stress depends on species, besides type and concentration of polyamines (Ali *et al.* 2009). Zapata *et al.* (2004) studied effects of salinity on polyamines in varied plant species like *Spinacia oleracea*, *Lactuca sativa*, *Cucumis melo*, *Capsicum annum*, *Brassica oleracea*, *Beta vulgaris*, and *Lycopersicon esculentum* and found that the polyamine titer is greatly altered by salinity with distinct increase in spermidine and spermine contents. Duan *et al.* (2008) witnessed an increase in enzyme arginine decarboxylase, ornithine decarboxylase, SAM decarboxylase, and diamine oxidase activities, as well as free spermidine and spermine and soluble-conjugated and insoluble-bound putrescine, spermidine, and spermine contents in the roots of cucumber cultivars following salinity stress. Wei *et al.* (2007) studied the effect of salt stress on polyamine contents in leaves of *Solanum melongena* grafted plants employing salt tolerant *Solanum torvum* as rootstock. The polyamine increase in grafted plants protects the grafted plants against salt stress by triggering higher activities of antioxidant enzymes like superoxide dismutase, peroxidase, and ascorbate peroxidase and glutathione reductase enzymes. Kim *et al.* (2010b) reported a decrease in spermidine but an increase in spermine under salinity in Chinese cabbage. Upreti and Murti (2010) found significant alterations in root polyamines in grape rootstocks by salinity stress with tolerant rootstock showing greater increases in spermidine and spermine as well as ABA. The high increase in polyamine helped plants in maintaining high root-shoot biomass ratio and high K-Na ratio in the tolerant rootstock. Similarly, Anjum (2008) reported that citrus rootstock, *Cleopatra mandarin*, shows better growth and chlorophyll efficiency under salinity stress which accumulated high spermine and carbohydrates but low chloride ions in leaves and roots under salinity. Application of exogenous polyamine is efficient in manipulating the levels of endogenous polyamine during stress, and the stress protection effect is rendered through the maintenance of membrane integrity, regulation of gene expression for the synthesis of osmotically active solutes, reduction in ROS production, and controlling of the accumulation of Na^+ and Cl^- ion in different organs. Anjum (2011) reported that spermidine treatment to NaCl stressed plants reduced salinity-induced decline in leaf number, chlorophyll content, Fv/Fm, net photosynthetic rate, and N content and reduced Na^+ contents of the plants. Additionally spermidine improved cellular Ca^{2+} and Mg^{2+} contents in salinity-treated plants. In pomegranate plants, putrescine and spermine are effective in reducing the stress-induced decline in growth rate without causing major alteration in Na+, Cl-, and K+ contents of roots and apical and basal leaves (Amry *et al.* 2011). In tomato, spermidine is responsive in improving drought resistance by increasing gas exchange parameters and reducing internal CO_2 concentration by preventing stomatal closure and stimulating CO_2 uptake (Zhang *et al.* 2010). Polyamines are also implicated in plant performance in flooded soil. Yiu *et al.* (2009) witnessed alleviation of flood stress in onion by putrescine through reduction in superoxide radicals and H_2O_2 (Yiu *et al.* 2009). Jia *et al.* (2010) reported that application of spermidine to cucumber roots enhances ATP production and alleviates flooding responses by enhancing aerobic respiration and declining fermentation metabolism. Shi *et al.* (2009) also observed that the putrescine application is efficient in the alleviation of stress-induced reduction of gas exchange variables of cucumber subjected to root-zone hypoxia by enhancing nitrate reductase activities. Thermo tolerance by polyamines is also possible by regulating heat stress-induced inhibition of photosynthetic efficiency. Upreti *et al.* (2012) reported that high accumulation in spermidine and spermine in flower buds and flowers of capsicum cultivars is an important attribute of thermo tolerance as found by lower decline in abscission of floral parts. The increase polyamine also downregulates ethylene production leading to reduction in abscission of reproductive organs under high temperature. Genetic modifications of the polyamine biosynthetic pathway are useful to establish the function of polyamines in plant responses to abiotic stress.

11. Triacontanol: Triacontanol is a natural plant growth regulator found in epicuticular waxes. It is used to enhance the fruit production. Quite a number of researchers have reported the TRIA mediated improvement in growth, yield, photosynthesis, protein synthesis, uptake of water and nutrients, nitrogen fixation, enzymes activities and contents of free amino acids, reducing sugars and soluble protein. Expectedly, TRIA enhances the physiological efficiency of the cells and, thus, exploits the genetic potential of plant to a large extent. Shinde *et al.* (2008) reported that application of triacontanol at 300, 500, and 700 ppm at flowering, pea and marble stage of fruit development in mango cv. Parbhani Bhusan and showed that spray of 700 ppm triacontanol significantly given maximum length (10.9 cm), breath (8.9 cm), volume (336.6 cm^3), weight (330.4 g), mesocarp (69.9 %) and lowest proportion of endocarp (12.0 %).

12. Paclobutrazol: Paclobutrazol is probably the most widely used in the production of fruit crops because of its wide range of efficacy and moderate to long-lasting response. Applications of paclobutrazol, particularly when delivered as a spray, delay flower development and reduce flower size. Paclobutrazol is absorbed by roots and stems, and to a lesser extent, by leaves. Therefore, it can be applied as a spray, drench, or bulb or young plant dip. Sprays are more effective when they penetrate plant canopies so that there is contact with stems. The post-harvest application of a small amount of paclobutrazol to the soil significantly promotes flowering and fruiting in the following year. Therefore, early and proactive applications are strongly recommended, and late applications should generally be used as a last resort. Hussein *et al.* (1998) observed that application of paclobutrazol @ 150 ppm as foliar spray and 6 g /tree as soil drench reduced vegetative vigour like shoot length, shoot thickness, internode numbers etc. in order to improve tree productivity and fruit quality and also had significant positive effect on yield/tree in fig cv. 'Sultania'. Early flowering and fruiting was also recorded by application of paclobutrazol in mango (Sao Jose and Rebounces, 2000). Similarly, Lichev *et al.* (2001) found that application of cultar (25% paclobutrazol) significantly inhibited the annual shoot growth and improves photosynthetic activity which may increase yield in cherry. Application of paclobutrazol, a gibberellin biosynthesis inhibitor, induced stress protection, and application of GA_3 to the dwarf phenotype reversed the inherent stress tolerance. Sankhla *et al.* (1989) described that the soil drenching treatment of paclobutrazol is important in minimizing water stress injuries in the fruits of ber trees, while Still and Pill (2003) found that foliar applications or seed priming with paclobutrazol is effective in improving drought tolerance in tomato seedlings. Likewise paclobutrazol improves plant water status in apple (Swietlik and Miller 1983), strawberry (Navarro *et al.* 2007), and peach (George and Nissen 1992) under drought conditions.

13. Chlorocholine Chloride (CCC): Chlorocholine Chloride (CCC) is gibberellins biosynthesis inhibitor involved in the inhibition of cyclization of geranylgeranyl pyrophosphate to copyallyl pyrophosphate. The chemical control of the plant growth to reduce the size through the use of plant growth regulators is a common practice to make a plant more compact and commercially more acceptable. Application of 3000 ppm CCC at 15 leaf stage in grape was found to be highly effective in increasing the yield / vine. Yield increased on account of growth retardant was mediated through increased number of cluster / vine (Shikhamany and Reddy, 1989). Brahmachari *et al.* (1995) studied effect of foliar spray (one before flowering and one a month after fruit set) of NAA, PCPA, 2,4,5-T,GA3, Kinetin and CCC in 6 years old guava cv. Sardar and observed that spray of 250 and 500 ppm CCC has enhanced fruit set as well as improved weight and quality of fruit.

Recent attempts at deciphering the role of phytohormones in abiotic stress tolerance:

a) **Hormones help in pollen development under cold stress**: Abiotic stress can affect any developmental stages, but the reproductive stage is perhaps most critical. If stresses are applied at the reproductive stage, it may damage the whole plant and ultimately reduce the crop yield. Further, among the various reproductive stages, pollen development is crucial and is affected by abiotic stresses such as drought, cold, and heat, leading to reduced crop yield (Zinn *et al.*, 2010). In pollen development, pollen meiosis is the stage most sensitive to cold, which causes pollen sterility and also reduces anther dehiscence, pollen load to the stigma, pollen germination, and pollen tube growth (Thakur *et al.*, 2010). Hormones such as GA and ABA are considered major signals for cold-induced pollen sterility and an understanding of molecular mechanism of pollen development under stress and non-stress conditions can be beneficial for hybrid seed production by producing novel sterile plants.

b) **Phytohormones alleviate high-temperature stress** : Crop productivity is decreasing in many parts of the world under high temperatures. Chhabra *et al.*, 2009 performed an experiment to test the effect of various hormonal concentrations on heat-stress effects and observed that both growth-promoting and growth-retarding hormones mitigated heat-stress effects. They also reported that the most effective concentration of kinetin was 100 ìmol L^{-1} followed by 50 ìmol L^{-1} and that soaking seeds in ABA delayed 50% seedling mortality by 1 h and 50 minute at 0.5 and 1.0 ìmol L^{-1} concentrations.

c) **Hormone responsive protein mediated stress responses**: In plant stress response, phytohormones play a vital role, and some proteins can be used for communication. Bhaskar *et al.* (2015) reported that C1-(cysteine rich protein family) domaincontaining proteins play a part in plant hormone-mediated stress responses.

Conclusions

Plant bioregulaters are vital components of plant growth and development under abiotic stresses. The stress conditions alter their levels which help in plant adaptation through their responses on stomatal functioning, plant water balance, nutrient allocations, and source-sink transitions, besides maintaining antioxidant status. There is either an increase or decrease in the endogenous PBRs levels in plants under stress conditions, and responses are cultivars, stress duration, and stress intensity dependent. While stressed plants invariably showed an increase in ABA and decrease in cytokinins, the gibberellins, auxin, ethylene, and polyamine levels show variable responses to abiotic stress factors. Significant progresses have been made over the last few years in understanding processes regulating the biosynthesis and metabolism of naturally present PBRs and their associated roles in signaling mechanisms. The success in elucidating roles of

PBRs in stress tolerance both at cellular and molecular levels helped in showing positive effects of exogenous application of PBRs and their synthetic substitutes and of compounds capable in modifying PBRs metabolism in improving stress tolerance in wide range of crop species. However, commercial benefits of such results are yet to be established because of gaps in understanding the physiological basis of their actions. Bacteria and mycorrhiza are efficient in producing PBRs and have potential to manipulate their endogenous levels for eliciting growth responses in plants and inducing stress tolerance. Efforts are needed to exploit their benefits as alternate tool in the management of abiotic stresses.

Future Directions

A wide-range of chemical and hormone-based PBRs has been used for enhancing stress tolerance and yield of different crops under field condition. Most of the studies related with PBR are currently limited to seedling level experiment or small-scale field trials and hence, their large scale evaluation in different agro-climatic zones along with various quality controls on grain nutritional quality, soil and human health are required to support their use at wider scale. Considering the changing and unpredictable climatic condition, future research may be directed to develop combined or mixture based bioregulator technology containing mixture of different bioregulators which can tackle multistress conditions in a better manner. Thus, area of PBR based research is still immature and can be explored in great detail with the final aim to develop a highly versatile and effective technology that can be directly applied on farmer's field to ensure sustainable agriculture.

References

Afzal, I., Basra, S. and Iqbal, A. 2005. The effect of seed soaking with plant growth regulators on seedling vigor of wheat under salinity stress. *J Stress Physiol Biochem.* 1: 6–14.

Ashraf, M.Y., Asshraf M, Akhtar M, Mahmood K, Saleem M. 2013. Improvement in yield, quality and reduction in fruit drop in Kinnow (*Citrus reticulata* Blanco.) by exogenous application of plant growth regulators, potassium and zinc. *Pakistan J. Bot.* 45:433-440.

Bhaskar, R.V., Mohanty, B., Verma V., Wijaya, E.and Kumar, P.P. 2015. A hormone-responsive c1-domain-containing protein at5g17960 mediates stress response in Arabidopsis thaliana,PLoS ONE 10: e0115418.

Bhujbal D. S., Naik, D. M. and Kale, S. A. 2012. Effect of growth regulators on growth and yield of sapota. *Asian J. Hort.* 7(2):351-353.

Brahmchari, V.S., Mandal, A.K., Kumar, R., Rani, R., Kumar, R., and Rani, R. 1995. Effect of growth substance on fruit-set and physico-chemical characteristics of 'Sardar' guava (*Psidium guajava* L.). *Recent Hort.* 2(2):127-131.

Bray, E.A., Bailey-Serres, J. and Weretilnyk, E. 2000. Responses to abiotic stresses. In: Buchannan B, Gruissem W, Jones R (ed) Biochemistry and molecular biology of plants. American Society of Plant Biology, Rockville, MD, pp 1158–1203.

Burman, U., Garg, B.K. and Kathju, S. 2004. Interactive effects of thiourea and phosphorus on cluster bean under water stress. *Biologia Plantarum*, 48: 61-65.

Chhabra, M.L., Dhawan, A., Sangwan, N., Dhawan, K. and Singh, D. 2009. Phytohormones induced amelioration of high temperature stress in *Brassica juncea* (L.) Czern & Coss, Proceedings of 16th Australian Research Assembly on Brassicas, Ballarat, Australia, 10–14 September, 2009.

D'Souza, S.F., Nathawat, N.S., Nair, J.S., Radhakrishna, P., Ramaswamy, N.K., Singh, G. and Sahu, M.P. 2009. Enhancement of antioxidant enzyme activity and primary photochemical reactions in response to foliar applications of thiols in water stress in pearl millet. *Acta Agronomica Hungarica* 57: 21–31.

Du, H., Wu, N., Chang, Y., Li, X., Xiao, J. and Xiong L. 2013. Carotenoid deficiency impairs ABA and IAA biosynthesis and differentially affects drought and cold tolerance in rice. *Plant Mol Biol*. 83:475–488.

Du, H., Wu, N., Fu, J., Wang, S., Li, X., Xiao, J. and Xiong, L. 2012. A GH3 family member, OsGH3-2, modulates auxin and abscisic acid levels and differentially affects drought and cold tolerance in rice. *J. Expt. Bot*. 63: 6467–6480.

Duan, J., Guo, Li, S, and Kang, Y. 2008. Exogenous spermidine affects polyamine metabolism in salinity-stressed Cucumis sativus roots and enhances short-term salinity tolerance. *J Plant Physiol.* 165:1620–1635.

Dunlop, J. R. and Binzel, M. L. 1996. NaCl reduces indole acetic acid levels in the roots of tomato plants independent of stress induced abscisic acid. *Plant Physiol.* 112:379–384

Fahad, S., Hussain, S., Bano, A., Saud, S., Hassan, S., Shan, D., Khan, F.A., Khan, F., Chen, Y.T., Wu, C., Tabassum, M. A., Chun, M.X., Afzal, M., Jan, A., Jan, M.T. and Huang, J.L. 2015 a. Potential role of phytohormones and plant growth-promoting rhizobacteria in abiotic stresses: consequences for changing environment. *Environ. Sci. Pollut. Res*. 22: 4907–4921.

Fahad, S., Nie, L., Chen, Y., Wu, C., Xiong, D., Saud, S., Hongyan, L., Cui, K. and Huang, J. 2015 b. Crop plant hormones and environmental stress, Sustain. *Agric. Rev*. 15: 371–400.

Fariduddin, Q., Yusuf, M., Ahmad, I. and Ahmad, A. 2014. Brassinosteroids and their role in response of plants to abiotic stresses. *Biol Plant* 58:9–17.

Farooq, M., Wahid, A., and Lee, D. J. 2009. Exogenously applied polyamines increase drought tolerance of rice byimproving leaf water status, photosynthesis and membrane properties. *Acta Physiol Plant* 31:937–945.

Finkelstein, R., Reeves, W., Ariizumi, T. and Steber, C. 2008. Molecular aspects of seed dormancy. *Ann. Rev. Plant Biol.* 59:387–415.

Garg, B.K., Burman, U. and Kathju, S. 2006. Influence of thiourea on photosynthesis, nitrogen metabolism and yield of cluster bean (*Cyamopsis tetragonoloba* L.) under rainfed conditions of Indian arid zone. *Plant Growth Regulation,* 48: 237–245.

Ghosh, S.N., Bera, B. and Roy, S. 2012. Influence of plant growth regulators on fruit production of sweet orange. *J. Crop & Weed*. 8(2):83-85.

Hayat, Q., Hayat, Iran, M. and Ahmad, A. 2010. Effect of exogenous salicylic acid under changing environment: a review. *Environ. Exp. Bot.* 68:14–25.

Hussein, M. A., Mostafa, F.M.A., Abdel-Allah, A.Y. and Ahmed, N.M. 1998. Effect of paclobutrazol on vegetative growth and fruiting of 'Sultania' fig cultivar. *Aust. J. Agril. Sci.* 29(3):189-200.

Iqbal, M., Khan, M.Q., Jalal-ud-Din, Khalid and Rehman M. 2009. Effect of foliar application of NAA on fruit drop, yield and physico-chemical characteristics of guava (*Psidium guajava* L.) red flesh cultivar. *J. Agric. Res.*, 47(3):259-269.

Jisha, K.C., Vijayakumari, K., Puthur, J.T. 2013. Seed priming for abio-tic stress tolerance: an overview. Acta Physiol. Plant. 35, 1381-1396.

Keskin, B.C., Saikaya, A.T., Yuksel, B. and Memoh, A.R. 2010. Abscisic acid regulated gene expression in bread wheat. *Aust. J. Crop Sci.* 4:617–625.

Khalid, S., Malik, A.U., Khan, A.S. and Jamil, A. 2012. Influence of exogenous applications of plant growth regulators on fruit quality of Young 'Kinnow' Mandarin (*Citrus nobilis* × *C.deliciosa*) Trees. *Int. J. Agric. Biol.* 14(2):229-234.

Kim, T.H., Bohmer, M., Nishilmura, N. and Schorder, J.I. 2010. Guard cell signal transduction network: advances in understanding abscisic acid, CO_2 and Ca^{+2} signalling. *Ann. Rev. Plant Biol.* 61:561–569.

Kong, J., Dong,Y.,Xu,L.,Liu,S.,and Bai, X. 2014..Effects of foliar application of salicylic acid and nitric oxide in alleviating iron deficiency induced chlorosis of *Arachis hypogaea* L. *Bot. Stud.* 55:9. doi:10.1186/1999-3110-55-9

Kumar, R., Bakshi, M. and Singh, D.B. 2012. Influence of plant growth regulators on growth, yield and quality of strawberry under U.P. sub tropics. *Asian J. Hort.*, 7(2):434-436.

Lichev, V., Berova, M. and Zlatev, Z. 2001. Effect of cultar on the photosynthetic apparatus and growth of cherry tree. *Bulgarian J. of Agri. Sci.* 7(1):29-33.

Lockhart, J. 2013. Salt of the earth: ethylene promotes salt tolerance by enhancing Na/K homeostasis. *Plant Cell* 25:3150.

Maggio, A., Barbieri, G., De Raimondi, G. and Pascale, S. 2010. Contrasting effects of GA_3 treatments on tomato plants exposed to increasing salinity. *J. Plant Growth Regul.* 29:63–72.

Min, X.J. and Bartholomew, D.P. 2005. Effects of flooding and drought on ethylene metabolism, titratable acidity and fruiting of pineapple. *Acta Hortic* 666:135–148.

Nathawat, N. S., Rathore, V. S., Meel, B., Bhardwaj, S. and Bhargava, R. 2018. Exogenous sulphydryl improves membrane stabilization, photosynthesis and antioxidant systems in *Vigna aconitifolia* L. under water stress. *Proceeding of the national Academy of Sciences, India, Section B: Biological Sciences* 88 (3): 875-885.

Nathawat, N.S., Nair, J.S., Kumawat, S.M., Yadava, N.S., Singh, G., Ramaswamy, N.K., Sahu, M.P. and D'Souza, S.F. 2007. Effect of seed soaking with thiols on the antioxidant enzymes and photosystem activities in wheat subjected to water stress. *Biologia Plantarum*, 51: 93–97.

Nathawat, N.S., Rathore, V.S., Meel, B., Bhardwaj, S. and Yadava, N.D. 2016. Enhancing yield of clusterbean (*Cyamopsis tetragonoloba* L. Taub) with foliar application of sulphydryl compounds under hot arid conditions. *Experimental Agriculture*, 52 (3): 418-433.

Nawaz, M.A., Ahmad, W., Ahmad, S. and Khan, M.M. 2008. Role of growth regulators on preharvest fruit drop, yield and quality in Kinnow mandarin. *Pakistan J. Bot.*, 40(5):1971-1981.

Onaha, A., Higa, M., Naksone, F. and Ikemiya, H. 2001. Effect of ambient temperature and defoliation on flower bud induction with chemical in pineapple plants. *Japanese J. Trop. Agri.* 45(1):1-7.

Parihar, G.N., Sahu, M.P. and Joshi, N.L. 1998. Nitrogen, sulphur and thiourea nutrition of pearl millet (*Pennisetum glaucum* (L.) R. Br.). II. Effect on yield and yield components. *Annals of Arid Zone,* 37: 59-67.

Pastor, V., Luna, E., Mauch-Mani, B., Ton, Flors, V. 2013. Primed plants do not forget. *Environ. Exp. Bot.* 94: 46-56.

Pillay, I. and Beyl, C. 1990. Early responses of drought resistant and susceptible tomato plants subjected to water stress. *Plant Growth Regul.* 9:213–219.

Ramaswamy, N.K., Nathawa,t N.S., Nair, J.S., Sharma, H.R., Kumawat, S.M., Singh, G., Sahu, M.P. and D'Souza, S.F. 2007. Effect of seed soaking with sulphydryl compound on the photochemical efficiency and antioxidant defense system during the growth of pearl millet under water limiting environment. *Photosynthetica*, 45: 477–480.

Ramburn, N. 2001. Effect of girdling and growth retardants on flowering and fruiting of litchi in Mauritius. *Acta Hort.* 558:229-232.

Reddy, P.A. and Prasad, D.M. 2012. Effect of plant growth raguletors on fruit characters and yield of pomegranate (Punica granatum L.) cv. Ganesh. *Int. J. P. A. E. Sci.* 2(2):2331-4490.

Sahu, M.P. and Singh, D. 1995. Role of thiourea in improving productivity of wheat (*Triticum aestivum* L.). *Plant Growth Regul..*, 14: 169-173.

Sahu, M.P. and Solanki, N.S. 1991. Role of sulphydryl compounds in improving dry matter partitioning and grain production of maize (*Zea mays* L.). *Journal of Agronomy and Crop Science,* 167: 356-359.

Sahu, M.P., Kumawat S.M., Ramaswamy, N.K., D'Souza, S.F. and Singh, G. 2006. Sulphydryl bioregulator technology for increasing wheat productivity. Research Bulletin RAU-BARC, 1-56.

Sakhabutdinova, A.R., Fatkhutdinova, D.R., Bezrukova. M.V. and Shakirova, F.M. 2003. Salicylic acid prevents the damaging action of stress factors on wheat plants. *Bulg. J. Plant Physiol.* 314–319.

Sao Jose, A.R. and Reboucas, T.N.H. 2000. Use of Paclobutrazol in Mango orchard in southwest region, Bahia state, *Brazil. Acta Hort.* 509:713-715.

Satisha, J, Prakash, G.S., Murti, G.S.R. and Upreti, K.K. 2005. Response of grape genotypes to water deficit: root, shoot growth and endogenous hormones. *Indian J. Plant Physiol.* 10:225–230.

Shibasaki, K., Uemura, M., Tsurumi, S. and Rahman, A. 2009. Auxin response in Arabidopsis under cold stress: underlying molecular mechanisms. *Plant Cell* 21:3823–3838. doi:10.1105/tpc.109.069906

Shikamany, S.D. and Reddy, N.N. 1989. Effects of growth retardants on growth, yield and quality in grape cv. Thompson seedless. *Ind. J. Hort.* 46(1):31-38.

Shinde, B.N., Pawer, B.R. and Kalalbandi, D.B.. 2008. Effect of chemicals and growth regulators on physical characters of Parbhani-Bhushan mango. Karnataka *J. Agric. Sci. 21(2):*318-319.

Sobeih, W.Y., Dodd. I.C., Bacon, M.A., Grierson, D. and Davies, W.J. 2004. Long-distance signals regulating stomatal conductance and leaf growth in tomato (*Lycopersicon esculentum*) plants subjected to partial root-zone drying. *J Exp. Bot.* 55:2353–2363.

Srivastava, A.K., Nathawat, N.S., Ramaswamy, N.K., Sahu, M.P., Singh, G., Nair, J.S., Paladi, R.K. and D'Souza, S.F. 2008. Evidence for thiol-induced enhanced in situ translocation of ^{14}C – sucrose from source to sink in *Brassica juncea. Environmental and Experimental Botany,* 64: 250-255.

Stamm, P. Ramamoorthy, R. and Kumar, P.P. 2011: Feeding the extra billions: strategies to improve crops and enhance future food security. *Plant Biotechnol. Rep.* 5:107-120

Thakur, P. Kumar, S. Malik, J.A. Berger, J.D. and Nayyar, H. 2010. Cold stress effects on reproductive development in grain crops: an overview. *Environ. Exp. Bot.* 67: 429–443.

Thompson, A.J., Barry, J.A., Mulholland, J., McKee, J.M.T., Howard, W., Hilton, H.W., Horridge, J.S., Farquhar, G.D., Smeeton, R.C., Smillie, I.R.A., Black, C.R. and Taylor, I.B .2007. Overproduction of abscisic acid in tomato increases transpiration efficiency and root hydraulic conductivity and influences leaf expansion. *Plant Physiol* 143:1905–1917.

Upreti, K.K. and Murti, G.S.R. 2004a. Changes in electrolyte leakage, chlorophyll concentration, endogenous hormones and bulb weight in onion in response to water stress. *Trop Agric.* 81:127–132.

Upreti, K.K. and Murti, G.S.R. 2004b. Leaf growth and endogenous hormones under water stress and stress recovery in French bean. *J. Plant Biol.* 31:61–64.

Upreti, K.K., Srinivasa, Rao, N.K. and Jayaram, H.L. 2012. Floral abscission in capsicum under high temperature: role of endogenous hormones and polyamines. *Indian J. Plant Physiol.* 17:207–214.

Veselova, R.G., Farkhutdinov, D., Veselov, S. and Kudoyarova, G.R. 2006. Role of cytokinins in the regulation of stomatal conductance of wheat seedlings under conditions of rapidly changing local temperature.*Russ. J. Plant Physiol.* 53:756–761.

Wolters, H. and Jurgens, G. 2009. Survival of the flexible: hormonal growth control and adaptation in plant development. *Nat. Rev. Genet.* 10: 305–317.

Wyatt, S.E., Rashotte, A.M., Shipp, M.J., Robertson, D. and Muday, G.K. 2002. Mutations in the gravity persistence signal loci in Arabidopsis disrupt the perception and/or signal transduction of gravitropic stimuli. *Plant Physiol.* 130:1426–435.

Zhang, J., Jia, W., Yang, J, Ismai,l A.M. 2006. Role of ABA in integrating plant responses to drought and salt stresses. *Field Crops Res.* 97:111–119.

Zhang, Q, Li, J., Zhang, W., Yan, S., Wang, R., Zhao, J., Li, Y., Qi, Z., Sun, Z. and Zhu, Z. 2012. The putative auxin efflux carrier OsPIN3t is involved in the drought stress response and drought tolerance. *Plant J.* 72:805–816.

Zhao, D., Shen, L., Fan, B., Yu, B., Zheng, Y., Lv, S. and Sheng, J. 2009. Ethylene and cold participate in the regulation of LeCBF1 gene expression in postharvest tomato fruits. *FEBS Lett.* 583:3329–3334.

Zhao, Y. 2010. Auxin biosynthesis and its role in plant development. *Ann. Rev. Plant Biol.* 61:49–64

Zhou, R., Cutler, A.J., Ambrose, S.J., Galka, M.M., Nelson, K.M., Squires, T.M., Loewen, M.K., Ashok, S., Jadhav, A.S., Ross, A.R.S., Taylor, D.C., Suzanne, R., Abrams, S.R. 2004. A new abscisic acid catabolic pathway. *Plant Physiol.* 134:361–369.

Zinn, K. E., Tunc-Ozdemir, M., Harper, J.F. 2010. Temperature stress and plant sexual reproduction: uncovering the weakest links. *J. Exp. Bot.* 61: 1959–1968.

Zottini,M.,Costa,A.,DeMichele,R.,Ruzzene,M.,Carimi,F.,andLoSchiavo,F. 2007. Salicylic acid activates nitric oxide synthesis in Arabidopsis. *J. Exp.Bot.* 58: 1397–1405.doi:10.1093/jxb/erm001.

7

Mass Multiplication of Quality Planting Material in Horticulture Crops

D.K. Sarolia, R.K. Meena, Ramesh Kumar, D.P. Singh and J.S. Gora

Introduction

Multiplication and perpetuation of plants through sexual and asexual means at desired quantum and quality is one of the important aspects of successful horticultural development. For successful propagation, stock production and growth of plants some facilities in the nursery, field or greenhouses are required for mass multiplication of the saplings. Horticultural crops included annuals, biennial and perennial crops so careful planning is require for the quality and genuinely. We known the various methods of propagation i.e., seed, cuttings or grafting and budding can be used for faster and true to type sapling production (Aldriance and Brison, 2000). Plantlets propagation mainly depends on season in open conditions, but for round the year material multiplication require hi-tech propagation structures like greenhouses, cold frames low polyethylene tunnels, sun tunnels, shade houses (lath houses), tissue culture facilities, hot frames and net house (Singh *et al.,* 2003; Ratha Krishna *et al*., 2014). Field management for mass multiplication of saplings require partial modification and refined in nursery beds preparation (raised bed in high rainfall area and sunken bed in arid region), mother bed preparation (bed with mixture of media, where generally seeds are sowing in enclosed area and all viable and vigours seedlings at early stage may be shifted in poly bags for further growth) and poly bag nursery (saplings of horticultural crops are produced in poly bags of 150-200 gauge and different sizes *i.e.*, 25 x10 cm (ber, khejri), 15 x10 cm (phalsa and pomegranate, karonda) and 30 x17 cm (kinnow, mosambi, aonla, sapota, mango) etc. Propagated plants should be healthy, vigorous and free from disease, weed, insect-pests as well as grafted plants should be equal in size of scion – rootstock, proper in shoot: root ratio and appropriate in morphological standard of scion according to varieties and species.

Present Status in India and Global Scenario

A total of 1661 rated nurseries and about 30 reputed tissue culture units and 27 model floriculture centre's were running across the India up to the end of plan period. The Working Group on Horticulture and Plantation Crops has projected the total requirement of planting materials of fruits and plantation crops as more than 2500 million at a modest growth rate of 4% per annum. In the segment of fruit crops alone, the demand for the grafted planting material was 8,359,632 by the year 2011-12; which may increase to more than 12 million by the year 2019-20.

The union Government number 23 and 24 has been made for the registration of nurseries in India. All the states have not enacted horticulture nursery act only 8 states have adopted nursery act, 9 states some system of registration/ monitoring exists but, 13 states there is no nursery act at present. Rajasthan state pass nursery bill 2008 and for quality fruit plant supply ensure through registered society RAJHANS (Rajasthan Horticulture Nursery Society) with the following responsibility assigned:

- Mother plants assessment by scientists
- District level committee to monitor, supervise and to register private nurseries
- Production, procurement & supply of quality material
- New mother plant blocks from ICAR Institutes/SAU's

National horticulture board recognition of service and quality planting material board stated a voluntary system of recognition of horticulture nurseries in 2008 and implemented of successfully across the country by rating 1604 fruits nurseries and 29 vegetable nurseries.

Region wise accredited nurseries of India and top three states

Region	Accredited nurseries (%)	Top 03 State
North	43	Himachal Pradesh> Jammu & Kashmir>Rajasthan
South	8	Karnataka> Andhra Pradesh> Kerala
East	22	West Bengal >Chhattisgarh>Orissa
West	27	Gujarat>Madhya Pradesh> Maharashtra

Mission on integrated development of horticulture (MIDH) launched in the year 2014 with the aim to develop horticulture sector and at the end of 2016-17 established 1092 number of commercial horticulture project in the area of 3046 hectare.

During normal grafting period, the success was at par in winter months, whereas it was significantly more under polyhouse (51.13-77.8 per cent) as compared to open field (8.87-17.8 per cent). The economic evaluation of plastic house for the year round grafting revealed that a net return of Rs 39,295 from 75 m^2 low-cost polyhouse is obtained in four months.

Gaps in Prevailing System

Use of quality seeds alone could increase vegetable production by 15-20 per cent. Inadequate availability of quality seed and planting material are the important deterring factors in development of sound horticulture industry. During the past decade, significant achievements have been made by plantation development and an area of > 6.7 Lakh hectare has been brought under horticultural crops. Fruits and vegetables account for nearly 90% of total horticulture production in the country. Ease of availability of quality seed and planting material (QPM) at reasonable costs, however, is a challenge. It is of special significance especially in fruit crops which has a long gestation period and effects are known only in later stages. The impact of investments in horticultural research can be realized only if farmers have better access to high quality seed and planting materials of the improved varieties. Quality comprises many aspects where four key attributes are explicitly identified: genetic, physical, physiological and health quality. Clearly, such efforts targeting quality seedling production will go a long way in promoting sustainable horticulture. National horticulture board a nodal agency that voluntarily recognize nurseries for producing quality planting material across the country, status report showed this agency accredited 1607 nurseries out of 2665 applications received during the year 2016-17 (www.nnb@)gov.in).

There are several constraints in the existing system of plant propagation. There are several private nurseries operating in the country playing important role in multiplication of planting material of horticulture crops and many of them follow traditional methods and lack adequate infrastructure and sell plant material of unknown pedigree. Of many other constraints, un-availability of standardized root stocks and non- maintenance of healthy stocks of elite varieties are worth mentioning. Guidelines prepared by National Horticulture Board for Recognition of Horticulture Nursery not only provides information for procedure of recognition but also provide technical information for setting up of fruit plant nursery. It is hoped that this system will fill up the existing gap and go a long way to establish a network of quality horticulture nurseries across the country. Even though all efforts presently only 30-40% demand of planting material is being met by the existing infrastructures (NHB web site).

One primary survey was conducted in the project NAIP regards to yearly demand of grafts and observed that more than 8.4 million fruit saplings in demand on which mainly mango, papaya, apple, grape, pomegranate, guava etc. and great demand of tissue cultured planting material in fruit crops like banana, pomegranate, strawberry etc. banana crop maximum acreage covered under tissue cultured plants (more than 95 %) and survey source showed about 31.5 lakh plantlets demand per annum (NAIP e-Publication, 2012).

Future Scope and Research Need

Mass multiplications of quality planting material in horticulture crops have lot of scope and opportunities. Wide gap exist in demand and supply of the planting material. Several low cost techniques need to be developed regards in lieu of mass multiplication of quality planting material, but cost minimization still has scope. Prim requirement is genuine, diseases free, healthy, vigorous and true to type planting material production as well as educational and training programmes needs to be strengthened for the development of human resource and sustainable progress in this field.

For the sustainable growth of the horticulture research is only the path and its prioritize areas may be

- Virus indexing of nursery stocks most essential in citrus, papaya, grape etc.
- Compatibility and adaptability of root stocks mandatory in mango, citrus, grape, sapota, ber etc.
- Maintained certified bud wood / scion bank.
- Accelerated growth techniques protocol for precocious attainment of desired planting material standards.
- Standardization of tissue culture protocol for perennial fruit crops.

Advanced Approaches in Mass Multiplication of Saplings in the Arid Region

- **Pro/plug tray nursery**: under this technique large number of vegetable seedlings and rootstocks of fruit crops can be produced annually in low cost poly houses, environmentally controlled greenhouses and under poly trenches by controlling optimum environmental conditions for seed germination and growth of the seedlings. A plug is a sapling which is produced in a small volume of medium in small individual cell or group of cells of different sizes ranging from 50 cc to 250 cc. Plug flats can be filled either manually or mechanically with a growing media and seeds are sown in to each cell. These plug trays are placed on stands in greenhouses under mist or fog conditions at optimum temperature and humidity. Plug trays can be initially put for 7-10 days in seed germinator or germinated seeds can also be sown in the plug trays filled with growing media. For production of plug transplants the seed of desired crop should be taken of high quality and vigour (Singh *et al.*, 2017).

- **Mist propagation of plants:** The plants produced through tissue culture or the soft wood leafy cuttings do not survive, when transferred to open field conditions, primarily due to excessive evaporative loss of water from above ground portion of plants/ cuttings. The excessive loss of water from leaves of cuttings and from the tender tissue culture plants can be prevented by use of misting. The mist propagation involves spraying of water in the form of mist, which is either done continuously, or intermittently to maintain high humidity. The aim of misting is to maintain continuously a film of water on the leaves, thus reducing transpiration and keeping the cuttings turgid until rooting takes place. In this way, leafy cuttings can be fully exposed to light and air because humidity remains high and prevents damage even from bright sunshine. Mist also prevents disease infection in cuttings by way of washing off fungus spores before they attack the tissues. While the leaves in this process must be kept continuously moist, it is important that only minimum water should be used. This is because excessive water leaches out nutrients from the compost which may cause starvation. Moreover, a directly injurious effect on the cuttings may occur from over watering. Hence, it is necessary to utilize nozzles capable of producing a very fine mist. The mist unit is controlled by a timer, operating a magnetic solenoid valve and is set in a way to turn on the mist for 3-5 seconds to wet the leaves and turn off for some time and when the leaves are dry, the mist is again turned on. In general, the mist may have any of the 5 control mechanisms timer, electronic leaf, thermostat and timer, screen balance and photoelectric cell. The two types of timers are used in mist unit, one turns "on" in the morning and "off' at the night and the second operate during day hours to produce an intermittent mist. Usually 6 seconds "on" and 90 seconds "off"", in electronic leaf, a plastic with two terminals, is placed under the mist along with cuttings. The alternate drying and wetting of the terminals breaks the current which in turn control the solenoid valve. A thermostat controls the temperature of the mist. In screen balance control mechanism, a stainless steel screen is attached to a lever with mercury switch. When mist is on, water is collected on the screen and when weight of water is more, it trips the mercury switch. The photoelectric controls are based on the relationship between light intensity and the transpiration rate. The mist unit can be set up in a glasshouse or in a polyethylene tunnel. Usually, it is set up on the propagation beds with 1.2m width. The layout of the misters is very important. The water supply and quality is equally important to have uniform misting. The water should have good pressure and it must be free from salts. The optimum pH of water to be used in mist unit is 5.5 to 6.5. Hard water or alkaline water may be avoided as it may block the nozzles of the mist chamber and it may accumulate on the young leaves of the cuttings and thus inhibit the growth and development of roots. While installing mist propagation unit, all the jets should be at equal height. Further, it is essential that a well-drained rooting medium is used and there should be a proper provision for adequate removal of excess water (Hartmann *et al.*, 2001).

- **Tissue (Micro propagation) culture***:* This *in vitro* technique where suitable explants multiply through axillaries shoot proliferation (shoot tip, merited tip, nodal culture, stool shoot, proliferation of *pseudocorms*, mini tubers), adventitious shoot formation (internodal section, leaf pieces, fragmented shoot apices, immature inflorescence, bulb scale), micro grafting (Shoot/meristem tip) and somatic embryogenesis (cell, nucellus tissue, callus, shoot tip) steps. Propagation of ornamental plants, banana, and strawberry by this method is gaining popularity.
- **Low cost energy efficient hardening technology:** this technique include, covering of pots by perforated poly ethylene bags, use of polyhouse or glasshouse equipped with fan/pad and fogging/misting devices, application growth retardant such as paclobutrazole or ancymidol have been reported to improve acclimatization of plantlets. However, these techniques are either inefficient or capital and energy intensive and not applicable to all agro-ecological conditions. Looking in to the inherent hardening problem of tissue culture plants, a three step low cost energy efficient hardening facility for tissue culture plants for arid agro ecosystem was developed at Central Institute for Arid Horticulture, Bikaner.

Major environmental regime under three step hardening system

S.No.	Steps of hardening	Temperature range(°C)	Relative Humidity (%)	Light intensity range (lux)
1	Acclimatization hood made of polycarbonate sheet with ventilation device kept inside culture room	25±2	75-95	3000-5000
2	Evaporative cool cham*ber* covered by insect proof net, thatches of local grass and shaded by agronet of 50 per cent light intensity.	25-30	70-80	5000-10000
3	Low cost polyhouse /shade house equipped with intermittent fogging device (250 watts capacity).	25-35	60-80	10000-20000

Source: Singh *et al.*, 2009 and Sarolia *et al.*, 2018

- **Standardized approaches in multiplying the horticulture plants:** Horticulture crops (fruits, vegetables, ornamental, flower, plantation and medicinal-aromatic) are propagated both by seed and vegetative methods. Sexually propagated plants show great variation especially due to cross-pollination. Whereas, vegetative method of propagation avoids genetic variation. Vegetative method of propagation includes cuttings, layering, grafting, budding and propagation by other special organs like bulb, corm, rhizome, tuber, tuberous root, sucker, runner, offset, crown etc. Each plant is propagated by specific method at specific time period. One method of propagation and time may not be suitable for other plants.

Methods of propagation in arid and semi arid fruit crops.

S.N.	Name of fruit	Methods of propagation	
		Standardized	Commercial
1.	Acid lime	Seed, budding, air layering	Seed
2.	Aonla	Budding	T-budding/ Patch budding
3.	Bael	Root cutting, budding, layering	Patch budding
4.	Ber	Budding (T, I, ring and patch)	*In-situ* budding (T-budding) / Polytube budding method
5.	Custard apple	Seed, Soft wood grafting (SWG), budding	T-budding/ inarching
6.	Date palm	Seed, offshoots, tissue culture	Off shoot, tissue culture
7.	Fig	Cutting, budding and air layering	Hard wood cutting
8.	Grape fruit	Seed and budding	T-budding
9.	Guava	Air layering and budding	Stooling, inarching, Soft wood grafting
10.	Jamun	Seed and budding	Soft wood grafting
11.	Ker	Seed and suckers	Suckers/ seed
12.	Khejri	Seed and patch budding	*In-situ* Patch budding
13.	Karonda	Seed and cutting	Seeds/ hard wood cutting
14.	Lemon	Seed, cutting and layering	Air layering
15.	Mandarin	Seed and budding	T/shield budding
16.	Mango	Inarching, side and veneer grafting, budding	Inarching, softwood grafting
17.	Mulberry	Cutting and budding	Stem cutting/ shield budding and cutting
18.	Olive	Budding, grafting	T-budding, tissue culture
19.	Papaya	Seed	Seed
20.	Phalsa	Seed and cutting	Seed
21.	Pilu	Suckers	Root suckers/layers
22.	Pomegranate	Budding (chip, patch and forkert), layering (air, ground and pot),	Hard wood cutting, air layering & tissue culture
23.	Pummelo	Seed and budding	Seed, T-budding
24.	Sapota	Air layering, inarching and soft wood grafting	Inarching
25.	Sweet orange	Seed and budding	T- budding
26.	Tamarind	Seed and soft wood grafting	*In-situ* soft wood grafting
27.	Wood apple	Root cutting, budding and seed	Seed/*In-situ* budding

Source: Vision ICAR-CIAH-2030 and Bose *et al.*, 1997

- **Registration of nurseries:** Government procurement facilitated may be posted on Government websites, free trainings to nurserymen, accessibility to quality propagation material, subsidies for fertilizers and pesticides and free soil testing. Recognition of Horticulture Nurseries has been proposed to facilitate, promote and monitor production and trade of quality planting materials as well as over all creditability will increase (www.nhb.gov.in).

Potential Benefits of Adoption of Advance Approaches

Planting material production mostly based on assumptions of demand of plants and availability of various production inputs like raw material and labour. It would therefore be wise to carry out market survey to anticipate demand of various plants in the coming season before production of plants in a nursery enterprise begins. The fixed and variable costs associated with establishment of nursery and expected return by saplings production activity will be helpful to assess the net returns of this activity. It is estimated that less than 10 per cent of nursery established face losses during the initial three years period only. This loss can be minimized by use of hi-tech approaches in mass production of saplings and assured marketing of the genuine planting material at reasonable price to the farmer's communities. Government agencies may also play an important role for accreditation of such production units and provide subsidies at initial establishment stage. By following standardized approach of budding in khejri as example expected economics was calculated from proper running nursery of Central institute for arid horticulture, Bikaner (Sarolia *et al*., 2017).

Variable cost for 30,000 sapling preparation of *Khejri*

Particulars	Required quantity	Rate (Rs.)/ unit	Cost (Rs.)
Cost of poly bag	55 thousands	1.0/-	55,000
Seed	15	800/-	12,000
FYM/ Compost	10 trolley	1500/-	15,000
Soil/ pond silt	20 trolley	600/-	12000
Growth regulators	5 g	600/ g	3000
Trench preparation/Bag filling/ arrangement labour	55000 bags	1/- bag	5,000
Insecticide-fungicide & other agrochemicals	12 months	-	15,000
Watering, seedling maintenance cost	12 months	5000/month	60,000
Budding/grafting labour on successful budding	30,000	8/-	2,40,000
Sprout removal/Shifting/ hardening labour	30,000	1/-	30,000
Miscellaneous	-	-	5,000
Total	4,52,000		

Expected returns per year from *Khejri* nursery

Particulars	Quantity and rate	Gross returns (Rs.)	Net returns (Rs.)
Sale of budded khejri saplings	(24,000 thousands) Rs.35 each	8,40,000	3,88,000

* Mortality of saplings in nursery before sale 10-20 %

Potential benefits needs to be coordination among three important aspects of nursery enterprise which are production of plants, rearing of plants and sales of plants. Adoption of frontier approaches in production of quality planting material will assist maximum success and less mortality of the plants that definitely enhance benefit by reducing cost as per potential of the unit.

Conclusion

Mass multiplication of saplings depends upon kind of crop and its variety, demand and future prospectus of the crop. Thus, four P's i.e., Product, Price, Promotion and Placement linkage very much important for successful nursery business. Nursery enterprise to be a profitable venture, meticulous planning is required in the way of improving the availability of healthy planting material of improved/ recommended varieties. The infusion of latest technologies has become essential for increased productivity. The demand for horticultural produce is accelerating with passing time. Unless uniform planting material of desired type is available, increased productivity levels cannot be achieved. Hence, adoption of frontier technologies like hi-tech nursery for raising plantlets has to be encouraged. The responsibility of providing food and nutritional security to aspiring populace, coupled with projected growth rate of 7 per cent can be met by various hi-tech inventions of which hi-tech nursery is the stepping stone. Educational and training programmes need to be strengthened for the development of human resource and sustainable progress in this direction.

References

Aldriance GW and Brison FR 2000. Propagation of horticultural plants. Mc Grow Hill Book Company. Inc, New York.

Bose TK, Mitra SK, Sadhu MK and Das P 1997. Propagation of Tropical and Sub tropical Horticultural Crops. IInd Edition, Naya Prakash, Calcutta.

Gunjal Surya (COPI, NAIP e- publication) Resource book on horticulture nursery management, NAIP, ICAR :1-268. www.fdcm. nic.in/PDF/ Horticulture %20plant%20nursery.pdf .

Hartmann H.T., D.E. Kester, F. Davies, and R. Geneve. 2001. *Plant propagation: Principles and practices (7th Edition).* Prentice-Hall, New York.

ICAR-CIAH vision-2030 2011. Mother plants & Propagation techniques standardization : 12.

Ratha Krishnan, P., Rajwant K. Kalia, Tewari, J.C. and Roy,M.M. 2014. Plant Nursery Management: Principles and Practices. Central Arid Zone Research Institute, Jodhpur, 40 p.

Sarolia D K., Meena R K. and Gora J S. 2017. Mass Multiplication of Quality Planting Material in Horticulture Crops for Higher Productivity. In: Compendium winter school on Doubling income through advance approaches for fruits and vegetables in the arid region published by P. L. Saroj, B. D. Sharma & M. K. Jatav at ICAR-Central Institute for Arid Horticulture, Bikaner during (28 Oct-17 Nov): 302-308.

Sarolia DK., Samadia DK., Choudhary BR. and Singh D. 2018. Production of quality seed and planting materials. CIAH/Tech./Publication No.75 Page (1-38). ICAR-Central Institute for Arid Horticulture, Bikaner (Indian). 138.

Singh D., Singh U V., More T A. and Awasthi O P. 2009. Hi-tech propagation of arid fruit crops. CIAH/Tech./Publication No.42 Page (1-51). ICAR-Central Institute for Arid Horticulture, Bikaner (India)

Singh R R., Meena L K. and Paramveer Singh. 2017. High-tech nursery management in horticultural crops: A way for enhancing income. *Int.J.Curr.Microbiol.App.Sci*. 6(6): 3162-3172. doi: https://doi.org/10.20546/ijcmas.2017.606.372.

Singh, H. P., Singh, Gorakh; Samuel, J. C., Pathak, R.K. (Eds), (2003). Precision Farming in Horticulture, NCPAH, DAC, MOA, PFDC, CISH, PP. 1-354.

8

Tissue Culture of Fruit Crops for Quality Planting Materials

Dhurendra Singh, Kamlesh Kumar and Ajay Kumar Verma

Introduction

Indian horticulture sector contributes about 38% to the agriculture Gross Value Added (GVA) making very significant contribution to the Indian economy. Apart from ensuring nutritional security of the nation, it provides alternate rural employment opportunities, diversification in farm activities, and enhanced income to farmers. India is currently producing about 283.5 million tones of horticulture produce which has surpassed the food grain production, that too from much less area (24.2 mha. for horticulture against 125 mha. for food grains). Productivity of horticulture crops is much higher compared to productivity of food grains (11.96 tones/ha against 2 tones/ha.). India has emerged as world leader in the production of a variety of fruits like mango and banana and is the second largest producer of fruits and vegetables. Besides, India has maintained its dominance in the production of spices, coconut and cashewnut. Among the new crops, kiwi, gherkins, kinnow, date palm and oil palm have been successfully introduced for commercial cultivation in the country. Launch of National Horticulture Mission has spurred the production and productivity of horticultural crops. Productivity of horticultural crops has increased by about 32% between 2004-05 and 2015-16. Increase in demand for horticultural produce due to greater health awareness, rising income, export demands and increasing population poses the challenge for further increasing the production and productivity of horticultural crops. The issue of climate change has thrown up greater uncertainties and risks, further imposing constraints on production systems. Therefore, continuous crop improvement programmes are the way forward for sustainable fruit production with greater opportunity of value addition of surplus produce.

India is rich in horticultural crops and their biodiversity which show the opportunities for the application of biotechnology to improve the quality of product, large scale multiplication of planting materials both for domestic and international market. To enhance the total production or productivity of any crop will require rapid area expansion and replacement of variety with new hybrids, genotypes or newly developed varieties. Therefore, quality planting materials is basic input for cultivation of horticultural crops in arid zone. Due to prevalent adverse climatic condition, the availability of quality planting materials of many arid horticultural crops remains short in supply which is commonly propagated through conventional technique. Now tissue culture or micro-propagation technique has been proved an alternative technique of biotechnology for rapid and mass multiplication of horticultural crops under adverse climatic conditions and many fruit crops can be successfully propagated through this technique. Micro-propagation is the most popular application of the biotechnology and a number of protocols have been developed for many important horticultural crops. It is now turning in to an industry for round the year supply of the planting materials. Even in India there are about 80 small and big commercial tissue culture companies, half of them concentrated on ornamentals. Banana, pomegranate and strawberry are the largest micro-propagated fruit crops. Besides micro-propagation of varieties, a large number of rootstock material is also multiplied in many developed countries. Micro-grafting is employed to develop disease fee mother stock/progeny. This technique has many advantages such as production of disease free planting materials, propagation of true to type plants, round the years production of the plants, production of transgenic and somaclonal variants with improved characteristics.

Application of Tissue Culture Techniques in Fruit Crops

Tissue culture/ micro-propagation has become an important part of commercial propagation for many fruit crops with following advantages

1. Mass Propagation of Specific clones/Variety/Hybrid

a) Plants whose natural rate of multiplication is relatively low for example date palm, banana, pomegranate and ker

b) New cultivars high in market demands for getting market opportunity quickly.

c) Cultivars whose high value makes micro-propagation a viable alternative to conventional methods. For example date palm, banana, orchids and many foliage plants

d) Propagation to difficult- to- root. For example many fruit crops, root stock genotypes.

2. Conservation of Endangered species. For example Elite Genotypes of Fruits

3. Production of disease free plants

a) Elimination of virus for example citrus, and strawberry.

b) Elimination of disease e.g. banana, citrus, pomegranate, guava, papaya etc.

c) Include micro-propagation as the initial step in production for foundation clone system

d) Facilitate the distribution of the commercial material in the international trade

4. Year Round Nursery Production

5. Germplasm Conservation

The potential importance of natural gene pool to meet the future need of crop improvement by introducing specific characters such as biotic and abiotic stress resistance can not be underestimated. However, the number of wild species and their natural habitats are disappearing rapidly, as a result of introduction of highly bred modern varieties, growing urbanization and an increased pressure on land for cultivation. This lead to the erosion of the natural germplasm to such an extent that there is a general fear that potentially valuable germplasm is being lost irretrievably. In plant improvement it is necessary to facilitate the assimilation of germplasm collection in working collection for the use of the breeders. The process of genetic erosion necessitates measure that germplasm must be conserved in such a manner that there are minimal losses or chances of genetic variability of a population. The conventional methods of germplsam conservation are proven to possible catastrophic losses due to (i) Attack by pest and pathogens (ii) Climatic disorders (iii) Natural disastrous and (iv) Political and economic causes. Besides this, the seeds of many important fruit plants such as mango, litchi etc. may lost their viability in a short time under conventional storage system. NBPGR, New Delhi has developed National Facilities of Plant Tissue Culture (NFTCR) and is maintaining large in-vitro germplasm collection of banana, phalsa, bael, Jackefruit, pomegranate etc.

Disadvantages of Micropropagation

Micro-propagation on commercial scale faces some physical, financial, technical and marketing problems. These include:

1. Capital intensive and sophisticated facilities are required.
2. Qualified and trained personnel of plant sciences are required.
3. Specialized art and technique are required.
4. Problem of contamination in cultures
5. Lack of assured marketing

Types of Systems for Regeneration of Plantlets by Tissue Culture

Micro-propagation is defined as the true-to-type propagation of selected genotypes using in vitro techniques. Four basic methods are used to propagate plants in vitro depending on the species and cultural condition, micro-propagation can be achieved by:

1. Enhanced Auxiliary Shoot Proliferation (shoot culture)

Suitable explants

- Shoot tip
- Meristem tip
- Nodal culture
- Stool shoot
- Proliferation of pseudocorms
- Minitubers

2. Adventitious Shoot Formation

Suitable explant

- Internode section
- Leaf pieces
- Fragmented shoot apices
- Immature inflorescence
- Bulb scale

3. Micrografting

- Shoot/meristem tip

4. Nonzygotic Embryogenesis

Suitable explant

- Cell
- Nucellus tissue
- Callus
- Shoot tip

Developmental Stages in Micro-propagation

Micro-propagation is the art and science of plant multiplication *in-vitro* success of micropropagation largely depend on the stages of stock plant preparation and subsequent culture preparation, establishment, morphogenesis and acclimatization of developed plants lets. Each stage is subjected to a specific cultural, nutritional, hormonal and environmental condition. Based on the various research information, it is now established that there are five major stages (o to 4) are critical for execution of successful plant regeneration.

Stage 0: Donor plant Selection and Preparation

Explant quality and subsequent responsiveness in vitro are significantly influenced by the phytosanitary and physiological conditions of the donor plant (Debergh and Maene, 1981; Read, 1988). Prior to culture establishment, careful attention is given to the selection and maintenance of the stock plants used as the source of explants. Stock plants are maintained in clean, controlled conditions that allow active growth but reduce the probability of disease. Maintenance of specific pathogen-tested stock plant under condition of relatively lower humidity, and use of drip irrigation and antibiotic sprays have proved effective in reducing the contamination potential of candidate explants. Such practices also allow excision of relatively larger and more responsive explants often without increased risks of contamination.

Numerous practices are employed to increase explants responsiveness by modifying the physiological status of the stock plant. These practices include the following: (1) trimming to stimulate lateral shoot growth; (2) pretreatment sprays containing cytokines or gibberellic acid; and (3) use of forcing solution containing 2% sucrose and 200 mg/L 8- hydroxyquinoline citrate for induction of bud break and delivery of growth regulators to target explant tissues (Read, 1988). Currently, information on the effect of other factors such as stock plant nutrition, light, and temperature treatments on the subsequent in vitro performance of meristem explants is lacking.

Stage 1: Establishment of Aseptic Cultures

Initiation and aseptic establishment of pathogen eradicated and responsive terminal or lateral shoot meristems explants are the goal of this stage. The primary explants obtained from the stock plants may consist of surface disinfested shoot apical meristems or meristem tips for pathogen elimination or shoot tips from terminal or lateral buds. The presence of microbial contaminants can adversely affect shoot survival, growth, and multiplication rate. Bacteria and fungal contaminants often persist within cultured tissues that visually appear contaminant free. Consequently, it is essential that stage I cultures be indexed (screened) for the presence of internal microbial contaminants prior to serving as sources of shoot tip and nodal explants for stage II multiplication.

The following facators may affect successful stage I establishment of meristem explants:

(1) Explantation time

(2) Position of the explant on the stem

(3) Explant size

(4) Polyphenol oxidation.

Time of explanation can significantly affect explant response *in- vitro*. In deciduous woody perennials, shoot tip explants collected at various times during the spring growth flush may vary in their ability for shoot proliferation. Shoot tips collected during or at the end of the period of most rapid shoot elongation exhibited weak proliferation potential. Explants collected before or after this period is capable of strong shoot proliferation in vitro (Brand, 1993). Conversely, the best results are obtained with herbaceous perennials that form storage organs, such as tubers or corms, when explants are excised at the end of dormancy and after sprouting.

Explants also exhibit different capacities for establishment in vitro depending on their location on the donor plant. For example, survival and growth of terminal bud explants are typically greater than lateral bud explants. Often similar lateral meristem explants from the top and bottom of a single shoot may respond differently *in-vitro* in woody plants exhibiting phasic development , juvenile explants typically are often more responsive than those obtained from the often non responsive mature tissues of the same plant. Sources of juvenile explants include the following:

(1) Root suckers

(2) Basal parts of mature trees

(3) Stump sprouts

(4) Lateral shoots produced on heavily pruned plants.

The excision of primary explants often promotes the release of polyphenols and stimulates polyhenol oxidase activity within the damaged tissues. Oxidized polyphenol products often blacken the explant tissue and medium. Accumulation of these products can eventually kill the explants. Procedures used to decrease tissue browning include

(1) Use of liquid medium with frequent transfer

(2) Adding antioxidants such as ascorbic acid or polyvinylpyrrolidone (PVP)

(3) Culture in reduced light or darkness.

(4) Season of explantation.

There clearly is no one universal culture medium for establishment of all species. However, modifications to the Murashige and Skoog (Murashige and Skoog, 1962) basal medium formulation are most frequently used. Cytokinins or auxins are most frequently added to stage I media to enhance explant survival and shoot development (Hu and Wang, 1983). The types and levels of growth regulators used in stage I media are dependent on the species, genotype, and explant size.

Knowledge of the specific sites of hormone biosynthesis in intact plants provides insight into the relationship between explant size and dependence on exogenous growth regulators in the medium. Endogenous cytokinins and auxins are synthesized primarily in root tips and leaf primordial, respectively. Consequently, smaller explants, especially cultured apical meristem domes, exhibit greater dependence on medium supplementation with exogenous cytokinin and auxin for maximum shoot survival and development (Shabde and Murashige, 1977). Larger shoot tip explants usually do not require the addition of auxin in Stage I medium for establishment. Rapid adventitious rooting of shoot tip explants often provides a primary endogenous cytokinin source. Most stage I media are agar-solidified and supplemented with at least a cytokinin (Wang and Charles, 1991). The most frequently used cytokinin are N^6 – (2- isopentenyl) - adenine (2-ip). Due to its low cost and high effectiveness, the cytokinin BA is most widely used (Gaspar et al., 1996). Substituted urea compounds, such as thidiazuron, exhibit strong cytokinin –like activity and have been used to facilitate the shoot culture of recalcitrant woody species (Huetteman and preece, 1993).

Many types of auxins are used. The naturally occurring auxin indole-3-acetic acid (IAA) is the least active ,whereas the stronger and more stable compounds á-naphthalene acetic acid (NAA), a synthetic auxin and indole-3-butyric acid (IBA)., a naturally occurring auxin, are more often used. Stage I medium PGR levels and combinations that promote explant establishment and shoot growth but limit formation of callus and adventitious shoot formation are selected.

A commonly held misconception is that primary explants exhibit immediate and predictable growth responses following inoculation. For many species, particularly herbaceous and woody perennials, consistency in growth rate and shoot multiplication is achieved only after multiple subculture on Stage I medium. Physiological stabilization may require from 3 to 24 months and four to six subcultures. Failure to allow culture stabilization, before transfer to a Stage II medium containing higher cytokinin levels, may result in diminished shoot multiplication rates or production of undesirable basal callus and adventitious shoots multiplication rates or production of undesirable basal callus and adventitious shoots. With some species, the time required for stabilization can be reduced by initial culture in liquid medium.

In many commercial laboratories, stabilized cultures, verified as being specific pathogen tested and free of cultivable contaminants, are often maintained on media that limit shoot production to maintain genetic stability. These cultures, called mother blocks, serve as sources of shoot tips or nodal segments for initiation of new Stage II cultures.

Stage II: Proliferation of Axillary Shoots

Stage II propagation is characterized by repeated enhanced formation of axillary shoots from shoot tips or lateral buds cultured on a medium supplemented with a relatively higher cytokinin level to disrupt apical dominance of the shoot tip. A subculture interval of 4 weeks with a three to eightfold increase in shoot number is common for many crops propagated by shoot culture. Given these multiplication rates, conservatively, more than 4.3×10^7 shoots could be produced yearly from a single starting explant.

Stage II cultures are routinely subdivided into smaller cluster, individual shoot tips, or nodal segments that serve as propagules for further proliferation. Additionally, axillary shoot clusters may be harvested as individual un rooted Stage II microcuttings or multiple shoot clusters for ex vitro rooting and acclimatization. Clearly, Stage II represents one of the most costly stages in the production process.

Both source and orientation of explants can affect Stage II axillary shoot proliferation. Subcultures inoculated with exhibit higher multiplication rates than lateral bud explants. Inverting shoot explants oriented explants per culture period in some species.

The number of subcultures possible before initiation of new Stage II cultures from the mother block is required depends on the species or cultivar and its inherent ability to maintain acceptable multiplication rates while exhibiting minimal genetic variation and off –types (Kurtz et al., 1991). Some species can be maintained with monthly subculture form 8 to 48 months in Stage II. In contrast, in some ferns (Nephrolepsis) , as few as three subcultures may only be possible before the frequency of off-types increases to unacceptable levels.

Increased production of off-types is often attributed to production of adventitious shoots particularly through an intermediary callus stage (Jain and De Klerk, 1988). Stage II cultures, originally regenerating from axillary shoot, clusters after a number of subcultures on the same medium. These so-called mixed cultures can develop without any morphological differences being apparent. Selecting only terminal shoots of axillary origin for subculture, instead of shoot bases, decreased the frequency of off-types including the segregation of periclinal chimeras.

Selection of stage II cytokinin type and concentration is made on the basis of shoot multiplication rate, shoot length, and frequency of genetic variation. Although shoot proliferation is enhanced at higher cytokinin concentrations, the shoots produced are usually smaller and may exhibit symptoms of hyperhydricity. Depending on the species, exogenous auxins may or may not enhance cytokinin-induced axillary shoot proliferation. Addition of auxin in the medium often mitigates the inhibitory effect of cytokinin on shoot elongation, thus increasing the number of usable shoots of sufficient length for rooting. This benefit must be weighed against the increase chance of callus formation.

The possibility of adverse carryover effects on the survivability and rooting of plantlets in Stage IV should be evaluated when selection a Stage II cytokinin. For example, with some tropical foliage plants species, the use of BA in stage II can significantly reduce Stage IV plantlet survival and rooting to as low as 10% (Griffis et al., 1981). Use of Kin or 2-ip instead of BA yields survival rates in excess of 90%. in some plants the adverse effect of BA on Stage IV survival and rooting has been attributed to production of an inhibitory BA metabolite (Werbrouck et al., 1995). Conceivably, switching the Stage III rooting in some species.

Stage III: Pretransplant (rooting)

These steps are characterized by preparation of Stage II shoots or shoot clusters for successful transfer to soil. The process may involve:

1. Elongation of shoots prior to rooting
2. Rooting of individual shoots or shoot clumps
3. Fulfillilng dormancy requirements of storage organs by cold treatment
4. Pre-hardening cultures to increase survival.

Where possible commercial laborites have developed procedures to transfer Stage II micro cuttings to soil, thus bypassing Stage III rooting. There are several reasons for eliminating Stage III rooting. Estimated costs for Stage III range from 35 to 75% of the total production costs. This reflects the significant input of labor and suupples required to complete Stage III rooting. Considerable cost saving can be realized if Stage III is eliminated. Furthermore, it is often observed that in vitro formed root systems are largely nonfunctional and die following transplanting. This results in a delay in transplant growth prior to production of new adventitious roots.

For various reasons, however, it may not always be feasible to transplant Stage II microcutting directly to soil. Given the aforementioned limitations of Stage III rooting, Debergh and Maene (1981) proposed using Stage III solely to elongate Stage II shoot clusters prior to separation and rooting ex vitro. Elongated shoots may be further pretreated in an aqueous auxin solution prior to transplanting. Usually, Stage III rooting of hergaceous plants can be achieved on medium in the absence of auxins. However, with many woody species, the addition of an auxin (IBA or NAA) in Stage III medium is required to enhance adventitious rooting. Optimum auxin concentration is determined based on percent rooting, root number, and length. It is critical that the roots not be allowed to elongate to prevent root damage during transplanting. Care must be taken when selection an auxin. For example, use of NAA for Stage III rooting has been shown to decrease survival rates or suppress post transplant growth (Connor and Thomas, 1981).

Stage iv: Hardening of Plantlets

The ultimate success of micro propagation depends on the ability to transfer and reestablish vigorously growing plants from *in vitro* to greenhouse conditions. This involves acclimatizing or hardening –off plantlets to conditions of lower humidity and higher light intensity. Even when acclimatization methods are carefully followed, poor survival rates are achieved. Micro propagated plants are difficult to transfer in vivo condition due to following reasons:

1. Heterotrophic mode of nutrition
2. Poor control of water loss

Plants cultured in vitro in the presence of sucrose and under condition of limited light and gas exchange exhibit no or extremely reduced capacities for photosynthesis. Reduced photosynthetic activity is associated with poor stomatal functioning. During acclimatization, there is a need for plants to rapidly transition from the heterotrophic state for survival (Preece and Sutter, 1991). Unfortunately, this transition is not immediate. For example, in cauliflower, no net increase in CO_2 uptake is achieved until 14 days after transplantation. This occurs only following development of new leaves since the leaves produced in vitro in the presence of sucrose never develop.

Photosynthetic competency interestingly, before senescencing, these older leaves function as "lifeboats" by supplying stored carbohydrate to the developing and photosynthetically competent new leaves. This is not the rule with all micro-propagated plants since the leaves of some species become photosynthetic and persist after acclimatization.

A composite of anatomical and physiological features, characteristic of plants produced in vitro under 100% relative humidity, contributes to the limited capacity of micro-propagated plants to regulate water loss immediately following transplanting. These features include reductions in leaf epicuticular wax, poorly differentiate mesophyll, abnormal stomata function, and poor vascular connection between shoots and roots.

To overcome these limitations, plantlets are transplanted into a well-drained "sterile" growing medium and maintained initially at high relative humidity and reduced light (40 to 160µmol. $M^{-2} s^{-1}$) at 20 to 27°C. Relative humidity may be maintained with humidity tents, single tray propagation domes, intermittent misting, or fog systems. However, use of intermittent mist often results in slow plantlet growth following waterlogging of the medium and excessive leaching of nutrients. Transplants and acclimatized by gradually lowering the relative humidity over a 1 to 4 week period. Plants are gradually moved to higher light intensities to promote vigorous growth. For successful acclimatization of plant lets the following infrastructural facilities are required.

1. Pots of different sizes
2. Potting mixtures such as vermiculite, perlite, cocopeat and vermicompost
3. Hardening chamber fitted with temperature, photoperiod and humidity control devices

4. Poly house/green house facility with controlled temperature and humidity factors
5. Shade house/ Net house with fogging/misting devices
6. Shade house for further growth and development of plantlets in containers filled with a mixture of sand, soil and compost.

Somatic Embryogenesis in Fruit Crop

One of the most interesting areas of plant tissue culture is somatic embryogenesis and synthetic seed technology. Somatic embryogenesis is the development of embryos from vegetative (somatic) cells, tissues and organs under *in vitro* conditions. Somatic embryogenesis is the development of embryo from vegetative cells rather than embryos arise from fusion of sexual male and female gametes. In this process a bipolar structure is produced with a root-shoot axis and a closed independent vascular system. Steward *et al.* (1958) made first attempt in obtaining embryo like structures called as embryoids in carrot cell suspension when treated with coconut milk. This treatment was found to stop cell multiplication process and lead to differentiation of miniature embryo like structures. At the same time Reinert (1959) independently discovered the same phenomenon in carrot cells grown on agar using high concentrations of auxins. After these preliminary studies somatic embryogenesis has been successfully achieved in many plant species including important fruit crops such as mango, guava, ber, pomegranate, mulberry, banana, date plam *etc*. Somatic embryos develop through different physiological and morphological stages similar to zygotic embryos. Non-zygotic embryos can develop from plants vegetative cells, reproductive tissues, zygotic embryos or callus cells. The selection of explant is very crucial in obtaining embryogenic response. It is not necessary that all genotype of a particular crop will be responded equally. Furthermore, type, age and physiological condition of explant play an important role in responsiveness of the cultures to be embryogenic. The embryogenic response of plant is heritable and it can be transferred from embryogenic lines into non-embryogenic lines through sexual hybridization. Although, there are chances of transfer of some desirable or undesirable traits via sexual hybridization.

Applications of Somatic Embryogenesis to Propagation

1. Mass Propagation-synthetic Seeds

Unlike micro-propagation from shoot cultures, mass propagation via somatic embryogenesis has not reached commercial application. However, somatic embryos are a clonal form of propagation and this procedure has the potential for mass propagation under industrialized conditions at low cost per unit. This would allow the clonal propagation of normally seed-propagated crops analogous to the production of apomitic seedlings. To make this potential a reality, many technical problems of engineering the entire process of embryogenesis to seedling production and planting need to be addressed. Also, the price needs to be competitive with conventional seed production. In concept, these are synthetic seeds. A synthetic seed is a somatic embryo enclosed in a synthetic seed coat suitable for sowing. Problems with inherent genetic variability with in the system need to be better understood and careful field testing and monitoring for variability is required.

2. Genetic Improvement via Somatic Embryogenesis

Somatic embryogenesis can be used as a method to recover plants that have unique genetic properties. These include:

Variants recovered throughsomaclonal variation: Somatic embryogenesis could be useful for isolating somaclonal genetic variation within populations of cells. Somaclonal variability results because (a) variation is inherently present within the source explant, (b) the callusing system results in variation, or (c) it can be induced by mutagenic agents.

Plants modified through genetic transformation: Somatic embryos develop from a few cells (often single cells) (239). This makes them attractive targets for genetic transformation. This shows embryogeneic tissue transformed with a marker gene (GUS). Blue spots indicate integration of the new gene into the embryogeneic cells. Transformation occurred by using the gene gun. Regeneration of somatic embryos from these cells would yield a genetically transformed plant.

Haploids from anther culture: It is common to recover plants from anther culture by production of somatic embryos.

Cybrids development: that result from protoplast fusion often regenerate from somatic embryos.

Germplasm preservation: Germ plasm conservationis always a concern for the plant breeder. Somatic embryos could be a convenient method to store clonal germ plasm through low- temperature storage. For example, in tree improvement programs, 10 or more years might be required to evaluated individual plants for ornamental or timber quality. By that time, trees have entered a mature phase that makes them difficult to clonally propagate. To solve this problem, long-term storage to preserve the original embryogenic cell cultures by crypreservation in liquid nitrogen (-196°C) has been used.

3. Rejuvenation Through Somatic Embryogenesis

Rejuvenation in difficult to root species of fruit crops can be achieved through somatic embryogenesis process for their ease propagation. This is important advantage of restoration of regeneration capacity mature plants which are difficult to propagate through cutttings for nodal explant. For example, mature date palm, guava, ber have been regenerated through somatic embryogenesis using shoot tip or leaf explant. These somatic seedlings can be used as mother plant in a juvenile phase to start micro or macro propagation.

Steps in Somatic Embryogenesis

Somatic embryogenesis although is genotype specific and its requirement are different even variety to variety of a particular crop. Therefore, each variety requires specific protocol for the development of somatic embryoids and their conversion to plantlets. Nevertheless, basic steps are common which are as following:

1. Selection of responsive explants
2. Conditioning of explants for embryonic state
3. Differentiation and maturation of somatic embryos
4. Germination of somatic embryos to a plantlets
5. Transplanting

1. Selection of Responsive Explants

The selection of explant source material is a critical decision and may require a systematic analysis of the embryogenic potential of different explant sources within the plant. Explants differ in their genetic background and include :

1. Embryo or seedling tissue
2. Seed tissues other than the embryo
3. Mature vegetative tissue such as leaf flower and root tissue

a) **Somatic embryogenesis from embryo or seedling tissue.** The developing zygotic embryo (at various stages of development) or seedling tissue is an initial explant source with a high potential; embryos. This is a common pattern of somatic embryogenesis in a wide variety of plants. However, since the explant comes from the zygotic embryo, it has gone through recombination and the resultant somatic embryos are not duplicate copies of the mother plant. This is s good example for genetic transformation studies or for crops normally propagated by seeds because the zygote has already gone through extensive genetic selection.

The suspensor also shows somatic embrogenic potential. The embryonic-suspensor mass (ESM) appears at the very earliest stage that precedsembryo development. This tissue is prominent in conifers and can be used as an explant source to develop suspension cultures. Its appearance is white and mucilaginous, stains red with acetocarmine (0.10 per cent w/v), and fluoresces weakly under UV light as compared to green for embryogenic cells and yellow for nonviable cells.

b) **Somatic embryogenesis from seed tissues other than the embryo.** Explants in this category come from the nucellus or integuments of young ovules of some polyembryogenic species including citrus, mangosteen and mango, as well as some monoembryonic species (loquat, rubber, apple, pear, cacoa and grape). Since these explants are derived from this tissue reproduce the genotype of the mother plant and are similar to certain type of apomixis.

c) **Somatic embryogenesis from vegetative mature tissues.** Leaf tissue has been used to initiate somatic embryo production in a variety of crops including citrus, ber, grape and date palm . In some cases, organs such as roots and flower tissue can be used as explant sources.. Immature flower parts have also been used to establish somatic embryos culture and organogenesis in many fruit crops such as date palm, mulberry. This is important in those fruit crops whose vegetative propagation are either slow or difficult to propagate clonally like date palm, khejri, ker, bael *etc.*

2. Conditioning of Explants for Embryonic State

Culture environmental conditions requirements are often specific for optimum culture growth and differentiation. Culture conditions requirements of light and dark over time can be optimized through experimentation for better results. Culture in dark may necessary for preventing adverse effect of light on many plant biological processes, which may adversely affect the callus formation and embryogenic cell population. Culture in dark condition suppresses unwanted physiological reaction and tissues differentiation in explant tissue for example by limiting the photosynthesis and change of plastids into choloroplasts. Other steps such as germination of embryos may also inhibit. Likewise, improvement in embryogenic condition can be obtained by optimization of nutrient medium composition.

This is another important step of somatic embryogenesis protocol in which explants are cultured to express its potentiality for embryo formation. Several auxins such as 2,4-D, NAA, Picloram, Dicamba and cytokinins like BA, 2-ip, TDZ, kinetin are used in different combination and concentration that are varied plant to plant or genotype specific. Somatic embryos formation can take place through two different pathways (Sharp *et al.*, 1980) which are as following:

Direct embryogenesis: The embryos initiate directly from the explant tissue in the absence of callus proliferation. This occurs through "pre-embryogenic determined cells" (PEDC) where the cells are committed to embryogenic development and need only to be tissues. Clumps of embryo are developed on embryogenic centres on the explant generally from epidermal or subepidermal cell. Direct embryogenesis easily takes place in developing zygotic embryo and seedling tissue but this pathway has also been observed in leaf explants.

Indirect embryogenesis: In this pathway the cell and tissues which are not predetermined for directly formation of somatic embryos are taken as explant. This indirect method involves a callus development phase or cell suspension phase for embryo induction. Cell proliferation i.e. callus from explant takes place from which embryos are developed. The cells from which embryos arise are called embryogenically determined cells and forms embryos which are induced to do so, also called as "induced embryogenic determined cells" (IEDC). Here, specific growth regulators concentration and/or cultural conditions are required for initiation of callus and then re-determination of these cells into the embryogenic pattern of development. Callus formation is achieved at high concentration of auxins with basal medium. This results in increasing cells and callus capacity to form somatic embryos. This pathway requires systematic analysis of the embryonic potential of different explants under optimum culture condition, hormonal and medium requirements. Although embryogenic callus and cell suspension cultures generally appear to be uniform group of cells. However, on closure inspections these cultures are contain different cell masses. For suspension culture of banana, five different cell aggregates types were identified that differed in their potential to form somatic embryos. Thus, aggregates can be identified that are well organized into pro-embryogeniic masses (PEMs). These cells are generally structurally distinct from callus and emenable for mass production. PEMs are developed continue in suspension cultures until they are move to a stationary medium to develop into mature embryos. PEMs are often passed through sizing screens to get uniformity and synchrony of development.

3. Differentiation and Maturation of Somatic Embryos

In many cases, 2,4-D is essential to initiate somatic embryogenesis, but most species must be removed from the 2,4-D medium to permit further development of somatic embryos. Where somatic embryos form directly from the explant without callus formation, the entire original explant is moved to an auxin-free medium to permit further embryo development. Cells from suspension culture, or proembryo masses (PEMs), are shifted to an auxin-free medium, usually high in ammonium nitrogen. In some cases, the tissue sized by passing through screens (sieves) of various sizes to give the highest number of PEMs and synchronous development. Somatic embryos arise from clumps or small masses, develop polarity, and follow a pattern mimicking normal embryo development within the ovule. Development may be variable in rate and abnormal in appearance with secondary embryos developing from primary embryos. ABA may be added to the development medium to promote more normal somatic embryo development. Dehydration may also induce embryo maturity similar to that produced by *in vivo* embryo development.

4. Germination of Somatic Embryos to a Plantlets

Once a mature somatic embryo has reached an appropriate size, it is plated onto an agar medium without auxin. Shoots and roots may then develop. Some somatic embryos show a poor ability to convert to seedlings. However, the most common treatment is to partially desiccate somatic embryos prior to germination. A slow rate of drying enhances survival of embryos during desiccating. In cases, somatic embryos may exhibit dormancy that is broken by chilling stratification, gibberellic acid or cytokinin.

5. Transplanting

Once an adequate shoot and root have developed, the seedling can be transplanted to a green-house medium as with any seedling plant or treated with a "synthetic seed coat" and handled as a synthetic seed. It require three step hardening procedure.

Classical Examples of Somatic Embryogenesis in Fruits Crop

Somatic Embryogenesis in Pomegranate

According to R.R.Bhansali (1990) plantlets were regenerated from somatic embryos originatingfrom cotyledonary tissues of pomegranate (*Punica granatum*) throughmultiple somatic embryogenesis. Embryogenic cell clusters proliferated vigorously with regular sub-culturing after 20 d on RBM-II mediumcontaining 1 μM kinetin (KN), 2 μM benzylamino purine(BAP) and 5 μM 2,4-dichlorophenoxyacetic acid (2, 4-D). Developmental stages of somatic embryos were expressed on sub-culturing with a low level of 2, 4-D (2.5 μM). Embryogenic initialscells were small, round to oval, thick-walled, contained densecytoplasm which stained with acetocarmine and were usually attached to non-embryogenic cells. Embryo maturation was obtained on RBM-III and IV media to produce young seedlings on the initiationof the first long tap root.

Somatic Embryogenesis in Papaya

M.H. Chen *et al*. (1987) reported somatic embryogenesis was obtained from 3 month old root cultures. A medium containing half strength of MS organic salts, 160 mg/l adenine sulphate, 1.0 mg/l NAA, 0.5 mg/l kinetin and 1.0 mg/l GA_3 was optimal for embryogenesis. The callus maintained high regenerative capacity after 2 years of culture on this medium.

Bhattacharya and S.S.Khupse (2000) induced somatic embryos in immature embryos on MS basal medium supplemented with 3 mg/l of 2,4,5-T and incubated in the dark for period of 3-6 weeks. Loosely attached globular somatic embryos appeared from apical dome within 3-6 weeks of incubation. The development of somatic embryos passed through globular, heart and torpedo shape stages. Embryo continue to proliferate with regular subculture and remained morphologically competent for up to one year. Maturation of the embryos was achieved in medium supplemented with ABA (0.1 mg/l)

Somatic Embryogenesis in Banana

Micchelle *et al*. (2005) described method with cell suspension cultures of banana were obtained from indirect embryogenesis of male inflorescence of banana cultivar maca. Part of the calluses formed (40%) showed embryogenic characteristics (non friable, compact and yellow colour). The cell suspension originated from embryogenic calluses, contained clusters of small tightly packed cells with dense cytoplasms, relatively large nucleoli. After 4 months of culture, somatic embryos started to regenerate. The maximum number of regenerated plants was observed between 45 to 60 days after embryo formation.

Somatic Embryogenesis in Date Palm

Quarishi *et al.* (1997) studied the response of shoot tips, axillary buds, leaf sheath and meristematic tissues as explants on MS medium containing 30 gm sucrose, 0.4 mg thiamine, 100 mg myoinositol, 3 gm activated charcoal and 8 gm agar per litre supplemented with 2,4-D , NAA, BA and 2-iP. Good tissue proliferation obtained only by apical buds. Somatic embryogenesis nodules formed in callus when transferred to MS medium with 0.1 to 1.0 mg/l GA_3 and 0.1 mg/l 2-iP and plantlets developed directly from embryoids. Roots are induced on MS mediumcontaining 0.1 mg/l NAA with an ex-vitro survival of 70-75% when rooted plants of 8-12 cm in length were used.

Madhuri and Shenkon (1997) reported that a white friable callus was initiated from young leaf primordia of date palm on MS medium with 2,4-D (50 mg/l), kinetin(1 mg/l), 2-iP(0.5 mg/l) and glutamine(100 mg/l) after 8-10 weeks of incubation in dark. The callus was transferred on embryogenic medium MS ,

Biotin 0.01 mg + Ca Protothenate 0.1 mg + BA 2 mg + NAA + NOA 0.5 mg per litre showed induction of large number of somatic embryos which continuously developed and proliferated for more than one year. Further development of embryo into plantlets with two leaves and thin root took when mature embryos were transferred to differentiation medium containing adenine sulphate and keep under light intensity of 6000 lux. Adventitious roots were develop by cutting of main root and sub culturing then MS medium with 0.5 mg/l NAA. Light intensity at this stage reduced to 3000 lux. Plantlets were successfully transferred to pots containing a mixture of 1:1 vermiculite and peat moss.

Asemota *et al.* (2005) conducted an experiment on in vitro culture initiation on MS basal medium with 2,4-D. Callogenesis was observed in second subculture. Continuous subculture of the callus of the establishment medium at about the third sub culture from callus production, resulted in somatic embryo formation. The somatic embryos were then transferred to MS medium without hormones under light where they matured after about two subcultures and developed into shoots.

Somatic Embryogenesis in Mulberry

Agarwal *et al.* (2004) develop a procedure for optimum secondary embryogenesis and maturation of *Morus alba* somatic embryos. The primary somatic embryos were inoculated on MS medium with different plant growth regulators and complex additives in various combination and concentrations. Best results were obtained on the medium supplemented supplemented with 0.05 mg/l 2,4-D + 0.1 mg/l BAP and 6% sucrose, where percentage of secondary embryogenesis was 98.93 with 17% well matured cotyledonary cotyledonary embryos. On this medium, the embryogenic competence was maintained for more than a year by repeated subculture.

Limitations

1. There is a probability of somaclonal variation.
2. The method is rather difficult.
3. The chances of losing regenerative capacity become greater with repeated sub cultures.
4. Induction of embryogenesis is often very difficult or impossible with many plant species.
5. A deep dormancy often occurs with somatic embryogenesis that may be extremely difficult to species.

Crop/explant	Medium Requirement				Culture Conditions	Reference
	I	II	III	IV		
Pomegranate Cotyledanory tissues	½ MS medium + sucrose(30 gm/l) + glutamine + casein hydrolysate + inositol (300 mg/l each)	½ MS medium + 2,4-D (5-2.5 µM) + BAP(2µM) +Kinetin (2µM)	MS medium + 2,4-D (0.5 µM) + BAP (2µM) + Kinetin(2µM)	MS medium + BAP (2µM) + Kinetin (2µM)	23° ± 2°C RH 60%, 2000 lux	R.R.Bhansali (1990)
Date palm Shoot tips	MS medium + 2,4-D(453 µM) + 2ip (15µM) + BAP (13 µM)	MS medium + NAA (54 µM) + 2ip (148µM)	MS medium + 2,4-D (679µM) + 2ip (15 µM)	-	24 ± 3°C	Eshraghi *et al.*(2005)
Shoot apical meristem	MS medium + 2,4-D (100 mg/l) + 2ip (3 mg/l)	MS medium + NAA (0.1 mg/l) + 2ip (5 mg/l) + glutamine (0.2 gm/l)	MS medium + NAA (0.1 mg/l) + 2ip (1 mg/l)	-	27°C, Darkness	Asemota *et al.* (2005)
Mulberry Zygotic embryo	MS medium + 2,4-D (2 mg/l) + BAP (0.50 mg/l)	-	MS medium + 2,4 -D (0.05 mg/l) + BAP (0.1 mg/l) + Coconut water (10ml/l)	-	25 ± 2°C, 60 hr photo period	Agarwal *et al.* (2004)

Somatic embryogenesis protocols of some fruit crops
Micropropagation protocols through Enhanced axillary shoot proliferation (shoot culture)

Fruit crops	Source of explant	Sterilization technique	Media requirement for shoot	Media requirement for root
Ber	Nodal explant	0.1% Hg Cl_2 for 5 to 7 minutes	MS + BA (7.5) + NAA (0.1)	White's Medium + Agar (8g) followed by Dipping in IBA 25 mg for 48 hrs
Lasoda	Nodal explant	20% Na O Cl for 15 minutes or 0.1% Hg Cl_2 for 5 to 7 minutes	MS+ BA(2.0) +NAA (0.1)	MS+ AC+3mg NAA
Bael	Nodal explant, Root segments	0.1% Hg Cl_2 for 5 to 7 minutes or NaOCl (1:5) for 10 Minutes	MS + BA (2.5) +IAA (1.0)	Exvitro 2 cm shoot, IBA (10.0)
Ker	Nodal explant	0.1% Hg Cl_2 for 5 to 7 minutes	MS+ BA(2.0) +NAA (0.1)	MS + NAA+IBA (0.5+0.5)
Mulberry	Nodal explant	0.1% Hg Cl_2 for 10 minutes minutes	MS+ BA(4.0) +NAA (0.1)	MS +IBA (0.5)
Guava	Nodal explant	0.1% Hg Cl_2 for 5 to 7 minutes	MS, BA (2.0), NAA (0.1), PVP	½ MS, 0.1% AC, NAA + IBA (0.5+0.5)
Grape	Three node shoot tip or 2mm meristem tip for virus elimination	5 cm shoot tips, remove leaves, stir in 70% alcohol for 1 minutes, rinse four times	(1) ¾ MS, BA(0.1)(2)BA (0.2), NaH_2PO_4 (170), AD (80)	Reduced additives, IAA (0.1) or omit*in vivo* rooting slow to establish
Strawberry	Meristem tip and Nodal explant	3cm runner wash with DT 5 minutes, 1:10 NaOCl, wash two times	(1) MS + BA (1.0)(2) Same	MS, IBA (1.0) + CharcoalStandard or use *in vivo* rooting

Applications of Citrus Shoot-tip Grafting *in vitro*

The main application of shoot-tip grafting *in vitro* (STG) is to control graft-transmissible pathogen that requires the use of healthy trees in the new plantings. The routine application of STG using 0.1-0.2 mm shoot tips is very efficient for elimination of all citrus graft-transmissible pathogens from local or imported varieties. It has allowed the recovery of hundreds of healthy cultivars and the planting of hundreds of millions of healthy certified trees worldwide. Only in Spain about 140 million certified nursery plants propagated from micrografted plants have been planted. STG is also a very useful technique for regeneration of elite genotypes in several research areas. In vitro grafting for these purposes may be done using larger shoots (up to 1 cm). STG is being routinely used for the following purposes: (i) Regeneration of somatic hybrids from embryos difficult to germinate; (ii) Regeneration of plants from irradiated shoots to produce seedless varieties; (iii) Regeneration of plants from haploid embryos that are very difficult to germinate. STG was used to regenerate the 'Clemenules' haploid plant that has been used by the International Citrus Genome Consortium to sequence the whole citrus genome; (iv) Production of stable tetraploid plants of non-apomictic genotypes, which are very useful for triploid breeding; (v) Regeneration of transgenic plants from shoots that are very difficult to root in vitro. STG has become a routine application in citrus genetic transformation.

References

Al-Khayri, J.M. 2001. Optimization of biotin and thiamine requirements for somatic embryogenesis of date palm (*Phoenix dactylifera* L.). *In Vitro* Cell. *Dev. Biol-Plant*. 37: 453-456.

Bhaskar, N. and Smith, S. 1992. Somatic embryogenesis from shoot tip and immature inflorescence of Phoenix dactylifera L. cv. Barhee. *Plant Cell Reports*. 12 : 22-25.

Eshargi, P., Zarghami, R. and Mirabdulbaghi, M. 2001. Somatic embryogenesis in two Iranian date palm cultivars. Paper presented in II nd International Conference on Date palm (Al-Ain UAE, March 25-27, 2001).

George, E.F. and Sherington, P.D. 1984. Plant growth regulators. In Plant explants from coconut and date palms cultured *in vitro*. *Physiol. Plant*. 42 : 173-178.

José Juárez, Pablo Aleza, Luis Navarro 2015 Applications of citrus shoot-tip grafting in vitro International Citrus Congress - International Society of Citriculture SHS Acta Horticulturae 1065: XII 10.17660/*Acta Hortic*.2015.1065.79

Kecker, N.C., Solanki, K.R and. Joshi, S.P. 1989. Micropropagation of date palm (Phoenix dactylifera L.) cv. Khadrawy using tissue culture technique. *Ann. of Arid Zone*. 28 : 137-141.

Saker, M.M. and Bakheet, S.A. 2001. In vitro propogation of Egyptian date palm morphogenic responses of immature embryos . Bulletin of Faculty of Agricultural University, *Cario*. 49(2) : 203-214.

Sharma, D. R., D. Sunita and J.R. Chowdhury. 1984. Somatic embryogenesis and plant regeneration in date palm (*Phoenix dactylifera* L.) cv. "Khadravi" through tissue culture. I*ndian J. of Exp. Biol.* Vol. 22: 596-598.

Tisserat, B. 1979. Propagation of date palm *in vitro*. *Journal of Experimental Botany*. 30(6) : 1275-1283.

Vermendi, J. and Navarvo. 1996. Influence of physical conditions of nutrient medium and sucrose on somatic embryogenesis of date palm. *Plant Cell Tissue and Organ Culture*. 45(2) : 159-164.

Yadav NR, Singh J, Yadav RC, Chowdhury VK. and Chowdhury JB 1998. Somatic embryogenesis and plant regeneration from cell suspension caltures of Phoenix dactylifera L. cv. Khadrawy. *Physiol. Mol. Biol. Plants*. 4: 135-138.

9

Strategies for Developing High Density Planting in Semi Arid Fruit Crops for Higher Yield and Returns

Sanjay Singh, A.K. Singh, D.S. Mishra and V.V. Appa Rao

Introduction

Most of the technological advances in the recent past in Indian Horticulture have been done for favourable agro-climatic regions. However, the vast and diverse land-scape of hot desert, arid, semi-arid and tribal area is not yet receiving logical thought for resource based horticultural development (Samadia and Haldhar, 2017). Although, the semi-arid regions of the country have now became potential hub for aonla, bael, ber, pomegranate, jamun, guava, fig and custard apple. There has been a steady increase in area and production of these fruits particularly under aonla, ber, custard apple and pomegranate in the country as a result of identification and development of suitable varieties and production technologies. The average productivity of these fruits in India is, however, low as compared to many developed countries. The main reasons for low productivity are old and senile orchards, low yielding varieties, poor orchard management and inadequate technological up-gradation and adoption by the growers. Presently, the continuing decline in the availability of cultivable land, rising energy and land costs together with the increased demand of fruit and fruit products, have given thrust to the concept of high density planting (HDP) in fruit crops. High density planting gives earlier production and return per unit area, shortens juvenility, eco-friendly, provides efficient land use and better use of resources like light, water and fertilizers, efficient pesticides application, besides, in this system the harvesting becomes easy(Mishra and Goswami, 2016). Among the factors such as cultivar, rootstock, quality of planting material and cultivation practices contribute to high yield of fruit trees, however, the number of trees^{-1} is the most important factor which brings about radical increase in fruit production. Therefore, high density planting has great potential for increasing productivity in fruit crops (Mishra *et al.*, 2003). The success of this technology in most of the fruit crops is dependent on the use of methods to control shoot growth and maximize light interception as the trees begin to bear fruit. The main aim of HDP is to achieve the twin requisites of productivity by maintaining a balance between vegetative and reproductive growth without impairing the plant health and fruit quality. The underlying principle of HDP is to make the best use of vertical and horizontal space per unit time and to harness maximum possible return per unit of inputs and natural resources.

The conceptual background of high density planting in fruit growing was pioneered in temperate fruits and first planted at the end of the nineteen sixties, since then there is rise in establishment of commercial high density orchards throughout the world. HDP system is normally understood as a system in which a higher number of plants are accommodated per unit area in comparison to the conventional planting density. However, the exact limit of plant density to be termed as HDP is not yet well defined. It varies with growing region, species, variety, rootstock, agro-techniques adopted for a particular crop and return from the orchard. In India, high density plantings have successfully been demonstrated in aonla (Singh *et al.*, 2011), guava (Singh *et al.*, 2007, Kumawat *et al.*, 2014), pomegranate (Haneef *et al.*, 2014), mango (Gunjate *et al.*, 2009, Joglekar *et al.*, 2013), jamun (Singh *et al.*, 2017) and custard apple (Farre *et al.*, 1999).

There are a large number of minor fruits such as tamarind, phalsa, wood-apple, mulberry, karonda and lasoda which have diverse uses, besides being hardy and well-adapted to different agro climatic conditions and stress situation.

Present Status in India and Global Scenario

There has been much talk about the potential of high density planting in fruit crops, but few studies demonstrated the long-term economic benefits (Ram, 1996). In some of the cases, yield was declined after few years as trees began to crowd each other. The high-density orchard provides several times (8-9) higher yields than the traditional densities as demonstrated by Ram, 1996 in alternate bearing Dashehari mango in north India. Dashehari mango at 2.5 m x 3m (1,333 plants ha^{-1}) was raised under HDP with pruning and dehorning after the harvesting followed by paclobutrazol application and yield was secured every year. In an experiment at CISH, Lucknow, ultra high density system of planting in guava was standardized. The ultra high density orchard system of guava accommodates 5000 plants ha^{-1}, at a spacing of 2.0 x 1.0m and managed with regular topping and hedging during initial stages which helped in controlling tree size and getting higher yield. Bhagwa pomegranate may be planted at a distance of 2 x 2 m (2500 plants ha^{-1}) under drip irrigation and fertigation. Such orchards resulted in higher profitable yield (net return Rs. 5,96, 177 ha^{-1}) with quality fruits (Haneef *et al.*, 2014). Similar observations have been made by various workers in citrus, custard apple and ber.

Gaps in Prevailing System

The average productivity and per capita availability of fruits in India is, however, low as compared to many developed countries. The main reasons for low productivity are old and senile orchards, wider spacing, low yielding varieties, poor orchard management and inadequate technological up-gradation and adoption by the growers. Presently, the continuing decline in the availability of cultivable land, rising energy and land costs together with the increased demand of fruit and fruit products, have given thrust to the concept of high density planting (HDP) in fruit crops. High density planting gives earlier production and higher yield per unit area in comparison to the traditional planting at wider spacing (Table 1).

Table 1: Successful stories in high density planting in India (Saroj and Singh, 2018)

Crop	Spacing	Plants ha^{-1}	Yield (t^{-1})	% increase in yield over traditional system
Mango cv. Amarpali	2.5 x 2.5m	1600	19.20	250
Citrus (Kinnow)	1.8 x1.8m	3000	20.0	200
Guava	1 x 2m	5000	50.0	250

Concept of High Density Planting

High density planting system ensures better utilization of land, labour and solar radiation and higher yield in the initial years of planting, because of accommodating more number of plants per unit area. High-density orcharding appears to be the most appropriate answer to the problem of low productivity and long gestation period depriving the orchardist of early returns. High density planting by adopting suitable planting systems has been viewed as successful in fruit trees like guava, pomegranate and aonla. Considerations of the soil condition, planting geometry and manipulation in the spacing are the important means in obtaining higher production and productivity.

Tools for High Density Planting

Planting Distance and Dwarf Cultivars

The availability of a dwarf variety is the first and foremost prerequisite for establishing any high density orchard. Closer spacing and growth regulation by the training and pruning may be tried for a successful adaptation of this concept. Normal planting distance in square system of planting is maintained at 10 m x 10 m in different fruits. In hedgerow planting system, distance between row-to-row and plant-to-plant is kept 10 m and 5 m respectively, while in double hedgerow system, two rows of hedge are planted at half of distance (5 x 5 m).In case of paired planting system, pair of two plants at the distance of 5 x 5 m in between the rows and 10 m within row and 6 pair per plot are maintained, whereas in cluster planting system a cluster of four plants at 5 x 5 m apart are accommodated, hence four clusters per plot (30 x 30 m) are adjusted. This aspect needs to be standardized under various agro climatic conditions of the country. Jamun and khirni may be planted at the distance of 5x5 m distance by following light non selective pruning.

Selection of Hardy Fruit Crops

Many fruits like tamarind, custard apple, bael and wood apple provide better choice in these areas because of synchronization of fruit development with rains and deep root system which can utilize the moisture from deeper soil layers and better adoptability in poor lands.

Rootstocks

Drought tolerant and hardy root stocks may be tried especially in the low rainfall regions for reducing vulnerability to droughts, long dry spells during rainy season and other uncertainties which are progressively increasing due to climatic changes.

Crop	Dwarfing rootstocks
Ber	*Zizyphus rotundifolia, Z. nummularia*
Citrus	Troyer citrange, Flying Dragon (*Poncirus trifoliata*), Rangpur lime
Guava	*Psidium friedrchsthalianum, P.chinensis*, Pusa Srijan
Mango	Vellaikullamban

Proper Training and Pruning

Direction of plant in a desirable form is known as training. Young plants should be allowed 3-5 well spaced branches to develop in to the main scaffold structure of the tree. Framework of branches is allowed to develop above 60-100 cm from the ground level. Pruning is a tool to regulate tree size and shape to achieve a desired architecture of the canopy and also to reduce the foliage density by removing the unproductive branches of the tree. It is necessary to train the trees right from the nursery stage. The young plants transplanted in the field should be supported with bamboo stake to avoid damage of the bud union and to support main stem. During the first year, tree puts forth vigorous growth and many side branches emerge from the main shoot. All the secondary branches up to 75 cm from the ground level should be removed and straight and vigorous growing main trunk be developed. On the main trunk 3 to 5 laterals/ secondary branches most favourably located around the main stem, should be selected to make the proper framework. In the second year, the process should be repeated to develop tertiary branches on the secondary shoots. On each secondary branch 3 to 4 well-distributed tertiary branches should be retained. Efforts to keep the branches and tree upright should be made all the time through light pruning. Final balancing and correction of framework should be done in the third year so as to obtain a strong framework.

Objectives of Pruning Under High Density

- To restore shoot and root ratio.
- To prevent the formation of week crotches and to regulate the number and location of main scaffold limbs.
- To regulate growth, vigour and direction of shoots.
- To regulate tree canopy and fruiting pattern.
- Exposure to sun light and for improvement in fruit quality and yield attributes.

Moisture Conservation Through Mulching Under High Density Planting System

Mulching refer to the process of covering the soil surface with plastics, organic materials, stones etc. to reduce evaporation, to keep down weeds, salinity level and also to moderate wide fluctuations in diurnal soil temperatures. It controls external evaporability by cutting of solar radiation falling on the ground, exerts a decisive effect on earliness, yield and quality of the crop. The effectiveness of mulches in conserving moisture has generally been found to be higher under drought condition and also during early periods of plant growth when canopy covers remain scanty. Mulching is mainly employed for-

1. Moisture conservation in drought prone and rain fed areas.
2. Reduction of irrigation frequencies.
3. Reduction of soil erosion
4. Enhancing water conservation and increase water infiltration.
5. Moderation of soil temperature.
6. Soil solarization for control of soil borne diseases.

7. Harvesting of better quality produce.
8. Increasing overall crop production.

Low soil fertility and moisture content are the main causes for low production and productivity under rainfed conditions of western India. Organic mulches also improve the organic matter content as well as microbial flora in the soil. Paddy straw mulch or grasses are very effective to conserve the soil moisture during summer season under ranfed areas.

Rain Water Harvesting and Water Management for High Density Planting

In-situ moisture conservation can be achieved by keeping soil surface rough and contour trenching etc. The *exsitu* rainwater harvesting includes water storage tanks, gully bunding construction of chek-dams followed by open wells etc., life saving irrigation through drip may be given to the plants. For example, irrigation in jamun during ripening enhances the fruit quality and yield attributes.

Canopy Management in High Density Planting

Canopy management of the crop deals with the development and maintenance of their structure in relation to the size and shape for the maximum productivity and quality. The basic concept in canopy management of a perennial tree is to make the best use of the land and the climatic factors for an increased productivity. Tree vigour, light, temperature and humidity play a vital role in the production and quality of the fruits. The crux of the canopy management lies in the fact, as to how best we manipulate the tree vigour and use the available sunlight and temperature to increase the productivity and quality and minimize the adverse effects of weather parameters.

Principles of the Canopy Management

The crux of the canopy management lies in the fact, as to how best we manipulate the tree vigour and use the available sunlight and temperature to increase the productivity and quality and to minimize the adverse effects of the weather parameters. The proper sunlight exposure to foliage, fruits and prevailing temperature play vital role in photosynthesis, fruit bud differentiation, ripening and fruit quality etc. Some of the basic principles in canopy management are as follows.

- Maximum utilization of the light.
- Avoidance of the build up of micro-climate congenial for the diseases and pests.

- Convenience in carrying the cultural operations.
- Maximizing the productivity and quality.
- Economy in obtaining the required canopy architecture.

Sequence/ Steps for Pruning/ Canopy Management in new Plantations

Direction of plant in a desirable form is known as training. Young plants should be allowed 3-5 well spaced branches to develop in to the main scaffold structure of the tree. Framework of branches is allowed to develop above 60-100 cm from the ground level. Pruning is a tool to regulate tree size and shape to achieve a desired architecture of the canopy and also to reduce the foliage density by removing the unproductive branches of the tree. It is necessary to train the trees right from the nursery stage. The young plants transplanted in the field should be supported with bamboo stake to avoid damage of the bud union and to support main stem. During the first year, tree puts forth vigorous growth and many side branches emerge from the main shoot. All the secondary branches up to 75 cm from the ground level should be removed and straight and vigorous growing main trunk be developed. On the main trunk 3 to 5 laterals/ secondary branches most favourably located around the main stem, should be selected to make the proper framework. In the second year, the process should be repeated to develop tertiary branches on the secondary shoots. On each secondary branches 3 to 4 well-distributed tertiary branches should be retained. Efforts to keep the branches and tree upright should be made all the time through light pruning. Final balancing and correction of framework should be done in the third year so as to obtain a strong framework.

High Density Planting in Semi Arid Fruits

Aonla

Singh *et al* 2007 and 2011 conducted the experiment on high density planting in aonla cv NA-7 under raifed conditions of western India. Total five planting systems *viz.*, square (9 plants / plot), hedgerow (15 plants/ plot), double hedgerow (20 plants /plot), cluster system (16 plants/ plot) and paired (12 plants / plot) planting systems were followed. Net area of plot of the each treatment was 900 m^2 and the number of plants in one ha was computed as 100, 133, 166, 177 and 222 under the square, paired, hedgerow, cluster, and double hedgerow systems, respectively. Normal planting distance was maintained in square system of planting (10 x 10m). In case of double hedgerow system, two rows of hedge were planted at half of distance at 5m x 5m.

Yield

Yield per ha during 10th year was recorded in double hedgerow (220.09q/ ha), hedgerow (176q/ ha), cluster (154.26q/ ha), paired (123.71q/ ha) and square system of planting (98.49q/ ha) under rainfed conditions of western India.

Fruit quality

Fruit weight (48.25 g), fruit length (4.45 cm), fruit breadth (4.27 cm) were recorded maximum in the fruits of square system. The quality in terms of TSS (9.00^{0}Brix), total sugar (5.70 %) and vitamin C (510 mg/100g) were recorded the highest in double hedgerow system.

Economics

In all the planting systems *viz.*, double hedgerow, hedgerow, cluster, paired and square, maximum net return was recorded in double hedgerow planting system (Rs.150070/ha) followed by hedgerow (Rs.120800/ha), cluster (Rs.106508/ha), paired (Rs.85968) and square (Rs.68792/ha). The net return over square system was recorded 118.15, 75.60, 54.82 and 24.96 per cent in double hedgerow, hedgerow, cluster and paired system of planting, respectively.

Based on the observations on various aspects, the overall results on different high density planting systems demonstrated that the double hedgerow system of planting can be adopted by *aonla* growers of rainfed area for maximizing the productivity.

Nutrient Management in Aonla Under High Density

For obtaining optimum yield purely under rain fed condition, 10 kg Farm Yard Manure, 100g nitrogen, 75g phosphorus and 75g potash per plant was applied to one year old plant and this dose was increased every year in same proportion. Half dose of nitrogen and potash and full dose of phosphorus was applied in the last week of June just after first rain and remaining amount of N and K was given in the end of August. The manure and fertilizer were spread under the whole canopy of the plant and it was incorporated well in surface soil by spading.

Bael

Recently, high density planting has been consistently gaining popularity particularly with small land holdings in the Indian conditions. High density planting makes maximum use of land to achieve high yields in the early period of orchard life along with better orchard management. With an increase in planting density, the yield may increase, but beyond threshold density, quality may be deteriorated and may not be profitable in terms of economic return. Normally, the bael is grown in India at planting distance ranging from 8-10 m distance. Keeping in view the growth habit of varieties, bael can be planted from 5 to 8 m. Goma Yashi variety of bael is performing very well under high density planting (5x5 m distance) (Singh *et al*, 2012). It is spineless with semi-spreading drooping growth habit. Flowering is taken place during April-May. Fruit yield was 51.00 kg in 6th year with 1.31kg fruit weight, 37°B T SS, 0.026% acidity and 22 mg/100 g vitamin C.

Canopy Architecture in Bael

Juvenile plants are trained with the help of stakes so that they can grow straight. The growth of bael plants can be managed at desired size and shape by canopy architecture. Plants may be headed back at 1 to 1.5 m from the ground level to facilitate the emergence of new growth below the cut point. Three to four equally spaced shoots (limbs) are retained around the stem to form the main scaffold of the trees. These shoots are allowed to grow and pruned to 50% of their total length during 3rd year (April-May) to promote emergence of new shoots below the cut point. As the new shoots emerge, they are allowed to develop further. It is better to train the plant in modified leader system of canopy architecture and it should be completed within initial 3-4 years. Thereafter, selective pruning should be done to maintain the balance canopy to the tree.

Jamun

Jamun is a crop where high density planting may be adopted (Singh and Singh, 2012). Jamun fruits are used as an effective medicine against diabetes, heart and liver trouble. Owing to its varied uses, it has now emerged as a potential fruit crop of country (Singh *et al*., 2017). Most of the plantations of jamun are established through planting of seedling types. However, now farmers are showing much interest in the plantation of grafted planting material of Goma Priyanka jamun developed by the station. In the absence of proper canopy management practices orchards become overcrowded which pose problem for orchard access and for adequate light interception needed for optimum photosynthesis, flowering, fruit set and quality. Pruning has been used as a tool to regulate tree size and shape to achieve a desired architecture of the canopy and also to reduce the foliage density by removing the un-productive branches of the tree. To maintain the dwarf framework of the jamun tree, toping of main stem (4–6 meters) is needed which also facilitates easy harvesting of the fruits. The fruit bearing branches in jamun are present on the periphery of canopy throughout its height. Also all fruits on a tree do not mature at once. Therefore, at least 3-4 pickings are required for harvesting of fully ripened firm fruits from a tree. The jamun fruits are also delicate and have very thin peel; therefore, fruits are to be essentially harvested by hand picking. For this, it is necessary to climb up the jamun tree. Thus, harvesting in jumun becomes tedious and 25 to 30% fruits are spoiled due to mechanical damage during harvesting. The jamun being a non-climacteric fruit, fully ripened fruits essentially has to be plucked for getting desirable fruit quality. As hand picking is the only way to harvest fruits, managing tree canopy at suitable height is the only lasting option to have clean harvest of ripe fruits in economical manner. Pruning intensity of 25% annual extension growth has increased the number of flowering shoots (70.33%), panicle length (13.97), number of flowers/panicle (46.25), pollen viability (93.80%), pollen germination (37.29%), fruit set (45.74%), and yield (42.70 kg/tree) in jamun cv Goma Priyanka.

Guava

High Density Planting and Canopy Management in Guava

Guava needs constant training and pruning to provide desired tree shape. The plant is trained 3-4 months after planting. Initially single stem is allowed up to the height of 75-80 cm from the ground level. A framework of 4 branched should be established. The crotch angles between the branches and main stem should be wide enough to facilitate adequate light penetration and provide physical strength to support fruit load at maturity.

Singh *et al.*, 2013 reported that guava cv Allahabad Safeda were planted at 1.5 x 3, 3.0 x 3.0, 3.0 x 6.0 and 6.0 x 6.0 m distance to determine the effect of canopy management on tree growth, yield and quality attributes. Trees were topped to uniform height of 60-70 cm from the ground level, 2-3 months after planting to induce the emergence of new growth below the cut points. Three to four equally spaced shoots were retained around the stem to form the main scaffold limbs of tree and allowed to grow for 4-5 months after topping until they attain a length of 40-50 cm. The selected shoots were further pruned to 50% of their length (concurrent pruning) for inducing multiple shoots from the buds below the cut end. Newly emerged shoots were again allowed to grow up to 40-50 cm and pruned once again for emergence of new shoots to obtain the desired shape. The pruning operations continue during the second year after planting. Higher yield of 80.76 per cent was obtained under density of 3 x 1.5 m than 6 x 6 m.

Ultra high Density Planting in Guava

In order to enhance fruit production and minimize requirements for labour, CISH, Lucknow developed (Singh, 2005) an ultra high-density orcharding for guava growing, which accommodates 5000 plants ha^{-1} (1m x 2m) coupled with regular topping and hedging, particularly during initial stages. On an average 12.5t/ha fruits are obtained in first year. The yield increases up to 48 t/ha 5 year after planting in multi-shoot meadow-orchard system. Two months after planting, when the plants are well established, the plants are cut 30-40 cm above the ground level. New shoots emerge below the cut point. These shoots are allowed to grow 40 to 50 cm in 3-4 months. These shoots are then pruned up to 50 percent of their length for new shoots to arise below the cut point. These are again cut after 3-4 months for flowering and fruiting. Pruning is done three times in a year so that the plants are maintained at height of 1-1.5 m in form of hedge. After appearance of flower on new shoots, which develop from previous cut, semi-hard wood portion (50%) of these shoots, is pruned out in May for different induction of new shoots below the cut end. Renewed growth is initiated, flowers differentiate and original size of well-feathered tree is recovered by September-end. The ultra high density planting is one of the most efficient modern systems particularly in respect of easy cultural operations, very high and early yield low demand on labour (Singh, 2008, 2010). The production under high density comes in range of 26-55 t/ha,18-38 t/ha, 11-31 t/ha, and 6-27 t/ha, under 1.5 x 3.0 m, 3.0 x 3.0 m, 6.0 x 3.0 m and 6.0 x 6.0 m, respectively. The cost benefit ratio under 1.5 x3.0 m, 3.0 x3.0, 6. 0 x 3.0m and 6.0 x 6.0 m is 2.34, 2.13, 2.02 and 1.91, respectively. Time and extent of pruning must be standardized for the varieties and agro-climate. After 5 years of planting, the yield starts to reduce as a result of overcrowding of shoots in the upper canopy and the lower part becomes woody and deprived of light. Therefore, back pruning to a height of 50 cm from ground level is required after 5 years to develop new canopy and sustain the yield.

Citrus

The concept of high density planting in citrus has drwn attention of citrus growers in US, Japan, Australia, Israel, South Africa, brazil and Italy during the last three decades. In India, sweet orange and mandarin are planted with density of 250 to 350 plants per ha depending upon the variety, root stock and agroclimatic conditions. Under Gujarat conditions, sweet orange cv Sathgudi is performing well with a density of 400 plants per ha. In Karnataka, Coorg mandarin on trifoliate orange and Rangpur lime are planted with density of 400 and 275 plants per ha respectively. In Haryana, a plant desity of 375 per ha in Kinnow mandarin on Jatti Khatti root stock was reported to be successful. Control of tree size could be done by removal of terminal portion of branches destroying apical dominance and stimulating lateral bud growth resulting in a bushy and compact tree. Up right growth is more vigorous and less fruitful, hence horizontal growth is encouraged through selective pruning to get more yield per unit area.

Ber

Ber is one of the potential crops under semi arid ecosystem and high density planting is being practiced. It is of deciduous nature and regular annual pruning is done for emergence of flowers on currents seasons growth and also to control the tree canopy. Under irrigated arid conditions 5 x 5 m distance may be practiced. During the first 2-3 years after planting, Ber trees are trained to develop a strong frame work. After that, old growth is beheaded in March-April, keeping 1-2 nodes above the graft union to allow vigorous new growth. On main trunk, 3-4 well spaced branches are allowed when it is headed back. During second year, these main branches are clipped, retaining 3-4 secondary branches on each of them. This process is continued to develop tertiary branches. Not more than one growing shoot is retained at a node so that narrow crotches are avoided. The Ber, fruit is born in the axils of leaves on the young growing shoots of current season. Therefore, a regular annual pruning is essential to induce a good and healthy growth which will provide the maximum fruit bearing area on the tree. Some thinning out of branches is also necessary to avoid too much crowding which helps in reducing the incidence of pest and diseases. The best time for pruning for Ber is during the hot and dry season or when it sheds its leaves and becomes dormant. In Delhi, Haryana, Punjab, Rajasthan, and Uttar Pradesh a Ber tree becomes dormant during May when pruning should be done. In Maharashtra, Gujarat and Andhra Pradesh, pruning varies with ecological regions and objectives. Singh *et al.,* 1978 recommended pruning at 75 cm from the base at leaf fall to obtain highest yield.

Pomegranate

Training of pomegranate plants to single stem or multi stem can be practiced. In single stem, plant is allowed to grow a height of about one meter. Pinching of stem tip results in branching and 4-5 well-distributed shoots are usually allowed to grow. This training system is not considered as economical for commercial purposes. Multi stem training is adopted by allowing 3-5 stems at a hill imparting a bushy appearance to plants. Production of more ground sucker in this system is a common problem. Regular pruning in pomegranate is confined to removal of water shoots, weak crotches, dried and dead twigs and old spurs. Flowers and fruits are produced terminally on shoot spurs produced all along the slow growing mature wood and hence pruning should be done carefully not to reduce the fruit bearing area. Planting density is the most important yield contributing factor which can be manipulated to attain the maximum production per unit area in pomegranate. The optimum spacing is important for the maximum utilization of land and good income over a long period. Fruit yield per hectare increased without affecting fruit quality and a density of 1000 plants/ha gave

2.30 times higher yield and 2.44 times more profit compared with normal plant population of 400/ha. It has also been recommended that for higher yield for the first four to five years after planting, a distance of 5 x 2 m may be adopted and alternate plants may be removed afterwards maintaining a planting distance of 5 x 4 m. Bhagwa variety has been found suitable for high density planting under semi arid ecosystem.

Custard Apple

The traditional planting distance was 6 x 6 meter, now it may be planted at the distance of 5 x 5 meter to achieve more productivity per unit area. Removal of 25- 30 per cent annual growth during dormancy period was found effective to control plant vigour and to harvest better quality fruits.

Phalsa

Pruning is one of the important operations in phalsa cultivation and optimum pruning gives good crop. Plants can be pruned any time during December or January when they are dormant. The Phalsa plant is normally pruned a 0.9 to 1.2 m above the ground level in late winter to avoid the possibility of frost damage of the newly emerged shoots. Plants should be pruned to 100 cm from ground level under to get very good yield under semi arid ecosystem of western India.

Mango

Among the several reasons, less plant population per unit area is the key factor for low productivity of mango in India. The conventional system of planting accommodates 100 plants per ha. High density planting as adopted in florida and China during recent years appears to be the most promising one to overcome the prevailing productivity barriers. In india, 1333 plants per ha were accommodated at GBPUA and T, Pantnagar to get higher productivity per unit area. Dashehari and Kesar mango may be planted at the distance of 5x5 m (400 plants / ha) to increase the productivity per unit area.

Future Research Needs

There is considerable potential for the expansion of cultivation of semi arid fruits in India for doubling the income of the farmers. Therefore, some suggestions for future research priorities are given bellow.

1. The plant genetic resource (PGR) research needs to be undertaken on the classification of the genetic diversity through use of the morphological, biochemical and molecular techniques and suitable varieties should be developed for high density planting.

2. Model nurseries for the local supply of quality plant material should also be established.
3. Dwarfing rootstocks are considered to be best because they induce dwarfness and enable to facilitate easy management of the orchard. Such type of work may be initiated and be standardized for different agro climatic conditions.
4. Agro techniques like integrated nutrient management, high density planting system, canopy management and irrigation management should be standardized under different ecosystem of the country.

Conclusion

Light is the most important factor in the production of fruits, as it plays an important role in their growth and development. In view of this, opening of the tree canopies by proper training and pruning would help in maintaining the ideal canopies of the trees. High-density orcharding is the most appropriate answer to the problem of low productivity and long gestation period depriving the orchardist of early return.

References

Haneef, M., Kaushik, R.A., Sarolia, D.K., Mordia, A. and Dhakar, M. 2014. Irrigation scheduling and fertigation in pomegranate cv. Bhagwa under high density planting system. *Indian J. Hort.*, 71(1): 45-48.

Gunjate, R.T., Kumbhar, A.R., Thimaiah, I.M. and Amin, S.M. 2009. Growth and fruiting of some Mango cultivars under high density plantation in arid conditions of Gujarat (India). *Acta Hort.*, 820:463-467.

Joglekar, V., Chivate, D. and Pujari, K.H. 2013. High density planting technique in dry region for 'Kesar' mango cultivation – a Savlaj Pattern. *Acta Hort.*, 820:463-467. 992:233-235.

Kumawat, K.L., Sarolia, D.K., Kaushik, R.A. and Jodha, A.S. 2014. Effect of different spacing on newly planted guava cv. L- 49 under ultra high density planting system. *African J. Agric. Res.*, 9(51):3729-3735.

Mishra, D.S., Misra, K.K. and Tiwari, J.P. 2003. High density orcharding in tropical and subtropical fruits – a review. *In*: Singh, S.P.(ed.) *Advances in Horticulture and Forestry*, vol. 9. Scientfic Publishers (India), Jodhpur, pp.179-195.

Ram, S. 1996. High density orcharding in mango. *Res. Bull.* 122, DES, GBPUAT, Pantnagar. pp. 1-24.

Saroj and Singh, K.K. 2018. Density Concept of Orcharding. Int. J. Adv. Sci. Res. Mana., I:56-61.

Singh, A. K., Singh, Sanjay, Appa Rao, V.V., Meshram, D. T., Bagle, B. G. and Dhandar, D. G. 2007. Efficacy of planting system on productivity and economics of aonla cv. NA-7 under rainfed conditions. *The Horticultural Journal*, 20 (2): 67-70.

Singh, A. K, Singh, Sanjay, Appa Rao, V. V., Bagle, B. G. and More, T. A. 2011. Effect of high density planting systems on the productivity of NA-7 aonla (*Emblica officinalis* Gaertn) under rain fed conditions of semi-arid ecosystem. *Indian J. Hort.* 68:461-65.

Singh, G. 2005. Meadow orchard system in guava production. *Indian Hort.*, 50 (1): 17-18.

Singh, G. 2008. High density and meadow orcharding of guava. CISH, Lucknow, pp 1-20.

Singh, G. 2010. Development of meadow orchard in guava for higher production. *Prog. Hort.* 42 (2):129-133.

Singh Sanjay and Singh, A K. 2012. Enjoy new jamun variety. Indian Horticulture. May-June, Pp 9-11.

Singh, A. K. Singh, Sanjay, Joshi, H. K.and Singh, R.S. 2012. Goma Yashi to enrich fruit basket. *Indian Horticulture,* Sept.-Oct, pp6-8.

Singh, Sanjay, Singh, A K, Mishra DS and Appa Rao, V. V. 2017. Effect of shoot pruning on yield and fruit quality of jamun cv Goma Priyanka. *Indian J. Arid Hort.*, 12 (1-2):100-102.

Singh, V K, Soni MK and Singh, Anurag 2013. Strategies and prospects of high density plnting and canopy management in guava. National seminar cum workshop on canopy management and high density planting in subtropical fruit crops, October 22-24, 2013, CISH, Lucknow, pp 23-33.

10

Canopy Architecture Management for Higher Fruit Production

Vishal Nath, Sanjay Kumar Singh and Alemwati Pongener

Introduction

Canopy management is defined as managing plant architectural design as per need, using their inherent characteristics which is in accordance with a given set of conditions and resources to perform the plant maximum (Nath *et al*, 2014).The modern principles of sustainable fruit production systems focused on optimal use of natural resources (light, water, soil fertility) where the input (labour, energy, fuel, other material) and output (yield) are well balanced and the resources remain renewable.

Fruit orchards can be considered as artificial ecosystems converting the energy of photosynthetic active radiation (PAR) into edible, marketable fruits for human consumption. The core of this system is the bio-factory of leaves driven by the absorbed PAR. Several data support that the light interception is the basic factor for the modern orchard systems. Factors as LAI (Leaf area index), tree spacing, training and pruning systems, fruiting wood development, rootstocks and genotypes also have an important bearing on orchard efficiency.

The main objectives of canopy architecture is to develop strong tree framework, maximize utilization of light, accommodate more number of trees per unit area, maintenance of tree orchard uniformity and induce and sustain proper amount of well distributed fruiting terminals.

In order for the fruit production system to remain sustainable, the orchard system should favour the contributory role of an input of natural origin to the maximum possible extent, in improving the yield in the early years and provide high yields after the trees are mature until the planned age of the orchard. Recent development in the tree architecture has produced some promising tree training and pruning systems as well for high density fruit orchards.

The significance of single row system is well implicated in stone fruit orchards planted for manual handpicking. Planting orchard either as double or triple row or "bed" system in HDP is besetting with many disadvantages mainly caused by shading the tree rows and picking difficulties, high cost of planting material and the requirement of special equipment (Robinson, 2003). Nowadays the fruit trees in handpicked orchards are trained to conic-shaped canopy with one central leader or to V or Y shaped canopy. Orchard systems design into V or Y shaped canopies favours more light interception (Robinson, 2003, Cittadini *et al*, 2006), probably due to the higher planting and support system. Apart from temperate fruit trees, the modified central leader and open centre system can also be applied to tropical and subtropical fruit trees.

Principles of Canopy Management

Uniform exposure to sunlight throughout the volume of the canopy is necessary for abundant formation of flower buds, growth, size, and color development of fruits. Fruit yield relies largely on the light use efficiency of canopies, whereas fruit quality is primarily related to the light distribution among the canopies. Reducing leaf area and increasing the proportion of canopy gaps decrease shading effect. However, excess proportion of canopy gaps allows sunlight to be 'lost', falling on the orchard floor. A balance between excessive gaps that minimize wastage of sunlight energy and insufficient gaps to avoid shading effect is an integral role of canopy management. Well-formed canopies permits better aeration and sunlight exposure to foliage, fruits, improve the photosynthetic efficiency, fruit bud differentiation, fruit set, growth and quality *vis-a-vis* reduced microclimate buildup for pests and disease. Closed canopies impair the penetration of both direct and diffused sunlight into the enclosure formed by canopy walls and the ground surface. This will ensure an appropriate photosynthetic partitioning between shoot and fruit growth. Maintenance of fruiting organs within the restricted zones in space facilitates mechanization in shoot pruning and fruit harvesting. While designing the canopy architecture, basic management considerations like development of strong structural framework, maximizing light penetration in and outside the canopy, natural control of diseases and pest, number of fruiting terminals and crop loads to improve productivity and quality of fruits, ease in mechanization, cultural operations and crop regulation should be the central point.

Ensuring light Interception Through Pruning and Canopy Management

Light microclimate plays a pivotal role in influencing fruit yield, quality and because photosynthetic carbon fixation depends mainly on the sunlight captured by a tree or an orchard, fruit yield of healthy and well-watered trees is related to total light interception. Usually, shade reduces fruit mass and fruit quality attributes like color and soluble sugar and secondary metabolite concentrations. Shading decreases shoot photosynthesis and fruit temperature (Abbott, 1984) and may change the light spectrum (Awad *et al*, 2001). Local irradiance may also affect flower bud initiation and development and leaf attributes affecting photosynthetic capacity, namely specific leaf area (Palmer *et al*, 1992). Light influence fruit trees at both the inter tree and intra tree level. Planting geometry and spacing are the major factors influencing light interception at the inter tree level. Within a tree and for a given genotype, light interception is affected by pruning and training procedures (Johnson and Lakso, 1986). When attempting training and pruning of litchi trees, utmost attention must be paid to induce growth of fruiting shoots towards the outer canopy to make use of available light in the outer periphery. This can be further improved upon by providing duct space between two rows in rectangular system and between rows and plants in square systems (Nath *et al*, 2014).

Canopy management is vital to control tree growth pattern, tree shape and maintain high quality fruit production. A large canopy surface area well exposed to sunlight is desirable, since it increase light interception and trapping of solar energy. The green leaves harvest sunlight to produce carbohydrates and sugars, which are transported to the sites/sinks where they are most needed like buds, flowers and fruits. For efficient harnessing of solar energy, regular pruning of trees under high density planting, planting geometry and orientation of planting must be considered. The planting of litchi trees in East-West row orientation results in more shading as compared to North-South orientation of the trees. A reduction in light interception due to dense canopies and shaded fruiting zones will affects the flower primordia development of the following year's crop. The development of these primordia is influenced heavily by light transmittance, as shaded canopies have been shown to lower the development of these primordia, thereby reducing the following year's yields. Improved sunlight transmittance to as many leaves as possible will provide for the most efficient use of photosynthetic processes, maximize flower primordia development and increase yields (Hunter *et al*, 1995).

In this context, maintaining tree size through development of pruning strategies is crucial to reach and maintain the economic profitability along the orchard life span. A better knowledge of the cultivar-related growth, branching and fruiting habits, referred to as tree architecture, is of major importance in understanding the cultivar-dependent reactions to manipulation.

In many fruit crops, 15-20% increase in production with enhanced fruit quality has been witnessed by managing appropriate and efficient canopies particularly in short-statured trees. Small trees capture and convert sunlight into fruit production in a better way than larger ones hence it has direct impact on orchard efficiency. Therefore, early height control and tree canopy management should be practiced in fruit crops for doubling orchard income. In general, pruned plants perform well both under wider spacing and narrow spacing in terms of yield and quality traits. Since, yield maximization is the prime objectives, the plants under narrow spacing with pruning are found to be the best. Pruning were also found to have a rejuvenating impact on aged trees due to better light interception leading to better photosynthetic rate, better nutrient and water supply with reduced canopy and better quality yields. Though fruit yield during the initial years of pruning, sharply declines, but in the subsequent years, yields stabilized in both pruned and unpruned trees. It is a well-proven fact that yield decreases in the year in which pruning is performed, but if the effect of pruning is analyzed over several years, this adverse effect disappears because the tree compensates for the previous reduction in the non-pruning years, probably as a consequence of the accumulation of reserves and better lighting (Sauls, 2008; Yildirim *et al*, 2010). Initiation of fruiting was observed after third year of rejuvenation pruning in mango cv. Amrapalli with recorded yield greater than 60 kg/tree (Das and Jana, 2012).

Intrinsic and Extrinsic Factors Before Managing Tree Architecture

Every plant has inbuilt inherent characteristics which governs the architectural behavior like tree forms, angling of branches and pattern of branching, flushing pattern, dormancy pattern and bearing behavior, level of endogenous growth hormones, etc. In trees with relatively natural tree forms such as in the multiple leader, central leader, vertical axis, or slender spindle forms, light penetration through the canopy can be enhanced by providing many small openings in the foliage. While tree forms which requires severe geometric restriction of the canopy such as narrow hedgerows, tree walls, and A V or T forms, provision of fewer but large and permanent openings can be resorted to favour increased light penetration into canopies. Based on these factors, the principles and action for canopy architectural design are planned and decided. Some plants have shoot growth with extension of flushes (mango) while others have continuous flushing during active growth phase (litchi). Therefore, the architecture of each plant is decided according to their inherent growth characteristics. The canopy architecture concept for litchi tree includes suggested series of operations (cutting, bending, and angling) on the above ground portion of the plant for maximizing production of quality fruits per unit canopy area. Litchi tree normally has thick, straight and short trunk and branches often have 'V'-shaped crotches with narrow crotch angles (40° or less from the vertical) which are prone to breakage. Narrow crotches result in greater loss of limbs due to windstorm and heavy fruit load than crotches with larger angles. The strongest crotches are those in which branches grow up from the trunk at angles ranging from 40-90°and scaffold branches evenly spaced around the tree, with each main branch at least 90° from the next one.

Hedgerow planting system with a principle of single canopy and multiple root system is ideal for those fruit species which are responsive to pruning and bear on either current season shoots or spurs (guava, peach *etc*), whereas open centre system may be more useful for plants requiring more solar radiation for fruiting and colour development (litchi).

Practices of Canopy Management for Doubling Arid Fruit Production

The commercial cultivars of ber are spreading in nature; therefore, staking is absolutely essential during the first two years to train the tree properly according to modified leader system. Light annual pruning i.e., heading back of one-fourth of the previous year's growth is desirable to obtain heavy yield, good fruit size and better quality. The cultivar 'Umran' pruned on May 9 by retaining 6 buds while in 'Sanaur-2', it is headed back at 8 buds during the third week of April. The selective pruning of late flowering branches (21 October to 5 November) improves fruit size, weight and palatability rating of fruits in 'Umran'. Ber trees needs to be rejuvenated after the age of about 25-30 years. The main limbs of the trees are headed back to about 30 cm during second fortnight of May. The first thinning of shoots is done during fourth week of August by retaining 12-18 shoots depending upon the vigor of the plant. The second thinning of shoots is done during second fortnight of September (before flowering) to retain 8-12 branches tree^{-1}. Trees start giving commercial crop of excellent fruit quality during the third fruiting season (Bal and Gill, 2016).

Training and pruning in guava begins at an early stage of plant growth to develop single trunk trees with well spaced scaffold branches to from the strong frame work. Apical growth is to be controlled within the first year of planting. Trees are topped to a uniform height of 60-75 cm from the ground. As a result, new shoots emerge. About 3 to 4 equally spaced shoots are retained around the stem to form the main scaffold limbs of the tree. These shoots are retained the stem to from the topping until they attain a length of about 40-50 cm. The selected shoots are further pruned to 50 per cent of their length for inducing multiple shoots form the buds below the cut ends and again emerged new shoots. This is mainly done to obtain the desired shape. The pruning operation continues maximize fruiting during the second year after density planting. After two years, the short branches within the canopy produce a compact and strong structure. All the plants are confined to a hedge and May-June of every year (Pal *et al.*, 2016).

Canopy architecture and its management is one of the important operations in fruit tree for harnessing natural resources through utilization of proper space, nutrients, moisture, solar energy and aeration in the new as well as existing productive and old senile fruit orchards. The tree architectural management, for strong frame and size control, has become a priority for reducing production cost and increasing input use efficiency, yield and quality in various perennial tree fruit species. The strongest crotches are those in which branches grow up from the trunk at angles ranging from 40-90°. Scaffold branches should be evenly spaced around the tree, with each branch at least 90° from the next one. Canopy design and shape influence light interception with assured higher monetary returns to litchi growers. This management process is an essential tree management operation that starts from the first year of plant establishment which of-course varies with species and environments but enables the plant to produce high quality and quantity by providing proper framework, more fruiting area and utilization of available natural resources. Therefore, early height control, open centre canopy, strong limbs with wide crotch angle and more fruiting area and sustained progressive tree canopy architecture management are the important techniques, which should be practiced in fruit crops for higher returns to growers.

With the advancement of knowledge in area of tree architectural design, now the effect is being realized in larger prospective and almost all crops are coming under the umbrella of tree architecture. Canopy management will continue to provide science with new ideas and knowledge on complex interactions of plant canopies with the environment, and pave way for development of technologies with new training systems for sustainable fruit production. Hence, furtherance is needed in this area of research for establishing basic facts and logical explanation of its impact on production quantum, productivity and quality of fruits. Some specific future thrust areas of research are suggested as follows:-

- Standardization of tree architectural design and frame for range of fruits are needed.
- Optimum canopy for sustainable yield needs to be worked out for commercial cultivars.
- Influence of canopy architecture on fruit quality needs to be quantified for varying cultivars.
- Canopy architecture and its management for different planting systems needs to be worked out so that HDP for many fruit crops can be recommended for small and marginal orchardists.
- Tree architectural plan against natural calamities needs to be developed/ postulated.

References

Abbott, D.L. 1984. The apple tree: physiology and management. Grower Books, London, 90 p.

Awad, M.A., P. Wagenmakers and A. de Jager. (2001). Effects of light on flavonoid and chlorogenic acid levels in the skin of 'Jonagold' apples. *Sci. Hortic*. 88:289–298

Bal, J.S. and Gill, K.S. 2016. Canopy management, nodal pruning and planting density in Indian jujube (ber). *Acta Hortic*. 1116, 99-104

Das, B., Jana, B.R., 2012. Effect of canopy management on growth and yield of mango cv. Amrapali planted at close spacing. *J. Food Agric. Environ*. 10:328–331.

Cittadini E.D., van Keulen H., Peri P.L., Ridder N. (2006). Designing a 'Target-Tree' for Maximizing Gross Value of Product in Patagonian Sweet Cherry Orchards. International *Journal of Fruit Science*, **6**(3):3-22.

Hunter, J., H.P. Ruffner, C. Volschenk, and D. LeRoux. 1995. Partial defoliation of Vitis vinifera L. cv. Cabernet Sauvignon/99 Richter: Effect on root growth, canopy efficiency, grape composition, and wine quality. *Am. J. Enol. Vitic*. 3:306-314

Nath, V., Thakre, M., Pandey, S.D., Kumar, A. and Kumar, R. 2014. Canopy architecture for high density litchi (*Litchi chinensis* Sonn.) planting. *Acta Hortic*. 1029:287-293

Pal, Virendra , Naveen Chandra, Anant Kumar and Mukesh Kumar 2016. Response of pruning in canopy management & high density planting in guava orchard under western Uttar Pradesh. condition. South Asian *J. Food Technol. Environ*., 2(3&4): 458-464

Palmer, J.W., D.J. Avery and S.J. Wertheim. 1992. Effect of apple tree spacing and summer pruning on leaf area distribution and light interception. *Scientia Hort*. 52:303–312.

Robinson, T.L. 2003. Apple orchard systems. In: D.C. Ferree and I. J. Warrington (eds.). Apples: Physiology, production and uses CABI Publishing. Wallingford, Oxon, United Kingdom.

Sauls JW, 2008. Citrus pruning. Texas citrus and subtropical fruits. Available in http://aggie-horticulture.tamu.edu/citrus/pruning/L2308.htm [14 Mar 2008].

Yildirim B, Yesiloglu T, Incesu M, Kamiloglu M, Ozguven F, Tuzcu O, Aka Y, 2010. The effects of mechanical pruning on fruit yield and quality in 'Star Ruby' grapefruit. *J. Food Agric. Environ*. 8 (2): 834-838

11

Protected Cultivation Technology for Arid & Semi Arid Region

P.P. Singh, A.K. Verma, S.K. Maheshwari, D.K. Sarolia and M.K. Jatav

Introduction

Protected cultivation practices can be defined as a cropping technique where in the micro environment surrounding the plant body is controlled partially or fully as per plant need during their period of growth to maximize the yield and resource saving. Protected structures can play important role to minimize the impact of temperature fluctuation, over/under precipitation, fluctuating sun shine hour and infestation of disease and pest. The controlled environment can be utilized for growing early season /off-season vegetables and high value vegetables. This may become relevant to those farmers having small land holding who would be benefitted by a technology, which helps them to grow more crops each year and save them from adverse situation. Vegetable crops are very sensitive to climatic vagaries and sudden rise in temperature as well as irregular precipitation at any phase of crop growth can affect the normal growth, flowering, pollination, fruit development and subsequently decreased the crop yield. Apart from that high input cost and higher labor wages make vegetable cultivation less profitable under open field condition. Again, in the present scenario, more people are demanding early season/off-season vegetables and high value vegetables at higher price which are not possible to grow under open field situation.

Present status in world: In world, China has the largest area (2,760 thousand hectares) under greenhouse cultivation followed by Korea having 57.44 thousand hectares (Nair and Barche, 2014). The advent of protected cultivation technology in India materialized during the early nineties, post globalization.

India: India is the second largest producers of vegetable in the world, however as per medical council of India, the production is much less than the requirement if balanced diet are to be provided to every individual. The advent of protected cultivation technology in India materialized during the early nineties, post globalization. In India, the area under protected cultivation faced with constraints of land holdings, rapid urbanization, declining crop production, declining biodiversity and ever increasing population, demand for food, especially vegetables has increased manifold and protected cultivation has offered a new dimension to produce more in a limited area. Today Dutch protected cultivation is one of the most intensive farming systems in the world with high levels of output by using the latest technologies.

Table 1: Greenhouse area (ha) under world

Present status in world	Greenhouse Area (ha)
China	27,60,000
Korea	57,444
Spain	52,170
Japan	49,049
Turkey	33,515
Italy	26,500
Mexico	11,759
Netherlands	10,370
France	9,620
United States	8,425

Table 2: Leading states in protected cultivation in India

S. No.	State	Area (in ha)	Crops
1.	Maharashtra	8000	Carnation, gerbera, rose, capsicum
2.	Karnataka	1000	Roses, gerbera, carnation, vegetable seed production and nursery raising of vegetables
3.	Himachal Pradesh	700	Capsicum, carnation, gerbera, tuberose
4.	Punjab	500	Vegetable crops
5.	Uttarakhand	300	Gerbera, capsicum
6.	Tamil Nadu	100	Floricultural crops
7.	North-Eastern States	800	Floricultural and vegetable crops

Gaps in prevailing system: Energy efficient cultivation continues to be achieved through precision equipment and protocols. While, the gap between demand and supply of most of horticultural crops remains wide and country plans to double production of horticultural crops by 2022, protected cultivation technology holds the key to meet the targets green house. Protected cultivation technology requires very careful planning, maintenance and management about timing of production and moreover, harvest time to coincide with the shortage period of availability of vegetables and high market prices, choice of varieties adopted to off season environments, and able to produce higher and economical yields of high quality produce etc. Vegetable farming in agri-entrepreneurial models targeting various niche markets of the big cities of the country is regularly inviting attention of the vegetable growers for diversification from traditional ways of crop cultivation to such modern methods. Even the unemployed educated youths who are not attracted or interested in traditional agriculture are also showing good interest and can be further motivated for this kind of modern agricultural technologies.

Methods of Protected Cultivation

1. Green house/Polyhouse/Net house/Shade house
2. Low Tunnels/row covers
3. Plastic low tunnel
4. Lath houses
5. Hot beds
6. Cold frames
7. Cloches

Greenhouses

A greenhouse is a framed or an inflated structure covered with a transparent or translucent material in which crops could be grown under the conditions of at least partially or fully controlled environment and which is large enough to permit persons to work within it to carry out cultural operations.

Types of Greenhouses

To the extent of environmental control, the existing green house structures can be classified as:

- High cost (fully controlled greenhouse)
- Medium cost (partially controlled)
- Low cost (naturally ventilated)

Other Classifications

- The greenhouse can also be classified based on type of structures, type of glazing, number of spans, environmental control etc. The various types are as follows.

- **Classification as per type of structure**
  a. Quonset type
  b. Curved roof type
  c. Gable roof type
- **Classification as per glazing**
  a. Glass glazing
  b. Fiberglass reinforced plastic glazing
     i. Plain sheet
     ii. Corrugated sheet
  c. Plastic film
     i. Ultra violet stabilized low density poly ethylene
     ii. Silpaulin
- **Classification based on number of spans**
  a. Free standing or single span
  b. Multi span or ridge and furrow or gutter connected
- **Classification based on environmental control**
  a. Naturally ventilated
  b. Passive ventilation

Principle of Greenhouse

A greenhouse is generally covered with a transparent material such as polythene or glass. Depending upon the cladding material and its transparency major fraction of sunlight is absorbed by vegetable crops and other objects. These objects in greenhouse in turn emit long wave thermal radiations for which cladding material has lower transparency. With the result, solar energy is trapped and raises the temperature inside the greenhouse. This is popularly known as greenhouse effect. This rise in temperature in greenhouse is responsible for growing vegetable in cold climates. During summer months, air temperature in greenhouse is to be brought down by providing cooling device. In commercial greenhouses besides temperature-controlled humidity, carbon dioxide, photoperiod, soil temperature, plant nutrients etc. facilitate round the year production of desired vegetable crops. Controlled climatic and soil conditions provide an opportunity to the vegetable crops to express their yield potentials

• Components of Greenhouse

Protected cultivation has two major components of technology. One is the infrastructure involving frames, cladding materials, irrigation system, tools, implements, other engineering inputs and another is crop production technology.

Infrastructural inputs ensure optimal light, air temperature, water and plant growth requirements. This optimal aspect of climatic parameters involves simple to most advanced engineering inputs such as automation, etc. to regulate several parameters such as ventilation which is one of the most important components in a successful greenhouse production. A major problem with conventional designs of greenhouses is the concentration of heat within the covered structures, which needs to be either expelled or neutralized through energy-intensive cooling facilities. This problem has been overcome by designing naturally ventilated greenhouses where the temperature can be maintained at the desirable level without consuming any energy. Importance of cladding materials in protected cultivation can hardly be overemphasized. Their quality and cost are important besides certification. Micro irrigation and fertigation involves a lot of science and technology, demanding research for continuous improvement. It is the engineering aspect of the protected structure which provides plants optimal conditions to grow normally. Another aspect of protected cultivation is crop production technology which involves development of high-yielding varieties and hybrids of crops suitable for protected cultivation.

Lath House

Lath house provides outdoor shade and protect container grown plants from high summer temperature, high light intensity. They reduce moisture stress and decrease the water requirement of the plants and thus these are useful for protecting young seedlings. During summer, overhead sprinklers are used for cooling down the temperature inside of shade.

Cold Frame

These are available with metal or wooden frames with glazed or solid sides and are useful for hardening off plants.

Hot Bed

Reinforced polyester plastic is also used for frame. It is light in weight, easy to handle and admits light, which however, is likely to decrease with age. Hot beds are headed with the help of different heating systems like hot air, hot water and electricity, which are adopted depending upon availability and cost, systems.

Shade Nethouse

Shade nets are available in different shade percentages or shade factor i.e. 15%, 35%, 40%, 50% 75% and 90% (for example 35% shade factor means - the net will cut 35% of light intensity and would allow only 65% of light intensity to pass through the net).

Traditional Cloche

Wire frame with sheets of glass used for establishment of transplants and hardening off and overwintering of autumn sown crops. It helps in warming soil prior to sowing and excludes some pests e.g. birds and cats.

Vegetables Grown Under Polyhouse

Tomato

Beefsteak varieties: FA-574, FA-180 and FA-514

Big fruited varieties: Naveen, Arka Vardan, Vishal, Delphi, Astona, Cornos, Shanmon

Cluster type varieties: HA-646, FA-556 and FA-521

Cherry tomato varieties: T-56, NS Cherry-1 &2, BR-124, HA-818

Sweet Pepper

Red: Heera, Bharat, Indira, Bombay, Jamini, Pusa Deepti

Yellow: Tanvi, boyaton, Orobellee, Golden Summer, Nun-3020

Cucumber

Satis, Alamir, Nun-9729, Kian

Pusasanyog, Priya and Poinsett can also be grown inside polyhouse.

Production Technology of Tomato under Greenhouse

Greenhouse cultivation of vegetables offers distinct advantages of quality, productivity and favourable market prices to the growers. Vegetable growers can substantially increase their income by greenhouse cultivation of vegetables in off season as the vegetables produced in the normal season generally do not fetch good returns due to large availability of these vegetables in the market.

Climate Requirements

Temperature and light intensity affect fruit set, pigmentation and nutritive value of the fruits. Critical factor in setting fruits of the tomato is the night temperature, the optimum range being 16°C to 22°C. Fruits fail to set at 12°C or below. Fruit set is also reduced markedly when average maximum day temperature goes above 32°C and average minimum night temperature goes above 22°C. At temperatures below 10°C, red and yellow colours do not develop. The ideal range for development of both red and yellow colours is 18-25°C. Under greenhouse conditions tomato crop can be grown for long duration (10-12 months) by cooling the greenhouses during summer months (April to July) and by heating the greenhouses during peak winter (December and January) months.

Preparation of Soil, Nursery Raising and Planting

In northern plains the tomato crop is planted from first August to second week of September under greenhouse conditions. August or September planted crop is continued upto June or July under climate controlled greenhouse conditions. If the greenhouse is naturally ventilated then the crop can be grown up to April or May months. Nursery for greenhouse tomato is raised under protected structure mostly in soilless media in plastic trays to produce disease free and virus free seedlings. The seedlings are ready for transplanting within 28-30 days under soil-less media. Healthy seedlings are transplanted at a planting distance of 60 x 45 cm mostly under drip irrigation system for efficient use of water and fertilizers for long duration cultivation of tomato crop and low pressure drip irrigation system can also be used for greenhouse tomato cultivation. Transplanting is mostly done in the early morning or late evening time for better establishment of the seedlings.

Training, Pruning and Trellising

Greenhouse grown tomatoes needs regular trellising, training and pruning of plants. Single stem (main) should be retained by removing all side shoots or suckers that develop between leaf petioles and the stems. Usually in early stage shoots are removed by snapping them off, not cutting them by knife, blade or scissors, as diseases mainly the virus (T.M.V.) can be transmitted from one plant to other. Plants are supported by plastic or binder twine, loosely anchored on the base of plants with the help of plastic clips or directly by non-slip loop and to overhead support wires (11 to 12 gauge) running to the length of the row of the bed. Overhead wires are fixed normally 8 to 9 feet above the surface of the bed and are anchored firmly to the support structure. Twine should be wrapped clockwise around the vine as it develops with one complete swirl every three leaves. The vine should be supported by the twine under the leaves, not the stems of the flower truss or fruit clusters. Twine is not wrapped around the growing tip otherwise the tip may break. When the plants reached overhead supporting wires, untie the twine (or take down the twine) from the twine roll after unlocking it to take down the vines at least 2 to 3 feet in every 15-20 days gap. Vines with twines are moved in one direction with twine roll in one row on the overhead wires and in opposite direction in the adjacent row. The plants must be pruned and trellised on regular basis for 10-11 months life cycle of the tomato crop in greenhouse and as a result plants are 30-35 feet longer in their 10-11 months life cycle of cultivation. Clearly, the greenhouse will not be able to accommodate the plants vertically. As fruits mature on the lower parts of the vines, pinch off older leaves below the fruits. This will, provide air circulation, which helps to reduce the incidence of the diseases and opens vines up for spraying and harvesting. Removal of excess fruits will also result in larger tomatoes at harvest that can fetch good price.

Pollination Requirement in Greenhouse

Although tomato is highly self-pollinated crop but aided pollination of flowers is generally needed in the green house grown tomatoes due to limited air movement and high humidity. In several countries bumble bees are being used as a pollinator for the green house tomato crops. Bumble bees are the perfect pollinators, even under environmental stress condition (i.e. under low and high temperature conditions). But in several countries an electric or battery –powered vibrators are being used to vibrate flowers cluster above the area where they originate from the stem. The vibration will release sufficient pollens necessary for pollination. This practice is done twice a day (10:00 AM to 11.00 AM and 2.0-3.0 PM). Flowers are vibrated or shaken every day. Air from a mistflower also has been found effective in pollinating tomatoes inside the protected structures. If tobacco mosaic virus (T.M.V) has been a problem the vibrator should be wiped after each use with a clean cloth.

Fertigation

Application of 350:350:350 kg NPK/ha in the form of water soluble fertilizers through fertigation at weekly intervals is recommended for getting maximum productivity and best quality of tomato under cost effective greenhouse.

Cooling and Heating of the Greenhouses

Most of the climate controlled greenhouses are having evaporative cooling system. In northern plains of our country cooling is done from September to October and from April to June months. Evaporative cooling system is quite effective when the relative humidity in the atmosphere is below 40%. Heating of the greenhouses can be done from 15th of December to end of January, to increase the temperature of the greenhouse during nights, so that the temperature cannot fall below 14°C. Heating and cooling of the greenhouses are required during growing of tomato for a period of 10-12 months.

Harvesting and Post-harvest Handling of Fruits

Harvesting of tomato fruits is a continual process throughout the growing cycle. Generally, most of the varieties are ready for first picking 75 to 85 days after transplanting. Big size tomato (slicing tomato) fruits are harvested singly with attached calyx, and are graded and packed according to grades. Under greenhouse conditions tomato can yield 100- tonnes/ha. During the summer months harvesting should be done in the early morning or late evening to avoid post-harvest losses.

Plant Protection in Greenhouse

Mostly greenhouses are designed to minimize or eliminate insects and diseases problems so that plants can be grown pesticide free under many conditions. Generally, all four sides of greenhouses are covered with insect proof nylon net of 40 to 50 mesh size to prevent insects, including white flies, thrips, aphids etc. Preventing insects from entering the protected structures is the best way of controlling virus and insect problems. However, if required, one or two sprays of insecticides can be done on need base. One spray of metasystox @ 1.5 ml/ litre of water is done 10 days after transplanting of the crop. If there is severe problem of mites, we can spray dicofol @ 2.0 ml/litre of water. Sometimes white flies or thrips entered in greenhouses, yellow and blue trappers are used for trapping of such insects. For control of root knot nematodes and other soil borne pathogens, soil sterilization is most effective and it is done by applying the formaldehyde (37%), one month before transplanting of the crop. Sometimes if some plants are infested with T.M.V, such plants are to be removed immediately to avoid further spread of virus inside the greenhouse.

Protected Cultivation of Capsicum

Capsicum, also known as sweet pepper, bell pepper or Shimla Mirch is one of the popular vegetables grown throughout India. Capsicum is a cool season crop, but it can be grown round the year using protected structures where temperature and relative humidity (RH) can be manipulated. This crop requires day temperature of 25-300C and night temperature of 18-20C with relative humidity of 5060%.If temperature exceeds 3500C or falls below 12C, fruit setting is affected. The traditionally grown green capsicum, depending upon variety and season, usually yields 20-40 tons per hectare in about 4-5 months.

In greenhouse, the crop duration of green and coloured capsicums is about 7 - 10 months and yields about 80-100 t per hectare.

Nursery Raising

Good quality seeds are required for producing better seedlings. The seedlings are raised in pro-trays of 98 cells or cavities. About 16,000 to 20,000 seedlings are required to plant one acre for which 160-200 gm of seeds is required. The pro-trays are filled with sterilized coco-peat and seeds are sown, one seed per cell to a depth of ½ cm and covered with the same media. The filled trays are staked one above the other and covered with plastic sheets till germination of seeds. Seeds germinate in about a week's time after sowing. The trays are shifted to net house/ polyhouse and lightly watered. After 15 days of sowing, Mono ammonium phosphate (12:61:0) (3g/L) and 22 days after sowing 19:19:19 (3g/L) solution has to be drenched. The seedlings in protrays are drenched with COC 3g/L before transplanting. The seedlings will be ready for transplanting in 30-35 days.Spray imidacloprid @ 0.2 ml/L and chlorothelonil @ 1gm/L before transplanting of seedlings. Always add about 0.3 ml/L of wetting agent per liter of water with each spray of pesticide.

Land Preparation

The land should be thoroughly ploughed and soil should be brought to fine tilth. Well decomposed organic manure at the rate of 20-25 kg per sq. mtr is mixed with soil. On application is sufficient to grow three capsicum crops successively. Raised beds are formed after bringing soil to fine tilth. The bed size should be 90-100 cm wide and 15-22 cm height. Between the beds walking space of 45 cm to 50 cm need to be provided.

Fertilizer Application

A basal fertilizer dose of 20:25:20 NPK is required per acre and is applied to the beds uniformly before transplanting in the form of 80 kg calcium ammonium nitrate, 125 kg super phosphate and 32 kg murate of potash or 40 kg sulphate of potash.

- **Application of neem cake and microbial bio-control Agents**

Fifteen days before transplanting, neem cake has to be enriched with bio agents like *Trichoderma harzianam* and *Pseudomonas lilacinous*. Neem cake of about 200 Kg is powdered and slightly moistened. *Trichoderma harzianam, Pseudomonas lilacinous* and *Paecilomyces chilmdosporia* each of two kg are mixed thoroughly to the neem cake. The mixture is covered with wet gunny bags or dry grass and left for 8-10 days. Avoid direct exposure to sunlight and rainfall. After 10 days, this enriched mixture of neem cake and bio-agent along with 600 kg of neem cake has to be applied uniformly to the beds for an area of one acre. This is highly useful to reduce the problem of soil borne pathogens and nematodes. *Azospirillum* or *Azoctobacter* or VAM which is a nitrogen fixing bacteria can also be applied to the growing bed.

Mulching and Spacing

Black polyethylene non-recycled mulch film of 30-100 micron thick, 1.2 m wide, is used to cover the planting beds. Holes of 5 cm diameter are made on the polyethylene film as per the recommended spacing (45cm x 30cm). The planting beds are covered with the film by securing the edges of the sheet firmly in the soil. Mulching practice conserves water, controls weeds and reduces infestation of pests and diseases and results in higher yield and good quality produce.

Transplanting

The planting beds are watered to field capacity before transplanting. Seedlings of 30-35 days old are used for transplanting. Care should be taken to see that no damage is occurred to roots, while taking out the seedlings from individual cells of portray. Seedlings are transplanted into holes made in polyethylene mulch film at a depth of 5 cm. After transplanting, seedlings are drenched with 3 g/L copper oxy chloride or 3 g/L captan or 2 g/L copper hydroxide solution to the base of seedlings at the rate of 25-30 ml per plant. Watering the mulched beds daily during afternoon by using hose pipe for a week continuously is essential to avoid mortality due to heat trapped by mulch sheet.

Pruning and Training

Capsicum plants are pruned to retain four stems. The tip of the plant splits into two at 5th or 6th node and are left to grow. These two branches again split in to two giving rise to four branches. At every node the tip splits into two giving rise to one strong branch and one week branch. The pruning is done after 30 days of transplanting at an interval of 8 to 10 days, resulting in bigger fruits with better quality and high productivity. The capsicum plants can also be pruned to two stems and same level of yield can be maintained.

The main stem of plant is tied with four plastic twine to train along and tied to GI wire grid provided on the top of the plants. This is practiced after four weeks of transplanting. The new branches and plants are trained along the plastic twines.

Drip Irrigation and Fertigation

Drip irrigation is given to provide 2-4 liters of water per square meter per day depending on the season. Water soluble fertilizers are given through fertigation for entire crop growth period, starting from third week after transplanting. Fertigation is to be given twice a week as recommended in below.

- Required fertilizer dosage per fertigation (kg/ac)19:19:19 4 kg
- Potassium Nitrate 1.5 kg
- Calcium Nitrate 1.5 kg
- Capsicum crop is sprayed with water soluble fertilizers like potassium nitrate and calcium nitrate at every 3 weeks interval after 2 months of transplanting @ 3g/ l as foliar application.

General Management of Pest and Diseases

Pests and diseases remain the greatest challenge in Tomato production. Appropriate and timely management makes all the difference between good production, poor production or total crop failure. Proper identification of the pest and disease is critical in a control strategy. The general principles in pests and disease management include:

- **Practicing crop rotation.** Observe minimum two-year rotation program.
- **Planting resistant varieties** - Use certified disease-free seed treated with an approved fungicide to control seed rots and post emergence damping off
- **Field hygiene**-old crop should be removed from the fields, control weeds and crop debris since these are source of pests and diseases. Staking and pruning is also key to disease incidence reduction.
- **Using proper crop production practices** that provide the right growing conditions for plants (sufficient water and balanced fertilization), particularly when crops are young. Strong healthy plants are more likely to withstand pests and diseases.
- **Irrigation management**; poor irrigation timing and scheduling may lead to disease, overhead irrigation in the evenings can encourage early blight.
- **Ensure regular crop scouting** for pest and disease as well as weed and nutrient deficiencies. Proper pest and diseases identification is the first and critical step in their management. This helps to detect problems early and take control measures on time.

- **Agronomic practices:** It includes nutrient management, irrigation, support, pruning, weeding, pest and disease management, harvesting.

Harvesting and Yield

Early morning hours are best suited for capsicum harvest. Green capsicum can be harvested at 55 to 60 days after transplanting, yellow capsicum at 70-75 days whereas red capsicum at 80-90 days. Fruits can be harvested once in 3 to 4 days. Yellow and red fruits can be harvested when they have gained 50-80 per cent of the colour development. After harvest fruits should be kept in cool place and avoid direct exposure to sunlight. The fruits should be handled carefully by adopting clip harvest technique and scuffing should be minimized. The average yield of capsicum per acre is 30-40 tons.

Protected Cultivation of Cucumber

Cucumber known as *khira* in hindi, is an important summer vegetable crop. Green house cucumbers are parthenocarpic (produce fruit without fertilization of ovules) and the fruit are usually seedless and does not require peeling, when ready for harvest, cucumbers are usually eaten as in salads.

Seed requirement: The amount of seed will be determined by the spacing used. A plan density of 1-2.5 plants/m^2 is recommended for most regions.

Nursery management: Trays can be used for growing seedlings. Watering will be done lightly using a watering can and timed in the morning to avoid conditions conducive for the development of diseases. It will take about a 20 day before the seedlings are ready for transplanting.

Transplanting: The seedlings need about 20 days of growth before they are ready for transplanting. Transplanting is best done in the evening when the weather is cool. Transplant directly into already prepared holes Spacing ranges from 90x45 cm, or 90x60 cm depending on soil condition and water availability.

Irrigation: Maximum yields and fruit quality will be realized only if the plant receives adequate and timely moisture when fruit begins setting and maturing. Depending upon the soil type and growth conditions, approximately 25-50 mm of Water per week is needed to obtain high quality cucumbers.

Nutrient Management

For growing cucumbers in greenhouses with soil, add as much of the required calcium and phosphorus as possible as a base dressing, because these nutrients store effectively in the soil and their absence from liquid feeds prevents most clogging problems of the irrigation system. Provide calcium in the form of finely ground calcite lime @ 800kg/ha and phosphorus in the form of super phosphate

@ 250 kg/ha. Also, supply a good portion of potassium sulphate @500kg/ha and magnesium sulphate @ 250kg/ha. The ratio of potassium to magnesium in the soil should be 2:1. For supply of nitrogen, pre plant application of ammonium nitrate @150kg/ha. Make the final decision on the base fertilization after receiving the soil test results. To correct micro-nutrient deficiencies, foliar feeds can be applied alongside the regular pesticide applications. Avoid excessive Nitrogen; it leads to excess vegetative growth, poor fruit set, smaller fruits, hollow fruits and poor keeping quality.

Weed Control

Weed control in a soil system is very much like cultivation in the field. However, fumigation will reduce the requirement for weed control. Cultivation can begin when weeds and grasses are very small, and should be done as shallow as possible to reduce root damage. Cultivation is usually accomplished by running a roto tiller between the rows. Soilless systems normally do not require weed control programs because the system itself usually prevents weed growth.

Trellising, training and pruning: Several training systems exist for net house and greenhouse cucumber. The basic principle in developing a training system is to uniformly maximize the leaf interception of sunlight throughout the house. The selection of a system will largely depend on the greenhouse facility, the production system, and grower preference. The most common pruning system is vertical cordon, which allow only one main stem to develop by pruning all lateral vines the main stem is allowed to run vertically on support nylon cord or twine to the overhead crop support. Continue to remove all lateral vines on the main stem throughout the life of the crop. The fruits on the base 30 inches of the main stem should be pruned off as soon as they appear. This allows the plant to vigorously produce early vegetative growth that is essential for maximum fruit production. Fruits above the basal 24-30 inches of the main stem are then allowed to develop. V-cordon trained plants are known as the umbrella system. In this system, all lateral branches are pruned out as they appear until the main stem reaches the overhead wire. The growing point of the main stem is removed when one or two leaves have developed above the wire. Two lateral branches near the top of the plant are allowed to grow and are trained over the overhead wire resulting in these two branches growing downward. The growing point of each lateral is removed when nearly to the ground. Fruits are developed at the node of each leaf. The productivity of the laterals is generally less than the main stem.

Fruit Physiological Disorder

Crooking: A serious physiological disorder, results in decreased yields and reduced quality of greenhouse cucumbers. Curvature in fruit begins at an early stage—often when the ovary is less than ½ inch long-and remains throughout maturity. Slight curvature (up to 1 inch per 12 inches of fruit length) is tolerable in first-grade fruit, but excessive curving or crooking reduces market value.

Harvesting

Cucumbers are harvested as immature fruit when full length has been reached. At suitable harvest maturity, a jellylike material has started to form in the seed cavity. Cucumber production will be reduced if the fruit are left on the plant for too long. Cucumbers are hand harvested, normally 3 times per week, depending on the weather and growth stage of the plant.

Post-harvest Handling

Cucumbers lose moisture quickly and have the tendency to soften during storage. Marketable cucumbers should be sorted according to size and quality and individually wrapped in clear plastic. The optimum storage temperature for cucumbers is 10-12.5°C, at relative humidity of 95% RH. Storage or transit temperatures below this range should be avoided as this will result in chilling injury after 23 days.

Advantages Through Advance Approaches of Protected Cultivation in Arid Region

1. Round the year production of vegetables.
2. Adverse climate for production of vegetables can be overcome by different systems of protected production.
3. Multiple cropping on the same piece of land is possible.
4. Off season production of vegetables to get better return to growers.
5. Production of high quality and healthy seedlings of vegetables for transplanting in open field supporting early crop, strong and resistant crop stands.
6. Use of protected vegetable cultivation can increase production as well as productivity per unit of land, water energy and labour. It supports the production of high quality and clean products.
7. It makes cultivation of vegetables possible in areas where it is not possible in open conditions such as high altitudes deserts.

8. Disease free seed production of costly vegetables becomes easy under protected structures.
9. The potential of polyhouse production technology to meet the demand of producing good nutrition and healthy foods and quality vegetables free from pesticides can be fully exploited.
10. Controlled environmental conditions are used for early raising of nurseries, off-season production of vegetables, seed production and protecting the valuable germplasm.
11. Water and nutrients are the critical inputs whose optimization is essential in arid and semi arid regions. In view of this, the standardization of efficient fertigation technology will improve the water and nutrient use efficiencies increase the production potential of the crops.
12. The emphasis will be laid on the development of multi-strata cropping systems which will ensure income throughout the year as well as increase the cropping intensity and efficient utilization of natural recourses.
13. Arid and semi-arid regions have low incidences of pest and disease, however, with climate change, the incidence of pest and disease are increasing. This warrants development of integrated pest and diseases models using botanical pesticides so that the produce will find a place in global market.

Thurst Areas for Improvement in Protected Cultivation

- Identification of new crops /varieties suitable for cultivation inside the green house.
- Large scale motivation and training of educated unemployed youths in the field of protected cultivation
- Standardizing proper design of construction of polyhouse including cost effective and indigenously available cladding and glazing material.
- Computerized Control System, which includes various factors which are as required for the plant to maximize returns.
- Developing cost effective agro-techniques for growing of different vegetable crops in the different types of poly houses and lowering energy costs of the green house environment management.
- Major research activities on growing of vegetables under protected covers should be launched by ICAR and SAU's.

- Import of planting materials, structural designs and production technologies which are not relevant under Indian conditions should be stopped and in turn emphasis should be given to develop own F1 hybrid varieties so that seed are made available to the growers in time and at cheaper rates.
- Government initiatives/efforts in popularizing the greenhouse technology among the farming community of the country are to be strengthened.
- With the increase production of arid fruits and vegetables, the country will not only keep pace with increasing demand of fruits and vegetables but will also be in a position to export the value added products. The programme on successfully completion will also limit the process of degradation of natural resources to promote sustainability by standardization of low input and environmental friendly technologies.

Conclusion

As the population of India is increasing day by day, the demand for quality produce would also increase. But the productivity of vegetables in India is less compared to other leading countries in the world. Besides this, open field cultivation of vegetables causes low yield due to various type of biotic and abiotic stresses. All these necessitate the adoption of new technology which can increase the productivity per unit area along with high input use efficiency. One such technology is protected cultivation which has the potential of round the year production of vegetables with quality produce. Although it is an intensive technology and needs proper care and maintenance but can be very rewarding if adopted on commercial scale by the resourceful farmers near big cities of the country. The maximum return on investment can be achieved through protected cultivation of vegetables because it not only produces high quality yield but also gives higher market price to the growers.

References

Chakraborty, Hillolmoy and Sethi, Laxmi Narayan 2015. Prospects of Protected Cultivation of Vegetable Crops in North Eastern Hilly Region. *International Journal of Basic and Applied Biology*, 2(5): 284-289.

Chatterjee Ranjit, Mahanta Sandip And Pal P.K. 2013. Adoption status and field level performance of different protected structures for vegetable production under changing scenario. *The Indian Journal of Agricultural Science,* 1 (1): 11-13.

Kouser Parveen Wani, Pradeep Kumar Singh, Asima Amin, Faheema Mushtaq and Zahoor Ahmad Dar. 2011. Protected cultivation of tomato, capsicum and cucumber under kashmir valley conditions. *Asian Journal of Science and Technology, Vol. 1(4); 56-61.*

NAAS (National Academy of Agricultural Science) 2010. "Protected agriculture in North –West Himalayas" Policy paper no.47.

Nair Reena and Brache Swati 2014. Protected Cultivation of Vegetables: Present Status and Future Prospects in India. *Indian Journal of Applied Research,* 4 (6); 2249-555X.

P Ishfaq Akbar, M S Kanwar, M Saleem Mir, Anwar Hussain 2013. Protected Vegetable Cultivation Technology for Cold Arid Agro-ecosystem of Ladakh.*International Journal of Horticulture* ,3(19): 109-113.

Singh,D.K.and K.V.Peter 2014. Protected cultivation of Horticultural Crops, published by New India Publishing Agency New Delhi-110034

Sanwal S.K.,Patel K.K. and Yadav D.S. Vegetable production under protected conditions in Leh region : problems and prospects. *ENVIS Bulletin, 12(2): Himalayan Ecology.*

Shweta, Bhatia S.K. Bhatia and Malik Manu 2014. *Protected Farming*, 2(1) pp. 21-25.

Wani, K. P., Singh, P. K., Amin, A., Mushtaq, F. and Dar Z. A. 2011. Protected cultivation of Tomato, Capsicum and Cucumber under Kashmir Valley Conditions. *Asian Journal of Science and Technology 1(4): 56-61.*

Yadav ,R.K., Kalia, P., Choudhary, H. and Zakir Husaian and Brihamadev 2014. Low cost polyhouse technologies for higher income and nutritional security. *International Journal of Agriculture and Food Science Technology*, 5(3):191-196.

12

Production of Seed Spices with Saline/ Alkali Soils and Water in Arid and Semi-Arid Regions

O.P. Aishwath, Pankaj Kumar and Rajesh Jalwania

Introduction

Under arid and semi-arid regions of the world, major problems are salinity, alkalinity and calcareousness, which seems further aggravate with changing climate scenario. Besides that most of the developing countries lies in this zone with flanging population resultant shrinking down of land resource and good quality of irrigation water. Globally area under saline/sodic soils is about 953mha covering 8% of the land surface. Resent estimates indicate that 6.7 mha area in India is affected by soil salinity and alkalinity and about 25% of ground waters are sodic and/or saline in the country. There are two important approaches for utilizing saline, alkali or calcareous environment. The first one is to modify the environment so as to suit the plant or modifying the plant or substitute the plant species so as to suit the environment. First approach has been used more extensively because it enables the plants to respond better not only to water but also to other production inputs and also not far sustainable. However, second option is cost effective and most sustainable.

Accumulation of excessive salt in root zone leaves adverse effect on general agricultural crops and or crop failures sometimes, while crop diversification using seed spices are the option for such conditions. Most of the seed spices particularly cumin, fennel, dill, fenugreek, ajwain, celery, anise, caraway and coriander (and their cultivars) have some degree of salinity and alkalinity tolerance. However, among the seed spices, cumin is having high degree of salinity tolerance. In case of sodic or calcareous environment, fennel is most tolerant followed by dill as well as these crops are the remedy for such degraded environment by bringing down the sodicity and calcareousness. In some cases, germination, growth, yield and quality of seed spices improved by saline sodic environment. Besides eco-friendly use of these soils, phytoremediation of saline, sodic and calcareous soils by seed spices seems most sustainable.

Though the sizable area is under seed spices production and their demand is increasing at the international level. However, not much attention was paid yet for their expansion in salt affected areas as theses have some inherent capacity to tolerate such environment. The degree of tolerance of these crops with salinity depend on some environmental factors include temperature, wind, humidity, light, and air pollution. The informations available on the aspect are compiled here so as to boost the production in such degraded environment by improving cropping intensity and resource use efficiency.

Anover View of Salinity, Alkalinity and Saline-alkaline Environments

a) **Saline:** The dominant exchangeable cations in saline soils are calcium and magnesium. Saline soils commonly have visible salt deposits on the surface. Salts in the root zone can reduce crop yield by making it difficult for roots to extract water from the soil. As salts increase in soil, osmotic potential increases and water to moves from plant tissue to resultant wilting takes place in plants even when soil moisture is adequate in root zone.

b) **Sodic/alkali:** Sodic soils are high in exchangeable sodium compared to (Ca and Mg). EC is less than 4 dS m^{-1}. Soil pH is often greater than 8.5 and can be as high as 10 or more. High exchangeable sodium percent (ESP), high pH, and low divalent cations force to disperse the soil. The dispersion of soil particles destroys soil structure and prevents water infiltration by clogging pore in soil. In most cases, sodic soils have a black color due to dispersion of soil organic matter (SOC) and rarely have scattered or no vegetation. These soils are known as black alkali soils.

c) **Saline-sodic:** Saline-sodic soils are high in sodium *and* other salts also. They typically have EC greater than 4 dS/m (mmhos/cm), SAR greater than 13, and. Soil pH can be above or below 8.5. Saline-sodic soils generally have good soil structure and adequate water movement through the soil profile. They can have the characteristics of either a saline or sodic soil by depending upon the dominance of either sodium or calcium.

Possibilities and Limitations to Expend the Areas of Seed Spices under soil salinity, alkalinity and calcareousness

Coriander (*Coriandrum sativum*)

Soil sodicity delayed about one week in germination with the rise of ESP levels in coriander. Sodicity has more adverse effect on the formation of inflorescence than the vegetative growth. The crop could endure ESP 20 for seed and ESP 30 for biomass yields. Variety RCR-41 produced is fairly tolerant to higher sodicity. However, this cultivar is more susceptible to higher pH under calcareous environment. However, Hissar sugandh, CO-1 and CO-2 cultivars are fairly adoptive to such conditions. Coriander can tolerate salinity upto EC 4.0 dS m^{-1}. However growth, flowering and yield of coriander reduce 50% when salinity exceeded 6.0 dSm^{-1} and do not adversely affects below the EC 4.0 dS m^{-1}. The essential oil content of coriander increases from 0.30% to 84% with higher salinity. For the germination under saline environment Hisae Surbhi, Sudha and Sadhana cultivars are more tolerant. However, Rajaendra Swathi, GCR-1 & 2 and RCR-20 are least tolerant. Based on the performance under throughout growing period, RCR-446, GCR-2 and CO-2 are more salinity tolerant cultivars (Aishwath *et al*, 2010).

Cumin (*Cuminum cyminum* L.)

Based on survey at farmer's field, cumin can be grown successfully up to EC 13.5 dS-m^{-1}. However, sodicity causes adverse effect on cumin cultivation, when soil pH goes beyond 8.5. At high pH, absorption and translocation of micronutrients adversely affected leads to induction of chlorosis and morphological impairments. Among the seed spices cumin is highly tolerant to salinity from germination to overall growth period. Salinity does not affect the essential oil content marked a special characteristics of the crop (Table 1). Among some of the cumin RZ-341 is the most salinity tolerant cultivar followed by GC-2, RZ-223, RZ-19, and RZ-209 with respect to seed germination. Seed germinates even up to EC 20 dSm^{-1} (Aishwath *et al*., 2010).

Fennel (*Foeniculum vulgare* Mill.)

Fennel can be grown successfully on salt affected black soil with salinity of 8-10 dS m^{-1}. It also reported by some workers that threshold EC tolerance in fennel was 10 dS m^{-1} for trimmed shoots and 11 dS $m^{-1}$for untrimmed plants. Usually fennel can be grown on sodic soil having Exchangeable Sodium Per cent (ESP) 20. The variety HF-107 is most sodicity tolerant followed by Local BRS, NDF-9 and NDF-7 (Table 4). Therefore, the crop can tolerate medium level of sodicity up to ESP-30. Higher sodicity adversely affect the seed yield than foliage even low sodium and potassium leads to lower growth of crop. It also mobilizes calcium in sodic soil as well as phosphorus in calcareous and reclaims them could be boon for companion and subsequent crops. Consequently, this crop also reduces down the EC and pH of the saline and alkali soils, respectively. Highest salinity tolerance cultivars of fennel are GF-1 and RF-125 among the 13 cultivars with respect to seed germination. It is interesting to note that growing age of fennel improve the tolerance in plants than its juvenile and germination stage. Therefore, transplanting is one of the options for further advancement in salinity tolerance (Yadav *et al.*, 2014).

Fenugreek (*Trigonella foenumgraecum*)

After cumin, fenugreek can tolerate very high salinity with respect to germination. However, it cannot be grown at higher salinity as that of its germination. Moreover, this problem could be overcomes by supplementary use of Si as Na_2SiO_3, which decreases membrane permeability. Besides that inoculation of salt tolerant strains of rhizobia improves salt tolerance in fenugreek. Based on farmer's field survey, it can tolerate EC 2.0-3.0 dSm^{-1} depends on the types of salinity and plant part require for marketing. For foliage it can tolerate more salinity than for the seed purpose. Hissar Mukta is a highest salinity tolerant cultivar with respect to germination in a cluster of seventeen cultivars. Fenugreek crop is able to endure soil sodicity at germination stage but an adverse effect was noticed on emergence of branches $plant^{-1}$ rather than plant height in cultivars namely, HM-346, RMt-1, Hissar Sonali and Kalyanpur Selection (Table 2). Medium tolerance of sodicity observed in this crop and yield attributes were unaffected up to ESP 30. The variety Kalyanpur Selection produced greater biomass and seed yield than other genotypes up to ESP 30. The cation composition of plant revealed Na inclusion mechanism and showed narrow K/Na and Ca/Na ratio in the foliage indicating the potential of fenugreek crop to withstand medium level of sodicity. Under green house conditions, mitigates the negative effects of salinity on plant dry matter and chlorophyll content in fenugreek (Aishwath *et al.*, 2011).

Dill (*Anethum graveolens* L)

Dill can well perform with salinity EC 5 dSm^{-1} under heavy textured (saline black) soil and under light textured soil up to 12 dSm^{-1} (Table 3). Up to a certain extent, salinity improves the growth, yield (foliage and oil) and essential oil content in dill. Salinity up to EC 4.0 dSm^{-1} hastens the seed germination also. Hence, mild saline areas are the potential opportunities for both yield and quality of dill cultivation. It is also a good tolerant to alkalinity (~ pH- 9.5 to ~ pH- 9.8). Among the dozens of cultivars, CSS-1 is comparatively more adoptive to higher soil pH. Flowering stage is most sensitive for higher pH as compared to seed yield, plant height, primary and secondary branches. It almost equally performs as fennel in calcareous.

Celery (*Apium graveolens* L)

Celery had been classified as a halophyte as it occurs naturally in saline marshy habitats in Sweden, Algeria, Egypt, and Abyssinia. It also grows in brackish marshes near the sea in those countries. It comes up well at salinity level EC <11.0 dSm^{-1} and above that level about 50% yield reduction takes places, it shows the EC 11.0 is critical. However, it is bit sensitive for sodicity as symptoms of blackheart appears when it exposed to Na^+, Mg^{2+} and K^+ based salinity than Ca^{2+}. In addition to it leaf size, intermodal distances and stalk girth shrinks with sodicity or high pH resultant plant appears as bunchy top. For the germination point of view Ajmer Cel-7 genotype is more salinity tolerant and more than 50% seed germination takes place at EC 12 dSm^{-1}.

Ajwain (*Trachyspermum ammi*)

Ajwain could withstand up to ESP 18 and thereafter growth and yield parameters reduction take place. In *Ammi majus* increasing salt concentrations caused a significant reduction in the number of umbels, fruit yield $plant^{-1}$ and weight of 1000 seeds. Ajwain germinates well up to EC 8.0 dSm^{-1}. However, AA-84 is the most tolerant among available genotypes at Ajmer followed by AA-1 and AA-2 (Aishwath *et al.*, 2010).

Anise (*Pimpinella anisum)* and Nigella (*Nigella sativa*)

In anise genotypes, seed germination per cent was highest in Ani-13 at higher salinity and more than 50% germination attained up to EC 8.0 dSm^{-1}. However, most of the genotypes have tolerance to germination for salinity EC 2-4 dSm^{-1} and some of them are highly sensitive like Ani-14 and Ani-15 (Aishwath *et al.*, 2010).

Nigella can germinate well up to EC 8.0 dSm^{-1} and detrimental beyond that level. The thresh hold vale of nigella for salinity is EC 2.0 dSm^{-1} for it growth and development. For NaCl salinity, it can tolerate up to EC 7.5 dSm^{-1}. Nigella extract have an ameliorative effect on growth of beans under saline milieu. It also comes up well in mild calcareous soils. Though it comes up well up to pH 8.5 yet it is sensitive for alkalinity.

Possibilities and Limitations for use of Saline/Sodic Water in Seed Spices

Water is a universal solvent and carries huge variety of chemicals in it. Thus, underground water always contains some soluble salts. Continuous use of such water may affect the crop growth and soil properties depending upon the quality and quantity of salts present. The major constraints of underground water uses for irrigation are (i) Salinity and (ii) Sodicity.

In American and European countries fennel can grown with saline water having salinity around 3.8 dS m^{-1}. However, sweet fennel irrigated by 1000 ppm sea water gave a good result for growth and yield parameters. Under Indian conditions, fennel and dill can be cultivated profitably in the salt affected black soils by using saline water of EC 4.0 dS m^{-1}.Under arid conditions celery yield increased 10% in response to irrigation waters with EC_i varied from 4.2 to 5.4 dSm^{-1}, but decreased 10% when EC_i was with 6.2-8.0 dSm^{-1}. Celery and fenugreek have tremendous potential for cultivation with irrigation of underground saline waters having salinity up to 10 dS m^{-1} under calcareous soils of Haryana with alternate use of canal water. Nigella can tolerate salinity of irrigation water up to EC 4.0 dSm^{-1}. Under Hisar conditions, saline water having EC 4.0 dSm^{-1} can be used in fenugreek, while EC 8.0 dSm^{-1} in fennel and coriander.

Table 1: Effects of different salt concentration in growing media on yield and yield components of cumin in sand culture.

Salt level (mol m^3)	Grain yield (g m^2)	Biological yield	Umbels plant^{-1} (g m^2)	No of seeds plant^{-1}	1,000-seed weight (g)	Essential oil weight
Control	113.42 a	183.64 a	15.0 a	17.0 a	2.80 a	2.95 a
50	95.97 b	165.12 a	13.0 a	13.0 b	2.50 b	2.72 a
100	46.47 c	103.12 b	7.0 b	10.0 c	1.40 b	2.51 a
150	32.40 c	76.47 cd	4.0 c	8.0 cd	1.10 c	2.27 a
200	19.98 d	51.55 d	1.0 d	6.0 d	0.80 d	2.14 a

a, b, c, d: in each column means following the same letter is not significantly different at 5% probability level. (Nabizadeh, E., 2002)

Table 2: Interaction effect of sodicity levels x varieties of fenugreek (yield g m^{-2})

Variety × Sodicity	10 (7.8)	20 (8.6)	30 (8.5)	40 (8.9)
HM-346	753.3	553.3	445.0	108.3
RMt-1	758.3	333.3	318.3	173.3
Hissar Sonali	756.7	506.7	426.7	141.7
Kalyanpur Selection	691.7	750.0	645.0	180.0

ESP=Exchangeable sodium percent (Garg, 2012)

Table 3: Mean dry weight of dry leaves, flowers, seeds and percentage yield of essential oil of dill under different salinity treatments.

Salinity (dSm^{-1})	Dry Weight (g)			Essential oil (%)	Essential oil yield (mg/pl)
	Leaves	flowers	seeds		
0	0.195b	1.291	0.648	0.518	3.657b
4	0.278a	1.275	0.808	0.523	4.045ab
8	0.309a	1.405	0.900	0.440	4.323ab
12	0.298a	1.491	0.842	0.574	4.834a

(Ghassemi-Golezani *et al.*, 2011).

Table 4: Effect of sodicity on yield parameters of fennel (mean of 2 years)

Variety	Umbel $plant^{-1}$	Umbell $ateumbel^{-1}$	Seed yield ($kg\ ha^{-1}$)	Test weight 1000^{-1} seed (g)
HF-107	43.12	24.20	1956.6	5.35
NDF-9	51.45	21.11	1577.3	5.23
Local BRS	35.02	17.84	170.94	5.25
NDF-7	46.20	17.16	1078.0	5.03
CD (P<0.05)	NS	NS	476.0	0.22
Treatment				
ESP 10	60.10	19.91	2316.6	5.76
ESP 20	50.89	18.14	2075.8	5.23
ESP 30	41.81	19.44	1587.5	5.26
ESP 40	37.84	18.89	1200.0	5.10
ESP 50	29.09	24.0	721.7	4.75
CD (P<0.05)	15.08	NS	469.0	0.64
Interaction (ESP × Variety)	27.53	19.85	1032.5	0.72

ESP=Exchangeable sodium percentage (Garg, 2012).

Table 5: Salinity/alkalinity tolerance limit in seed spices

S.No	Name of crops	EC dSm^{-1} (soil/water)	Remark
1	Coriander	4.0	Sensitive to higher pH in calcareous soil.
2	Cumin	13.5	Sensitive to higher pH in calcareous soil.
3	Fennel	8.0-10.0	Tolerance improves with advancing age of crop. Sensitive to very high pH in calcareous soil.
4	Fenugreek	2.0-4.0	Seedling stage is more tolerant than the later.
5	Dill	5.0-12	Depends upon the soil texture and salinity type. It is also tolerant to high pH (9.8)
6	Celery	9.0-11.0	Bit sensitive to sodicity.
7	Ajwain	6.0-8.0	Mild sensitive to high pH in calcareous soil. Medium tolerant to sodicity.
8	Anise	6.0-8.0	This tolerance is for germination.
9	Nigella	2.0-7.5	Depend upon the types of salinity. Mild sensitive to high pH in calcareous soil.

Aishwath *et al.* (2013)

Plate 1: Fennel intercropped with garlic in saline-alkali soils at Gauhana, UP

Plate 2: Cumin in saline soil (EC 13.5) at Ranudhar, Rajasthan

Plate 3: Fennel in calcareous soil with high pH at Bilara, Rajasthan

Plate 4: Coriander, fennel and fenugreek with saline water at Hisar, Haryana

Conclusions

Most of seed spices mainly cumin, fennel, dill, celery, ajwain, coriander, fenugreek and anise have some degree of tolerance to salinity, alkalinity and calcareousness (Table 5). Cumin has high degree of salinity tolerance among these spices. In sodic or calcareous situation, fennel and dill crops are the best suited. Seed spices are also the remedy for such degraded environment by brings down the salinity, sodicity and calcareousness. In some cases, germination, growth, yield and quality of seed spices improve by saline and sodic locale. Besides phytoremediation of saline-sodic and calcareous soils by seed spices, eco-friendly use of these soils proves the most sustainable and a bonus for farmers of the area.

References

Aishwath, O.P., Mehta, R.S., Meena, R.S., Anwer, M.M., Dagar, J.C., Yadav, R.K., Lal, K. and Sethi, M. 2010. Seed Germination Study of Fenugreek Cultivars with Extended Exposure to Hyper Saline Conditions. *Journal of Soil Salinity and water quality*, 2(2): 95-101.

Aishwath, O.P., Mehta, R.S., Meena, R.S., Vishal, M.K., Khan, M.A., Singh, R.K. and Anwer, M.M. 2010. Seed spices: A solution of Problematic Soils (Hindi). *Kheti* 63(6):33-34.

Aishwath, O.P., Singh, B., Mehta, R.S., Yadav, R.K. and Meena, R.L. 2013. Seed Spices fetches more in saline/sodic soil and with poor quality water. *Indian Horticulture*, 58 (6): 33-35.

Aishwath, O.P., Singh, H.R., Velmurugan A., and Anwer, M.M. 2011. Analysis of Soil Suitability Evaluation for Major Seed Spices in Semi-arid Regions of Rajasthan using Geographic Information System. *International J. Seed Spices*, 1(1) 29-37.

Aishwath. O.P., Mehta, R.S., Lal, G., Anwer, M.M., Dagar, J.C., Yadav, R.K. and Singh, R. 2010. Growing seed spices a boon for problematic areas. *Indian Horticulture, 55(4):* 19-21.

Garg, V.K. 2012. Sodicity affects growth, yield and cation composition of fennel (*Foeniculum vulgare* Mill.). *Journal of Spices and Aromatic Crops*, 21 (2): 130–135.

Garg, V.K., 2012. Response of fenugreek (*Trigonella foenum-gracecum* L.) to sodicity.*Journal of Spices and Aromatic Crops*, 21 (1): 25–32.

Ghassemi-Golezani, Zehtab, K., Salmasi, S. and Dastborhan, S. 2011. Changes in essential oil content of dill (*Anethum graveolens*) organs under salinity stress. *Journal of Medicinal Plants Research,* 5(14): 3142-3145.

Nabizadeh, E. 2002. Effect of salinity on cumin growth and yield. *Iranian Field Crop Res.* 1: 20-29

Yadav, R.K., Meena, R.L., and Aishwath, O.P. 2014. Salinity tolerance of coriander, fennel and fenugreek seed spices under semi-arid conditions of northern India. *Journal of Soil Salinity and Water Quality* 5(2): 114-118.

Conclusions

References

13

Spices, Medicinal and Aromatic Plants in Degraded Arid and Semi-Arid Regions

O.P. Aishwath, Pankaj Kumar, R.C. Balai and Rajesh Jalwania

Introduction

According to FAO estimates, about 50% total geographical area in India is under various degraded hazards and about 2.1 million ha of land is getting degraded and deforested annually (Sehgal *et al.* 1998). Out of that underground water in arid and semi-arid areas is saline or sodic in nature. While spices, medicinal and aromatic plants may be a boon for these areas, which not only withstand under adverse conditions but also could be productive and helps in mitigating these problems. However, other crops are remain less or unproductive in these areas.

Spices are low volume and high value crops and have tremendous export potential, which is witnessed by several invasions faced by this Indian peninsula during ancient time. Medicinal and aromatic plants have unequally market potential with the world demand of herbal products growing at the rate of 7 percent per annum (Aishwath and Tarafdar, 2007). Seed spices and aromatic plants are the natural source of flavour, perfumes and fragrance widely exploited by essential oil industries in the world. Owing to considerably diversity of edaphoclimatic conditions ranging from temperate to tropical and arid to sub-humid that prevails in India and an array of medicinal and aromatic flora occurs in wild state. Many exotic essential oil bearing plants, including mints, citronella, geranium, patchouli, lavender and chamomile have been naturalized in India. As the country has emerged as one of the main producers of seed spices and some of the essential oils, theses plants are significant foreign exchange earners. Some of the aromatic plants especially grasses can sustain both biotic as well as abiotc stresses, even in the areas unsuitable for conventional crops (Aishwath *et al.*, 2008). Most of the seed spices crops are seasonal crops and comes up well with limited moisture and nutrients. Therefore, underutilized natural resources can be utilized profitably by crop diversification.

Spices, Medicinal and Aromatic Plants in Saline Alkali Soil Resources

About 7.2 million ha soils are saline and alkaline out of which 3 million ha is alkali in India. Due to poor physical properties, excessive exchangeable sodium and high pH, most of these lands support a very poor vegetation cover. Most of these areas are under arid or semi-arid regions. This waste area is not utilized properly by the traditional farming. There is lot of scope for utilization of our valuable soil resource by seed spices and aromatic plants for the cultivation of these high-value crops with marginal inputs.

i) **Saline soil:** Dill can be grown up to EC 3.3 dSm^{-1}(Sharma *et al.*, 2001). In another study on dill with salinity shows that it comes up well with EC 4-5 dSm^{-1}(Rao *et al.,* 2001 and Singh *et al.*, 1995). The beneficial effect of salinity on lemongrass (*Cymbopogon citrates*) and Java citronella (*Cymbopogon winterianus*) was observed at EC 5 dSm^{-1}as compared to lower orits higher levels of salinity (Patra & Singh 1995; Pal *et al* 1989). Lemon grass is a very hardy crop which grows successfully on cultivable waste land and the salt affected soil having EC up to 11 dSm^{-1}. The essential oil of German chamomile (*Matricaria chamimilla*) popularly known as blue oil can withstand up to salinity 12.0 dSm^{-1} without leaving any adverse effects on yield.Palmarosa comes up well in various textural types of soils. The salinity upto 5.5 dSm^{-1} found beneficial with respect to its productivity. Based on the pot culture study, there was no adverse effect on yield and quality of palmarosa observed upto EC 11.5 and beyond that soil salinity may adversely affect the yield and quality of palmarosa. Fig. 1 enunciates that vetiver, jamarosa and citronella could withstand up to salinity 12, 12.5 and 5.5 EC dSm^{-1}, respectively (Singh and Anwar, 1985, Patra and Singh, 1995). In aromatic trees, various species of eucalyptus have wide range of salinity tolerance ie 3.0 to 20.0 EC dSm^{-1} (Fig. 2) as reported by Gupta and Gupta (1997). Some of the aromatic plants are salt includer type salinity tolerant has potential for hyper accumulation of salt resultant helps in reclaiming the saline soil such as palmarosa is an hyper salt accumulator (Aishwath and Pal, 2004). Palmarosa grown on the soil having EC 4.8 dSm^{-1}, after two years of cropping, the EC of the soil reduced down to 0.64 dSm^{-1}(Prasad and Singh, 1998).

ii) **Alkaline soil:** Fennel performs well on heavy textured soils and can also be cultivated on salt affected black soil with salinity of 8-10 dS m^{-1} (Singh, 1999). Fennel and coriander grown on partially reclaimed sodic soil having Exchangeable Sodium Per cent (ESP) 20 with various N and P levels yielded well and large accumulation of Na in fennel stover indicating its tolerance to sodicity. Reduction in soil pH and ESP and increase in exchangeable calcium indicates its capability for reclaiming sodic soils (Garg *et al.*, 2000). Dill (*Anethem graveolens*) can tolerate the pH up to 8.6 (Farooqi, and Sreeramu, 2001). It has been reported that aromatic grasses such as, palmarosa (*Cymbopogon martinii* Roxb.Wats.) and lemon grass (*C. flexuosus* Steud. Wats.) could successfully be grown on moderately alkali soils having pH up to 9.0 and 9.5, respectively, while vetiver (*Vetiveria zizanioides* L. Nash) which withstands both high pH and stagnation of water, could successfully be grown without significant yield reduction on highly alkali soils (Fig. 1&3). In some cases the yield and quality of vetiver was better on sodic soil (Patra & Singh, 1995, Gupta & Gulati, 1965 and Farooqi & Sreeramu, 2001). These grasses not only produce essential oils used for industrial purpose but also ameliorate the soil (Dagar *et al.* 2004). Palmarosa, lemon grass and vetiver grown for two years on sodic soils having pH 10.6, 9.8 and 10.5, respectively, the reduction in pH was noticed as 9.4, 8.95 and 9.50 in each soil, respectively (Prasad and Singh, 1998; Patra *et al.*, 2002 and Anwar *et al.*, 1996). Japanese mint (*Mentha arvensis*) also comes up well in slightly alkaline soil than acid sedentary soil (Prasad & Chattopadhyay 1999). Among the aromatic grasses like palmarosa, lemongrass, vetiver, citronella and jamrosa, the jamrosa is a high pH tolerant (pH=10) and citronella is least (8.5). However, under well supplied organic matter conditions (Fig. 3), even pH up to 9.7 has no detrimental effect on growth, yield and oil content in citronella plant as reported by Gupta and Gulati (1965). In case of amrette (*Abelmoschus moschata*), chamomile (*Matricaria chamomilla*), and Lavender (*Lavandula species*) can tolerate upper pH level as 8.6, 9.0 and 8.4, respectively (Farooqi, and Sreeramu, 2001). An another study it was found that dill could be grown up to soil salinity pH 8.8 with slightly yield reduction (Sharma *et al.*, 2001). German chamomile can produce 2.5-3 t ha^{-1} of flower yield in alkaline soil having pH 9.5 and also have the high sodium uptake efficiency could help in improvement of sodic soil (Patra and Singh, 1995). Seed spices like cumin, anise and nigella are fairly tolerable to higher pH (Fig. 4).

Spices, medicinal and aromatic plants with poor quality irrigation water

Underground water always contains some soluble salts. Continuous use of such water may leave some harmful affect the crop growth and soil properties depending upon the quality and quantity of salts present. Accumulation of salt in soil profile depends upon the texture, structure and the depth of the soil (Kelly, 1963 and Aishwath and Pal, 2004), which directly affects the yield and composition of the crops (Aishwath and Pal, 2000). The major constraints of underground water uses for irrigation are (i) Salinity and (ii) Sodicity.

Fennel can be cultivated profitably in the salt affected black soils by using saline water of 4 dS m^{-1}. It is also revealed that the water of sub-marginal quality with salinity of 4 dS m^{-1}can be used without any significant yield reduction in dill. It is found to tolerate a pH of 8.6 (Rao *et al.*, 2001) Usage of saline irrigation water (EC_{iw} 8-10 dS m^{-1}) on highly calcareous saline soils, *Eucalyptus tereticornis* and aromatic grass species vetiver (*Vetiveria zizanoides*) showed great promise by yielding 1.19 to 1.73 t ha^{-1} root biomass. Lemon grass (*Cymbopogon flexuosus*), Palmarosa (*Cymbopogon martinii*), Ocimum (*Ocimum sanctum*), Celery (*Apium graveolens*), Mint (*Mentha piperita*), and Metha (*Trigonella foenum-graecum*) were irrigated either with saline or canal water alone or with the alteration of canal and saline water. These species have tremendous potential for cultivation with irrigation of underground saline waters having salinity up to 10 dS m^{-1}. Among the aromatic grasses, vetiver was found the most promising, followed by palmarosa and lemon grass (Dagar 2007). Irrigation water having water EC 10.0 and EC 16.0 dSm^{-1} does not have adverse effect on herb, essential oil yield and quality of lemongrass and palmarosa, respectively (Patra *et al.*, 1992). German chamomile is also tolerant to irrigation water salinity up to 8 dSm^{-1} and decreased the yield beyond that level of salinity (Patra & Singh 1995). In *Cymbopogon martinii* the detrimental effect of 4 EC $dSm^{-1}$was not observed in sandy soil, while same EC level was detrimental for palmarasa on sandy clay loam and sandy loam soils (Aishwath *et al.*, 1990 and Aishwath and Pal 2004).

The irrigation water applied in sandy loam and sandy clay loam soil having SAR 9.1 leaves adverse effect on the yield of palmarosa without any effect on oil content. However, at same level of SAR no adverse effect observed on yield of palmarosa in sandy soil (Aishwath and pal, 2004). Up to seven numbers of irrigations with sodic (RSC 10.0) water, there was no adverse effect on growth and yield of ambrette, palmarosa and lemongrass. At the initial stage of establishment of kewada has problem with this sodic water, but after 3-4 years of its establishment no adverse effect on growth was observed with sodic water (Aishwath, 2005a). However, some of the aromatic plant like ginger, lemongrass and jasmine are sensitive for sodic water (Aishwath, 2005b). The adverse effect of high SAR (10.0) of irrigation water did not appear on growth and yield palmarosa. However, beyond that level the growth, yield and uptake of the N, P, K, Ca and Mg reduced (Pal *et al.* 1993).

Spices, medicinal and aromatic plants in acid soils

Almost all the seed spices are fairly tolerable to mild soil acidity and alkalinity while coriander, fenugreek and celery could comes up well even up to 5.5 pH level (Fig. 4). It has also been reported that coriander comes well with the soil pH 6.5-8. Cumin can be grown with pH ranging from 6.8 to 8.3. While, fennel could be grown in heavy clay to sandy loams soils with pH of 6.5-8.0. In case fenugreek, it is cultivated on well drained loams or sandy soils and the acceptable pH rang is 6.0-8.0. The heavy loam or clay with pH 6.0-7.5 is most suitable. Celery thrives well in acid soil but not beyond pH 5.5 and rich in organic matter. Micro and secondary nutrient are essential otherwise plant shows deficiency symptoms as Boron- Cracked stem, Ca - Black heart and Mg-Chlorosis. Anise requires sandy loam to clay loam soil and pH 6.0-8.5 is well suited to this crop. For nigella, well drained sandy to heavy clay soil acid to alkaline pH is suitable for optimum plant growth. Caraway requires neutral to slightly alkaline soil having pH 6.5-7.5 is most suitable. It shows that these crops have wide adaptability with respect to soil reaction could be fitted into any adverse soil conditions (Aishwath *et al*, 2008).

Spices, Medicinal and Aromatic Plants in Eroded Lands

Out of total land area, about 50per cent is under degraded and cultivated land subjected to serious hazards of water, wind, shifting cultivation, water-logging, salinity and alkalinity and changing water courses. (Narayana and Babu 1983). While seed spices crops and aromatic plants can combat against these problems of erosion on sloppy land. The cultivation of seed spices and aromatic plants like *Anethum sowa*and *Ocimum kilimandscharicum* recommended in degraded habitat orchards as inter-crop (Gupta *et al.* 1998 & Narayana 1998).Aromatic grasses on sloppy lands have wide potential for the extraction of oil. In *Cymbopogon species*, *Cymbopoganmartinii* produced oil yield of 68.3kg/ha/ year followed by *C. winterianus*, 41.3kg/ha/year. At many a places, Cymbopogons are utilizing for dual purpose to control the gully erosion and oil extraction in Uttar Pradesh, Madhya Pradesh, Rajasthan and Gujarat (Verma and Chinnamani, 1998). Similarly, Eucalyptus globules and Eucalyptus robusta utilized in humid regions for dual purpose to control the soil erosion and stabilization of gully erosion as well as essential oil extraction (Ramamohan Rao, 1998). Lavender (*Lavandula officinale*) and rose (*Rosa domascena*) could also be grown in sloppy and erodes lands or terraces in hill areas (Shiva *et al.*, 2002).

Spices, Medicinal and Aromatic Plants in Nutritionally Eroded Lands

Fast mineralization of soil organic matter and release of nutrients in soil leads to losses of nutrients through water runoff in arid and semi-arid areas. It has been estimated that India is loosing about 8.2 million tonnes of nutrients annually (Singh 1998). In these areas major portion of soil is composed of sand (>85%) resulting lower retention power of nutrient, high infiltration rate (9-16 cm hr^{-1}) and poor moisture storage capacity. Lower fertility and moisture stress which offer a limited choice for increasing crop production under varying rainfall situation.

Coriander, cumin, fenugreek, Aromatic grasses, peppermint, davana, lavender and geranium are suitable for nutritionally eroded soil and some of them have association with VAM leads to effective utilization of available nutrients and water by them. It is well known fact that grasses can squeeze the soil by extracting the nutrients with their complex root system. Here, vast potential of aromatic grass could be harvested by utilizing nutritionally eroded soil with limited water (Aishwath *et al.*, 2008). Some of the seed spices and aromatic plants have lower requirement of nutrient can be grown on lower fertile soil (Table 1). It has also been reported that VAM simulate plant growth by physiological effects other than by enhancement of nutrient uptake or by reducing the severity of diseases caused by soil borne pathogens (Dehne 1982). The responsiveness of VAM to the host specific seed spices and aromatic plant is given in table 2. Recently, possibilities have been explored for the cultivation of rose scented geranium (*Palargonium sps*) in semi-arid conditions by multiple cuttings/ harvesting (Kothari *et al.*, 2004).

Spices, Medicinal and Aromatic Plants with Deficit Moisture Stress (Arid Areas)

The arid regions of the world occupy very extensive areas of the continental and masses and constitute nearly one – half of the land surface of the Earth. The major portion of the African continent and a great part of the Asian continent are arid.

Vegetation in the dry zones is very sparse. The general landscape is desolate and barren. Herbs appear only during a short period of the year when the conditions become favorable. The shrubs, herbs and trees have various means of storing water. These adaptations include increased ability to store water in their succulent stems or leaves; thickening of the leaf cuticle or reduction of leaf surface or the entire absence of leaves to lower the transpiration rate, and the ability to survive as a seed through many years of aridity. Drought – resisting plants may be divided into two groups: Succulents and Ephemerals. Seed spices are arid and semi-arid crops and come up with limited water conditions, while dill, ammi, Anise, fennel, papaver crops are fairly tolerant to moisture stress among the seed spices. The aromatic plant recommended for arid areas are ferula, artemisia, sage, lemongrass, guggal and Myrrh Recent research on *Ammi visnaga* and *A. majus*, *Pimpinella anisum,Foeniculum vulgare,Ferula asafetida*, *F. foetida*, Artemisia proved to be adaptive in dry areas. Some artemisias are of medicinal interest and a few are prized for their volatile oils. The chief source of santonin, in the U.S.S.R., is A. cine Berg. Growing in abundance in Russian, Turkestan and Persia, other species which have been found to contain santonin include *A. mexicana* Willd., A. neo – mexicana Woot, and *A. wrightii.A. grey*, found in America and *A. gallica* Willd. in East Germany, France, England and Scotland, and A. maritime L., a fairly widely distributed species spread from England to as far east as Chinese Mongolia. The content of santonin in the American and English species is too low to warrant commercial extraction. *A. fragrans* Willd. and *A. parviflora* Roxb. occurring in Afghanistan have also been reported to contain santonin. *A. maritime* growing in certain areas of Kashmir (India) and Kurram (Pakistan) has been found to contain substantial amounts of santonin (1 – 2 per cent), which is commercially extracted for aromatic and medicinal purpose. Lemongrass or Indian Verbena (*Cymbopogon citrates, C. schoenanthus*and C. *nardus*) southern parts of the eastern Egyptian desert and which yields a volatile oil are adaptive to water stress. Myrrh is still used to some extent in pharmacology as an aromatic stimulant; fenugreek (*Trigonella foenumgraecum* L.) contains diosgenin content could serve for the demi-synthesis of cortisone derivatives and sex hormones are come up well under water stress (Chopra *et al*., 2003).

Spices, Medicinal and Aromatic Plants Under Excess Moisture Stressor Water Logged Conditions

There has been no accurate estimate of the area afflicted in India under water logging. Framji (1974) reported that nearly 3.5 million hectares areas become waterlogged with the rise in water table in the root zone. Recently, Sehgal *et al*., (1998) reported about 20 million hectares area is under flooding causes water logging.

Most of the seed spices are susceptible for the water logging condition. However, water logging or submerged conditions can be utilized for the production of *Acorus calamus*, while, cardamom (*Elettaria cardamomum*) is best suited in swampy areas for its ideal production. Clove can withstand for a shorter period under water logging conditions. In case of kewada (*Pandanus species*), it is highly tolerable to water logging conditions for a longer period. The mint species *Mentha arvensis* is moderately tolerable to water logging for 8-10 days provided low atmospheric temperature. However, vetiver is capable of prolong water stagnation and also yielded with full potential (Shiva *et al.*, 2002).

Spices, Medicinal and Aromatic Plants in Metal Contaminated Soils

With our industrial progress, we are also progressing to contaminate our soil and water resource by various metals and heavy metals causes serious problem to the living ecosystem and their environment. Aromatic crops can provide economic return and metal-free final product, the essential oil. The study in Bulgaria using coriander, sage, dill, basil, hyssop, lemon balm, and chamomile grown at various distances from the smelter revealed that aromatic crops may not have significant phytoremediation potential, but growth of these crops in metal contaminated agricultural soils is a feasible alternative. Herbage essential oil yields of basil, chamomile, dill, and sage were reduced when they were grown closer to the contaminants. Metal removal takes place from the site with the harvestable plant parts was as high as 180 g ha^{-1} for Cd, 660 g ha^{-1} for Pb, 180 g ha^{-1} for Cu, 350 g ha^{-1} for Mn, and 205 g ha^{-1} for Zn. It also reported that high concentrations of heavy metals in soil or growth medium did not result in metal transfer into the essential oil. Of the tested metals, only Cu at high concentrations may reduce oil content (Valtcho *et al.*, 2008). No detectable amount of Cd, Cu, or Pb was found in the oils plant species Peppermint, basil, and dill. These can be grown in soils enriched with Cd, Pb, and Cu medium without risk for metal transfer into the oils, and without significant alteration of essential oil composition that may impair marketability (Valtcho*et al.*, 2006). Zheljazkov and Jekov (1995) investigated by Furnace Atomic Absorption (GFAA) and Inductively Plasma (ICP) techniques. The content of Cd, Pb, Cu, Mn, Zn and other trace elements determined in essential oils and plant extracts from the genera *Rosa*, *Lavandula*, *Mentha*, *Salvia*, *Ocimum*, *Foeniculum*, *Coriandrum*, *Anethum*, *Hyssopus* and *Rhus*. On the other hand the heavy metal content in oils from different regions were compared to heavy metal content in some oils obtained from species grown on heavily polluted soils and under severe atmospheric pollution. Results, obtained from the two apparatuses were very similar. They found that the concentration of the most hazardous heavy metals in all of the tested oils and plant extracts was very low, near the detection limits of the used apparatuses. It was concluded that most of the essential oil and medicinal crops could be successfully grown on heavy metal polluted soils and under atmospheric pollution as substitutes for some other edible crops.

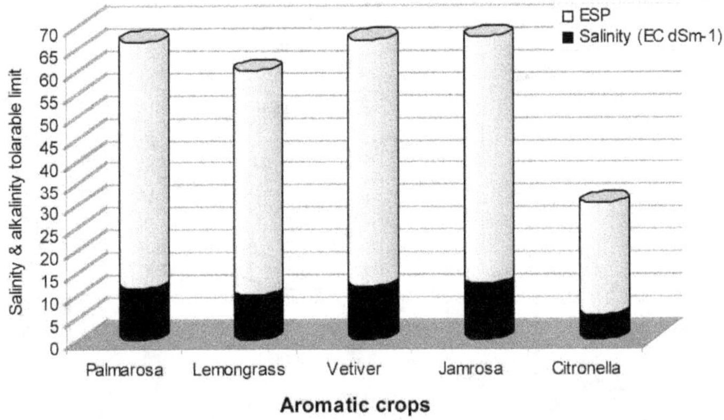

Fig. 1: Adaptability of aromatic crops with salinity and alkalinity.

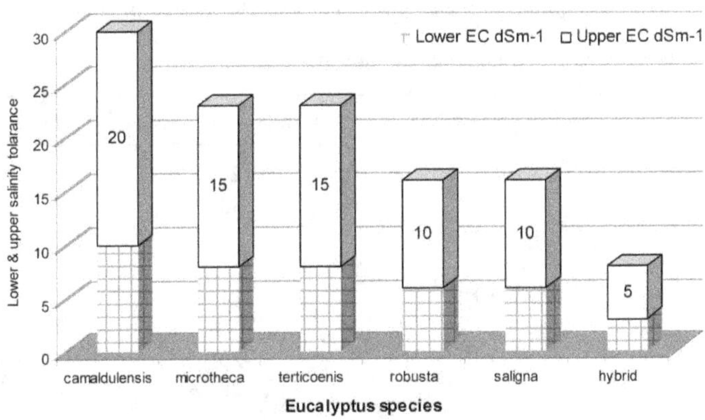

Fig 2: Salinity tolerant limit of various species of *Eucalyptus sps*.

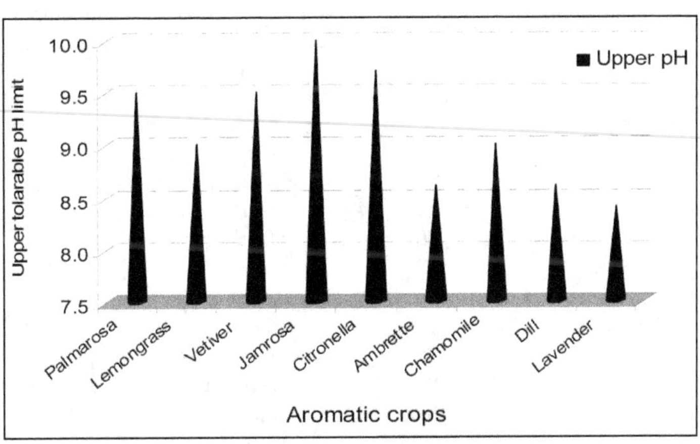

Fig. 3: Upper permissible limit for adaptability of aromatic crops with pH.

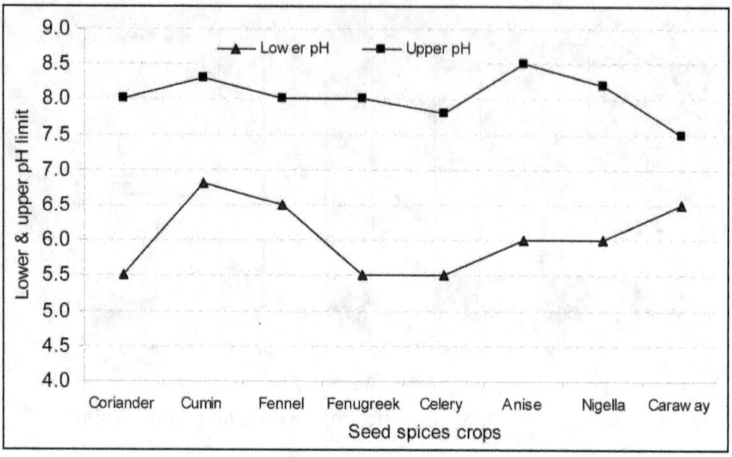

Fig. 4: Adaptability of seed spices crops with various pH ranges.

Table 1: Nutrient removal by some spice and aromatic crops

Crop	N uptake (kg ha^{-1})	P uptake (kg ha^{-1})	K uptake (kg ha^{-1})
Menthol mint	109	28	129
Ctronella	181	33	255
Lemongrass	160	32	194
Palmarosa	430	62	339
Geranium	180	24	137
Coriander	30	4	36
Davana	179	26	171
Japanese mint	109	28	129
Pappermint	42	—	—

Source: Prakasa Rao, 1992 and 1993

Table 2: Per cent colonization of VAM fungi in some spices, medicinal and aromatic plants

Crop	Per cent colonization	Crop	Per cent colonization
Basil	63	*Mentha arvensis*	46
Citronella	75-91	*Mentha piperita*	100
Coriander	32	*Mentha spicata*	100
Cymbopogon citratus	80	*Mentha citrata*	40
Cymbopogon flexuosus	72	Musk	83
Cymbopogon jwarancusa	50	Palmarosa	41-98
Cymbopogon khasianus	76	Patchouli	43
Davana	86	Rose-scented geranium	22-59
Fenugreek	92	Sacred basil	77
Lavender	80	Vanilla	57
Lemongrass	21-75	Vetiver	42

(Aishwath and Tarafdar, 2007)

References

Aishwath, O.P. 2005a. Appraisal of underground water in four consecutive years at NRC MAP, Boriavi (Gujarat) and impact of its irrigations on medicinal and aromatic plants. *Ecology, Environ. & Conserv.* 11: 373-380.

Aishwath, O.P. 2005b. Coefficient of variation and correlation coefficient in underground water quality parameters in and adjoining municipal area of Boriavi, Gujarat. *Pollution Res.* 24: 9-17.

Aishwath, O.P.and Pal, B. 2000. Performance of palmarosa (*Cymbopogon martinii* var Motia) under various soil texture and salinity levels. *Indian Perfumer*, 44: 285-290.

Aishwath, O.P. and Pal, B. 2004. Mineral composition of palmarosa (*Cymbopogon martinii*), soil EC and pH as influenced by saline water irrigation in texturally different soils. *J. Guj. Soc of Agron. Soil Sci.* 4: 42-47.

Aishwath,O.P.,Pal, B.and Singh, A.K. 1990. Effect of soil texture and water salinity on the growth, dry matter production, nutrients uptake and oil content of palmarosa (*Cymbopogon martinii*). *J. Agri. Sci. Res.* 32: 45-47

Aishwath, O.P. and Tarafdar, J.C. 2007. Role of organic farming in medicinal and aromatic plants. In: *"Organic Farming and Sustainable Agriculture"* (Eds. Tarafdar *et al.*, 2007). Publisher - Scientific Publisher, Jodhpur. pp 135- 185.

Aishwath, O.P., Vashishtha, B.B. and Malhotra, S.K. 2008. Scope of production of aromatic plants in arid and semi-arid regions. In: Proc. *National Workshop on Spices and Aromatic Plants*. Held on 6-7 February, 2008 at ARS (RAU) Mandor, Jodhpur, Rajasthan. pp 10.

Anwar, M., Patra, D.D. and Singh, D.V. 1996. Influence of soil sodicity on growth oil yield and nutrient accumulation in vetiver (*Vetiveria zizanioides*). *Ann. Arid Zone*, 35:49-52.

Chopra, L.C., Abrol, B.K. and Handa, K.L. 2003. Medicinal plants of the arid zone. Publisher Today & Tomorrow printers and Publishers. pp11-55

Dagar, J.C. (2007). Alternative land uses in areas with water scarcity. In, On-farm land and water management (A Compilation of National Training Course) pp 154-165.

Dagar, J.C., Tomar, O.S., Kumar, Y. and Yadav, R.K. 2004. Growing three aromatic grasses in different alkali soils in semi-arid regions of northern India. *Land degradation and development.* 15: 143-151.

Dehne, H.W. 1982. Interaction between vesicular arbuscular mycorrhizal fungi and plant pathogens.*Phytopathol.* 726: 1115-1119.

Farooqi, A.A. and Sreeramu, B.S. 2001. Ambrette, Chamomile, Dill, Lavender, and Palmarosa. In: *cultivation of medicinal and aromatic plants*. Pub. Univ. Press (India) Ltd. Hyderabad. pp 295-423.

Framji, K.K. 1974. International Commission on Irrigation and Drainage bull. pp 36-41.

Garg, V.K., Singh, P.K and Katiyar, R.S. 2000. Spices and aromatic plants. In: Proc. *Centennial conference on spices and aromatic plants* (20-23 September, 2000), Kalicut, Kerla, India. pp 133-138.

Gupta, R. and Gulati, B.C. 1965. Salt tolerance in *Cymbopogon nardus* (Lin) Rendle (Citronella cylon type) possibilities of commercial cultivation in *ushar* lands. Proc. Sem. *Sea salt and Plants*, held at Bhavnagar. pp 100-106.

Gupta, R.K., Bhardwaj, S.P.and Mittal, S.P. 1998. Management of degraded lands in the Himalayas. In, *Technologies for waste land development* (Eds Abrol I P and Dhurva Narayana, V.V.). Pub. Information Div. ICAR, KAB, New Delhi pp 196-208.

Gupta, S.K. and Gupta, I.C. 1997. Crop tolerance to saline conditions. In: *Management of saline soils and water*. Scientific Publishers, Jodhpur, India. pp 99-117.

Kelly, W.P. 1963. Use of saline irrigation water. *Soil Science*, 96: 305-391.

Kothari, S.K., Vijay Kumar, Y., Bhattacharya, A.K.and Ramesh, S. 2004. Harvest management in rose-scented geranium (*Pelargonium sps.*) for higher essential oil yield and superior quality under semi-arid tropics. *Archives of Agron.&Soil Sci.*,50: 593-600.

Narayana, V.V. Dhurva, 1998. Land degradation due to water erosion. In, *Technologies for waste land development* (Eds Abrol, I.P. and Dhurva Narayana,V.V.). Pub. Information Div. ICAR, KAB, New Delhi pp 190-213.

Narayana, V.V. Dhurva and Babu, R. 1983. Estimation of soil erosion in India. *J. Irri. Drain. Engng.* ASCE 109: 419-434

Pal. B, Singh, A.K., Singh, P.K.and Singh, R.P. 1993. Effect of saline water on herb, oil yield and composition of palmarosa (*Cymbopogon martini* Var Motia). *Indian Perfumer,*37: 267-269.

Pal, B., Singh, K.K., Jadaun, S.P.S. and Parmar, A.S. 1989. Effect of water salinity on herb and oil yield and composition of some *Cymbopogon species*. *Indian Perfumer*33: 196-199.

Patra, D.D., Anwar, M.and Singh, D.V. 1992. Herb yield essential, oil yield and nutrient accumulation by palmarosa under saline water irrigation. Abstract of technical paper presented at the conference on symposium on Environmental Soil Science, Canadian Society of Soil Science, August 8-13, Alberta, Canada pp 36.

Patra, D.D. and Singh D.V. 1995. Utilization of salt affected soil and saline sodic irrigation water for cultivation of medicinal and aromatic palnts. *Current Res. on Med. and Arom.Pl.* 17: 378-381.

Patra, D.D., Prasad, A., Anwar, M., Singh, D., Chand, S., Ram, B., Katiyar, R.S. and Kumar, S. 2002. Performance of lemongrass cultivars intercropped with chamomile under sodic soil with different levels of gypsum application. *Comm. Soil Sci. Pl. Analy.*, 33: 1707-1721.

Prakasa Rao, E.V.S. 1992. Fertilizer management in medicinal and aromatic crops. In: *Non traditional Sectors for fertilizer Use* (Eds Tandan, H.L.S). Fertilizer Development and Consultation organization, New Delhi, India. pp. 73-93.

Prakasha Rao, E.V.S. 1993. Nutrient management in some important tropical aromatic crops present and future. *Indian Perfumer*.37: 35-39.

Prasad, A.and Chattopadhyay, A. 1999. Growth and yield of Japanese mint (*Mentha arvensis*) in acid sedentary and slightly alkaline soil. *Indian Perfumer,*43: 9-14.

Prasad, A. and Singh, D.V. 1998. Effect of organic amendments on yield of palmarosa (*Cymbopogon martinii*Roxb. Wats) var Motia and chemical properties of sodic soil. *Indian perfumer*, 42: 21-26.

Ramamohan Rao, M.S. 1998. Approaches to management of denudated lands in peninsular, India. In, *Technologies for waste land development* (Eds Abrol, I.P. and Dhurva Narayana, V.V.). Pub. Information Div. ICAR, KAB, New Delhi. pp 232-244.

Rao, G.G., Nayak, A.K., Chinchmalatpure, A.R. and Khandelwal, M.K. 2001. Yield of dill (*Anethum graveolens*) on saline black soils of different unirrigated farm sites of Bhal area in Gujarat. *Indian Journal of Agricultural Sciences,*71: 711-712.

Rao, G.G., Nayak, A.K., Chinchmalatpure, A.R. Singh, R. and Tyagi, N.K. 2001. Resource characterization and management options for salt affected black soils of agro-ecological regionV of Gujarat state.pp 48-53.

Sehgal, J.L., Saxena, R.K. and Pofali, R.M. 1998. Degraded soil-their mapping through soil surveys. In, *Technologies for waste land development* (Eds Abrol, I.P. and Dhurva Narayana,V.V.). Pub. Information Div. ICAR, KAB, New Delhi pp 1-20.

Sharma P.C., Mishra, B., Singh R.K., Singh, Y.P. 2001. Variability in the response of spinach, fenugreek and coriander to alkalinity and salinity stresses. *Indian Journal of Plant Physiology,*6: 329-333.

Shiva, M.P. Lehri, A. and Shiva, A. 2002. Introduction, Calamus, Cardamom, Clove Kewada Mint and Vetiver. In. *Aromatic and Medicinal Plants*. Publisher, International Book Distributors, Dehra Dun, India. pp 67-307.

Singh, D.V. and Anwar, M. 1985. Effect of soil salinity on herb and oil yield and quality of some *Cymbopogons species*. *J. Indian Soc. Soil Sci.* 33: 362-365.

Singh, R. 1999. Lavaniya jal aur mrda mai sounf ke kheti. Phal phool. Oct-Dec. pp 30-35.

Singh, R., Bhargava, G.P., Singh, R. 1995. Response of safflower (*Carthamus tinctorius*) and dill (*Anethum graveolens*) to soil salinity. *Indian Journal of Agricultural Sciences,* 65: 442-444.

Singh, R.P. 1998. Land degradation problems and their management in semi-arid tropics. In, *Technologies for waste land development* (Eds Abrol, I.P. and Dhurva Narayana,V.V.). Pub. Information Div. ICAR, KAB, New Delhi. pp 125-136.

Valtcho, D. Z., Lyle E. C. and Baoshan, X. 2006. Effects of Cd, Pb, and Cu on growth and essential oil contents in dill, peppermint, and basil.*Environmental and Experimental Botany,* 58: 9-16.

Valtcho, D.Z., Lyle, E. C., Baoshan, X., Niels, E. N. and Andrew, W. 2008. Aromatic plant production on metal contaminated soils. *Science of the Total Environment*, 395: 51-62.

Verma, V. and Chinnamani, S. 1998. Rehabilitation of revine lands. In, *Technologies for waste land development* (Eds Abrol, I.P. and Dhurva Narayana, V.V.). Pub. Information Div. ICAR, KAB, New Delhi. pp 263-272.

Zheljazkov, V. and Jekov, D. 1995. Heavy metal content in some essential oils and plant extracts. In: *Proc. International Symposium on Medicinal and Aromatic held on 1 November 1995 at Amherst, USA,Published in Acta Horticulturae 2006.* pp 426.

14

Fruit Cultivation in Salt-Affected Soils: Constraints and Management Options

Prabodh Chander Sharma and Anshuman Singh

Introduction

Salinity is an encompassing term used to describe saline and sodic soils, and the marginal quality groundwater unsuitable for crop irrigation. A soil is considered to be 'salt-affected' when either excess soluble salts or excess exchangeable sodium or both adversely affect the soil properties and plant growth. Depending on the values of soil saturation paste extract electrical conductivity (EC_e), pH (pH_s) and exchangeable sodium percentage (ESP), salt-affected soils (SAS) are grouped into saline ($EC_e \geq 4$ dS m^{-1}, $pH_s < 8.2$ and high ESP< 15) and sodic ($EC_e < 4$ dS m^{-1}, $pH_s > 8.2$ and ESP >15) categories. Under certain conditions, both excess soluble salts and exchangeable sodium may be present resulting in saline-sodic conditions ($EC_e \geq 4$ dS m^{-1}, $pH_s > 8.2$ and ESP >15)(Sharma and Singh, 2015). Based on FAO/UNESCO soil map of the world (1970-1980), the global extent of SAS was earlier computed to be 831 Million ha (M ha) consisting of 397 M ha saline and the remainder 434 M ha sodic soils. Although this estimate is still widely used, recent studies suggest comparatively higher global extent of salinity and the associated problems; especially in irrigated arid and semi-arid regions across the world. Not withstanding the discrepancies in such estimates, it is certain that salinity continues to impose substantial costs on global food production and environmental sustainability.

Weathering of rocks and primary minerals is the dominant natural process responsible for salt accumulation in the soil and groundwater. Deposition of wind-borne salts and ingress of sea water also contribute to soil salinization to varying extents. Sea water intrusion is likely to heighten in the coming decades due to climate change impacts and intensive pumping of the coastal aquifers putting the coastal lands at increased risk of salinization. Naturally formed saline and sodic soils have been reported from about 100 countries across continents indicating the ubiquitous presence of salts in the soil and water; albeit in small to moderate amounts. Term 'primary salinity' is used to designate such naturally formed SAS where use of salt tolerant cultivars often in conjunction with doable agronomic interventions is a *sine qua non* to harness their productivity. In contrast, 'secondary salinity' refers to the salinization of agricultural lands due to anthropogenic activities (Shabala and Munns, 2012; Sharma and Singh, 2015).

Excessive and indiscriminate irrigation at the neglect of drainage and land clearing for crop production are the drivers of secondary salinization in irrigated and dry land regions, respectively. High susceptibility of irrigated lands to waterlogging and salinization is attributed to prolonged use of saline water for irrigation, seepage from canals and water channels and deep drainage beyond the root zone. Salt removal by the crops is often insignificant such that salts present in the irrigation water keep accumulating unless appropriate leaching and drainage interventions are implemented to remove them below the root zone. Salt tolerant crops have nearly half the leaching requirement compared to salt sensitive crops suggesting that development of crops and cultivars capable of enduring salinity and related problems without appreciable reductions in economic yield can greatly lessen the management expenses by curtailing the costs of irrigation and drainage. In drylands, replacement of deep rooted perennial vegetation with shallow rooted annual crops proves conducive to the percolation of much larger amounts of rainfall below the root zone resulting in wetter and salty soil profiles. Perennial tree- and pasture-based farming systems are suggested to contain the dryland salinity problem (Shabala and Munns, 2012).

A recent investigation suggests that globally over 1100 M ha area suffers from salinity. In South Asia including India, nearly 52 M ha area is salt-affected. It further reveals that a vast proportion of global SAS (~85%) have only slight to moderate constraints while the rest 15% severely affected lands pose complex and often unmanageable obstacles to crop production (Wicke *et al*., 2011). This observation points to the fact that relentless salinity onslaught has affected new agricultural areas in spite of ongoing reclamation efforts. It also underscores the need for precise delineation of SAS to identify the easily manageable pockets where feasible solutions can lead to tangible outcomes. It is increasingly becoming evident that salinization of newer areas has surpassed the reclamation efforts in several salt affected countries. In India, for example, an estimated 2.0 M ha salt-affected area has been reclaimed while 6.73 M ha still suffers from salinity and nearly 16 M ha is predicted to become salinized by 2050. It leads to the conclusion that reactive measures put into effect after salinization has attained damaging levels have become, by and large, irrelevant and that focus should shift to devise the ways and means that ensure simultaneous improvements in crop yields and soil quality with the minimal water and energy usage. While reclamation efforts need to be further strengthened, innovative tools for sustained returns from saline/partially reclaimed lands and for arresting the salinization of new areas are urgently needed.

Fruit Cultivation: A Driver of Socio-economic Development

Global fruit and vegetable production increased by about 3% over the last decade. In 2011, total global production of fruits was around 640 Million tonnes. Continued increase in production is largely driven by the area expansion in Asia (FAO, 2013). India is a leading producer of horticultural crops and ranks second in the fruit production globally. In the past few years, horticultural production has steadily increased. Surprisingly, India's total horticultural production surpassed food grain output during 2012-13 and 2013-14. Horticulture sector plays a pivotal role in the socio-economic development. Compared to the majority of field crops, horticultural crops are much more remunerative and provide handsome returns to the growers. Currently, horticultural crops together contribute ~29% to India's agricultural GDP and support ~20% of the agricultural labour force from ~13% of the total cropped area. Concerted R&D efforts and favourable policy initiatives have created a congenial environment for the sustained growth of Indian horticulture industry (ICAR-IIHR, 2015). Commercial fruit and vegetable production is seen as a viable option to harness the potential of marginally productive and undulating lands for poverty reduction and higher farm incomes provided that production, post-harvest and marketing constraints are adequately dealt with (Braidotti, 2013).

In addition to high incomes to the growers, horticultural production being a labour-intensive activity also contributes to poverty reduction by providing assured employment opportunities to the rural youth and women. Regular consumption of fruits and vegetables is known to improve the human health. According to World Health Organization, low consumption of fruits and vegetables contributes to 1.7 million deaths worldwide (FAO, 2013). In contrast to prescribed daily minimum intake of 400 g (5 daily servings with an average serving size of 80 gm) of fruit and vegetables (excluding potatoes, cassava and other starchy tubers) for the prevention of micronutrient deficiencies and chronic health problems such as heart diseases and cancer, current average fruit and vegetable intake in India is abysmally low, *i.e.*, 3.5 servings day^{-1} consisting of 1.5 servings of fruits and 2 servings of vegetables. Notwithstanding the fact that proportion of household expenditure on total food items has decreased in both rural and urban households, the proportion of income spent on fruits and vegetables has not changed much and the average Indian diet continues to be skewed towards cereals with fruits and vegetables constituting only 9% of the total calorie intake (Mukherjee *et al.*, 2016).

Constraints Hampering the Growth of Indian Fruit Industry

Despite huge benefits in terms of higher incomes, employment generation and poverty alleviation, development of Indian fruit industry is hampered by a myriad of constraints. Low productivity of fruits, ascribed to the poorly managed low density and senile orchards, compared to other major fruit growing countries is a cause for concern. With the exception of banana and papaya, average fruit productivity stands at 12.0 tha^{-1}. Most of the fruit crops are inherently susceptible to biotic and abiotic stresses. On an average, insect-pest, disease and weed problems lead to 20% yield reduction which can reach up to 80% in cases of severe infestation (ICAR-IIHR, 2015). Climate change and its anticipated long-term impacts, *inter alia*, higher atmospheric temperature and CO_2 concentration, intense heat waves, glacial ice melt, reduced river flows and sea level rise could further accentuate these problems. Climate change induced changes in tree phenology and reproductive behaviour, shifts in production belts, decrease in irrigation water availability and more losses due to abiotic stresses and pest-diseases outbreaks are increasingly becoming noticeable. Furthermore, continual shrinkage of productive farmlands and fresh water has necessitated the development of doable technologies for extending fruit cultivation in degraded areas suffering from constraints such as low water availability and salinity (ICAR-IIHR, 2015). Although demand for processed and value added products has consistently increased over the years, value addition is limited to only about 2% of the total fruits produced hampering the export competitiveness vis-à-vis other major fruit producers like China, Brazil and the United States (Bung, 2017). Post-harvest losses in fruits and vegetables are alarmingly high; 20-30% of the total production is wasted due to poor linkages between the field and fork resulting in huge monetary losses (~Rs. 30,000 crore year^{-1}) (Arumugam and Manikandan, 2011). It is due to these reasons that despite being a leading producer, India is also a net importer of fruits and vegetables (Mukherjee *et al.*, 2016). Consistent with the goal of doubling the farmers' income, Government of India has recently initiated many programmes to make farming a rewarding activity. In this regard, 'Mission for Integrated Development of Horticulture (MIDH)' was launched in 2014-15 to promote the holistic development of horticulture in the country by reducing the cost of cultivation, increasing the per unit yields and ensuring remunerative prices to the growers (midh.gov.in).

Irrigation-induced Salinity: A Rising Threat

Although net sown area in India has remained static around 141 Million ha over the past few decades, unabated increase in human population has necessitated agricultural intensification to meet the growing food needs. Concerted efforts towards technology development have led to quadrupling of food grain production surpassing nearly threefold increase in human population from 361 million to 1140 million over 1951-2011. Nonetheless, most of the growth in agricultural productivity has come from irrigated areas while rainfed regions continue to languish (Srinivasa Rao *et al.*, 2015). Although irrigation has played a critical role in sustaining the food production, on-farm irrigation mismanagement has come at the expense of severe land degradation. Irrigation-induced salinization has emerged as a severe environmental problem in many areas of Punjab, Haryana, Rajasthan and Uttar Pradesh important for the national food security. There is ample evidence that 'Punjab pattern of agricultural development', origins of which are traced back to the Green Revolution period (mid 1960s-mid 1980s), has virtually become unsustainable. Severity of the problem can be gauged by the fact that continual decline in farm incomes due to stagnant crop harvests, higher costs of cultivation, depleting soil fertility, secondary salinity and increasing debt burden has eventually led to rampant cases of farmer suicides in the region (Chand, 1999; Gill and Singh, 2006). In addition to large tracts of fertile Indo-Gangetic Plains affected by salt and waterlogging problems (Sharma *et al.*, 2016a), salinity has kept farming on tenterhooks in many other areas.

Twin menaces of salinity and waterlogging are also increasingly coming to fore in different fruit growing areas of the country. Salinity has emerged as serious hurdle to commercial viticulture in semi-arid areas of peninsular India having nearly 90% grape area of the country. Bhargava *et al.* (2006) found that groundwater salinity (EC_{IW}) rose from average 0.84 dS m^{-1} during 1980-1981 to 1.11 dS m^{-1} during 1999-2004 in the grape growing areas of Maharashtra and Karnataka. Out of the total samples tested, 49% had EC_{IW} <1.0 dS m^{-1} considered safe for grapevines, 30% had EC_{IW} between 1.0-2.0 dS m^{-1} at which growth is suppressed, 18% had EC_{IW}>2.0 dS m^{-1} where rootstock use becomes absolutely essential, and 3% showed EC_{IW}>4.0 dS m^{-1} at which salt tolerance of the commonly used rootstocks breaks down. A subsequent study also confirmed steadily deteriorating irrigation water quality leading to higher salt accumulation in vineyards in Pune, Sangli, Solapur and Nashik districts. In 2013-14, average irrigation water salinities (EC_{IW}) were 1.6 and 2.3 dS m^{-1} in Solapur and Sangli, respectively, compared to the corresponding values of 1.3 and 1.6 dS m^{-1} in 2007-08 (Mohan, 2015). High salinity in soil and groundwater, drought and strong winds are the major limitations to mango production in the Saurashtra region of Gujarat where traditionally grown Kesar variety has gained reputation in the global export market. Techniques such as drip irrigation using rain water and mulching have made it possible to raise profitable high density mango orchards under such conditions (Nathwani, 2014). Use of potential salt tolerant rootstocks is also being explored to tide over the salinity problem; especially in coastal lands suffering from sea water ingress (Balan, 2017). Commercial

Kinnow cultivation in Fazilka district of Punjab and the adjoining areas is threatened by the rising watertable. In areas with shallower water tables (<0.5 m), considerable increase in soil salinity (EC_2= 1.13-2.84 dS m^{-1}; pH= 8.51-9.05) has led to large scale Kinnow decline (Fig. 1). Even in areas with relatively deeper water tables (~ 1.5-2 m), 'apparently healthy' Kinnow plantations seem to be at risk. Unabated decline has enhanced the livelihood vulnerability of the farmers who previously fetched remunerative returns from the sale of Kinnow fruits. A survey conducted in ber growing areas of Haryana revealed that salinity is a severe impediment to profitable ber cultivation in areas having saline groundwater (~3.0-10.0 dS m^{-1}). Salinity caused marked reductions in fruit yield with saline irrigated trees invariably tending to bear small crops compared to the trees irrigated with either good quality or less saline water. For example, fruit yield of 20 year old trees of Gola variety decreased from average 0.2 t $tree^{-1}$ to 0.1 t $tree^{-1}$ when irrigation water salinity (EC_{IW}) increased from ~6.0 dS m^{-1} to ~10.0 dS m^{-1} under similar agro-climatic conditions. Development of waterlogged sodic lands and the repeated instances of resodification in Sharda Sahayak canal command, if left unaddressed, can spell doom for the flourishing mango industry of Lucknow district (ICAR-CSSRI, 2017). Climate change induced rise in sea level can accentuate the problem of sea water ingress in many coastal areas of country where different fruits and plantation crops are grown on a commercial scale (Sharma and Singh, 2015).

Fig. 1: Salinity-induced Kinnow decline in Fazilka district of Punjab, India (Photo: Anshuman Singh).

Effects of Salt Stress

Salinity retards water absorption by the plants by raising the osmotic pressure of the soil solution. Salinized plants facing 'physiological drought' fail to extract the available water. Osmotic stress induced water deficiency in saline soils is strikingly similar to the effects of drought. Osmotic stress constitutes the first phase of growth inhibition in salt stressed plants while salts still remain outside the plant. Reduced rate of cell expansion, decrease in stomatal conductance and low availability of CO_2 eventually limit the rate of photosynthesis (Shabala and Munns, 2012). Irrigation creates favourable conditions for crop growth by alleviating the osmotic stress. Irrigated soils have high soil moisture content and low salt concentration while the reverse is true for the dry saline soils (Abrol *et al*., 1988). Second phase of salt stress begins with the entry of salts inside the plants. In absence of effective salt exclusion by the roots, salts accumulate to the toxic levels causing a range of ion specific physiological abnormalities. Soils with the high levels of Na^+ exhibit the reduced availability of K^+, Ca^{2+} and NH_4^+.Excess Na^+ results in the depolarization of plasma membrane hampering the passive uptake of essential cations such as K^+. Again, Na^+ not only competes with K^+ for major binding sites in the key metabolic processes such as protein synthesis but also increases K^+ efflux from cytoplasm (Shabala and Munns, 2012). As an essential micronutrient, Cl^- modulates the enzymatic activities, serves as a co-factor in photosynthesis, stabilizes membrane potential and regulates the cell turgor and pH. However, in high concentrations, Cl^- becomes toxic. Cl^- concentrations as low as 4-7 mg g^{-1} prove limiting to the salt sensitive species whereas adverse effects in tolerant species appear at relatively higher Cl^- levels (15-50 mg g^{-1}). Root exclusion and restricted translocation of Cl^- to shoots are necessary to prevent salt injury symptoms in citrus and grapes (Tavakkoli *et al*., 2010). In salt stressed plants, continued accumulation of reactive oxygen species (ROS) causes the oxidative stress. Osmotic stress induced decrease in stomatal conductance and excessive Na^+ accumulation in the cytosol together impair the photosynthetic machinery so that light captured by the photosynthetic pigments is only partially utilized. This accelerates ROS production in leaves leading to lipid peroxidation, DNA and protein denaturation, and ion leakage. ROS activate a certain class of K^+ channels resulting in heavy K^+ loss from the cytosol (Shabala and Munns, 2012).

Salt Stress Symptoms

Excess salts in the growing medium hamper plant growth in two ways. Reduced water uptake, *i.e*., 'water-deficit effect of salinity' leads to the initial reductions in the growth rate. Shortly, salt ions enter inside the plant in the transpiration stream and interfere with key physiological processes further reducing the growth. This is referred to as the 'salt-specific' or 'ion-excess effect of salinity'. In fruit plants, salinity suppresses the plant height, emergence of new branches and leaves while hastening the senescence of older leaves. In most of the cases, salt injury symptoms initially appear as the yellowing, scorching and chlorosis of the leaf margins. If salt stress is not relieved, leaves soon become chlorotic, develop necrotic spots and eventually shed off the plant. Salinized plants show significant reductions in the fresh and dry weights of shoots and roots. In certain crops like guava, plant height and stem girth may be relatively less affected by salinity but branch emergence and leaf production significantly decrease giving the salinized plants an upright and sparse look (ICAR-CSSRI, 2016; Fig. 2). In contrast to the annual plants, which tend to become relatively more salt tolerant with age, majority of the fruit crops become salt sensitive as they grow older. It is possibly due to gradual translocation of the salts stored in roots to leaves coupled with a slower growth rate in the older plants. Moreover, salt sensitive species like citrus and stone fruits often absorb excessive amounts of salt ions in soils otherwise considered to be normal for other crops. In some fruit crops, use of salt excluder rootstocks may decrease Na^+ and Cl^- toxicity implying that osmotic stress will mainly be responsible for most of the deleterious effects in the composite plants (Singh *et al.*, 2016).

Depending on species, either ion toxicity or osmotic stress may account for the salt-induced injury. Grapevine growth seems to be highly sensitive to water deficit than to an equivalent salinity level suggesting that besides predominant organic osmolytes (tartrate and malate) salt uptake also contributes to osmotic adjustment in the grape leaves (Cramer *et al.*, 2007). Higher NaCl levels (80 and 160 mM) reduced the plant height, leaf number, total leaf area and dry matter by >50% compared to control in *Ziziphus spinachristi* L., a wild fruit species distributed in Asia and Africa. At 160 mM NaCl, leaf Na^+ and Cl^- levels increased by 81- and 21-fold, respectively, over control but leaf N, P and K concentrations were not significantly affected indicating the role of Na^+ and Cl^- in osmotic adjustment (Sohail *et al.*, 2009). Addition of NaCl into irrigation water increased the leaf and root Na^+ and Cl^- concentrations in *Citrus limonia* seedlings and olive (cv. Arbequina) cuttings causing decline in plant dry mass production, stomatal conductance and net photosynthesis. Growth reduction was apparently due to ionic effects and not due to osmotic stress since both the species maintained a leaf pressure potential higher than in control leaves (Melgar *et al.*, 2008). In most of the fruit crops, shoot growth is more sensitive to salt stress than root growth. In certain species such as avocado, however, root growth is more affected. In avocado plants treated with 25 mM NaCl, leaf biomass production, leaf initiation and elongation rates decreased by 19.5%, 12%, and 5%, respectively, while root volumetric growth and root elongation rate reduced by 65% and 75%, respectively (Bernstein *et al.*, 2004).

Fig. 2: Salt stress symptoms in guava cv. Allahabad Safeda planted under shallow watertable saline conditions at Nain Experimental Farm, Panipat. Note the progressive loss of leaves, sparse growth and reduced fruiting and plant mortality with increasing salinity (in clock wise direction; Photo: Anshuman Singh)

Mechanisms of Salt Tolerance

Regulation of salt uptake may occur either at the whole plant or at cellular and molecular levels. At the whole plant level, both halophytes and glycophytes tend to achieve salt tolerance by the selective ion uptake, preferential loading of K^+ into xylem and retention of Na^+ in the basal stem and root tissues (Munns *et al.*, 2002). The relative contribution of these mechanisms to salt exclusion is essentially a genotype-dependent phenomenon. Salt exclusion assumes great significance as plants transpire 30-70 times more water than they use for cell expansion. In non-excluder genotypes, shoot salt concentrations can shortly peak to the toxic levels (Munns, 2002). Salt exclusion is especially desirable in perennial fruit crops which retain the leaves for much longer periods than annual crops having short-lived leaves. In perennial species, high shoot: root ratio, high intrinsic growth rate and absence of an apoplastic pathway in roots are the features which reduce the rate of salt loading into the transpiration stream. In relatively more salt tolerant species, there would be little retranslocation of absorbed Na^+ or Cl^- in the phloem to prevent the salt export to the actively growing shoot tissues. Halophytes also possess specialized cell types called salt glands or bladders for salt excretion (Munns, 2002; Munns *et al.*, 2002).

At the cellular (organelle) level, plants sequester the excess salts present in the cytoplasm in the cell vacuole, a process called ion compartmentation, to protect the metabolic enzymes from salt damage. Both Na^+ and Cl^- inhibit the enzymatic activities at concentrations above 100 mM. After Na^+ and Cl^- have been sequestered in the cell vacuole, a variety of inorganic and organic solutes accumulate in the cytoplasm (*i.e.*, osmotic adjustment) to balance the osmotic pressure of the ions in the vacuole. In comparison to salt exclusion, ion compartmentation and salt excretion, osmotic adjustment requires more energy use to synthesize the solutes. Again, accumulation of compatible organic solutes requires much more energy compared to that needed for inorganic ions (*i.e.*, 3.5, 34, 41 and 50 moles of ATP for Na^+, mannitol, proline and glycine betaine, respectively; Munns, 2000).

At the molecular level, different ion channels and transporters regulate the net movement of salts across the cell membranes. Under normal conditions, plants maintain a high cytosolic K^+/Na^+ ratio. Higher extracellular Na^+ concentrations, however, favour the passive transport of Na^+ over K^+. As hydrated ionic radii of Na^+ and K^+ are strongly similar with each other, ion channels often fail to discriminate between them. Na^+ ions can be transported into the cell through low- and high-affinity K^+ transporters (Blumwald, 2000). Low-affinity K^+ channels with low K^+/Na^+ selectivity are more adversely affected by Na^+ ions. Salt stressed plants must maintain more high-affinity K^+ channels for adequate K^+ uptake (Zhu, 2007). Three types of low affinity K^+ channels- inward rectifying channels (KIRCs), outward rectifying channels (KORCs) and voltage-independent cation channels (VICs)- are known. In contrast to voltage dependent KIRCs and KORCs having a high K^+/Na^+ selectivity ratio at physiological K^+ and Na^+ concentrations, VICs exhibit a relatively high Na^+/K^+ selectivity and are implicated in Na^+ uptake at high salt levels (Blumwald, 2000). Integrated membrane proteins (IMPs) present in plasma membranes play a crucial role in different cell functions including control of the solute movement across the membranes. High-affinity potassium transporters (HKTs) are a class of IMPs occurring only in plants. While some HKT proteins are highly selective for Na^+, others exhibit affinity for K^+. Moreover, the selectivity of certain HKT proteins may change depending on the ionic environment (Waters *et al*., 2013).

Available evidence suggests that in many species higher Cl^- concentrations, and not Na^+, cause the toxicity symptoms. Nonetheless, little is known about Cl^- uptake and translocation mechanisms in plants. Some of the key traits that control Cl^-transport processes in plants include reduced net xylem loading, intracellular compartmentation and efflux of Cl^- from roots. Restricted loading into xylem prevents excess Cl^- accumulation inthe shoot tissues. In some species, Cl^- is preferentially sequestered in leaf epidermis to lessen the toxicity to the photosynthetically active mesophyll cells. In contrast to extensively studied K^+ channels, there is paucity of information on genes and proteins involved in anion movements in plants. Recent studies have shown that several traits may contribute to reduce Cl^- loading into shoots (Cl^- exclusion); previously considered to be controlled by a single dominant gene. A few genes and proteins directly or indirectly regulating Cl^- flux in plants include the aquaporins (regulating the water flow in roots) and ATP binding cassette (ABC) transporters. Transcriptome analysis of two citrus genotypes differing in Cl^- exclusion revealed the possible involvement of several anion transporter families (Teakle and Tyerman, 2010).

Assessment of salt tolerance

Salt tolerance of a given plant species or genotype can be assessed using two criteria, *viz*., percent reduction in 'biomass' or 'yield' production under saline and normal conditions over a period of time. Salt tolerance of a crop can be assessed by plotting its relative yield as a continuous function of soil salinity. Maas and Hoffman (1977) proposed that this response curve could be represented by two line segments, *viz*., '*a tolerance plateau with a zero slope*' and '*a concentration-dependent line*' whose slope indicates the yield reduction per unit increase in salinity. The point at which the two lines intersect designates the threshold, *i.e.* the maximum soil salinity (EC_e) that does not reduce yield below that obtained under normal conditions. It is pertinent to mention, however, that threshold salt tolerance reported by Maas and Hoffman (1977) serves only as a guideline to understand the relative salt tolerance among crops as absolute tolerance varies with the agro-climatic conditions, genotypes and management practices. Furthermore, data in most of the cases represent the salt tolerance of a single cultivar or few cultivars; they do not necessarily represent the salt tolerance of species. In case of woody fruit crops, another limitation is that these data are applicable when Na^+ or Cl^- excluder rootstocks are used or when these ions do not predominate in the soil. Despite these discrepancies, the data are useful in showing the wide range of salt tolerance across species, and in that yield has a different pattern of response than vegetative biomass (Munns *et al*., 2002).

Salinity Management Options in Fruit Crops

Several options can be exercised for expanding the area under fruit crops in salt-affected regions (Table 1). It appears to be a safe option to raise the fruit orchards in lands having moderate salinity and relatively deeper water tables as majority of them are harmed by the elevated salt and moisture levels in the root zone. In lands where water table lies below 2 m, and preferably at 4-5 m depth from the surface, there will be least possibility of damage as direct contact between deep growing tree roots and saline water will not occur. Upward movement of water brings the dissolved salts to the root zone. Plants suffer from severe salt stress when the water table is <1.5 m below the soil surface. Depth of water table is a major factor governing the rate of evaporation; in areas where water table depth is 1.5-3.0 m evaporation will be minimal (Salama *et al.*, 1999). In waterlogging free degraded sodic lands having assured fresh water supplies feasible interventions like gypsum application could prove very effective (Singh *et al.*, 1997). In either case, availability of quality planting material of salt tolerant genotypes is desirable as it can considerably reduce the dependence on other means of salinity management. Although such 'salt-affected lands of horticultural relevance' provide ample scope for crop diversification through fruit crops, there are constraints that need to be addressed. Third, and the challenging, option relates to putting the waterlogged salty lands under fruit crops. Over the years, it has become difficult to reclaim and utilize such lands even through time tested management practices including gypsum application and drainage such that they lie barren for extended periods of time. An 'ensemble of technologies', *inter alia*, land shaping, storage of rain/canal water, salt tolerant cultivars, mulching and drip irrigation is suggested to enhance the economic value of such lands.

Table 1: Context-specific choice of salinity management practices in fruit crops.

Watertable depth	Soil type	Main limitations	Strategies
Deep (> 2m)	Saline	Marginal to moderate salinity, osmotic stress and specific ion toxicities	Leaching, use of salt tolerant cultivars, drip irrigation
	Sodic	High pH (9.0-10.0), degraded soil structure, water stagnation and ionic imbalances	Improved planting methods; gypsum application (5-10 kg pit^{-1}); selection of drought and salt tolerant cultivars
	Partially reclaimed sodic	Moderate pH (~8.5), unfavourable soil structure and susceptibility to resodification	Integrated fruit-based farming systems for resource recycling, integrated crop management practices and micro-irrigation
Shallow (< 2m)	Saline	High to very high salt stress and related hazards like hypoxia	Land shaping, raised bed planting, storage of canal/rain water for irrigation, drip irrigation and salt tolerant cultivars
	Sodic	Structural problems, impeded drainage, wetter soil profiles and oxygen depletion	Inversion of soil column, integrated crop-fish module, planting on pond dykes, rainwater harvesting and drip irrigation

Different techniques available for salinity management in fruit crops are briefly discussed under the following heads:

Improved Planting Techniques

Waterlogging free sites are always desirable for raising fruit trees. Of late, however, growing land scarcity and food security concerns are making it difficult to spare the lands under staple crops for horticultural production. In large parts of north-western India suffering from poor soil health and stagnant crop yields, farmers seem to be reluctant to switch over from rice-wheat system to the more remunerative horticultural crops. This state of affairs has necessitated the development of appropriate planting techniques to revive the productivity of waterlogging prone salty lands through fruit crops. Experiments conducted under shallow salinewatertable conditions indicate the possibility of commercial cultivation of guava (cv. Allahabad Safeda) and bael (cv. NB-5) in saline soils irrigated with saline water (EC 3.0-4.0 dS m^{-1}). Planting on raised beds and the amelioration of planting pits with FYM considerably enhance the salt tolerance in young plants [Fig. 3]. Simple land shaping interventions have also made it possible to raise high value horticultural crops in waterlogged sodic lands of Central Indo-Gangetic Plains and the coastal saline lands otherwise lying unproductive. Land shaping interventions are based on the premise that inverting less saline deeper soil profiles upside down would make the surface soil congenial for crop growth by lowering the water table and improving the soil properties (ICAR-CSSRI, 2016). Ten different fruit species were evaluated in a highly sodic land (pH ~10.0). Planting was done in the auger-holes and pits containing 5-20 kg gypsum. Increasing gypsum dose had a positive effect on tree growth. After 7 years, Indian jujube, jamun, guava, aonla and karonda were found to be suitable species for such soils (Dagar *et al.*, 2001). Auger planting holes are especially effective in promoting water and air flux and root growth in calcareous sodic soils having a sub-soil hardpan.

Fig. 3: Raised bed planting of bael cv. NB-5 and guava cv. Allahabad Safeda under shallow saline water table conditions of Nain Experimental Farm, Panipat (left) and tractor mounted auger for piercing the hard pans in sodic soil (right) (Photo: Anshuman Singh).

Amendments, Microbial Inoculants and Plant Growth Regulators

In addition to gypsum, the preferred amendment for sodic soils, other materials have also been successfully used to alleviate salt injury in fruit crops. Amelioration of planting pits with sand and FYM (20 kg each pit^{-1}) considerably improved the properties of a moderately sodic soils (pH ~9.0) resulting in better establishment of litchi cv. Rose Scented (Saxena and Gupta, 2006). Fruit plantations also lead to steady improvements in the properties of sodic soils. Among different fruit and agro-forestry land uses in a semi-reclaimed sodic soil, the minimum (6.8) and the maximum (9.5) soil pH at 1.5-2.0 m depth was noted in litchi and *Eucalyptus* plantations. The highest soil organic carbon storage (133 t C ha^{-1}) was observed in the guava land use. Long-term improvements in soil properties under fruit trees may be due to *in situ* decomposition of the litter, deeper root growth and nutrient mineralization in the root zone (Sharma *et al.*, 2016b). Arbuscular mycorrhizal (AM) fungi alleviate the detrimental effects of salinity by stabilizing photosynthesis, regulating the ion and water uptake by roots and by enhancing the accumulation of compatible solutes and antioxidant compounds. In AM symbiosis, the fungus forms an appressorium on the root surface and enters the root cortex by extending its hyphae. The hyphae form arbuscules and vesicles in the cortex. Inoculation with a mixture of two AM fungi (*Glomus intraradices* and *Glomus mosseae*) significantly improved the growth in NaCl stressed (50 mM) Cleopatra and Alemow seedlings compared to the salinized but non-inoculated plants (Navarro *et al.*, 2011). Rhizospheric and endophytic bacteria (CSR-G-1, CSR-B-2, and CSR-B-3) have been found promising to enhance salinity tolerance in different horticultural crops. Their growth enhancing and bioameloiration properties are due to higher activities of superoxide dismutase, phenyl alanine lyase, catalase, peroxidase, phenols, and proline in the treated plants (Damodaran *et al.*, 2014).

Of late, interest in plant growth regulators as a tool for salinity management has considerably increased. Endogenous hormones regulate important physiological processes linked to salt tolerance. Under stress conditions, indole acetic acid (IAA) increases water use efficiency by enhancing the stomatal resistance to transpiration. Gibberellins modulate ionic balance and endogenous abscisic acid levels in the salinized plants. Cytokinins such as kinetin reduce the severity of salt stress by up-regulating endogenous polyamine levels (Rajkumar *et al.*, 2016). Paclobutrazol (PBZ), an anti-gibberellin growth retardant, has been used in many fruit crops to overcome the salt induced oxidative stress and other adverse effects. It improves the salt tolerance by regulating the plant water relations, promoting the activities of proline and antioxidant enzymes and by protecting the photosynthetic pigments (Hu *et al.*, 2017). Salicylic acid, an endogenous plant growth regulator, is also known to improve abiotic stress tolerance in plants in trace amounts (0.05 mM) mainly by stimulating the transient accumulation of IAA and ABA (Shakirova, 2007). Despite these beneficial effects, it remains to be seen whether such growth regulators exert a similar influence under field conditions as most of such experiments have been carried out under relatively controlled environment.

Nutrient Management

Majority of the fruit crops being sensitive to excess Na^+ and Cl^- ions, sodium and chloride free fertilizers are recommended to prevent any adverse impacts on orchard productivity and health. Similar to normal soil conditions, nutrient management plans under saline conditions should ideally be based on soil and leaf tests and should take into account the specific nutritional requirements at a particular crop stage. Consistent with the fact that appropriate nutrition management is a key to sustaining tree growth, fruit yield and quality, specific fertilizer management guidelines have been developed for the crops irrigated with saline water. In both Na^+ and Cl^- sensitive crops, potassium sulphate (K_2SO_4), and not potassium chloride, should be used. Fertigation with water soluble fertilizers should be done wherever possible to save the nutrients and to prevent the groundwater contamination. Some nitrogen and calcium supplements have been found to partly alleviate the salt induced injury in fruit crops.

Salt Excluder Rootstocks

Reduced salt uptake by the plant roots can considerably enhance the plant growth and yield under saline conditions. Rootstocks that either fully or partially exclude Na^+ and Cl^- ions and/or preferentially accumulate K^+ may be used for conferring salt tolerance to the scion cultivars. Of late, a few breeding projects seek to combine traits such as high water use efficiency, drought tolerance and salt tolerance into a single rootstock so that they can simultaneously tolerate multiple stresses. Certain genetic factors govern the salt tolerance of a given rootstock. For example, polyembryonic mango rootstocks often outperform the monoembryonic ones in saline soils (Bright *et al*., 2001). Tetraploid citrus seedlings may be more salt tolerant than diploid genotypes. Genome duplication leads to physiological and anatomical changes resulting in differential capacities for mineral uptake and transport. Under two NaCl levels (40 and 80 mM), diploid *Citrus macrophylla* seedlings accumulated more Na^+ and Cl^- in leaves than tetraploids. Leaf K^+ concentrations declined in diploid but not in the tetraploid seedlings obviously due to differences in nutrient absorption and translocation (Ruiz *et al*., 2016). Fine root turnover, a trait unique to citrus relative *Poncirus trifoliata*, enables the plants to continuously produce fine roots to remove the excess salt ions. This mechanism, critical in delaying the ion accumulation, seems to be transferable to related species through inter-generic hybridization (Tozlu *et al*., 2000). In citrus, Na^+ and Cl^-exclusion mechanisms are independent traits suggesting that a good Na^+ excluder would not necessarily be an effective Cl^- excluder and vice versa (Sykes, 1992).

A list of salt tolerant rootstocks identified in selected fruit crops is furnished in Table 2. This list, however, is only illustrative as several such rootstocks have been reported in other fruit crops. Moreover, ongoing researches are expected to unravel many more potential salt excluders in the future. It is worth mentioning that salt exclusion capacity in some of the widely used rootstocks breaks down when salinity exceeds the critical threshold. For example, grape rootstock Dog Ridge was introduced in India to cope up with salinity and drought stresses. Initial results suggested that non-grafted Dog Ridge roots could tolerate up to 6.5 dS m^{-1} NaCl-induced salinity. Later on, however, its ability to sustain prolonged saline irrigation became questionable spurring interest in other rootstocks (Sharma *et al*., 2011). Furthermore, Dog Ridge induces scion vigour, uneven bud sprouting and dead wood formation resulting in reduced bud fruitfulness and low fruit yield in Thompson Seedless vines (Satisha *et al*., 2016).

Table 2: Salt excluder rootstocks identified in fruit crops.

Crop	Rootstock(s)	Finding	Reference
Mango	Gomera-1	Salt stressed (EC_e ~2.5 dS m^{-1}) Osteen scion on Gomera-1 had much lower leaf Na^+ and Cl^- than on Gomera-3.	Duran Zuazo (2003)
	M. zeylanica	*M. zeylanica* had higher CO_2 assimilation and root K^+/Na^+ ratio but lower leaf/root Na^+ than *M. indica* 13-1 at 60 mM NaCl.	Schmutz (2000)
Citrus	Cleopatra mandarin	Excluded Cl^- at the lowest salinity (NaCl induced osmotic potentials of -0.10, -0.20 and -0.35 MPa) but not at higher levels; Sour orange and Rough lemon were not able to exclude Na^+/Cl^- even at -0.10 MPa.	Zekri and Parsons (1992)
	Rangpur lime	A good Cl^- excluder; salttolerance capacity of citrus rootstocks decreases in the order: Cleopatra> Rangpur lime> SB812 (*C. sunki* x *P. trifoliata*)> x639 (*C. reshmi* x *P. trifoliata*).	Boman *et al*. (2005)
Guava	Crioula	Salinity drastically suppressed seedling emergence, growth and	Sá *et al*. (2016)

Contd.

		biomass production in Paluma and Ogawa, but Crioula was least affected.	
Grape	140Ru, St. George and Schwarzmann	Strong Cl^- excluders; Ramsey (syn. Salt Creek) is not a strong Cl^- excluder as believed but simultaneously tolerates drought and moderate salt stress.	Fort and Walker (2011)
Almond	Hansen 536 & Bright's	A long term (20 y) trial revealed that peach-almond hybrids (Hansen 536, & Bright's) were far superior over peach rootstocks (Halford, Lovell, Nemaguard, and Nemared) with regard to Na^+ and Cl^- exclusion.	Doll *et al.* (2013)
Pistachio	*P. atlantica*	*P. atlantica* outperforms *P. vera* with regard to low Na^+ and Cl^- uptake, higher K^+/Na^+ and Ca^{2+}/Na^+ selectivity and ability to resume normal growth and fruiting after stress relieve.	Mehdi-Tounsi *et al.* (2017)

Salt tolerance is a complex biological trait governed by several genes whose expression is highly influenced by the environmental variables. For example, in citrus species, around 60 different traits seem to directly or indirectly influence the salt tolerance (Raga *et al.*, 2016). Thus, despite significant strides in deciphering the physiological regulation of salt stress, genetic and molecular mechanisms imparting salt tolerance in fruit crops remain elusive. This has impeded the progress in developing salt tolerant rootstocks through conventional breeding techniques.

Many studies suggest that genetic transformation could be a very effective tool to enhance salt tolerance by transferring the genes from other species. Preliminary results clearly indicate that stable expression of appropriate transgenes can remarkably increase the salt tolerance even in highly salt sensitive species. *Agrobacterium*-mediated integration of vacuolar Na^+/H^+ antiporter gene (AtNHX1) from *Arobidopsis* into kiwifruit genome considerably enhanced the salt tolerance of transgenic plants (up to 200 m mol l^{-1} NaCl). Transgenic lines outperformed wide-type plants with regard to osmotic adjustment and antioxidant activities (Tian *et al.*, 2011). Constitutive expression of a *Vitis vinifera* dehydrin (DHN) of the YSK_2 type (VvDhn) allowed for a more vigorous growth in transgenic tobacco plants than in wild-type (WT). Transgenic lines were able to germinate under 300 mM mannitol. Adult plants expressing VvDhn also displayed improved tolerance to both drought and salt stresses (Jardak-Jamoussi *et al.*, 2016). Genetic transformation of lemon rootstock *Citrus macrophylla* with *Arabidopsis* gene CBF3/DREB1A led to better growth in salinized transgenic lines in comparison to wild-type plants (Alvarez-Gerding *et al*., 2015). In spite of such encouraging results, most of the transgenic development programmes are in infancy stage and a range of biological, regulatory and public safety concerns need to be adequately addressed before the actual field application of the genetically transformed genotypes.

Irrigation Techniques

As saline and sodic soils are often also underlain with low quality water, water use efficient techniques are recommended to allow the sustained use of saline water without impairing the soil properties in the long run. In many salinity affected parts of India, about 25 % of the groundwater contains either excess dissolved salts or residual sodium carbonate rendering it unsuitable for irrigation. Nonetheless, farmers' grappling with acute fresh water shortages have no other choice but to irrigate their crops with such water further exacerbating the salinity and waterlogging problems. In many areas of Haryana, fruit growers construct the cemented farm ponds to store the rainfall/canal water for conjunctive use with saline water (Fig. 4). Available evidence suggests that this could be a sustainable practice in areas having light textured soils and moderate rainfall to leach the salts accumulated in the previous season. Adoption of drip irrigation and simple agronomic interventions such as mulching can further enhance the viability of this practice.

Fig. 4: Cemented pond for storing the fresh canal water in a rainfed saline field at Hisar, India (left) and a drip irrigated vineyard in South Australia (right) (Photo: Anshuman Singh).

Despite the fact that water is a critical resource that needs to be sustainably used, unsustainable anthropogenic practices have caused continued depletion of fresh water. Globally, 70% of the annual available water is used in agriculture to irrigate only 17% cultivated area that provides 40% of the global food production. Consistent with the fact that even small savings (~15%) in agricultural use can more than double available water for domestic and industrial use, counties like Israel have made onerous efforts to maximize the water use efficiency by the large scale adoption of micro-irrigation practices (Barak, 2014). Drip irrigation offers a number of advantages such as improved soil-water regime, partial soil wetting, and minimum salinity hazard to plants, water savings, weed control and higher yields. Very high impact of drip technology in Israel is evident from the fact that of the 1,129 Million cubic meters of water used by agriculture annually nearly 30% is wastewater and 16% saline water applied through drip (Megersa and Abdulahi, 2015). Despite multifarious benefits and policy thrust, restricted adoption of drip irrigation in India is worrisome. A study conducted in 9 Indian states revealed that small farmers see micro-irrigation as a capital intensive technology suited only to the large farmers having access to capital and technical knowhow. As a consequence, only about 12% of the potential drip irrigation area and 8% the potential sprinkler area is covered in the country with large variations across states. In order to promote the uptake of drip irrigation, reductions in capital cost of the system, provision of technical support for operation and maintenance, relaxation of farm size limitation in providing subsidies and creation of a single state level agency have become necessary (Palanisami and Raman, 2012).

Conclusion

Horticulture is increasingly being seen as a sunrise sector with huge growth potential. In many developing and underdeveloped countries, horticultural crops are being promoted to achieve the intertwined goals of sustainable food and nutritional security, equitable growth and rural prosperity. In the last few decades, India has made rapid strides in the production of horticultural crops. Globally, India is the second largest producer of fruits. Despite huge socio-economic and environmental benefits, the growth of Indian fruit industry continues to be hampered by different constraints. Low productivity of fruits vis-à-vis other major producers, huge post-harvest losses and negligible presence in global export market are some of the factors hindering India from becoming a leader in the global fruit trade. Majority of the fruit crops are highly sensitive to biotic and abiotic stresses with about three-fourths of the fruit crops listed known to be highly sensitive to salinity and the related soil constraints. As further expansion of area under fruit crops seems possible only in marginal situations, this paper attempts to shed light on the properties of the marginally productive salt-affected soils, mechanisms underpinning salt tolerance and doable interventions for growing fruit crops under saline conditions.

Taking a serious note of the agrarian distress adversely affecting the farmers' livelihoods, Government of India has set the target of doubling the farmers' income by the year 2022. Efforts to improve the land productivity and assured remunerative returns to the farmers will be the keys to realize this goal. Previous attempts to enhance the farm profits have unwittingly led to land degradation in different parts of the country. Accordingly, agricultural technologies capable of providing higher incomes without jeopardizing the environmental sustainability are needed. Despite many efforts, farm diversification through horticultural crops remains sluggish. Given the fact that future expansion in area under fruit crops will only be possible in marginal areas suffering from problems such as salinity and fresh water scarcity, feasible interventions such as improved planting methods, use of amendments, salt tolerant rootstocks and drip irrigation are suggested to enhance the salt tolerance in high value fruit crops. Recent innovations such as genetic engineering provide a new opportunity to develop multiple stress tolerant genotypes capable of simultaneously tolerating two or more stresses; albeit with caution. Continual refinements in existing technologies are equally important to unlock the economic potential of saline lands through fruit crops.

References

Abrol, I. P., Yadav, J. S. P. And Massoud, F. I. 1988. *Salt-affected soils and their management,* FAO Soils Bulletin 39. Rome: Food & Agriculture Organization of the United Nations.

Alvarez-Gerding, X., Espinoza, C., Inostroza-Blancheteau, C. and Arce-Johnson, P. 2015. Molecular and physiological changes in response to salt stress in Citrus macrophylla W plants overexpressing Arabidopsis CBF3/DREB1A. *Plant Physiology and Biochemistry*, 92: 71-80.

Arumugam, R. and Manikandan, M. 2011. Fermentation of pretreated hydrolyzates of banana and mango fruit wastes for ethanol production. *Asian Journal of Experimental Biological Sciences*, 2: 246-256.

Balan, P. 2017. Kesar farmers to reap fruits of Israel technology. http://timesofindia.indiatimes.com/city/ahmedabad/kesar-farmers-to-reap-fruits-of-israel-tech/articleshow/59451835.cms

Barak, N. 2014. Drip irrigation- Israeli innovation that has changed the world. http://www.maala.org.il/wp-content/uploads/2014/06/course-13-nettafim.pdf

Bernstein, N., Meiri, A. and Zilberstaine, M. 2004. Root growth of avocado is more sensitive to salinity than shoot growth. *Journal of the American Society for Horticultural Science*, 129: 188-192.

Bhargava, B. S., Kalbhor, J. N., Deshmukh, S. U. and Sharma, J. 2006. Deteriorating ground water quality used for irrigating grapes. *Indian Journal of Horticulture*, 63: 235-239.

Blumwald, E. 2000. Sodium transport and salt tolerance in plants.*Current Opinion in Cell Biology*, 12: 431-434.

Boman, B. J., Zekri, M. and Stover, E. 2005. Managing salinity in citrus. *Hort Technology*, 15: 108-113.

Braidotti, G. 2013. Fruit and vegetables boost economic health. http://aciar.gov.au/files/node/14983/03fruitandvegetablesboosteconomichealth_pdf_13169.pdf.

Bright, J., McAlister, S. and Renfre, R. 2001. Rootstock effects on mango productivity. *Horticulture Technical Annual Report* 2000-2001.

Bung, P. 2017. Indian Fruit processing Industry: Import and Export Analysis. *DHARANA-Bhavan's International Journal of Business*, 6: 72-86.

Chand, R. 1999. Emerging crisis in Punjab agriculture: Severity and options for future. *Economic and Political Weekly*, A2-A10.

Cramer, G. R., Ergül, A., Grimplet, J., Tillett, R. L., Tattersall, E. A., Bohlman, M. C., Delphine Vincent, D., Sonderegger, J., Evans, J., Osborne, C., Quilici, D., Schlauch, K. A., Schooley, D. A. and Cushman, J. C. 2007. Water and salinity stress in grapevines: early and late changes in transcript and metabolite profiles. *Functional and Integrative Genomics*, 7: 111-134.

Dagar, J. C., Singh, G and Singh, N. T. 2001. Evaluation of forest and fruit trees used for rehabilitation of semiarid alkali-sodic soils in India. *Arid Land Research and Management*, 15: 115-133.

Dalton, F. N., Maggio, A. and Piccinni, G. 2001. Assessing the effect of solar radiation on plant salt tolerance as deûned by the static and dynamic indices. *Plant and Soil*, 229: 189-195.

Damodaran, T., Rai, R. B., Jha, S. K., Kannan, R., Pandey, B. K., Sah, V., Mishra, V.K. and Sharma, D. K. 2014. Rhizosphere and endophytic bacteria for induction of salt tolerance in gladiolus grown in sodic soils. *Journal of Plant Interactions*, 9: 577-584.

Doll, D. A., Duncan, R. A., Hendricks, L. an Micke, W. 2013. Performance of six almond rootstocks with long-term exposure to sodium and chloride. In: Proceedings of *VI International Symposium on Almonds and Pistachios, 1028*, pp. 273-277.

Duran Zuazo, V. H., Martinez-Raya, A. and Aguilar Ruiz, J. 2003. Salt tolerance of mango rootstocks (*Mangifera indica* L. cv. Osteen). *Spanish Journal of Agricultural Research*, 1: 67-78.

FAO. 2013. Statistical Yearbook- World Food and Agriculture. Food and Agriculture Organization of the United Nations Rome.

Fort, K. and Walker, A. 2011. Breeding salt tolerant rootstocks. *FPS Grape Program Newsletter*, 9-11.

Gill, A. and Singh, L. 2006. Farmers' suicides and response of public policy: evidence, diagnosis and alternatives from Punjab. *Economic and Political Weekly*, 2762-2768.

Hu, Y., Yu, W., Liu, T., Shafi, M., Song, L., Du, X., Huang, X., Yue, Y. and Wu, J. 2017. Effects of paclobutrazol on cultivars of Chinese bayberry (*Myrica rubra*) under salinity stress. *Photosynthetica*, 55: 443-453.

ICAR-CSSRI. Annual Report 2016-17. ICAR-Central Soil Salinity Research Institute, Karnal-132001, India.

ICAR-IIHR. 2015. Vision 2015. ICAR-Indian Institute of Horticultural Research (Indian Council of Agricultural Research), Bengaluru, India.

Jardak-Jamoussi, R., Zarrouk, O., Salem, A. B., Zoghlami, N., Mejri, S., Gandoura, S., . Khiari, B., Mliki, A., Chaves, M., Ghorbel, A. and Pinheiro, C. 2016. Overexpressing *Vitis vinifera* YSK 2 dehydrin in tobacco improves plant performance. *Agricultural Water Management*, 164: 176-189.

Maas, E.V. and Hoffman, G.J. 1977. Crop salt tolerance- current assessment. *Journal of Irrigation and Drainage Division, ASCE* 103: 115-134.

Megersa, G. and Abdulahi, J. 2015. Irrigation system in Israel: A review. *International Journal of Water Resources and Environmental Engineering*, 7: 29-37.

Mehdi-Tounsi, H., Chelli-Chaabouni, A., Mahjoub-Boujnah, D., & Boukhris, M. (2017). Long-term field response of pistachio to irrigation water salinity. *Agricultural Water Management*, 185: 1-12.

Melgar, J. C., Syvertsen, J. P., Martínez, V. and García-Sánchez, F. 2008. Leaf gas exchange, water relations, nutrient content and growth in citrus and olive seedlings under salinity. *Biologia Plantarum*, 52: 385-390.

Mohan, M. D. 2015. Assessment of Irrigation Water Quality for Grape Gardens in Western Maharashtra. M. Sc. Thesis, MPKV, Rahuri, Maharashtra.

Mukherjee, A., Dutta, S. and Goyal, T. M. 2016. India's Phytonutrient Report: A Snapshot of Fruits and Vegetables Consumption, Availability and Implications for Phytonutrient Intake. Academic Foundation, New Delhi, p. 100.

Munns, R., Husain, S., Rivelli, A. R., James, R. A., Condon, A. G., Lindsay, M. P., Lagudah, E. S., Schachtman, D. P. and Hare, R. A. 2002. Avenues for increasing salt tolerance of crops, and the role of physiologically based selection traits. *Plant and Soil*, 247: 93–105,

Munns, R. 2002. Comparative physiology of salt and water stress. *Plant, Cell and Environment*, 25: 239-250.

Nathwani, P. 2014. Reliance Industries Limited- Reliance Mangoes from Jamnagar. http://www.parimalnathwani.com/reliance-industries-limited-reliance-mangoes-from-jamnagar.

Navarro, J. M., Andujar, S. , García-Olmos, B., Rodríguez-Morán, M., Morte, A. and Pérez-Tornero, O. 2011. Arbuscular mycorrhizal fungi influence the response of citrus rootstock seedlings to salinity. *Acta Horticulturae*, 922: 245-252.

Palanisami, K. and Raman, S. 2012. Potential and Challenges in Up-Scaling Micro-Irrigation in India- Experiences from Nine States. IWMI-Tata Water Policy Program, http://www.iwmi.cgiar.org/iwmi-tata/PDFs/2012_Highlight-20.pdf.

Raga, V., Intrigliolo, D. S., Bernet, G. P., Carbonell, E. A.and Asins, M. J. 2016. Genetic analysis of salt tolerance in a progeny derived from the citrus rootstocks Cleopatra mandarin and trifoliate orange. *Tree Genetics and Genomes*, https://doi.org/10.1007/s11295-016-0991-1.

Rajkumar, Singh, Anshuman and Yadav, R. K. 2016. Nursery management in fruit crops in salt-affected soils. In: Quality seed production, processing and certification of selected field and vegetable crops in salt affected areas. Training Manual, ICARCSSRI, Karnal, India. P. 125-131.

Ruiz, M., Quiñones, A., Martínez-Alcántara, B., Aleza, P., Morillon, R., Navarro, L., Primo-Millo, E. and Martínez-Cuenca, M. R. 2016. Effects of salinity on diploid (2x) and doubled diploid (4x) Citrus macrophylla genotypes. *Scientia Horticulturae*, 207: 33-40.

Sá, F. V. D. S., Nobre, R. G., Silva, L. D. A., Moreira, R. C., Paiva, E. P. D., & Oliveira, F. A. D. (2016). Tolerance of guava rootstocks under salt stress. *Revista Brasileira de Engenharia Agrícola e Ambiental*, 20: 1072-1077.

Salama, R. B., Otto, C. J. and Fitzpatrick, R. W. 1999. Contributions of groundwater conditions to soil and water salinization. *Hydrogeology Journal*, 7: 46-64.

Satisha, J., Somkuwar, R. G., Sharma, J., Upadhyay, A. K., & Adsule, P. G. (2016). Influence of rootstocks on growth yield and fruit composition of Thompson seedless grapes grown in the Pune region of India. *South African Journal of Enology and Viticulture*, 31: 1-8.

Saxena, C. K. and Gupta, S. K. 2006. Uniformity of water application under drip irrigation in litchi plantation and impact of pH on its growth in partially reclaimed alkali soil. *ISAE Journal of Agricultural Engineering*,. 43: 1-9.

Schmutz, U. 2000. Effect of salt stress (NaCl) on whole plant CO_2 gas exchange in mango. *Acta Horticulturae*, 509: 269-276.

Shabala, S. and Munns, R. 2012. Salinity stress: physiological constraints and adaptive mechanisms. In: *Plant Stress Physiology* (S. Shabala Ed.), CAB International, pp. 59-93.

Shakirova F.M. 2007. Role of Hormonal System in the Manifestation of Growth Promoting and Antistress Action of Salicylic Acid. *In*: Salicylic Acid: A Plant Hormone (Hayat S. and Ahmad A. Eds). Springer, Dordrecht.

Sharma, D. K. and Singh, A. 2015. Salinity research in India-achievements, challenges and future prospects. *Water and Energy International*, 58: 35-45.

Sharma, D. K., Singh, A. and Sharma, P. C. 2016a. Role of ICAR-CSSRI in sustainable management of salt-affected soils—achievements, current trends and future perspectives. In: The Proceedings of the 4th International Agronomy Congress (November 22-26, 2016a), New Delhi, India, p. 91-103.

Sharma, D. K., Singh, A., Sharma, P. C., Dagar, J. C. and Chaudhari, S. K. 2016b. Sustainable Management of Sodic Soils for Crop Production: Opportunities and Challenges. *Journal of Soil Salinity and Water Quality*, 8: 109-130.

Sharma, J., Upadhyay, A. K., Bande, D. and Patil, S. D. 2011. Susceptibility of Thompson Seedless grapevines raised on different rootstocks to leaf blackening and necrosis under saline irrigation. *Journal of Plant Nutrition*, 34: 1711-1722.

Singh, A., Kumar, A., Yadav, R. K., Datta, A. and Sharma, D. K. 2016. Growth and mineral nutrition in salt stressed guava (*Psidium guajava* L. cv. Allahabad Safeda). *Journal of AgriSearch*, 3: 21-25.

Singh, G., Dagar, J. C. and Singh, N. T. 1997. Growing fruit trees in highly alkali soils—a case study. *Land Degradation & Development*, 8: 257-268.

Sohail, M., Saied, A. S., Gebauer, J. and Buekert, A. 2009. Effect of NaCl salinity on growth and mineral composition of *Ziziphus spina-christi* (L.) Willd. *Journal of Agricultural and Rural Development in the Tropics and Subtropics*, 110: 107-114.

Srinivasa Rao, C. H., Lal, R., Prasad, J. V. N. S., Gopinath, K. A., Singh, R., Jakkula, V. S., Sahrawat, K. L., Venkateswarlu, B., Sikka, A. K. and Virmani, S. M. 2015. Potential and challenges of rainfed farming in India. *Advances in Agronomy*, 133: 115-172.

Sykes, S. R. 1992. The inheritance of salt exclusion in woody perennial fruit species. *Plant and Soil*, 146: 123-129.

Tavakkoli, E., Rengasamy, P. and McDonald, G. K. 2010. High concentrations of $Na^{+}$and Cl^{-}ions in soil solution have simultaneous detrimental effects on growth of faba bean under salinity stress. *Journal of Experimental Botany*, 61: 4449-4459.

Teakle, N. L. and Tyerman, S. D. 2010. Mechanisms of Cl transport contributing to salt tolerance. *Plant, Cell & Environment*, 33: 566-589.

Tian, N., Wang, J. and Xu, Z. Q. 2011. Overexpression of Na+/H+ antiporter gene AtNHX1 from Arabidopsis thaliana improves the salt tolerance of kiwifruit (*Actinidia deliciosa*). *South African journal of botany*, 77: 160-169.

Tozlu, I., Moore, G. A. and Guy, C. L. 2000. Effects of increasing NaCl concentration on stem elongation, dry mass production, and macro-and micro-nutrient accumulation in Poncirus trifoliata. *Functional Plant Biology*, 27: 35-42.

Waters, S., Gilliham, M. and Hrmova, M. 2013. Plant high affinity potassium (HKT) transporters involved in salinity tolerance: structural insights to probe differences in ion selectivity. *International Journal of Molecular Sciences*, 14: 7660-7680.

Zekri, M. and Parsons, L. R. 1992. Salinity tolerance of citrus rootstocks: effects of salt on root and leaf mineral concentrations. *Plant and Soil*, 147: 171-181.

Zhu, J. K. 2007. Plant salt stress. *Encyclopedia Of Life Sciences,* John Wiley & Sons, Ltd. (www.els.net), p. 3.

15

Date Palm Cultivation in Arid Ecosystem

R. S. Singh and Ramkesh Meena

Introduction

Date palm (*Phoenix dactylifera* L.; Family- Arecaceae) also known as *Khajoor*, *Kharek* is an ancient important fruit tree of semi-arid and arid regions of the world. It grows well under poor desertic soils due to its hardy plant characteristics and deep root system. It is also suitable fruit trees for growing under saline irrigation conditions. It is well known fact for date palm that its feet in the water and its head in sun. It requires dry hot climate for growth and development of fruits. In arid region, crop production is risky preposition, where date palm cultivation contributes in achieving food security, high nutritive value food, crop diversification, desertification control, higher income generation and foreign exchange earnings. It can be grown at such places where adequate irrigation facility is available besides other dry hot climate conditions (Chandra *et al,* 1992; Singh and Dhandar, 2007). Better returns from date plantation can be achieved through good management practices besides adequate marketing of produce. The quality produce and its marketing will help to increase the income from date palm cultivation.

Date fruit is a highly nutritious and favorite fruit throughout the world. Fruits of date cultivars viz. Halawy, Khalas, Khuneizi, Chip chap, Braim and Barhee are eaten as fresh (*doka* stage). Besides fresh consumption, several value added products dry date, pind, beverages, jam, paste, chutney, etc are prepared from fruits. The date-palm fruits have been recognized for high nutritive and energy food. It is known to produce more well mineralized, highly flavored and nutritious food per hectare than any other fruit of even temperate zone. The mineral content of dates amounts to about 2 per cent of cured fruits on weight basis. It is a good source of iron, potassium and fair source of calcium but a poor source of phosphorus (Nixon, 1966).

This fruit can supplement the dietary needs of the desert people where very less nutritious food is available. Dates are cured as dry dates (*Chhuhara*) and soft dates (*Pind Khajoor*), which has great demand and market in the country. Date fruits provide abundant quantities of sugar, iron, potassium, calcium and nicotinic acid. Small amounts of protein, copper, magnesium, chlorine, and sulphur, Vitamins A, B_1 and B_2 are also present in date pulp. The flesh of dates with a moisture content of 20 per cent contains 60-65 per cent sugar, about 2.5 per cent fiber, 2 per cent protein and some what less than 2 per cent fat, mineral matter and pectin substances, provides about 3,550 calories per kg (Nixon and Carpenter, 1978). Fresh 100g date fruit contains 59.2 g moisture, 1.2 g protein, 0.4 g fat, 1.7 g minerals, 3.7g fiber, 33.8 g carbohydrates, 22.0 mg calcium, 38.0 mg phosphorus while 100 g dry dates (*chhuhara*) provides 2.5 g protein, 2.1g minerals, 3.9g fibers, 75.8g carbohydrates, 120 mg calcium, 50 mg phosphorus, 7.30 mg iron, 0.02 mg riboflavin, 0.9 mg Niacin, 3.0 mg vitamin C (Gopalan *et al*, 1985).

Date-palm leaves are used for preparation of paper, baskets, mats and brooms. The tree trunk provides timber for making furniture. The date palm tree also furnishes shelter, clothing, fuel, building material, fiber, starch, sugar, oil, wax and many other products. Wine is also prepared from date fruits. It has also several medicinal properties. Date tree is also mentioned on coins/currency as symbol in Arab countries. It has religious significance also in Muslim religion. Due to better keeping quality of dry and soft dates and their increasing demand, it does not face the problem of domestic consumption. Besides, *doka* dates, *chhuhara* (dry dates) and *pind khajoor* (soft dates) are in high demand for raw eating (Vishal Nath and Singh, 2003). Date fruits are also used for making squash, pickles, chutney, beverage and syrup. Its delicious fruits at doka (khalal) stage are eaten as fresh and used for processing to different value added products (Singh *et al*, 2004). Caramel colour has been produced from dates which can be used in food and pharmaceuticals industries. Moreover, all plant parts of date palm are useful since its history of cultivation (Chao and Krueger, 2007) in several ways to make handicraft items. Its seed kernel is also used for making cattle's and fish feed.

Present Status in India and Global Scenario

Date palm requires specific climate for flowering, fruiting and fruit quality hence, in India it is mainly grown in the states of Rajasthan, Gujarat, Punjab, Haryana, and Tamil Nadu however, the exact area and production data is not available. The extremely dry hot areas comprising Jaisalmer, Barmer, Hanumangarh, Sri Ganganagar, Bikaner, Churu, Nagaur and Jodhpur in Rajasthan, Bhatinda, Fazlika, Abohar in Punjab, Sirsa, Hisar, Mahendergarh, Bhiwani in Haryana and Anjar,

Bhachau, Mandvi, Mundra, Bhuj, Kachchh region in Gujarat are the potential areas for date palm cultivation in India.

The major dates growing areas in India are Kachchh region (Gujarat), Rajasthan and certain parts of the Punjab, Haryana as well as Tamil Nadu to some extent. In Kachchh region, there are more than 1.9 million palms, the majority of them grown from seeds and offshoots, providing a huge biodiversity for crop improvement programmes. Now, tissue culture plantation is coming very well in the region particularly of cv. Barhee (Muralidharan *et al,* 2008). The performance of cv. Barhee was excellent in Kachchh region in comparison to other genotypes with respect to fruit yield and quality. The maximum return Rs.992.70 per palm was observed in variety Barhee due to sweet in taste and soft at Khalal stage (Ramdevputra *et al*, 2009). Tissue culture plants of Local elite type are also produced by Tissue culture Lab at AAU, Anand and KCS Ltd., Mundra, Gujarat for sale purpose and these are being planted by farmers. Though, the cost of planting material is very high @2500 -3750/- per plant depending on varieties.

Table 1: Area and production of date palm in India

S. No.	States	Area (in ha)	Production (MT)
1.	Rajasthan	850.0	800.00
2.	Gujarat	18847	173997.66
3.	Punjab	8.0	—
4.	Haryana	2.0	—
5.	Tamil Nadu	809.0	—

In Rajasthan, earlier area under date palm cultivation was about 10-20ha, only now it has increased by planting of imported tissue culture plants in about 833.43ha in districts of Jaisalmer, Barmer, Jodhpur, Sri Ganganagar, Hanumangarh, Churu, Pali, Sirohi and Bikaner (Table-2). Further, date palm is also planted in about 20ha area at research farms, Bikaner, Surat garh, Jetsar, Jodhpur, and Jaiselmer. At Bhojka (Jaisalmer), plants of cultivars viz., Barhee, Khadrawy, Khuneizi, Khalas, Medjool, Zamali, Saggai and male palms (cv. Madsari and Ghanami) have been planted during 2009-10 and have started flowering/ fruiting. At Khara Farm, Bikaner total of 5,268 plants have planted with same cultivars during August 2010 by M/s Atul Ltd, under aegis of Rajasthan Horticulture Development Society, Jaipur. For promotion of date palm in the state, Government of Rajasthan is providing subsidy on tissue culture plating materials and drip irrigation systems. On farmer's field in the following districts of western Rajasthan, date palm plantation has also done and is started flowering/ fruiting.

Table 2: Area of date palm cultivation in Rajasthan (2015-16):

S. No	Districts	Area(in ha)	Varieties
1.	Sriganganagar	153.13	Barhee, Khuneizi, Khalas, ,Medjool,
2.	Hanumangarh	120.00	Barhee and Medjool,
3.	Churu,	9.00	Barhee,
4.	Bikaner	210.00	Barhee, Khadrawy, Khuneizi, Khalas, Medjool, Zamali, Saggai and male plants
5.	Jaisalmer	189.00	Barhee, Khadrawy, Khuneizi, Khalas, Medjool, Zamali, Saggai and male plants
6.	Barmer	64.50	Barhee, Khuneizi, ,Medjool,
7.	Jhunjhunu	3.50	Barhee, Khuneizi, Khalas, ,
8.	Nagaur	29.50	Barhee, Khuneizi, Khalas, ,
9.	Jodhpur	33.00	Barhee, Khuneizi, Khalas, ,
10.	Pali	7.00	Barhee, Khuneizi,
11.	Sirohi	3.80	Barhee,
12.	Jalore	11.00	Barhee, Khuneizi, Khalas,

In Kachchh region of Gujarat, the maximum date palms are natural and as grove in depressions, on riverbank, *nala* and boundary of the farms. In Gujarat the major area of date palm cultivation is in the coastal area of Bhuj, Anjar, Khedoi, Mundra, Mandavi, Gandhidham, *etc*. During the year 2015-16, the area of date palm cultivation in state has increased to 17658 ha with production of16,5632 MT fruits with productivity of 9.38 MT/ha during 2015-16 and 18847 ha and 173990.6MT (2016-17) (Table- 3). The tissue culture plants of cultivar Barhee imported from UK through UAE have been planted on farmers' field in Kachchh district and started fruiting after 3-4 years of planting. In these plantations, suckers production is not allowed by the growers for better growth and early fruiting. The entrepreneur Mr. Rahul Gala Shah founded a Society in 2006 for date palm farming. KDDC is an association of farmers registered under a society act (1960) based in Bhuj, Gujarat and is the largest organization of Barhee date growers in India. A total of 25,000 offshoots/ tissue culture plants were procured from Israel, Jordan and United Arab Emirates (UAE). Now, date palm cv. Barhee plantation has spread over throughout Kachchh and is started production of fruits. Further, the date palm planting material is also spread over in other states from Gujarat.

Table 3: Area and production of date palm in Kachchh region of Gujarat (2015-16):

S. No.	Taluka/districts	Area (in ha)	Production (MT)
1.	Abdasa	470	4376
2.	Anjar	2257	27049
3.	Bhachau	677	5503
4.	Bhuj	1216	10521
5.	Lakhpat	209	1946
6.	Mandvi	1051	8985
7.	Mundra	10448	96471
8.	Nakhtrana	706	5773
9.	Rapar	490	3762
10.	Gandhidham	134	1248
	Total	17658	165632
	Productivity (MT/ha)	9.38	

The area under tissue culture plantations have increased in Kachchh region, Gujarat state under Government financially assisted a programme. A private organization M/s Kutch Crop Services Ltd., Gajod C/ o Excel Crop Care Ltd., Mundra has supplied date palm tissue culture plants either imported cv. Barhee or produced Local elite type plants at Tissue culture laboratory, Mundra.

In Punjab, the area under date palm cultivation is limited and few plantations have made through off-shoots and seedlings. The very best dates known can be grown at Montgomery, Multan, Khushab, Sindh and Lahore (now in Pakistan) and parts of Ludhiana district (in India). The old date plantations are in Abohar, Fazlika and Bhatinda. About 11 acre area of date palm is located at RFRS, PAU, Abohar/ some farmers have planted date palm through offshoots supplied by PAU, Abohar centre. The tissue culture plantation of date palm is very limited in Punjab.

In Haryana, the farmers of arid parts in the districts of Sirsa, Hisar, etc. have initiated planting of tissue culture plants of cv. Barhee. However, few old date palm plantations exist in Hisar, Haryana. The climatic conditions in some parts of Haryana are suitable for date cultivation.

Tamil Nadu is an important state in southern India where climate, soil and irrigation facilities in certain coastal regions such as Tirunelveli, Tuticorin and Ramanathpuram districts favour date palm cultivation. About 809 ha in Tamil nadu are cultivated with Arabian dates. The first plantlets for commercial growth were procured over 15 years ago at a cost of INR 3,250. (USD 54)/palm from Saudi Arabia. Mr. Nizamuddin is the first and foremost person of date palm cultivation in South India. He brought from Saudi Arabia cultivars like Aguah, Makthumii, Sukkary, Katima and Muscat and planted them in Dharmapuri District. Out of 100 seedlings only 7 survived including 2 males. After 14 years an average yield of 150 kg of dates per palm per year is obtained (Dates India, 2011).

At present, in India date palm production is not meeting our domestic demand and maximum dates are imported from date-growing countries of the world. India imports about 311575 MT date from UAE, Pakistan, Egypt, etc, of worth 183169,000 US $. Under National scenario, import of dates can be reduced up to some extent by expansion in area and quality production by planting of date palm in suitable areas of Rajasthan, Gujarat, Punjab, Haryana and Tamil Nadu states.

Table 4: Import of dates in India.

Units	2007	2008	2009	2010	2011	2012	2013
Quantity (MT)	253341	230926	298423	193467	256295	247300	311575
Value 1000US $	72,359	81318	99059	95043	141713	112649	183169

(Source: FAO Stat database)

Further, India also exports a small quantity of dates and processed products to neighboring countries like Nepal, Myanmar, Bhutan, Maldives and Sri Lanka amounting 635,000 US $.

Table 5: Export of dates from India.

Units	2007	2008	2009	2010	2011	2012	2013
Quantity (ton)	284	1303	1769	3115	627	442	531
Value 1000US $	194	546	1792	6494	1134	584	635

(Source: FAO Stat database)

At International level, commercial date palm growing countries of the world are Iraq, Saudi Arabia, Algeria, Iran, Egypt, Morocco, Tunisia, Pakistan and Sudan besides many other Near East and North African countries. Date palm has played a vital role in the history of mankind by providing food and shelter to millions of people besides economy of the country (Zaid, 1999). Date fruits are major source of food and income of local population in the Middle East and North Africa and play significant roles economy, society and environment in these countries. Pakistan is the fourth largest date producing country, contributing 11% to the world production. In India date palm production is not meeting our domestic demand and maximum dates are imported from date-growing countries of the world. However, India also exports small quantity of dates to neighboring countries like Nepal, Myanmar, Bhutan, Maldives and Sri Lanka.

The commercial date palm growing countries of the world are Egypt, Iraq, Saudi Arabia, Pakistan, Algeria, Iran, Morocco, Tunisia, and Sudan besides many other Near East and North African countries. Pakistan is the fourth largest date producing country, contributing 11% to the world production. Now, it is on the sixth rank in date's production in the world .The date palm cvs Begum Jangi and Dhakki are popular and high yielding varieties in Pakistan. The other countries which are producing dates are Libya, Tunisia, Sudan, Muscat, Oman, United States of America, Behrain and Mauritiana. Dates are also produced in small quantities in Spain, Chad, Mexico, Yemen, Israel and Somalia. These countries have about 60 million palms out of which Iraq alone has 20 million and producing 80% of total world's production. Apart from above a few date palm plantations exist at Peru, Brazil, Argentina, Namaland Territory (South Africa), Australian desert in Queensland and Northern Territory, Cyprus, Ethiopia and Tanzania.

Table 6: Countries production of date palm in world (2013)

S. No.	Countries	Production (tonne)	Value (Million $)
1.	Egypt	1328468	766.974
2.	Iran	1148500	553.460
3.	Saudi Arabia	1095158	543.916
4.	Algeria	848199	433.178
5.	Iraq	676111	273.794
6.	Pakistan	526749	269.012
7.	Sudan	437835	223.604
8.	Oman	308400	133.038
9.	United Arab Emirates	237864	79.159
10.	Tunisia	195000	89.620
11.	Libya	170224	63.347
12.	China, mainland	153803	76.605
13.	Morocco	111924	38.000
14.	Yemen	54197	27.678
15.	Israel	45195	23.081
16.	Turkey	33232	16.971
17.	Qatar	31182	11.286
18.	Kuwait	31180	16.971
19.	USA	27669	11.117
20.	Mauritania	22482	—

Constraints

The major constraints in expansion of date palm in our country are availability of quality planting materials on reasonable cost. At present, the planting material is available @2500 to 3700/- per plant depending on variety of plant which is too high rate from input cost point of view.

Standardization of agro-techniques for different agro-ecological conditions and sufficient irrigation water is required for quality fruit production.

Irrigation is absolutely necessary for date palm production, because it is grown in hot and dry, low rainfall areas. The water requirement of date palm is also high, under such circumstances date palm cultivation could not be extended without using water use efficiency devices in hot arid zone.

At present, farmers of arid region are facing the problem of availability of pollens due to less number of male palms and delay in flowering pattern since last 2-3 years. Ghanami male plants are better than AL-Ain City (Madsari) in respect of pollen production as informed by the date growers in arid region. Further, male plants population is also very less (2-3%). Hence, the farmers are purchasing pollen from available sources @ 8000/- to 10,000/- per kg. This input cost can be reduced by planting of male palms on the boundary of farm/ fields.

In tissue culture -derived date plants, morphological variation, stunted growth in plants has also reported by farmers and poor fruit set problem in some of the variety plants. Morphological abnormalities in tissue culture- derived plants have also reported in literature by many workers.

Another constraint faced by farmers is, no proper market and storage facilities of fruits in the region. Further, there is a need to develop processing unit in the date growing areas for proper utilization of *doka* fruits and to explore export potential. More emphasis should be given on storage, marketing and post harvest management aspect to achieve more income from the date palm plantations in arid region.

Techniques for doubling income from date palm cultivation

- Grow high yielding varieties
- Cropping system for proper utilization of land
- Inputs Management for production of quality fruits
- Cleaning, Grading and packaging
- Processing for value added products
- Storage
- Marketing
- Exploitation of export potential/market

Genetic Diversity and Status of Germplasm

All the commercial date cultivars have developed through selection of chance seedlings based on local needs. A rich genetic diversity is available in costal belt of Kachchh region (Gujarat) in India which should be exploited. In other parts of country, seedlings are less in population. At CIAH, Bikaner, 64 genotypes/ cultivars have been conserved in National field gene bank collected from different available sources (Annon, 2016-17) and evaluated for growth performance and fruiting characters including three germplasm introduced from ICARDA, Amman, Jordan during the year 2015. One late maturing, red berry colour and rain tolerant genotype has marked/identified at CIAH, Bikaner (Annon.2016-17). Further, SDAU, DRS, Mundra, (Gujarat); CAZRI, Jodhpur, SKRAU, Bikaner, PAU, RFRS, Abohar, (Punjab); Central State Farm, Jetsar, Sri Ganganagar, Surat garh Farm, has also maintained and evaluated the date palm varieties (table-7).

Table 7: Status of date palm germplasm in India

S.No.	Centre	Nos.	Suitable varieties/genotypes
1	ICAR-CIAH, Bikaner	64	Halawy, Khalas, Zahidi, Khuneizi, Chipchap
2.	DRS, SKRAU, Bikaner	35	Halawy, Barhee, Khalas, Zahidi, Khuneizi,
3.	RFRS, PAU, Abohar, Punjab	34	Halawy,Barhee, Medjool, Zahidi,
4.	DRS,SDAU, Mundra, (Gujarat)	18	Halawy, Barhee and Local red type
5.	CAZRI, Jodhpur	18	Halawy
6	Centre of Excellence, Bhojka, Jaiselmer, Rajasthan	09	Barhee, Khadravi, Khalas, Jamli, Saggai, Khuneizi, Medjool and 02 male cultivar.

A large number of varieties and some promising selections (Sel.- 9, Sel.-13, Yaqubi, Kotho, Trofo, Bhukso, Gulchati, Madhepura, Khedoi-7, Sopari, Saidy,) have been made in our country from natural populations existing in Kachchh region. Some selections have also been made by the farmers of the Kachchh region. Further, these yellow and red berry colour types are suitable for making different processed products (Singh *et al*, 2004; Pareek and Sodagar, 1986). An elite type of green colour, sweet berry at *doka* stage has been identified in Kachchh region at farmer's field. M/s Kachchh Crop Service (KCS) has also identified elite type date palm on farmers field and producing tissue culture plants of these genotypes.

Varieties

More than 1000 varieties/genotypes are available in the world. The cultivars (34) imported from different countries have been successfully grown at Date palm Research Station, SKRAU, Bikaner but only Halawy, Barhee, Khuneizi, Medjool, Zahidi, Khalas and Sewi have shown better performance. In Kachchh region, the performance of variety Barhee is found most suitable and is covering maximum area under cultivation (Ramdevputra *et al*, 2009). Under Jodhpur condition, cultivar Halawy is found suitable for early maturity, ripening and better yield. There is no hybrid developed in date palm and maximum available varieties have developed through selection.

The varieties have been classified according to fruit maturity (early/medium/late) and uses. The descriptions of some important varieties are as follows:

Halawy: An early maturing, regular bearing high yielding and suitable variety to grow in large areas for fresh fruit and processing. Its fruits are yellow and sweet in taste at doka stage with an average weight 9.6g; size 3.56 x 2.20 cm and TSS 32-40^{0}brix. It produces about 70- 80 kg *doka* fruits/tree/year.

Barhee: Its origin is Basrah, Iraq. It is a medium to late maturing cultivar and good for fresh consumption. In this variety, sucker production is low. The fruit at khalal stage (doka) has excellent flavour, yellow and attractive fruit colour and sweet in taste. The stone is very small in size with high pulp: stone ratio. It has high market potential in the world's market. The yield potential of this variety is 80-120 kg per plant which goes up to 200 kg per palm.

Medjool: The origin of this variety is Morocco from where it was introduced to the other countries of the world. It is No. 1 variety in the world and very high yielding cultivar due to its bigger attractive colour berries (size 4.5 x 3.2 cm). It is also known as Mejhool. It also matures late and it is highly suitable for dry date productions. Its doka fruits are astringent in taste. The colour of berry is yellow-orange at doka stage. It has high commercial value and most important variety for export purposes. Average yield is 80-100 kg/palm.

Khalas: It is a medium maturing type cultivar. Its berry is big in size at doka stage and sweet in taste. The yellow colour doka fruit is suitable for fresh eating and processing. This variety is medium to heavy yielder (70-80kg/tree).

Khuneizi: It is an early to mid season maturing variety. The colour of berry is dark red and suitable for fresh consumption at doka stage. The doka fruits are very sweet in taste and flavour. The yield potential of this variety is 50-60 kg/tree but market price is high due to earliness and sweetness.

Zahidi: It is a late maturing variety and suitable for pind preparation. Its doka fruits are astringent in taste and solid with hard skin. The size of berry is small, attractive yellow colour but it is a heavy yielder (125 kg doka fruits/tree). The bunch of this variety is compact with 20 to 25 berries per strand.

Shamran: It is a mid season to late maturing and high yielding variety. The average fruit weight is 6.5 g and fruit size of 3.50 x 2.18 cm. The colour of berry is light yellowish red and astringent at doka stage. This variety is grown commercially. It is suitable for processing in to dry dates and other value added products. It is a prolific bearer. The yield potential of the variety is 80-100 kg *doka* fruits/ tree.

Chipchap: This variety has been introduced in India from Iraq during 1997 and suitable for fresh as well as for processing purpose. Its fruits are yellow, bigger in size and sweet in taste at doka stage because of less astringency. It is medium in yield (60-70 kg/tree) and maturity period is also less in arid region. It is suitable for dry date preparation.

Braim: This variety has been introduced to India from Iraq during 1997. It is medium maturing variety suitable for fresh consumption and processing. The berry is yellow in colour, juicy and sweet in taste at doka (Singh *et al., 2005*). It is medium fruit yielding variety (50-60 kg/tree).

Khadrawy: The plants of this variety are dwarf in growth. The fruits at *doka* stage are greenish yellow in colour and oblong in shape with broad apex. The average fruit weight is 7.7 g and fruit size of 3.10 cm x 2.10 cm. It has pulp thickness of 0.66 cm having 30.0 per cent TSS at *doka* stage. The fruits are highly astringent at full *doka* stage and are badly spoiled due to rains at the time of ripening. This variety is suitable for both soft dates as well as dry dates. Average yield potential is 50-60 kg *doka* fruits per/palm.

Climatic and soils

The climatic factors which decide the commercial possibility of date palm production are: (a) long dry summer and sufficient heat unit accumulation for development and ripening of fruit, (b) sufficient water resources for irrigation, and (c) production technology suitable for different agro-climatic zones. The heat unit available at some of the locations in arid zone counted from the time of flowering (February) indicates that the most of these regions are suitable for date palm cultivation (Pareek, 1984). For successful date fruit maturation, nearly 3000 °C heat units are required, most of which accumulate by July in all these locations. The rate of maturity of berry and quality depend upon maximum temperatures (45- 48 °C). The rainfall pattern also influences the cultivation of date palm. The annual average rainfall is less (240 mm) and erratic under Bikaner condition. However, it can be grown from 150- 450 mm annual rainfall.

Date palm thrives well in light to heavy soils, which have 2.0 to 3.0 m soil depth and good drainage system. The soil of Thar Desert is suitable for date palm. The pH of soil from 8.0 to 8.6 is suitable for date palm cultivation. Deep sandy loam and sandy soils has been considered to be the best for date plantations, which is available in parts of Rajasthan, Haryana, Punjab and Gujarat States. It can tolerate salinity up to maximum level ($30dSm^{-1}$).The water logging conditions for longer period are not suitable for date palm trees. The soils of coastal belt of Gujarat are suitable for growing dates. Due to very little vegetation cover on soil surface prone to dryness and absence of clay particles, these soils are susceptible to wind erosion and form shifting sand dunes. These dunes are hazardous if unstabilized. Because of high permeability of sandy soils, infiltration of rainwater is very rapid and run – off is minimal even after heavy rainfall.

Propagation techniques

The propagation technique in date palm *viz.*, seeds, offshoot/suckers (traditional methods) is in practice and the recently developed tissue culture techniques are being popular for mass multiplication.

Seeds: In date palm variability and sex ratio is the main problem by seed propagation and 50:50 ratio of male and female plant has been observed in seedlings. The germination of seeds in date palm have been observed about 80-90 per cent when seed sowing take place in the month of July- August. Moreover, growth in seedlings is very slow at initial stages. It has been observed that seed soaking enhanced germination and seedling growth by pre-treatments of seeds with cow dung for 24 hours in *Phoenix* species (Singh and Bhargava, 2009). Seedlings varied considerably with respect to vegetative and reproductive characteristics. However, date palm propagation through seedlings can provide an ideal material as some times seedlings palms are outstanding in performance (Chandra *et al.* 1992). In Kachchh region, the seedlings are used for planting on farm boundary due to lack of planting material (Singh *et al*, 2004).

Suckers/offshoots: Presently, date palm is mainly multiplied through suckers / offshoots. The offshoots bear fruits 2-3 years earlier than seedlings. But production of suckers is a limiting factor because a palm tree produces about 10-20 suckers in whole life. It is a main constraint in increasing area under date palm cultivation. Survival of offshoots is about 40-50% under field conditions. Work on initiation of roots in suckers and better survival is also very important for increasing the area of cultivation in date palm. Sodagar *et al* (1996) observed that the best time of planting is April – May for higher survival and rapid growth of plant in Kachchh region. At the time of removal of suckers from mother palm, leaves should be pruned and tie up.

The offshoots should be well rooted and of proper weight for better survival in the field. Well rooted offshoot of medium weight (range 8-12 kg) is better in survival performance in the field compared to poorly rooted small sized or bigger sized offshoots (Singh *et al*, 2003). The offshoots weighing between 8 and 12 kg have been reported to give maximum rooting (Mohammed, 1978). High offshoots, which emerge from stem, perform poorly because of weak root development. Such offshoots of cultivar Medjool even after dipping in IBA (1000 or 3000 ppm) and planting in nursery failed to strike roots under arid irrigated conditions at Abohar, Punjab. However, in an experiment when such offshoots were covered at the base with moist moss grass enclosed in polythene and attached to mother palm treated similarly with wooden shavings, there was 44.9, 7.5 and 22.0 per cent increase in average root length, root thickness and root weight, respectively (Vij *et al.*, 1977). The treatments of suckers with organic such as cow dung solution, cattle urine etc also enhanced the survival of plants.

Tissue culture techniques: Date palm is multiplied through micro-propagation, which is also called *in vitro* propagation. (Tisserat, 1979, Zaid, 1999). For mass multiplication, tissue culture technique is useful in date palm and its protocol varies from variety to variety. By both organogenesis and embryogenesis plantlets have been raised in Gulf countries. In our country, date palm tissue culture work is limited. However, at AAU, Anand, protocol for date palm has been developed and tissue culture plants are being produced. Two or three private company at Mundra, Hyderabad, is also producing tissue culture date palm plants in India. The micro propagation technique has many advantages in date palm cultivation like:

- Production of genetically uniform plants which is a limiting factor at present situation,
- Propagation of healthy selected female cultivars.
- No seasonal effect on plants because they can be multiplied under controlled conditions in the laboratory throughout the year.
- Clones to be propagated from elite cultivars already in existence, or from the F1 hybrids or previous selections, and seed-only originated palms.
- Ensure an easy and fast exchange of plant material.
- Economically reliable when large production is required
- Large scale multiplication.

The sucker and tissue culture plants were evaluated for growth and production under hot arid conditions. The physiological performance of tissue culture plants is at par with those of sucker plants (Bhargava *et al.*, 2014). Tissue culture is the only tool to produce a large number of plantlets in date palm to meet out the requirement. It was noticed under evaluation that sucker production is more in tissue culture raised plants. However, for proper vegetative growth and early fruiting, suckers are not allowed by the growers and removed from time to time. In India, groves of seedling date palm are found on the coastal belt in Kachchh region of Gujarat.

Planting methods

The layout and preparation of pits is very important aspect for planting of suckers or tissue culture plants of date palm. The suckers/offshoots require lot of efforts and care for proper establishment. The pits should be at least 1x1x1 m in size and filled with equal proportion of well rotten manure, pond silt and top 30 cm fertile soils. For effective control of termites, Methyl Parathion or Fenval dust (100 g/pits) should be added with filling mixture. Besides this, Neem cake 2 kg per pit can also be applied. Planting at distance of 8 x 8 m in square system (156 plant/ha) is found suitable for date palm (Singh and Dhandar, 2007) cultivation in hot arid region. However, in Kachchh region of Gujarat, date palm suckers are planted at 9 x 9 m or 10 x 10 m distance. Date palm being dioecious in nature requires plantation of male plants for proper pollination along with the female plants. These should be at least 5 per cent of the total plantation.

At the planting time, suckers should be treated with Carbendazim @ 0.1 % solution to check fungal infections. Planting of date palm suckers should be done either in the spring season *i.e.* during February-March or during August – September in arid regions. It may also be dipped in to IBA 1000 ppm solution for better root initiation before planting. Mertia *et al* (2010) reported that transplanting of offshoots in the month of September is more beneficial with higher survival, 90% as against 65% in the offshoots transplanted in the month of March-April. However, May is the best time of planting of suckers for better survival and growth of plant in Kachchh, Gujarat (Sodagar *et al*, 1996). The offshoots should be planted in such a way that their tops get straight or they lean very slightly if at all towards the south. Too deep planting of offshoots may be damaged by irrigation water. The crown of the planted offshoots should be at least 10-15 cm above ground level so that the irrigation water does not damage it.

Irrigation

Irrigation plays an important role in establishment of suckers during initial stage. Water stress conditions during this period may be detrimental for survival of suckers because at the time of plantation suckers have generally poor root growth and it can not take moisture from deeper soil layer. After the planting, the suckers should be irrigated daily during the first month of plantation and thereafter these may even be irrigated at weekly intervals. However, it is known that water requirement of date palm is very high particularly after fruit set till the growth and development of the fruits. The irrigation should be given at 7-10 day interval because high temperature prevailing from April to June when fruiting takes place in date palm. During winter season, irrigation to date palm plantations may be given at an interval of 20 -30 days. Drip irrigation is useful in date palm tissue culture plantations for better growth and fruiting. The soil should be kept moist up to the depth of 2.0-2.5 m .The saline water can be used for irrigation purposes as it is salt tolerant plant.

Manures and fertilizers

Date palm plantations should be provided with adequate manure, which not only results in better growth of trees but also improves physical properties of the soil and productivity. Application of nitrogen, phosphorus and potassium may be helpful in better vegetative growth and fruit yield of better quality. The application of annual dose of 50-60 kg/ tree farm yard manure and 400 g N/ palm/year is found suitable to obtain maximum yield in cv. Khadrawy under irrigated conditions of Abohar, Punjab. The phosphorus and potash should be given @ 250 –500 g per tree according to age of plant. In Kachchh, Gujarat, the manure and fertilizers is applied @10 kg FYM and 100 g N, P and K per plant /year with increasing age of plant up to 10 years (Patel *et al*, 2001). By reducing input cost on irrigation and fertilizers, income from orchards can be enhanced. For this, fertigation technique should be used.

Application of annual dose of 50-60 kg manure or 1-2 kg ammonium sulphate per bearing tree in two equal splits, one during the month of January – February and the other during August–September has been recommended. It also has suggested that application of farm yard manure at the rate of 10 t/ha which works out as 33 kg of nitrogen, 21.75 kg of P_2O_5 and 50 kg or K_2O per hectare..In Kachchh, date palm orchards or groves are recommended to be manured with 30 kg farmyard manure in August-September and 1 kg ammonium sulphate in Jan. – Feb. to each palm (Pareek and Sodagar, 1986). However, these are seldom manured except the intercrops of fruits, vegetable and other crops. However, under Bikaner conditions, application of 1500 g N / tree / year was found optimum with respect to better yield and fruit quality in full-grown date palm cv. Halawy.

Training and pruning

Being monocotyledonous, date palm trees have single stem and generally do not branch and require pruning. Very old leaves must be removed, as these are less efficient in photosynthesis. Unwanted leaves should be cut down before manuring period *i.e.* September - October. Cultivars, which have short stalks, should be pruned more severely so as to provide ventilation to bunch. It is also found that pruning date leaves leaving 6 leaves per bunch and also removal one-third strands from centre have given good quality date production. It is suggested that nearly 60-80 / leaves palm are required for optimum productivity.

Inter-cropping

The intercrops may be utilized to grow inter crops without disturbing the physiology of date palm trees when trees are planted at wider spacing. At Abohar, citrus plantations *viz.* mosambi, kinnow gave good results under trees shade of date palm. Sapota tree is planted in interspaces of date palm orchards in Kachchh region (Singh *et al*, 2004). In initial years of establishment, different crops viz. cluster bean, moth, cowpea can be taken in between rows of plantations. The crops cultivation under trees shade is economical and suitable for per unit production. During *rabi* season, interspaces can be effectively utilized for growing barley, mustard, taramira crops in arid regions where very limited irrigation is available. The cultivation of taramira, an oil seed crop, was found suitable as rain-fed crop during *rabi* season. Shade loving crops like turmeric, ginger, colocasia *etc*, has grown as inter crops under date palm plantations at Abohar and it is remunerative.

Spathe emergence and flowering

Flowering takes place from January to March months depending upon agro-climatic conditions of the region and varieties. It is dioecious in nature and the identification of sex (male or female) of seedling palm trees is possible only after the trees start flowering and fruiting. About 5-10 % male plants should be planted for pollination of female plants. In north India, flowering starts after winter season *i. e.* with the start of spring season. Under Bikaner conditions, spathe emergence starts from last week of January and completed in February. It has been observed that prolonged winter delays the spathe emergence. At Abohar, flowering takes place by the middle of March. Flowering in male palm trees was observed 15 days earlier than female trees, yet fresh pollens were available in abundance for pollination. Mertia and Vashishtha (1985) have observed variation in spathe emergence in different years in cultivar Halawy. The variation in spathe emergence/opening and pollen production was observed in male palm trees in hot arid region.

Pollination and storage of pollens

The pollination in female palm trees is done manually and for that strands of freshly opened male inflorescence is cut from the tree and 3-4 strands are placed between the strands of female flower cluster in an inverted position as soon as flower anthesis in female takes place. The second method of pollinating female clusters consists of dusting of pollens on it within 2 days of anthesis. The pollens may be collected in abundance as a creamy powder from the freshly opened inflorescence. It should be done in morning hours for obtaining better berry set. Pollens storage of date palm can help in pollination of female palm trees where spathe opening takes place much after flowering in male trees. About 10-20 g pollen is required for pollination of a tree. An elite male palm (CIAH/DP/M-03) has marked/identified for better pollen production (644g/ tree). The pollens can be stored at room temperature or in a refrigerator in a dry air tight container which can be used in later part of pollinating period. The pollen stored in refrigerator (9°C) gave higher fruit set (42.9%) as compared to that stored at room temperature (36.8%) (Gupta and Thatai, 1980). The pollens can be stored either in refrigerator or at room temperature for a period of four weeks after collection without any loss in viability but thereafter viability decreases under room temperature. Under refrigeration (4^{0C} temp.), the pollens can be stored for one year but viability of pollen is decreased. The cryopreservation is best method for long term storage without any significant effect on pollen viability (75-84%) which depends on time of storage (Maryam *et al,* 2017). Due to lack of pollens, growers are purchasing pollen @ Rs. 8,000/- to 10,000/- per kg from available sources. The production and marketing of pollen can also become as source of income.

Fruit growth and development

The fruit development in date palm passes through various developmental stages. Berry initiation was earliest in Sewi cultivar (Ist week of April) followed by Medjool, Bint-Aisha and Sedami cultivars (II week of April). It was latest in cultivars viz. Zahidi, Barhee and Nagal Hilali (II to IIIrd week of April) under Bikaner conditions during the year 1987 (Chandra *et al.,* 1992). After pollination, the fruit set stage is known as Hawabuk, when un-pollinated or unfertilized flowers are dropped. The influence of pollens on fruit set, yield and quality have reported in date palm (Nasim *et al*, 2017).

The first stage of fruit growth is known as pre-doka or gandora or Kimri stage in which the berries are hard and remain green in colour. The fruits at this stage attain growth and size. Second stage is known as *doka* or *khalal* stage in which the fruits are fully grown, hard and the berries attain yellow, red or its combination depending to varieties characteristics. This stage may further be subdivided in two stages viz. early *doka* stage (colour changes to yellow reddish) and late *doka* stage (colour fully changes to yellow/red). Early *doka* stage may also be called as colour break stage.

The third development stage during ripening of *doka* berries is known as *dang* or rutab stage which is characterized by softening of date berries. When the berry becomes completely translucent or soft, it is known as pind khajoor. The dang to early pind/full pind stages can be obtained in extremely dry areas like Jaisalmer and western parts of Barmer, Bikaner and Jodhpur (Pareek, 1984). The final stage of fruit maturity is pind or tamer when berries fully ripen and soft. Pind (tamer) stage maturity and harvest of berry is rare in all varieties under Indian condition due to early monsoon. Maximum fruit is harvested at *doka* stage.

Fruit Ripening

Berry matures when berry colour turn either into yellow or red according to cultivars from green colour. In general, yellow and red colour cultivars are available for cultivation. One genotype has observed green berry colour even at full maturity stage in Kachchh region of Gujarat. Application of ethrel has been found to improve quality of date berries at colour break stage. Ripening was advanced by 20 or 28 days by spray of 1,000 ppm ethephon in Hayani and Khadrawy, respectively. It also increased berry weight, reducing and total sugars. Meena *et al* (2013) also reported spray of 1000 ppm ethephon is suitable for early ripening in date palm cultivars under arid conditions.

Plant Protection

Like other fruit crops, date palm orchards are also not free from the attack of diseases and insects. In fact, survival of offshoots is a limiting factor for large-scale establishment after planting which may due to several reasons like proper weight of sucker, availability of roots, climatic conditions, etc.

Diseases

***Graphiola* leaf spot:** The major disease of date palm plantation has been identified as *Graphiola* leaf spot, which is caused by *Graphiola phoenicis* and is characterized by a numerous dark coloured black pustules with full of yellow spores which are formed on both sides of pinnae and on the rachis. It occurs in almost all date palm plantations of Kachchh regions during the month of December and January (Muralidharan *et al*., 2008). In infected trees fruiting structure emerges as small, black coloured sorri which are abundant on apical pinnae and open to linerate masses of yellow spores. Spray of Copper oxy-chloride @ 0.4% solution was also found effective to control *Graphiola* leaf spot.

Besides leaf spot disease, fruits of date palm are also spoiled by *Aspergillus* sp. This type of spoilage has been found to be more in cultivar Medjool where the attack of this fungus penetrates deep up to the stone of the fruit resulting in blackening of pulp. Application of Blitox @ 0.2 % after fruit set may help in minimizing the incidence of diseases. For the control of such problem, suckers should be taken from healthy palm. The offshoot should be of proper weight i.e. between 8-10 kg with well-developed root system. After planting, suckers should be drenched with carbendazim 0.1% or Ridomyl 0.1% at monthly interval to check crown rot diseases.

Insects

Scale insects, red palm weevil and termites are major insects, which have really posed a problem for date palm cultivation in arid region.

Termites

Termites (*Microcerotermes diversus*) feed on the root zone of newly transplanted suckers and gradually tunnel their way upwards and galleries are formed inside the main trunk. The tree may ultimately wither and die away. It is the main problem in establishment of plants in arid region. Endosulphan dust 50g/pit or Methyl parathion dust (2%) should be mixed in the filling mixture in each pit before planting to check the termites attack. After planting, apply Chloropyriphos-20EC @ 0.1% at the time of irrigation to control termites.

Scale insects: For control of this insect, removal and burning of all the infested leaflets in case of severe incidence is required. Spraying with Malathion (0.08%) can control the pest population. Sprays of monocrotophos (0.03%), dimethioate (0.05%), parathion (0.05%) or diazinon (0.02%-0.04%) are very effective when spraying is done at an interval of 20 days.

Red palm weevil (*Rhynchophorus ferrugineus*): It attacks only on stem and the grubs are harmful. Due to hollowing of trunk from inside, the top of young tree dries up and rots giving out bad smell and finally trees breaks and drops off. Red palm weevil infestation is found in Kachchh region. For control of insect population, dead and infected palms should be destroyed. Aggregation pheromone ferrugineol is found effective to control red palm weevil.

Bird management

The period of fruit growth and development from May to June coincides with the period when there are almost no crops available for birds for feeding in the areas where date can be grown in Rajasthan. The dates bunches at the time of ripening are very much attacked by the birds like bulbuls, sparrows, mynas, parrots, purple sunbird and crows. The number of damaged fruits is more during the later half of June to July when the fruits are in full doka, dang and pind stages. For control of the birds, individual bunches should be covered with wire netting or Grow cover bags (degradable material) after thinning of bunches. It provides maximum ventilation for proper fruit development though it is costly but durable for 4-6 years. However, covering of bunches with other degradable material is cheap and seasonal. The protection from damage can also be done by covering bunches with wire gauge (3 x 3 mm mesh) at the start of *doka* stage till harvesting.

Harvesting and Yield

Date palm trees starts commercial bearing after 5-6 years of planting till to even more than 50 years age of trees. The fruit yield varies according to age, cultivar and growing sites. Variation in fruit yield 89-125kg per plant was reported in date palm cultivars by Mertia *et al* (2010) under extreme hot arid region. On an average 50 to 60 kg *doka* fruits of good quality can be harvested from a full-grown tree however, fruit yield may go up to 100- 150 kg/palm with adoption of proper cultural practices and plant protection measures. Maximum date palm fruits are harvested at *doka* stage in Indian arid condition due to early rains.

Post harvest management

Date fruits are utilized as fresh for raw eating due to delicious, sweet in taste and good flavour at *doka* (Khalal). However, *doka* fruit of some cultivars are astringent in taste and these are not used as such for fresh consumption. Processing involves curing of dates for preparation of dry dates i.e. *chhuhara* and soft dates (*pind khajoor*). Dry date can be prepared through dipping of *doka* fruits for 8-10 minutes in boiling water and then drying. Recovery percentages vary from cultivar to cultivar. *Doka* dates are used for preparation of a number of processed products like jam, chutney, pickle, RTS, beverage and bakery products. A large number of processed products of date palm are available in the world market including wine. Processing of date includes cleaning, grading, heat-to use date products, derived products such as syrup, juice, liquid date sugar. It has been reported that alcohol, wine .organic acids are made from date palm fruits (Zohra- Ashraf and Zohreh Hamidi-Esfahani, 2011). Date products can be preserved for a long period by using preservatives and citric acid. Delicious biscuits were prepared by adding 10-20% date fruit pulp powder and they were observed rich in protein in comparison to control (Singh *et al,* 2013).

Future scope and research needs

- Introduction of rain tolerant, early maturing genotypes/varieties
- Introduction of suitable genotypes/cultivars of high yield and quality
- Production of quality planting material on affordable cost
- Development of roots in small and aerial suckers
- Development of pollen bank and Pollination techniques
- Plant protection
- Post harvest management
- Storage, Marketing and export

Conclusion

It can be concluded that growing of date palm is profitable venture in hot arid region supported with proper marketing, storage, processing facilities. For doubling income from date palm production, proper land utilization, increase in fruit yield, marketing and reduction in inputs cost are very much essential. Technical knowledge/ guidance on various aspects of production technology are also needed to the growers. Because, at present the cost of planting material *i.e.* tissue culture plants is very high. The grading and packaging for proper marketing of produce is required to fetch better income from date palm. All varieties fruits are not used for fresh eaten hence, processing of date fruits is also essential to enhance income from date palm production and employment opportunities in the region.

References

Anonymous, 2016-17. Annual Report, ICAR- Central Institute for Arid Horticulture Bikaner, 162 p.

Bhargava, R.; Singh, R.S. and Vashishtha, B.B. 2014. Studies on physiological parameters in tissue culture and sucker plants of date palm *(Phoenix dactylifera* L.) cultivars. *Indian Jour. of Arid Hort.,* 9 (1&2) : 91-95.

Chandra, A., Chandra, Anju and Gupta, I.C. 1992. Date palm Research in Thar Desert. Scientific Publishers, Jodhpur 116p.

Chao, C.T. and Krueger, R. 2007. The date palm (*Phoenix dactylifera*); an overview of biology, varieties and genetics. *Hort Science*, 42: 1077-82.

Dates India. 2011a. Date palm in Tamil Nadu. http://www.datesindia.com/about.php Dates India. 2011b

Gopalan, C.; Ramasastri, B.V. and Balasubramanium, S.C.1985 Nutritive Value of Indian Foods, National Institute of Nutrition, ICMR, Hyderabad, 87p.

Gupta, M.R.and Thatai, S.K 1980. Receptivity of pistillate flower in date palm cultivars Shamran. Jour. of Research, (P.A.U.), 17(4): 366-368.

Maryam, Jaskani,M.J and Naqvi,S.A. 2017.Storage and viability assessment of date palm pollen. Methods Mol.Biol. 1638: 3-13.

Meena, Ramkesh, Yadav,P. K. and Singh, R.S. 2013. Response of date palm cultivars to varying concentration of ethephon on yield and yield attributing. *Prog. Agric*., 13(1):1-3.

Mertia, R.S. and Vashishtha, B.B. 1985. A note on performance of date palm cultivar Halawy at Chadan, Jaiselmer. *Annals of Arid Zone*, 24: 263-64.

Mertia, R.S. ;Birbal and Kumawat, R.N. 2010 . Performance and field management of six prominent cultivars of date palm (*Phoenix dactylifera*) in extreme arid region of Thar desert, India, *Acta Hort*. 882: 65-68.

Mohammed, S. 1978. Problems in date palm propagation. *Indian Hort*. 23(3): 15-18.

Muralidharan, C. M., Tikka, S.B.S. and Piyus Verma 2008, Date palm cultivation in Kachchh, Tech. Bull.No. 02/2008 S.D. A.U. Mundra, Kachchh,36p.

Nasim, Mohammadi, Rastgoo,Sasan and Mahmood Izadi 2017.The strong effect of pollen source and pollination time on fruit set and the yield of tissue cultured-derived date palm (*Phoenix dactylifera* L.) cv. Barhee. *Scientia Horticulture*, 224: 343-350.

Nixon, R.W. 1966. Growing dates in the United States. Agric. Inf. Bull. USDA no. 207, pp 56.

Nixon, R.W., Carpenter, J.B. 1978. Growing dates in the United States, Agr. Inf. Bull. No. 207, USDA, Washington, D.C., 63p.

Pareek, O.P. 1984. Date palm growing potential of Indian arid zone. *Indian Horticulture,* 2-5 & 8.

Pareek, O.P. and Sodagar, N.N. 1986. Date palm grooves in Kachchh. *Indian Hort*., 31: 21-27.

Ramdevputra,M.V.,Butani,A.B.,Savaliya,J.J.Punsuriya,A.G. and Kanzaria, D.R. 2009. Performance of date palm cultivars in Kachchh region .*The Asian Jour. Hort.,* 4(!):181-183.

Singh, R.S., Samadia, D.K. and Vashishatha, B.B. 2004. Genetic diversity in date palm *(Phoenix dactyliferra* L.). *Indian Horticulture*, 49 (3): cover page II & 24.

Singh, R.S. and Dhandar, D.G. 2007. *Khajoor ki Kheti*, Bulletin published by Directorate of Extension, Min. of Agric. (Govt. of India), New Delhi. 21p.

Singh, R.S., Bhargava, R. and Dhandar, D.G. 2003. Status and prospects of Date palm propagation, *Current Agric*., 27(1-2): 9-16.

Singh, R.S., Bhargava, R. and Dhandar, D.G. 2005. Performance of Date palm introductions under hot arid conditions. *Indian Jour. Plant Genetic Res*., 18 (1): 65-66.

Singh, R.S. and Bhargava, R. 2009. Effect of Seed treatments on Germination and Growth behaviour of *Phoenix* species under hot arid conditions, *Journal of Tropical Forestry*, 25: (1) 42-48.

Singh, R.S., Garima Pal, Bhargava, R. and Sharma B.D. 2013. Utilization of Date fruits for making delicious Biscuits *Annals of Arid Zone*, 52(2) :125-127.

Sodagar, N.N.; Murlidharan, C.M. and Ramdeoputra, M.V. 1996. Date palm propagation problems and prospects with special reference to Kachchh. *Jour. Applied Hort.*: 2:102-05.

Tisserat, B.H. 1979. Tissue culture of date palm. *Jour. Hered*. 70: 221-222.

Vishal nath, and Singh, R.S.2003. Boost up date cultivation. *Intensive Agric.,* 41:3-7.

Vij, V.K.; Kalra, S.K. and Bajwa, M.S. 1977. Studies on rooting of high offshoots in date palm, *Punjab Hort. Jour*., 17 (3&4): 135-138.

Zaid, A. 1999. Date palm Cultivation, FAO, Rome. 238p.

Zohra- Ashraf and Zohreh Hamidi-Esfahani, 2011. Date and Date Processing: A Review. *Food Review International*, 27:101-133.

16

Flower Regulation in Pomegranate (*Punica Granatum* L.) Under Hot Arid Climate

R. Kumar, P.L. Saroj, B.D. Sharma and Ramkesh Meena

Introduction

Pomegranate (*Punica granatum* L.) is an important fruit crop of hot arid region of the India. It is grown commercially in the states of Maharashtra, Gujarat, Karnataka, Andhra Pradesh, Tamil Nadu, Madhya Pradesh and Rajasthan. India is one of the largest producers of pomegranate in the world. The pomegranate area in arid region is increasing at faster rate due to its incessant demand, drought hardy nature, low cost management and wider adaptability to arid soils. In Rajasthan it is mainly grown in Jaipur, Ajmer, Alwar, Tonk, Sriganganagar, Tonk, Pali, Kota, Jalore, Banswara, Sawai Madhopur, Bhilwara, Jhunjhunu and Sirohi districts in 12,000 ha area (Rathi, 2017). The pomegranate plants flower and provide fruits throughout the year. In Rajasthan pomegranate is harvested in all three seasons during March-April, Aug.-Sep. and Nov.-Dec. However, it needs to be forced into rest period for prolific harvest at a given time. The basic objectives of flower regulation in pomegranate is to manipulate the natural flowering and fruiting in such a way that it produce high yield of quality fruits in desired season with sustainable use of farm resources. Transition of flower from vegetative phase to reproductive phase is critical phase in pomegranate flowering, which determine production efficiency. The differentiations of vegetative tissue to reproductive tissue decide flowering intensity and production efficiency. Withholding of irrigation (moisture stress), plant growth substances, nutrient content along with canopy management are major horticultural interventions for flower regulation which influence flowering. It is based on principles that pomegranate bears three kinds of flowers namely hermaphrodite, intermediate and male flowers (Fig.1) on both new and old growth in three flushes i.e. *ambe bahar* (Feb.-March), *mrig bahar* (June-July) and *hasta bahar* (Sep.-Oct.).

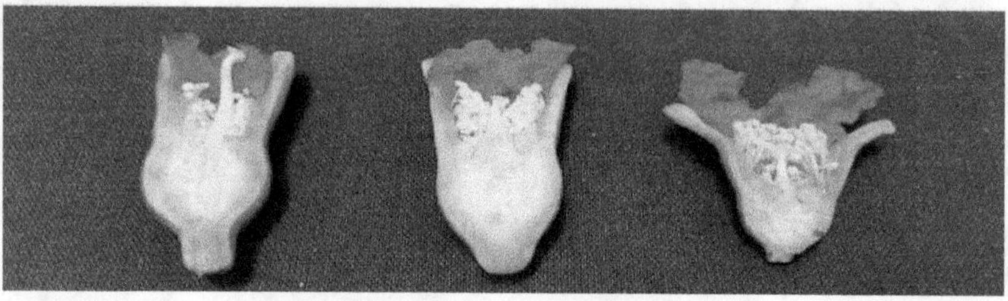

Fig. 1: Hermaphrodite, intermediate and male flowers in pomegranate

Under temperate climate, pomegranate sheds their leaves in winter and behaves as deciduous but in sub-tropical and tropical climate it behaves as an evergreen or semi deciduous. It thrives best under hot dry summer and cold winter under irrigated condition. Pomegranate flowers continuously when watered regularly. Under irrigated conditions plants continue bear flowers and produce small crop irregularly at time of the year, which may not be desirable economically. To avoid this trees are subjected to flower regulation/crop regulation/bahar treatment (Fig. 2). In this treatment, the irrigation is withheld one to two months prior to the bahar. This facilitates the shedding of leaves. The trees are then medium pruned 40-45 days after withholding irrigation. Ethrel is used (1-2 ml/l) as defoliant to remove leaves. The recommended doses of fertilizers are applied immediately after bahar treatment and irrigation is resumed. This leads to profuse flowering and fruiting. The fruits are ready for harvest 4-6 months after flowering. In hot arid climate, there are three flowering seasons, viz., February –March (*ambe bahar*) June-July (*mrig bahar*) and September-October (*hasta bahar*). The choice of flowering/fruiting is regulated taking into account the irrigation water source, market demand and pest, disease incidence in a given locality. The fruits of *ambe bahar* are ready for harvest in the month of June to September. As the fruit development takes place during dry months, they develop attractive colour and quality. Owing to dry weather, the incidences of pest and disease attack are very less. Therefore, *ambe bahar* can be taken in areas having assured irrigation water. The *mrig bahar* crop is harvested in the month of December to January. Usually this bahar is favoured as the flowering and fruiting period coincides with moisture availability period, and the crop is taken without much irrigation. As the fruits develop during the rainy season and mature during winter, the colour and sweetness of the fruit is affected. The fruits from *hasta bahar* become ready during the month of March to April and have very attractive rind with dark coloured arils. Since the availability of the fruits during this period very less, they fetch premium prices. Optimum water stress cannot be developed during this period as withholding of irrigation not possible because of rainy season. This leads to poor flowering and yield. The plant habit, flowering physiology and fruiting changes with the habitat. Most commercially grown varieties in India have advantage due to their free flowering nature. To utilize this beneficial characteristic, flowering regulation is practiced in pomegranate according to the need and choice of growers.

a. Withholding of irrigation b. Light pruning c. Application of Ethrel d. Dropped leaves on ground

e. Defoliated plant f. New foliage emergence g. New twigs h. Flowering initiation i. Profuse blooming

Fig. 2: Process of flower regulation in pomegranate

Fig. 2: Process of flower regulation in pomegranate

Plants integrate environmental signal for their growth and more specifically for flowering phenomena. In the flowering process, the most critical stage regulated by environmental factors is the floral transition. Infect, the differentiation of vegetative primordia to reproductive primordia is the crucial first step for the reproductive phase which determines the blossom intensity and yield capacity. In this context, floral transition is the key factor that determines the productivity especially in perennial fruit crops. The perennial fruit crops are constantly exposed to annual climate change and vagaries hence, assume greater significance in determining productivity. Different environmental signals influence the flowering are photoperiod, light quality, vernalisation and temperature where as horticultural factor influence the flowering includes moisture stress, plant growth regulators, nutrient status and canopy management.

Among the environmental factors, the most important is temperature, including the canopy temperature a plant experiences throughout its life, along with prolonged low temperature (vernalization), and day length (photoperiod). Finally, the activities of the flower inductive pathway genes come together on a small number of floral integrators. When the expression of these floral integrators exceeds a threshold limit, the plant starts flowering, which is an irreversible process. Apart from ensuring the reproductive success of a plant, the management of the vegetative-to-reproductive phase transition is also of considerable horticultural implication. Moisture and nutrient status also influence the maturity of flushing branches.

Keeping in view the flowering and fruiting period of different *bahar* along with their advantages and disadvantages (Table 1) flowering regulation is practiced in pomegranate.

Ambe bahar is practiced in South India and Dholka area of Gujarat where sufficient irrigation water is available (Patil and Karale, 1985). It is preferred in Maharashtra as the yield and quality of this *bahar* is better, whereas *mrig bahar*, crop is severely damaged by pests and diseases (Anonymous, 1981). *Mrig bahar* is taken in the North West arid region (Mann and Pareek, 1974) and drylands of South India where water is scarce during the hot and dry summer. Prasad *et al.* (1997) and Singh *et al.* (2003) also reported that due to hot weather during July-August and inferior colour development and quality of fruits during January–February, both *hasta bahar* and *ambe bahar* are not recommended to in arid and semi arid climate in pomegranate. They also recorded maximum cracking and deterioration in quality of fruits in advanced stages of harvesting during November–December that might be due to day night temperature fluctuation prevailed in arid zone. For better colour development, fruit set and maximum size fruit, usually *mrig bahar, i.e.* June-July flowering is preferred in arid zone. But this season causes cracking at maturity due to day night temperature variation (Lawande, 2000).

Hiwale (2009) reported that good crop was obtained under semi arid condition when September flowers were allowed to set because fruit development coincided with maximum moisture availability and cool climate leading to fewer incidences of insect pest and reduced cracking leading to improved quality of fruits. *Hasta bahar* is generally not taken because, during the preceding monsoon season in July-August, it is not possible to withhold irrigation to induce flowering in September. Flowering can be induced in pomegranate during *hasta bahar* by foliar spray of suitable growth hormones at 2-4 weeks after defoliation (Babu and Pal, 2014).

Importance of Stress in Flowering

The beneficial effects of stress are accentuated with the objective to manipulate the flowering physiology. Beneficial responses of stress may be profuse and uniform flowering and improved sex ratio and fruit set. A positive correlation was found between leaf proline content and the number of flowers produced (Powerwanto and Inoue, 1990). Conceptual model of flowering shows two distinct phases, a shoot initiation phase and an inductive phase which determines

Table 1: Flowering and fruiting period of pomegranate under different *bahar*/waves.

Bahar	Flowering period	Harvesting period	Advantages	Disadvantages
Ambe bahar	February to March	June to August	• Fruit mature during dry period having attractive colour and good quality fruit - suitable for export. • Low pest and disease incidences.	• Require assured irrigation facilities. • Poor growth and yield.
Mrig bahar	June to July	December to January	• Flowering and Fruiting coincides with maximum moisture availability periods, hence not require much irrigation. • Suitable for hot arid region.	• Flowering and fruiting coincides with rainy season. • Not favoured in the bacterial blight endemic area.
Hasta bahar	September to October	March to April	• Fruits attractive with dark colored arils. • Fetch high value due to less availability.	• Stress cannot be given during rainy season. • Poor flowering and yield.

that shoot become either vegetative or reproductive. Initiation of shoot growth affected different internal and external signals than induction. The buds which initiated in the inductive, low temperature or water stress environments produced generative shoots. Those initiated after the removal of inductive conditions produced vegetative shoots (Southwick and Davenport, 1986). According to Lovatt *et al.* (1988), the accumulation of ammonia during stress resulted in increased biosynthesis of arginine, polyamines and subsequently an increased rate of cell division. These physiological changes and the subsequent rapid increase in cell division are pre-requisites to flower initiation in citrus. Xie *et al.* (2015) studied the effect of water stress on vegetative growth and nitrogen metabolism of pomegranate seedling and reported that water stress reduced protein content and increased NH3-NH4 +, arginine and proline in leaves of pomegranate.

There are two methods practiced to expose plant against the stress, first by water reduced stress and second is the chemical induced stress. In first method, 50-70% leaf defoliation is reported and considered to be an optimum stage of stress while in second method 90-100% leaf defoliation is observed. Soil factors also affecting flowering as in heavy soil of South India 45 to 60 days water stress is given while in light sandy soil of hot arid zone 30 day stress is sufficient. Under such conditions, the success of flower induction is governed by the right quantum of water deficit stress as a function of combined effect of soil physico-chemical properties.

Role of Ethylene in Flower Regulation of Pomegranate

- When the plant is then exposed to ethylene, gene expression of cell wall degrading enzymes such as cellulase and polygalacturonase are activated which cause defoliation and impose stress.
- Ethylene perception is involved in the arrest of stamen development through the induction of DNA damage that may alter sex ratio in pomegranate.

Plant Architectural Engineering and Flower Regulation

i) **Bahar Pruning:** The pomegranate plant requires pruning throughout its life cycle. The pomegranate bears fruits on both current growth as well as short branches known as "spurs." The spurs remain productive for 3 to 4 years but with the advancement in age, they lose capacity to bear fruits and needs to be removed. These old spurs are pruned and new spurs are encouraged to develop. Pruning operation is carried out to modify natural habits of pomegranate plant which control unnecessary vegetative growth and regulate flowering and fruiting for optimum yields. At initial stage young plant requires light pruning while heavy pruning is done in old plants. Pruning technique are of two types:-

a) **Topping (External pruning):** Pruning is done to reduce height by removing the branches growing in external direction during winter (Dec.-Jan.) after fruit harvesting. Top level of fruit bearing branches is also pruned up to convenient heights for easy harvesting.

b) **Skirting (Internal pruning):** Removal of lower tree branches. This improves under tree access for irrigation, weeding, fertilizers application. Improve sunlight penetration inside and lower parts of the tree.

Physiological Factors Affected by Pruning

i. **Photosynthesis:** The sunlight intercepted by the tree canopy is the prime importance for the production of qualitative and quantitative yield.

ii. **Apical dominance:** The suppression of lateral bud growth by the terminal bud is known as apical dominance. Apical dominance dependent upon hormonal balance, explains many of the growth characteristics of plants and their responses to pruning.

iii. **Carbohydrate: Nitrogen ratio:** The balance between plant growth and fruitfulness depends upon a relationship between carbohydrates and nitrogenous compounds within the tree.

iv. **Flowering behaviour:** Pruning encourages more flow of nutrients and water to the remaining shoots which flower and numbers of flowers that develop to form fruits is increased consistently. As long as a balance between loss of potential bearing surface and the nutrient supply is kept at optimum pruning always encourages better flowering.

v. **Branches orientation:** The orientation of branches in space has a marked effect on growth and fruiting. A decrease in growth rate and an increase in flowering occur when branches bend to a horizontal position.

iv. **Thinning:** Under favorable environmental condition, pomegranate plant produces many flowers and fruits which causes bending and breaking off the branch. Hence thinning is required to remove some flowers and immature fruits in earlier stages of Bahar. It is a cultural practice of maintaining a specific number of fruits on tree for getting good quality and size of fruits. A full grown pomegranate tree produces 60- 100 fruits per tree under intensive management practices. Nearly 98% farmers insisted on higher yields so did not carry thinning operation in the orchards. Ultimately, it affects on the quality of fruits and so also makes trees weak that in turn minimizes the life span of orchard trees.

Nutrient management: Nutrients affect pomegranate plant growth, flowering, yield and quality. With each harvest considerable amounts of macro and micro-nutrients are being depleted which need to be replenished regularly. Hence, maintaining soil health is necessary for sustainable production. The requirement of fertilizers varies greatly as per stage of Bahar from flowering and fruiting. Time of application varies in bahar like May for mrig bahar, October for hasta bahar and January for ambe bahar.

Nitrogen assimilation and flowering: Role of nitrogen metabolism in regulating the induction of flowering has been considered as one of the valid hypotheses to explain the phenomenon of flowering. Up to 80 % of the nitrogen utilized during the spring growth and fruit set period is resultant from storage reserve of nitrogen (Kato, 1986). The timing of the normal soil application is such that it probably does not significantly contribute to the nitrogen requirement of the emerging flowering and vegetative flush. The regulation of flower initiation by the balance of carbohydrate to nitrogen or by a precise mathematical ratio of carbohydrate/nitrogen is recognized internationally as the Kraus-Kraybill hypothesis.

Rest period: Under hot arid condition, a rest period of about 2-3 months is recommended for better plant health, vigour and reduction of pathogen inoculums load. Sprays of Bordeaux mixture 1% alternated with COC (2.5g/l). During rainy season (*hasta bahar* crop) sprays at 15 days interval of Bordeaux mixture 1% alternated with sprays of streptocycline (0.5g/l) + COC (2.5g/l) to reduce pathogen load. Minimum irrigation should be given during rest period. Remove the water shoots regularly.

Improvement in fruit quality: The thinning of flowers/fruitlets during the selected bahar to an optimum number has helped in uniform maturity of good size fruits. The fruit size progressively increased when the number of fruits per tree was decreased and the gross income was the highest when 50 fruits were retained (Anon., 1981). However, it is advisable to keep around 80-100 fruits / tree depending on the age of plants, vigour of plants and management conditions. Improvement in fruit quality was observed by Raturi and Hiwale (1991) when deblossoming was carried out for obtaining a particular bahar. Sharma and Singh (2000) reported that juice percentage was more due to deblossoming in April month as compared to control in Ganesh and Kandhari cultivars of pomegranate.

Sun scald management: Develop proper canopy architecture to avoid the direct exposure of fruits to sunlight. Spraying Kaolin (5 %) during the hot summer months is useful in reducing the sunscald. Covering of plants with 35% shade net is effective in reducing sunburn. Fruit bagging with butter paper covers is also practical in minimizing fruit spoilage due to sun sunscald. White colour bags are more effective in reflecting sunlight and protecting the growing fruits.

Conclusion: Stress induced by reducing moisture and chemical application, plant growth regulators, nutrient and canopy management are major horticultural interventions which influence flowering ultimately affect yield. Pomegranate bears male, intermediate and hermaphrodite flowers on both new and old growth in different flushes. Plant may continue to bear flowers and produce low yield of inferior quality with unsynchronized maturity which may not be advantageous commercially. Therefore, flowering is regulated to encourage plentiful harvest at particular period considering irrigation facilities, pest & diseases incidence and market demand. This will leads to profuse flowering with improved sex ratio, fruit set, retention and ultimately high quality fruiting in desired season. The information on molecular mechanisms, particularly differential expression of genes involved in flower transition phase is very limited. The strategic and basic approach would be to genetically modify them which would be difficult. In this context, the new emerging trend is to take clues from chemical regulation of the target proteins and transcripts.

References

Anonymous. 1981. Annual Report. All India Coordinated Fruit Improvement Project, ICAR. pp.21-23.

Babu, K.D. and Pal, R.K. 2014. Induction of flowering in pomegranate through chemicals and hormones. Abstracts of the National Seminar cum workshop on physiology of flowering in perennial crops held at Central Institute for Subtropical Horticulture, Lucknow, 24-26 May 2014. pp.42.

Kato, T. 1986. Nitrogen metabolism and utilization in citrus. *Hort. Reviews,* 8: 181-216.

Lawande, K E. 2000. Pomegranate production technology. Lecture delivered during training on changing scenario in arid hort. held during 20-29 September 2000 at NRC for Arid Horticulture, Bikaner.

Lovatt, C.J. 1990. Stress alters ammonia and argimine metabolism. Polyamines and Ethylene. Biochemistry, physiology and interactions (ed. H.E. Flores, R.N. Arteca and J.C. Shannoneds) *Am. Soc. Pl. Physiol.* pp. 166-178.

Mann, H.S. and Pareek, O.P. 1974. Potentialities of growing fruit in the hot arid zone. *Indian Hort.,* 19: 1-3.

Patil, A.V. and Karale, A.R. 1985. Pomegranate. In : Bose, T.K. and Mitra, S.K. (eds.). Fruits : Tropical and Subtropical, Naya Prokash, Calcutta.

Powerwanto, R. and Inoue, H. 1990. Effects of air and soil temperatures in autumn on flower induction and some physiological responses of satsuma manciarin. *J. Jap. Soc. Hort. Sci.* 59 (2) : 207-214.

Prasad, R N., Bankar, G. J. and Vashishtha, B. B. 1997. Problems and prospects of pomegranate cultivation in arid region. Proceedings of symposium on recent advances in management of arid ecosystem, held during 3-5 March 1997 at Jodhpur, Rajasthan.

Rathi, A. 2017. Present status and future of pomegranate in Rajasthan. In: Souvenir, 2nd National Seminar cum Farmers fair- Pomegranate for health, growth and prosperity, April 28-30, 2017, ICAR-National research Centre for Pomegranate, Solapur-431255 (Maharashtra), pp.243-44.

Raturi, G.B. and Hiwale, S.S. 1991. Regulation of cropping in pomegranate under rainfed conditions. *Indian J. Hort.* 48:13-15.

Sharma, K.K. and Singh, N.P. 2000. Effect of mechanical deblossoming on yield and quality of pomegranate (*Punica granatum* L.). *J. Res. Punjab Agric. Univ.,* 37(3-4):203-5.

Singh, D. B., Sharma, B. D. and Bhargava, R. 2003. Effect of boron and GA to control fruit cracking in pomegranate *(Punica granatun).Current Agri.* 27(1-2): 125-7.

Southwick, S.M. and Davenport, T.L. 1986. Characterization of water stress and low temperature effects on flower induction in Citrus. *Plant Physiol.* 81:26-29.

Xie, S. Xiaopeng, Lu. Xiaoli, Zhao and Qun, Nie. 2015. Effect of water stress on vegetative growth and nitrogen metabolism of pomegranate seedling. *Acta Hortic.* 1089, 63-69.

17

Bael: An Emerging Crop of Wastelands

A. K. Singh, Sanjay Singh and P. L. Saroj

Introduction

Bael (*Aegle marmelos* Corr.) is an indigenous fruit of India belongs to family Rutaceae and it is commonly known as Bengal quince (John and Stevenson, 1979), *Bilva,* Indian quince, Golden apple, Holy fruit, *Bel, Belwa,* sriphal, *bilipatra* Stone apple and *Maredo* in India. The deciduous with trifoliate aromatic leaves is traditionally used as sacred offering to the 'Lord Shiva' and thus the tree is also known by the name Shivaduma (The tree of Shiva). It has tolerance to arid conditions as well as high rainfall. It is known from prehistoric time and mentioned its importance in different ancient and medicinal system *viz., Rigveda, Yajurveda, Atharvaveda, Charak Samhita and Sushrut Samhita, and Ayurveda, Siddha, Unani* medicines and religious Jain and Buddhist literature (Balunas and Kinghorn, 2005, Om Prakash,1961 and Kirtikar and Basu, 1935). It can be grown successfully in water scarce and rainfed areas of the country. Besides, occurring as natural wild plant of forest, it is commonly grown in the homestead gardens, backyards, and religious places particularly in the premises of the Lord Shiva temples and in the farmers' fields. In India, bael is being grown throughout the country except temperate region, but there is no systematic plantation of bael except Uttar Pradesh, Rajasthan and Gujarat where few progressive farmers have initiated growing bael economically (Dhankhar *et al*., 2011). Its fruit is highly nutritive and rich in vitamins, bioactive compounds, minerals, carbohydrate, *etc* (Gopalan *et al.,* 1964). Every part of the tree such as fruit root, bark, leaf and flower of any stage of maturity and ripening are important ingredients of several traditional formulations. Recently, advanced scientific techniques have reported that plant has various medicinal properties viz, anticancer, antibacterial, antifungal, antidiabetic, antioxidant, hepatoprotective, haemolytic, larvicidal and anti-inflammatory activities. The *sherbet* and marmalade prepared from fruit pulp is prescribed for diarrhea and dysentery. It is said that the ripe fruit is a tonic, an astringent and laxative, whereas unripe fruit is an astringent and digestive and good for heart and brain. The important medicinal properties of bael are antidiabetic, antimicrobial, anti-inflammatory, antipyretic, analgesic, cardioprotective, antispermogenic, anticancer and radio protective (Kumar *et al.,* 2010 and Vadivelan *et al*., 2011) The fruit is efficient to avoid malnutrition and health insecurity in the underprivileged. The ripen and unripe fruits can be used for preparation of various value added products such as jam, slab, squash, powder, candy, nectar, toffee, pickle, wine, *sherbet,* ice cream *etc.*

It is highly valued for the formulation of *Ayurvedic* medicines. The importance of bael fruit lies in its curative properties, which make the tree one of the most useful medicinal plants of India. All the parts of the tree, whether it is stem, bark, root, leaf, flower, seed oil or fruits of any stage of maturity and ripening are used in various *Ayurvedic* medicines (Singh *et al*., 2011a). An infusion of flowers is a cooling drink in Indonesia. The ripe fruit is eaten fresh and has a more demand from them, who wants it for therapeutic use. The fruit pulp can be used to prepare delicious cooling drink. The bael pulp is used as a base of various fruit products like squash, jam, slab, toffee, powder, nectar, RTS etc. Green bael fruits are used for preparing *murabba* (preserve). The leaves, roots, flower, fruit and other parts are known to have medicinal properties especially for stomach disorders and improvement in digestive system. The green bael fruit slices often are dried and stored for future use. The unripe or half ripe fruit is regarded as astringent, digestive stomachic and good for heart and brain (Kirtikar and Basu, 1935). The fruit is used in chronic diarrhea and dysentery, and is said to act as tonic for heart and brain. It is useful adjuvant, as it helps to remove constipation which hinders the healing ulcerated surfaces of intestine. The roots as well as bark are used in the form of a decoction as a remedy in melancholia, intermittent fever and palpitation of heart. Root has anti-amoebic and hypoglycemic properties. The leaves are bitter and used as febrifuge, poultice made of leaves is used for ophthalimia and ulcers. Fresh leaves are also used as remedy for 'dropsy, and beriberi associated with weakness of heart. The astringent rind of ripe fruits and bark are employed in dyeing and tanning. The timber is commonly used for making pestles of oil and sugar mills, for posts, shafts, axles and naves of cart. The shells of smaller fruits are used as snuff boxes. A yellow dye is extracted from the rind of unripe fruit. The diluted leaf juice is used for catarrh. The alkaloid aegelin present in the leaves are efficacious in asthma. Utilization of bael fruit has been reported by Roy and Singh (1979a).

Present Status and Global Scenario

It is a hardy plant and grows up to an altitude of 1,200 m altitude from sea level. It grows well in the dry forests on hilly and plain areas. It is a widely distributed plant and found in India, Ceylon, China, Nepal, Sri Lanka, Myanmar, Pakistan, Bangladesh, Nepal, Vietnam, Laos, Cambodia, Thailand, Indonesia, Malaysia, Tibet, Sri Lanka, Java, Philippines and Fiji. In India, it is distributed throughout the country. The wild plants are found growing in Sub-Himalayan tracts from Jhelum eastwards to West Bengal state in central and southern India (Dhankhar *et al*., 2011), but concentrated area under bael is in eastern parts of the Gangetic plains and nearby areas particularly in Uttar Pradesh, Bihar, West Bengal and Orissa. In Gujarat, bael trees are found growing naturally in the forest with great diversity (Sharma *et al*., 2013 and Singh *et al.,* 2011a). Apart from systematic orchards, bael trees are also planted in nutritional gardens, parks, temple gardens, roadsides for various purposes. It was introduced into Europe from India in 1759 (John and Stevenson, 1979). The area under bael cultivation is increasing particularly in arid and semi-arid region because of its ability to grow in diverse climate and helpful in controlling the various ailments of mankind.

Bael is very hardy, deciduous fruit tree and can thrive well in alkaline, saline and stony soils having pH range from 5.0 to10.0 where many trees fail to establish. It has wide range of adaptability to varied soil and climate. The extent of hardiness of bael plants has also been observed and the plants are giving good yield in rainfed conditions of western India having soil depth from 0.60 to 0.75 m, derived from mixed alluvial basalt, quartzite, granite and layers of limestone, and can tolerate salinity up to 9 dsm^{-1}and sodicity up to 29 ESP. However, well drained sandy loam soil is ideally suitable for *bael* cultivation. It has ability to tolerate to the extremes of temperature and soil moisture stress. Bael can be grown up to an altitude of 1200 m and is able to grow as low as -7^{0}C and higher temperature up to 48^0 C, but it is sensitive to frost and water logging conditions. However, plants tolerate high temperature and drought by shedding their leaves during summer (April-May).

Gaps in Prevailing System

Fruits are more vulnerable to adverse weather, leading to higher risk of failure. Rather than pay for crop insurance (where it is available), farmers prefer to simply avoid these crops. Farmers don't really have much option to store fruits for longer period. The need to sell immediately means that they are at the mercy of current market prices, unlike grains that can be held on to for a longer time. Fruits experience a much higher degree of price volatility than grains. Part of the reason for this is the high level of mismatch between demand and supply of fruits. Another reason is the inefficiency of markets in matching supply and demand in different parts of the country. And of course, their inherent perishable nature and lack of cold-chain is an additional worry. Lack of proper storage and transport facilities has yet another impact–spoilage of produce resulting in lower price realization due to poorer quality of produce by the time it reaches markets. Lack of trait specific varieties is being a medicinal fruit tree, non availability of genuine planting materials of improved varieties and mechanism for assessing the quality planting materials, lack of awareness about medicinal significance of bael, health benefits and nutritional security, lack of awareness to develop small scale unit for value added product, technologies to avoid frost injury and sunscald. Lack of adequate standards for quality produce to harvest export potential are the prevailing gaps in system.

Advance Approaches for Bael Production

Genetic Resource Management

Diverse germplasm of bael has been enormously collected from various states of India *viz.,* Uttar Pradesh, Bihar, Gujarat, Haryana, Madhya Pradesh and West Bengal by NDUAT, Faizabad, CIAH and its regional Station CHES, Godhra, CISH, Lucknow, CCSHAU, Regional Research Station, Bawal, CAZRI, Jodhpur, NBPGR Regional Station, Jodhpur (Malik *et al*., 2010), Jharkhand (Jana *et al*., 2014), Bihar (Rai *et al*., 1991), Rajasthan (Singh *et. al.,* 2015c) and Punjab (Bal, 2003). Characterization of bael genotypes has been undertaken at CHES, Godhra under rainfed hot semi-arid environment of Gujarat (Singh *et al*, 2014 a,b&g). Different genotypes of bael showed wide variations with respect to vegetative characters under rainfed conditions of western India (Singh *et al.,* 2008 &2014 a&c). Varieties also showed wide differences with regards floral traits (Singh *et al.,* 2008, 2014b). The yield and yield attributing traits of fruit of different genotypes, viz., fruit yield (40.50- 69.29 kg), fruit weight (0.43-4.25 kg), length (10.61-19.59 cm), width (9.40-22.00 cm) and fruit girth (29.10-70.00 cm) also showed variations. Physical composition of bael fruit exhibited wide variation in their shell weight (115.25- 560.05g), shell thickness (0.16-0.31cm), number of seed/fruit (90.34-212.25), total fresh seed weight (17.34-43.41 g), number of seed sacs (10.23-19.17), fibre weight (15.91-106.50g) and pulp weight/fruit (0.27-3.67 kg). The qualitative characters of fruit in terms of TSS mucilage, TSS pulp, total sugar, reducing sugar, non reducing sugar, vitamin C, total phenols, acidity and TSS to acid ratio ranged between 37.00-49.50° brix, 30.57-37.45° brix, 16.15-19.98%, 3.30-4.95%, 12.85-15.13%, 17.13-21.03 mg/100g, 2.34-2.75%, 0.30-0.49% and 68.88-124.83, respectively. Diversity of bael germplasm based on morphological characteristics of the fruit has been performed by Singh *et al.,* (2011b &2015a) and Rai and Misra (2005). The bael genotypes were found to have considerable genetic variability, demonstrating the importance of RAPD markers to analyze each genotype in a collection in order to maintain the germplasm for genetic improvement of bael efficiently.

Genetic resources of bael are being conserved at various field gene banks of ICAR institutes and State Agricultural Universities. Various collections are maintained at CCSHAU, Regional Research Station, Bawal (10), NDUAT, Faizabad (22), CIAH, Bikaner (16), CISH, Lucknow (44), GBPUAT, Pantnagar (10), CAZRI, Jodhpur (5) and at CHES, Godhra (181).

Varieties

In existing biodiversity of bael, two distinct types *viz.* the small fruiting type having acrid pulp, high seed and mucilage and the large fruiting type with thin shell, less seeds and mucilage and more sweet pulp have been broadly identified. In recent past, some promising varieties of bael have been developed through clonal selection at SAUs and ICAR Institutes/Centers. The important bael varieties have been characterized by Singh *et al.* 2008; Singh *et al.*, 2012b; Singh *et al.*, 2011a, Singh *et al.*, 2014a and Singh *et al.*, 2016d).

Goma Yashi

Yield 65.45 kg/plant during 8th year, fruit weight 1.00-1.62 kg, fruit size 13.00 cm x 12.50 cm, fruit girth 41- 45 cm, shell thickness 0.17cm, seed weight 25.37-32.50g, fibre weight 40.24-51.20 g, shell weight 180-210g, locules in cross section 13-15, pulp 68.00-72.25%, TSS pulp 35-39^{0}B, TSS mucilage 41-43^{0}B, acidity (0.29-0.34%) and vitamin C 19.00-22.00 mg / 100 g pulp. This variety matures during March and belongs to early maturing group and found ideally suitable under dryland semi-arid ecosystem and highly suitable for high density planting (Singh *et al.*, 2012a & Singh *et al.*, 2017).

Thar Divya

It starts ripening after 270 days of fruit setting under rainfed conditions of hot semi-arid ecosystem. The average yield/tree(kg) during 8th year, fruit weight (kg), fruit size (cm), fruit girth(cm), shell thickness (cm), number of locules in cross section (seed cavity), peel weight (g), pulp weight (kg), fibre weight (g), seed weight(g), total seed weight (g), total number of seed, TSS of pulp (^{o}brix), TSS of mucilage (^{o}brix), acidity (%),TSS/acidity ratio are 7.50, 1.62, 18.60x14.80, 45.80, 0.19, 14.72, 235.50, 1.30, 61.10, 0.13, 32.00, 120.75, 38.50, 51.00, 0.30 and 128.33 respectively. This variety matures during February and belongs to very early maturing group and can be grown successfully under rainfed semi-arid ecosystem (Singh *et. al.*, 2015b & Singh *et al.*, 2017).

Thar Neelkanth

It is superior genotype with desirable characters like earliness, compact growth, medium height, spiny, better yield with quality fruits having pleasant flavour with attractive colour of pulp. It started flowering and fruiting from 3rd year of budding. Average yield per plant 75.67 kg (8th year), average fruit weight 1.45 kg, fruit size 15.10 cm x 15.00 cm, fruit girth 47.30 cm, shell thickness 0.18cm, total number of seed 73, seed weight 0.21g, total seed weight 15.46g, fibre weight 110.17 g, shell weight 265.00g, locules in cross section 13-16, pulp 71.30%, TSS pulp 40.10^{0}B, TSS mucilage 51.50^{0}B, acidity (0.30%) and vitamin C 19.90 mg / 100 g pulp were recorded.The fruits of this genotype is having good flavour and aroma. It is highly suitable in draught prone dry land conditions and also suitable for sherbet, powder and squash making (*Singh et al.*, 2016c & Singh *et al.*, 2017).

Narendra Bael -5

It is having spreading growth habit with medium dense foliage. The fruits are medium in size, round, with smooth surface and very thin rind (0.16-0.17cm), straw yellow colour at maturity, low in mucilage, moderately fibrous and dull yellow pulp with low seed content. Excellent in taste and flavour, the fruits have 33^{0}Brix total soluble solids in pulp and 48^0 Brix in mucilage and can be used as fresh as well as for various post harvest products.

Narendra Bael -7

The fruits are large in size, round with smooth surface and have medium thick rind, yellow at maturity, low in mucilage and fibrous, and an attractive yellow pulp with low seed content. Sparse in bearing and fruit did not ripe uniformly. Fruits are good in taste and flavor and it is highly suitable for *murraba* and other processed products.

Narendra Bael -9

Fruits are medium to large in size, roundish-oblong with smooth surface with thick rind, light yellow at maturity, average in mucilage, moderately fibrous, slightly golden-yellow pulp with low seed content. The fruits are good to taste, slightly acidic and can be used as fresh as well as processed into several value added products. Its keeping quality for storage is very good.

Narendra Bael -16

The leaves are small dark green in colour. Fruits are small in size, round with rough surface and thick rind, greenish yellow at maturity, average in mucilage, and fibrous, yellow pulp with medium seed content. The fruits are good to taste, acidic and can be used for powder making as well as processed into several value added products. Its keeping quality for storage is good.

Narendra Bael -17

The leaves are large in size. Fruits are medium to large in size, roundish-oblong with smooth surface and thin rind, light yellow at maturity, average in mucilage, moderately fibrous, yellow pulp with medium seed content. The fruits are good in taste, and flavor, slightly acidic and can be used as fresh as well as processed into several value added products like powder and sherbet. Its keeping quality for storage is good.

Pant Aparna

It is dwarf in stature, almost thornless, precocious and heavy-bearer. The leaves are large, dark green and pear shaped. Fruit has globose shape, medium sized fruit; pulp yellow with thin shell. Mucilage, seed and fibre are low. Mucilage and seeds are enclosed in separate segments. Flavour and taste are very good.

Pant Shivani

It is an early season maturing and spineless variety having sparse foliage and fruiting. Trees are tall, vigorous, dense, upright growth having sparse fruiting. Fruit shape is ovoid, oblong and medium to large sized fruits. Colour of fruit is lemon-yellow and its storage quality is good. Shell is thin, lemon-yellow pulp with pleasant flavour and mucilage, seeds and fibre content are medium. Taste is very good and highly suitable for powder making.

Pant Sujata

It is a mid-season variety and fruit is globose shaped, slightly depressed at both ends with medium sized fruits. Fruit and pulp is light yellow. Storage life is better, rind is slightly thick, and seeds, mucilage and fibre are low. Its flavour is pleasant and taste is very good.

Pant Urvashi

It is a mid-season variety. Fruit is ovoid-oblong with medium to large fruits. Fruit is yellow, rind is medium to thin and pulp is light yellow. Fruit has 62.35% pulp with pleasant flavour. Seeds and mucilage are medium, fibre content low. Small to big sized fruits may be seen on the tree.

CISH-B-1

It is early maturing variety. The fruits are medium in size, oval-oblong, with smooth surface, yellow at maturity, low in mucilage and fibrous, an attractive yellow pulp comparatively high seed content. Fruit pulp is very good in taste and flavour with low mucilage. Keeping quality of fruit is very good and highly suitable for sherbet and powder making.

CISH-B-2

It is late maturing variety. The fruits are medium in size, obovate, smooth surface, and yellow at maturity and the rind is thick, low in mucilage and fibre, an attractive yellow pulp, with low seed content. Good in taste and flavor with very good keeping quality. Fruit does not ripe properly; hence an application of ethrel is required for uniform ripening.

Propagation

Raising Root Stocks in Polyethylene Bags

Rootstock raising in nursery beds and lifting of budded plants with earth ball in highly sandy soils is practically not feasible. Transportation of plants for a long distance may also cause high mortality particularly under hot semi-arid and arid environment. To reduce the time for raising root stock and to avoid damage during handling and transportation, polyethylene tubes and polyethylene bags may be used on commercial scale. Generally, polyethylene bags (25cm x 10cm) are used for raising the rootstocks. Small holes are made in the bottom and sides of polythene bags for drainage and aeration and filled with porous rooting medium or pot mixture for raising rootstocks. Generally, 1-2 seeds are sown in each poly bags and then placed in trench bed, so that it can be irrigated easily. Sometimes coiling of root is become problem, hence root pruner is also used for trimming of roots. Polyethylene bags of the size of 25 cm x10 cm, 25cm x 15 cm and 30 cm x 15cm as per the requirement are being used for raising the rootstocks in India. Seeds can also be sown in the raised nursery beds in open or playhouses and transplanting of very young seedlings (2-4 leaf stage) in the polythene bags gives 82.00% success. Growth of the rootstocks can be improved with the application of 1g/ litre urea solution. About 8-12 months old seedling of uniform size having stem of pencil thickness are used as rootstock for budding and grafting. Plants raised in the polythene bags can easily be transported to distant places with higher planting success.

Rootstocks Raised in the Field

For getting maximum survival and success of budding under rainfed conditions of hot semi-arid and arid ecosystem, sowing of seeds may be done directly in the field during monsoon (June-July) at demarcated location, and the seedlings become ready for *in-situ* budding during June, 10-11 months after sowing. If there is long gap between two spell of rain, plants should be irrigated as and when required. Proper irrigation is required to save the plants in the field during summer. Since no planting is involved in this case, the plants retain the roots properly and thus become more hardy and vigorous after *in-situ* grafting or budding.

Raising of Rootstocks in Nursery Beds

For raising of rootstocks, seeds are sown in nursery beds at 2 cm depth during March to May and watered regularly. The seeds have no dormancy and they germinate within 8-12 days after sowing. These seedlings can be directly transplanted in pit in the field at 3-5-leaf stage during rainy season for *in–situ* patch budding or can be done in nursery beds. The seedlings became ready after 6-8 months for *in–situ* patch budding under field condition. This method not only saves time but also ensures higher success (95%) under dryland conditions. Under Godhra condition, sowing is done during February-March so that the seedling should attain buddable size during May–June. The budded plants become ready for transplanting 1 month after budding (Singh *et al.*, 2012 b). Foliar spray of GA_3 (1000 ppm) has been found effective to accelerate the growth of seedling with maximum root and shoot ratio

Rootstock

Rootstock selection for vegetative propagation of bael is important, as it controls the vigour and equilibrium between yield and quality. Dwarfing rootstocks induce dwarfness and facilitate easy management of the orchard. Generally, seeds of *deshi* plants are used for raising root stocks in nursery, but it can also be grafted on the *Aegle fraeglegabonensis, Aeglopsis chevalier* and *Aegle paniculatum.* However, it is not commonly used for raising of bael plant. .

Asexual Propagation

Bael can be propagated asexually through budding, root suckers, grafting, and air layering and cuttings. For better success through layering, invigoration of the mother tree by heading back of few thick branches in April and layering in August with the application of IBA at 10000 ppm in lanolin paste should be done, while the highest rooting rates of 75-80% can be achieved using 5000 ppm IBA, etiolation and invigoration in ringed stem cuttings. Patch budding is the commercial method of propagation of bael.

Patch Budding

Promotion of Scion Wood

For getting vigorous and healthy scion shoots, the branches are detopped in April- May so that axillary shoots could arise below the cut portion and attain the length of 0.40 m to 0.60 m in two months, and can be used as scion shoot for better success and survival.

Selection of Bud Wood

Under rainfed conditions, scion shoots become available during the active growth (May onwards). The budsticks (1month old, 25 mm girth, current growth) with well swollen and recently matured buds (but still not open) are collected. The active growth period is indicated by easy and clear separation of the scion shoot from the wood of scion sticks. Patch budding is most common method of propagation of bael. Under rainfed conditions, patch budding gives more than 95% success rate when it is performed in May-June before onset of rain either in nursery under field condition (Singh *et al*., 2014e).

Soft Wood Grafting

Scion shoots are detached from the mother plant with the help of secateurs or sharp grafting knife for softwood grafting. For this, seedling rootstock is cut at 20-25 cm height and the top portion is removed. With the help of knife, 3-5 cm long vertical downward incision is made in the center of the rootstock. A sharp cut of same size is made on both the sides on the base of the scion shoot to make wedge shape. Thereafter, prepared scion is carefully inserted in vertical slit of the rootstock and tightly secured with the help of 200 gauge thick and 2 cm wide polythene strips. The polythene strips should carefully be removed after completion of the union. Maiti *et. al.,* (1999) reported that *bael* can be propagated successfully and among different grafting methods tried; whip grafting had given good response (70% success). Recently, softwood grafting during May-June (before onset of rain) has also been found successful with more than 85 per cent success in bael under semi-arid ecosystem of western India.

High Density Planting

Planting

Pits of 1m x 1m x 1m size should be prepared during March- April and left open during summer for solarization. The pits should be filled with 50:50 a mixture of FYM and top soil and it is also advisable to mix 125 g SSP and 150 g *neem* cake. Drenching of pit with Tafaban (chlorpyriphos) @ 2-3 ml/l should be done to avoid the termite attack. With the onset of the monsoon season (June- July), the pit soil started settling down and the seedlings grown in nursery should be planted in the pits during monsoon. Planting should be done during drizzling rain for better establishment particularly in rainfed areas. After planting, the soil around the seedlings should be firmly pressed to avoid formation of air pockets for better survival of the plant.

Planting Density

Recently, high density planting has been consistently gaining popularity particularly with small land holdings in the Indian conditions. High density planting makes maximum use of land to achieve high yields in the early period of orchard life along with better orchard management. With an increase in planting density, the yield may increase, but beyond threshold density, quality may be deteriorated and may not be profitable in terms of economic return. Normally, the bael is grown in India at planting distance ranging from 8-10 m distance. Keeping in view the growth habit of varieties, bael can be planted from 5 to 8 m. However, the variety Goma Yashi can be grown at 5m x 5m distance successfully because of its short nature of growth habit (Singh *et al.*, 2016b & Singh *et al.*, 2017).

Canopy Architecture

Juvenile plants are trained with the help of stakes so that they can grow straight. The growth of bael plants can be managed at desired size and shape by canopy architecture. Plants may be headed back at 1 to 1.5 m from the ground level to facilitate the emergence of new growth below the cut point. Three to four equally spaced shoots (limbs) are retained around the stem to form the main scaffold of the trees. These shoots are allowed to grow and pruned to 50% of their total length during 3rdyear (April-May) to promote emergence of new shoots below the cut point. As the new shoots emerge, they are allowed to develop further. It is better to train the plant in modified leader system of canopy architecture and it should be completed within initial 3-4 years. Thereafter, selective pruning should be done to maintain the balance canopy to the tree.

INM

Systematic work on its nutritional requirement particularly under rainfed conditions has not been reported on bael till now. The Fertilizers are applied in two split doses, 10 kg FYM, 75g nitrogen, 50g phosphorus and 50g potassium per year per plant is given to one year plant and this dose should be increased in same proportion up to the 10 years age for its better growth. FYM plus half dose of nitrogen and potassium and full dose of phosphorus should be applied just after rain while remaining doses should be applied in the last week of August under rainfed conditions of hot semi-arid ecosystem. However, use of organic based nutrient will be ideal for sustainable production of bael under dryland condition (Saroj *et al.*, 2006).

Irrigation

It is a hardy tree and it can be grown successfully without irrigation under rainfed condition, however, young plant should be irrigated need based particularly during summer at 15 days interval for better growth and development. In bearing orchards, irrigation should be given at monthly interval after rainy season. Regular soil working around the tree should be done at appropriate time to avoid loss of water through evaporation under rainfed conditions.

Mulching and Intercropping

Bael plants are susceptible to water logging and frost; proper care should be taken to avoid such conditions. Suckers appearing from rootstocks should be removed periodically. Under rainfed condition, application of organic mulch in tree basin is found very beneficial for successful cultivation of bael. Mulches not only conserve soil moisture, but also impart manifold beneficial effect on soil fertility and growth and yield of plant. Locally available and easily decomposable organic materials like paddy straw, maize straw, grasses, husk etc should be applied with 20 cm thickness after rainy season. Leaf litter of bael under the canopy is not only an effective mulch to retain soil moisture during summer but also improve the soil properties (Singh *et al.,* 2013a and Saroj *et al.*, 2006).

Orchard Floor Management

Inter cropping in newly developed bael orchard had no adverse effects on plant growth up to 5 years. Intercropping of guar with bael cv. Goma Yashi increased the fruit yield. Cover cropping with lobia, moth bean is found to increase water holding capacity of light soils as a result of increased organic carbon content in the soil. Bael +cluster bean, bael +bottle gourd, bael + Bhindi combination are ideal to generate extra income from bael orchard under rainfed semi-arid conditions (Singh *et al.,* 2011a & 2013a).

Bael Based Cropping Models and Crop Diversification

Under rainfed conditions, various fruit crop models can be adopted to minimize the risk and enhance the productivity. Bael based cropping model are found to be beneficial in term of income as well as risk of crop failure. Bael+ aonla + karond+ drumstick, bael+ chironji + fig + custard apple, bael+ khirni + phalsa + wood apple cropping models are useful to enhance the productivity of dryland tracts of the country. Lay out and plantation of these crops should be done at closer spacing with proper management of canopy so that more income can be generated (Singh *et al.,* 2017).

Green Manuring

Green manuring with leguminous crop in the basin of fruit tree is very beneficial for proper growth and development of plant under rainfed semi-arid environment. At Godhra, Gujrat, sunhemp (*Crotalaria juncea*) and mung (*Vigna radiata*) sown in the basin of bael tree during monsoon and incorporated in the basin in the first week of September which not only improved the physical and chemical properties of basin soil but also enhances the growth and development of tree (Singh *et al.*, 2017).

Management of Physiological Disorders

Fruit Drop

Fruit drop is a natural phenomenon, but its extent of damage is a matter of concern. The extent of fruit drop varied according to genotypes/varieties and locality. Fruit drop in bael may be due any one or combination of reasons. It depends on cultivar, maturity, extent of fruit set, disintegration of ovule, embryo abortion, physilogical imbalance, fruit cracking etc. Immature fruit drop (marble size) has also been observed. Sometimes cricket ball size fruits also fall down after cracking during rainy season. The extent of fruit drop in bael can be reduced effectively by adopting better orchard practices which include mulching with organic materials and proper soil nutrient management (macro–micro) and application of growth hormones like NAA (15-20 ppm/ litre) at pea size stage during July-August.

Sun Scald

Sun scald is one of the major disorders of *bael* under dryland conditions. It is manifested by turning of normal green shell into dark brown at the point where it is exposed to hot sun for maximum period during day hours. Sometimes the pulps of fruit beneath the shell also get affected due to moisture loss and irradiation. The temperature of sun scalded portion is increased by 8-10° C as compare to unexposed portion of the fruit. Mulching and canopy management and wrapping of fruit individually (cotton cloth, perforated polythene bags) are useful to reduce down this disorder up to some extent. Optimum moisture level in rhizosphere is also helpful in reduction of fruit sunscald.

Fruit Cracking

It is physiological disorder and its degree of damage depends according to genotypes/varieties and agro-climatic conditions. The cracking can be minimized by maintaining optimum soil moisture regime and by provision of wind breaks against hot desiccating wind side of orchard. Organic mulches like paddy straw, maize straw, *subabul* loppings can effectively be utilized in maintaining soil moisture of tree basin particularly during summer under rainfed conditions of hot semi-arid ecosystem (Singh *et al.*, 2017).

Harvesting and Post-harvest Management

Development of bael fruit follows a single sigmoid curve with three distinct period of growth. The initial growth is slow about one month followed by rapid increase for three months and then more or less stationary phase till the fruit are harvesting. Bael is one of the fruits that intact on the tree for long time. The fruit is categorized as climacteric fruit. In general the ripening of fruit is judged by the turning of the colour of the shell, flesh colour, TSS and aroma of the fruits. Ripe fruits can be made available 2-3 months ahead before natural ripening time by giving pre harvest treatment of 1000-1500 ppm ethrel and storage of fruits at 30° C after harvesting. It takes 18-25 days for fruits to be artificially ripened. Varieties CISHB-1 and Pant Shivani start ripening from the second fortnight of February; Goma Yashi during first fortnight of March; Pant Aparna and Pant Sujata in the second fortnight of March; NB-16, NB-17 and Pant Urvashi in first fortnight of April; NB-9 in second fortnight of April; NB-7, NB-5 and CISHB-2 in first fortnight of May under rainfed conditions of western India (Singh *et al.*, 2011b and Singh *et al.*, 2017)).

Proper care should be taken during harvesting. The fruits should be harvested individually with little portion of stalk and should not be allowed to drop on the ground, as the fruits are liable to develop cracks easily due to brittleness of peel, otherwise a minor cracks or damage on the peel can cause heavy spoilage during storage. To prevent fruit from falling on the ground, fruit picker is used for harvesting. Fruits take about 10-11 months after fruit set to ripen on the tree. Mature green fruit about 8 months after fruit set are ideal for making of preserve while ripe fruits are mostly used for beverages hence harvesting should be done at ripe stage (Singh *et al.,* 2013 a).

Yield

Bearing in budded plants starts during 3rd year and gives 80-160kg/ tree/year at the age of 10 years whereas seedling trees require 7-9 years to bear fruits, giving 120-300kg/ tree of smaller size fruits at the age of 12 years (Singh *et al.*, 2011a and Singh *et al.*, 2014 c). Yield and quality attributes of different varieties of bael during 7th to 8th year under rainfed conditions has been reported by Singh *et al.*, 2016d.

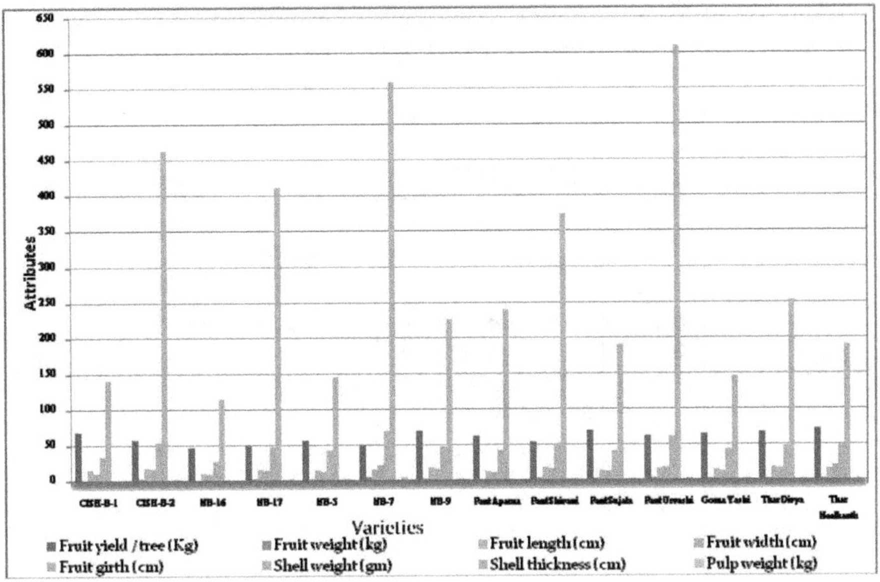

Fig.1: Yield and quality attributes of bael varieties under rainfed semi-arid condition

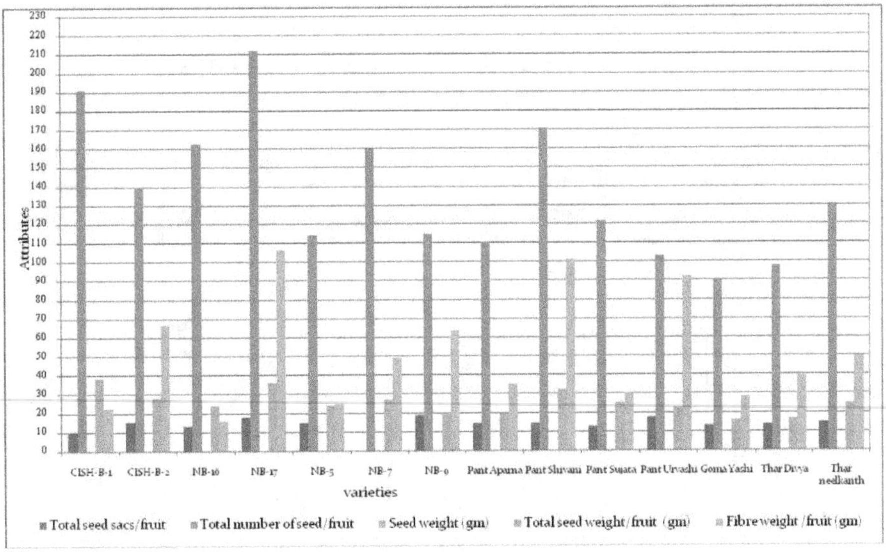

Fig. 2: Physical composition of bael varieties under rainfed semi-arid condition Singh *et.al.,* (2016d)

Table 1: Chemical composition of bael varieties under ranfed semi-arid condition

Varieties	Acidity (%)	Vitamin C (mg/100g)	Total phenol (%)	T.S.S pulp (Brix)	T.S.S. mucilage (Brix)	Total sugar (%)	Reducing sugar (%)	Non reducing sugar (%)	TSS: Acid ratio
CISH-B-1	0.34	19.75	2.75	33.50	46.28	18.00	4.80	13.13	91.76
CISH-B-2	0.43	17.37	2.83	34.00	40.75	17.34	4.25	12.04	79.06
NB-16	0.48	18.19	2.54	35.90	45.60	18.00	4.00	14.50	74.79
NB-17	0.32	18.28	2.43	34.78	48.52	16.60	4.50	13.00	108.68
NB-5	0.32	20.63	2.38	36.21	46.15	18.50	4.75	13.75	113.15
NB-7	0.36	19.78	2.47	34.50	49.00	16.15	3.30	12.85	84.88
NB-9	0.33	19.20	2.46	38.29	46.83	19.98	4.85	15.16	106.93
Pant Aparna	0.35	17.15	2.43	34.23	37.00	19.93	4.62	14.98	97.8
Pant Shivani	0.48	19.54	2.44	35.49	48.65	17.77	3.54	14.23	73.93
Pant Sujata	0.43	17.16	2.41	33.25	44.47	18.00	3.35	14.65	77.32
Pant Urvashi	0.49	17.59	2.34	36.44	49.50	16.15	4.00	14.00	74.36
Goma Yashi	0.31	21.03	2.35	37.45	43.43	18.75	4.12	15.00	124.8
Thar Divya	0.30	19.67	2.79	36.78	43.10	17.50	4.50	13.04	122.60
Thar Neelkanth	0.29	23.40	2.15	41.20	52.50	19.84	4.84	13.67	142.07
C D at 5%	0.043	1.78	0.23	3.45	4.32	1.77	0.39	1.34	10.21

Singh *et al.*, (2016d)

Storage, Grading and Packaging

There is no recommended practice for grading and packaging of bael fruits. Injured fruits should be sorted out and healthy ones are graded based on the size of fruits. Generally, fruits packed in gunny bags, baskets or wooden crates and sometimes they are transported without any packaging. It is desirable to use cushioning materials for packaging of bael fruits. Storage life of bael depends upon the variety and time of harvesting. Fruits harvested at maturity can be stored up to 12-18 days, whereas fruits harvested at ripe stage can be stored from 7-10 days under normal room temperature. In cold storage, Fruits can be stored for about 3months at 90°C and 85-90 per cent relative humidity (Singh *et al.,* 2011a & 2011b)

Management of Pests and Diseases

Bacterial Leaf Spots

It is characterized by water soaked spots on leaf surface and minute, round, water soaked (oily) grayish spots are seen on pericarp surface which shows raised margin and gradually increase in size and turn cankerous in look. Removal of more affected twigs followed by spray with Bordeaux mixture has been suggested to manage it. Spraying once or twice with streptomycin sulphate 250 ppm or Bordeaux mixture 1% at 12-15 days interval effectively control the disease (Singh *et al.,* 2011a & Singh *et al.,* 2014d)

Fungal Leaf Spots

It is characterized as small 4-8 mm, dark spots on leaf which generally start from leaf margin but can also be seen on any part of lamina. Concentric rings can also be seen around spots. Spray with mancozeb (0.2%) helps to manage the disease.

Fruit Rot

The disease is caused by *Aspergillus nidulans* and *A. nigerum*. The disease is somewhat serious during storage period of fruits. Pre harvest spray of Carbendazim (0.05%) and avoiding bruises to the pericarp during picking, storage and transport are suggested to manage the disease. It is better to wrap the fruits with phenol paper/newspaper and to line the containers with newspaper. Bamboo stick baskets and plastic crates are better for transport purpose.

Powdery Mildew

The disease is characterized by appearance of white floury patches on leaflets, especially on younger leaves which increase in size and cover entire lamina soon within 7-10 days (during November–December under Godhra condition).

Later the color of the colony turns slightly pinkish or grayish. The disease is caused by *Oidium* sp. Spray with Carbendazim 50 wp (Bavistin 0.1%) or wettable sulphur (0.2%) is found to be useful (Singh *et al.*, 2014d).

Gummosis

Oozing of gum is common in bael orchards. The disease is characterized by oozing of pale or amber coloured gummy substance initially from bark of lower portion of trunk and later on other branches also.To manage the disease, scrapping off the infected portion of bark with the help of a sharp knife followed by application of Bordeaux paste. Removal of highly infected twigs and incorporation of *Trichoderma viridae* propagules in the soil of rhizosphere of bael were found helpful to control the disease (Singh *et al.*, 2014d).

Pests

Generally, bael is free from serious pests and diseases, but few insect pests are known to cause damage the crop, especially when environmental conditions are very conducive to paste attack. Termite attack on new saplings is a major problem which results into defoliation and whiting of plants. Application of chlorpyriphos @ 2ml/litre has been found effective to control the termite. Chafer beetles or leaf eating caterpillars (lemon butterfly) cause damage to the plants and it can be controlled by 1-2 sprays of dimethoate @ 1.5-2.0 ml/litre at 15 days interval.

Value Addition

Bael fruit is not popular as table fruit due to its hard shell, mucilaginous texture and numerous seed and fiber content but popular as processed fruit. Bael fruit has been used widely from time immemorial for processing in the mature green form to prepare preserve. While from ripe fruit, bael sherbet is very common and useful as tonic.(Singh *et al.*, 2011c) Both mature and ripe fruits can be utilized for making of various post harvest products. Fruits can be processed into number of acceptable products like pickle, candy, slab, powder, toffee, squash, jam, RTS, ice cream, wine etc. (Singh and Roy,1984, Singh *et al.*, 2013 and Singh *et al.*, 2016a)

Potential Benefits of Adoption of Advance Approaches

Production and Productivity

Change in temperature and rainfall patterns with occurrence of natural calamities due to climate change are major threat to future bael cultivation. *In-situ* soil water conservation and rainwater harvesting are very important for its commercial cultivation in arid region. By adopting climate resilient varieties, high density planting, canopy architecture, intercropping, water harvesting techniques, mulching, organic farming etc., the production and productivity will be improved and area under bael cultivation will also be increased especially in arid and semi-arid region.

Socio-economic Impact

There will be great effect on socio-economic condition of the resource poor farmers by adoption of bael in farming system. There will be significant impact on labour engagement, farm income, social participation, farm expenditure, and increase in level of education resulting into better socio-economic condition of bael growers. However, socio-economic and bio-physical factors along with policy environment have larger contribution in explaining bael production and productivity and technology impact alone cannot be expected to bring a positive impact. Government intervention in price fixation of bael and formation bael farmers association will have the larger impact for commercialization and expansion of area under bael cultivation in resource poor and aberrant agro-climatic conditions.

Health and Nutritional Security

Large number of bioactive compounds has been isolated from various parts of the bael tree. Several compounds such as skimmianine, cineole, citral, citronellol aegelin, lupeol and marmesinin have been purified from bael leaves. The bioactive compounds isolated from bael fruits are marmelosin, luvangetin, aurapten psoralen marmelide. Marmelosin is most probably most therapeutically active compound present in fruit (0.03-0.37%) and vary according to variety and locality (Dixit and Dutt, 1932). Tannin present in the unripe fruit of this plant has astringent property and is excellent remedy for diorrhoea. Among the important bioactive compounds isolated from bark are fagarine, marmin, auropotins, umbelliferone, lupeol and skimmianine etc.(Chatterjee and Roy, 1959). Various chemical constituents *viz,* alkaloids, coumarins, steroid etc. have been isolated and identified from the different parts of the plants. Different parts of the plant viz., leaves, roots, stem, flower, fruit have been used in enthno medicine to exploit its' therapeutic properties including astringent, antidiarrhoeal, antidycentric, antipyretic, antiulcer, antidiabetes, antibacterial, antiviral antifungal, anticancer, analgesic, redioprotective because compound purified from bael have been proven to be biologically active against several major diseases (Maity *et al.*, 2009). They are also valued in *Ayurvedic* medicines.

Bael fruit is one of the most nutritious fruits. Analysis of the fruit gave the following values: 61.5g moisture, 1.8 g protein, 0.39 g fat, 31.8 g carbohydrates,

1.7 g minerals, 55 mg carotene, 0.13 mg thiamine, 1.19 mg riboflavin, 1.1mg niacin and 8.0 mg vitamin C per 100 g of edible portion (Gopalan *et al.,* 1985). No other fruit has such a high content of riboflavin. Tannic acid is only phenolic substance detected from bael fruits. Analysis of the leaves gave the following values (dry basis): crude protein, 15.13, ether extra, 1.54, crude fibre, 16.45, N-free extra, 52.83, ash, 14.05, calcium, 5.93 and phosphorus, 0.69 per cent. The seed yields oil 34.4% on dry basis and the fatty acid composition of oil as follows: palmitic, 16.6; stearic, 8.8; oleic 30.5; linoleic, 30.0; and linolenic, 8.1 per cent. Bael seed contains 62% protein and 3% each carbohydrate and ash (Banerjee and Maiti, 1980). The fruit yields two per cent dried, water soluble gum. Hydrolysis of the gum gave: galactose, 20.4, arabinose, 10.7, D-galacturonic acid, 25.2% and traces of rhamnose (Haskar and Kendulkar, 1961). Physico-chemical analysis of different cultivars has been reported by Roy and Singh (1978).

Environmental Security

Bael has substantial carbon sequestration potential for mitigation of climate change and has capacity to grow in variety of soil and climatic conditions. Carbon sequestration has gained interest in the recent past due to its impact on sustainable crop production. Carbon fixed by plant during photosynthesis and added to the soil in form of litter is the primary source of C in ecosystem. Bael has dense foliage with high photosynthesis capacity and can be grown under HDP, there is more chance of sequestrating the atmospheric carbon resulting into the reduction of CO_2 in atmosphere.

Future Scope and Research Needs

1. Existing biodiversity needs to be exploited for selection of better and trait specific genotypees.
2. Emphasis should be given on post harvest technology to develop export oriented value added products.
3. Establishment of small scale processing unit should be promoted for its commercialization.
4. Screening of genotypes against abiotic especially against drought and frost tolerance and moisture stress.
6. Standardization of high density planting and canopy architecture.
7. To create awareness about its medicinal significance among the masses.
8. Standardization of production technologies suits to arid and semi-arid region.

8. Bael based farming models should be developed for high yield and proper land utilization.
9. Being high therapeutic values due to presence of large number of bioactive compound in different parts of tree, an organic farming should be emphasized to generate better income.

References

Bal, J. S. 2003. Genetic resources of underutilized fruits in Punjab subtropics. *Acta Horticulturae*, 623: 325-331.

Balunas, M. J. and Kinghorn, A. D. 2005. Life Sciences, 78, 431-441.

Chatterjee, A. and Roy, S. K. 1959. Chemistry of extractives from hardwood. 1. Constituents of hard wood of *Aegle marmelos*. *J. Indian Chem. Soc.*, 36:267-69

Dhankhar, S., Ruhil, S., Balhara, M., Dhankhar, S. and Chhillar A. K. 2011. *Journal of Medicinal Plants Research*, , 5, 1497-1507.

Dixit, B. B. L and Dutt, S. 1932. The constitution of marmelosin. *J. Indian Chem. Soc.*, 9: 271-79.

Gopalan, C.B. N. Rama Sastri and Balsubramanian 1961. Nutritive Value of Indian Food. I.C.M.R, Hyderabad, India.

Haskar, C.N. and Kendurkar, S.G. 1961. Uses of bael gum. *Printindia*, 11(1): 135.

Jana, B. R., Das, B. and Singh, M. 2014. Conservation and performance of some bael (*Aegle marmelos* correa) genotypes under rain-fed ecosystem of eastern India. *International Journal of Information Research and Review*, 1(10): 133-134.

Johns, L. and Stevenson, V. 1979.The complete book of fruit. Anqus and Roberson Publications.

Kirtikar, K. R. and Basu, B. D. 1935 *Indian medicinal plants*, Vol-I Published by L. M. Basu, Allahabad.

Kumar, G., Karthik and Rao, L. K.V.B. 2010. *Journal of Pharmacy Research* 3, 539-542.

Maity Pallab, Hansda Dhananjay, bandopadhyay, Uday and Mishra, D. P.(2009).Biological activities of crude extracts and chemical constituents of bael (*Aegle mamelos* Correa). *Indian Journal of Experimental Biology*, 47:849-861.

Malik, S. K., Chaudhury, R., Dhariwal, O.P. and Bhandari, D.C. 2010. *Aegle marmelos* (Bael). In: Genetic Resources of Tropical Underutilized Fruits in India. NBPGR, New Delhi, pp. 18-24.

Om, Prakash 1961. Food and drinks in ancient India. Munshi Ram Manohar Lal. Orient Book Sellers and Publishers, Delhi.

Rai, D., Mishra, K. K. 2005. Studies on genetic divergence in Bael (*Aegle marmelos* Correa). *Indian Journal of Horticulture*, 62:152-154.

Rai, M., Dwivedi, R. and Gupta, P.N. 1991.Variability and potentials of identified germplasm in bael (*Aegle marmelos* Corr.). *Indian Journal of Plant Genetic Resources*, 4(2):86-92

Roy, S. K. and Singh, R. N. 1978. Studies on utilization of bael fruits (*Aegle marmelos* Correa) for processing. I.Physico-chemical analysis of different cultivars. *Indian Fd. Packer*, 32 (6): 3-8.

Roy, S. K. and Singh, R. N. 1979a. Studies on utilization of Bael fruit (*Aegle marmelos* Correa.) for processing II. Extraction of bael fruit pulp. *Indian. Fd. Packer,* 33 (1): 5-9.

Saroj, P. L., Singh, R. S. and Singh, A. K. 2006. Bael (*Aegle marmelos* Correa).In: Advances in Arid Horticulture-production technology of arid and semi-arid fruits (Saroj *et. al.*, Eds.), IDB, Lucknow, Pp.21-38.

Sharma, S. K., Singh, R. S. and Singh, A. K. 2013. Bael, Biodiversity in Horticultural Crops Vol.4 (Edt. K. V. Peter), Daya Publishing House, New Delhi pp. 285-300.

Singh A. K., Singh, Sanjay, Singh, R. S., Joshi, H. K., and Sharma, S. K. (2014a). Characterization of bael varieties under rainfed hot semi-arid environment of western India. *Indian Journal of Agricultural Sciences*, 84(10): 80-86.

Singh, A. K., Singh, Sanjay, Singh, R. S. and Makwana, P. 2014b. Phenology, floral biology and pollination in bael varieties under rainfed semi-arid conditions of western India. *Indian Journal of Arid Horticulture*, 9 (1&2): 84-90.

Singh, A. K. Singh, S. , Singh, R. S.and Makwana, P. 2016a. Organoleptic scoring of RTS prepared from bael (*Aegle marmelos* Corr.) varieties, *Indian journal of Agricultural Sciences*, 86 (5): 611-4.

Singh, A. K. Singh, S. and Makwana, P. 2015a. Intervarietal morphological variability in bael (*Aegle marmelos* Corr.) under rainfed semi-arid hot ecosystem of western India. *Current Horticulture* 3(2):3-9.

Singh, A. K. Singh, Sanjay and Joshi, H. K. 2013a. Improving socio-economic through rainfed bael. *Indian Horticulture July- August,* pp-14-17.

Singh, A. K. Singh, Sanjay and More, T. A. 2014c. Preliminary evaluation of bael varieties under rainfed conditions of western India. *Indian Journal of Horticulture,* 71 (2): 264-68.

Singh, A. K. Singh, Sanjay and P. L. Saroj 2017. Cultivating climate resilient bael for future, *Indian Horticulture*, 62(4): Cover II-43-45.

Singh, A. K. Singh, Sanjay, Joshi, H. K. Bagle, B. G. and More, T. A. 2008. Evaluation of bael genotypes for growth behaviour and floral traits under semi-arid ecosystem of western India. *The Horticultural Journal*, 21(3):140-42.

Singh, A. K. Singh, Sanjay, Joshi, H. K. and Singh, R. S. 2012a. Goma Yashi to enrich fruit basket. *Indian Horticulture,* 57(5):6-8.

Singh, A. K. Singh, Sanjay, Mishra D. S.and P. L. Saroj 2016b. More crop with minimal water *Indian Horticulture*, 61(6): 86-91.

Singh, A. K. Singh, Sanjay, Singh, R. S and Makwana, P. 2015b. Thar Divya: an early maturity variety of bael for dryland. *Indian Horticulture,* November-December pp.11-13.

Singh, A. K. Singh, Sanjay, Singh, R. S. and Sharma B. D. 2016c. Thar Neelkanth: a new bael variety. *Indian Horticulture,* 61(5):8-10.

Singh, A. K., Singh, Sanjay, Makwana, P. and Sharma, S. K. 2016d. Evaluation of bael germplasm under rainfed semi-arid environment of western India. *International Journal of Noni Research*, 11(1&2): 11-19.

Singh, A. K., Singh, Sanjay, Singh, R. S., Sharma B. D. and More, T. A. 2011a. The bael- fruit for dryland, CIAH Bulletin.

Singh, A. K., Singh, Sanjay, Singh, R. S. and Joshi. H. K. 2012b. Morphological variability of bael varieties under rainfed conditions of hot semi-arid environment of western India. *Indian Journal of Arid Horticulture,* 6 (1-2): 35-37.

Singh, A. K., Singh, Sanjay, Singh, R. S., and Makwana, P. 2014d. Reaping diseases and pest free bael. *Indian Horticulture*, 60(5):35-36.

Singh, A. K., Singh, Sanjay, Singh, R. S., and Sisodia, P. S. 2011b. The Bael, A potential fruit for future, Agro-tech Publishing Academy, Udaipur, Rajasthan.

Singh, A. K., Singh, Sanjay, Singh, R. S., Contractor, K. and Makwana, P. 2014e. *In-situ* patch budding for better establishment of bael in rainfed areas, *Indian Horticulture,* 59(5):24-25.

Singh, A. K., Singh, Sanjay, Singh, R.S. Joshi, H. K and Contractor, K. 2014g. Exploring biodiversity in bael for healthy and wealthy life. *Indian Horticulture*, 57(6): cover II&III. pp24-26.

Singh, R. N. and Roy, S. K. 1984. *The Bael*. I.C.A.R., New Delhi, p.25.

Singh, R. S., Bhargava, R. and Singh, A. K. 2013. Organoleptic rating of RTS from pulp of bael cultivars. *Indian Journal of Arid Horticulture,* 8(1-2): 62-64.

Singh, R. S., Meena S. R., Singh, A. K., Bhargava, R.and Sharma, b. D. 2015c.Variability in bael (Aegle marmelos Correa) germplasm collected from Rajasthan. *Indian Journal of Arid Horticulture*, 10 (1&2): 91-93.

Vadivelan, Umasankar,R. ,P., Dipanjan, M., Dhanabal, S.P., Shanish, A., Satish Kumar and Elanko, M.N. K. 2011. *Der Pharmacia Sinica* 2, 299-304.

18

Tunnel Technology of Cucurbit Cultivation in Hot Arid Region

B.R. Choudhary, A.K. Verma and Dhurendra Singh

Introduction

Indian arid zone is one of the largest subtropical deserts of the world of which 20% is arid and rest is semi-arid. Hot arid zone is spread over 31.7 million ha area and mainly confined in the states of Rajasthan, Gujarat, Andhra Pradesh, Punjab and Haryana. The major part of hot arid region of the country lies in western Rajasthan (19.62 mha) followed by North-western Gujarat (2.16 mha). It is characterized by high aridity index, extremes of temperature low and variable precipitation, very high wind velocity and dust storm, high soil pH, high infiltration rate, very limited availability of groundwater and saline ground irrigation water (More, 2010). Under such harsh climatic conditions cultivation of vegetable crops in open condition results in very low and poor-quality yield which fetches less return per unit area. By creating the suitable micro-climate for plant growth round the year cultivation of vegetables can be done under adverse climatic conditions. It is possible through protected cultivation which has tremendous scope in peri-urban areas of the country (Singh and Sirohi, 2004). It is the best way to increase the productivity and quality of vegetables, especially cucurbits. India is the second largest producer of vegetables in the world next only to China. Presently India has 178.17 million tonnes of total vegetable production, but the productivity and quality of most of the vegetable crops is very poor (17.40 t/ ha) due to several biotic and abiotic stresses due to open field cultivation. In Rajasthan, the total production of vegetables is very less *i.e.* only 1.81 million tonnes from an area of 0.17 m ha with a productivity of 10.66 t/ ha.

Production of cucurbits in India is mainly restricted to open field cultivation and most of the varieties grown are open pollinated in nature and are mainly grown in two seasons, as a summer and rainy season crop in northern plains of the country. The crops grown under open condition are always affected by several diseases and pests, resulting in low productivity and poor-quality produce. On the other hand, vegetable markets are flooded with these vegetables during their main season and sometimes the vegetable growers are even not able to recover the cost of cultivation spent on these vegetables. But the same vegetables are sold at very high prices during offseason in several parts of the country. The demand of off-season vegetables is increasing rapidly in several big cities of the country because of continuous increase in availability and continuous changing choice of consumers towards off season and high quality produce. There are different types of protected structures based on the cost and technical feasibility such as high cost (fully automated), medium cost (semi-controlled) and low cost (naturally ventilated, tunnels). Thus, high cost (fully automated), medium cost (semi-controlled) structures requires high input costs and good management practices compared to low cost structures, which have direct bearing on the economic viability of the production system. Before going for protected cultivation, knowledge about the techno-economic suitability of the protected structure or protected cultivation technology for the particular region is necessary for the growers. Therefore, it becomes important to adopt low cost protected structures to ensure maximum benefit against per unit of area and cost invested. Protected cultivation technologies can provide specific advantage over open field practices these advantage includes protection against both biotic as well as abiotic stress factors (Chandra, 1996).

Protected cultivation of horticultural crops is reasoned to be the next logical step to open field agriculture because one million hectares of area under protected cultivation is efficient to increase the production of vegetables alone by about 60-70 million tonnes. This can also create about one million additional jobs in the rural sector where unemployment problem is more acute (Chandra, 2001). The north Indian plains pose problems of cold winters and hot summers where the requirement is to moderate the winter and summer conditions. Besides, availability of suitable water in these areas is often highly inadequate. Under such climatic conditions a bit of protection from weather aberrations and biotic stresses greatly improves horticultural crop productivity (Chandra, 2004).

Looking to the increasing population, climate change, decreasing land holdings, increasing pressure on natural resources *i.e.* land & water and high demand of quality horticultural fresh produce we are forced to shift towards modern technologies of crop production like protected cultivation. Presently area under protected cultivation of horticultural crops is only around 40,000 ha and out of which large portion mostly in northern parts of India is not successfully being utilized for protected cultivation (Singh, 2014). Plastic tunnels had registered a significant growth of 60.5% for the years 2007 to 2012. The area protected by the plastic tunnels was 4.3% of the total area under the protected cultivation in 2012. Although promotion of protected cultivation will certainly help in creation of huge self-employments for unemployed educated youths and will also raise the national economy by sale of high quality produce in domestic and international markets. Production of vegetables under protected conditions not only provides high water and nutrient use efficiency but it can easily increase the productivity by 3-5 folds over open field cultivation of these crops under varied agro-climatic conditions of the country. This technology has very good potential especially in urban and peri-urban areas adjoining to the major cities which is a fast growing market for fresh produce of the country. Vegetable farming in agri-entrepreneurial models targeting various niche markets of the big cities of the country is regularly inviting attention of the vegetable growers for diversification from traditional ways of crop cultivation to such modern methods.

Despite the several constraints, the hot arid region of India offers very good opportunity for cultivation of cucurbits like, watermelon, muskmelon, longmelon, bottle gourd, ridge gourd, *tinda*, summer squash, cucumber, *etc*. The sowing of cucurbits is generally done in second fortnight of February for summer season crop. The flowering stage of February sown crop coincides with the prevailing high temperature and hot wind (*Loo*). These climatic conditions leads to high transpiration resulted in wilting of plants. Desert soil being sandy in texture gets warm soon which cause burning of pistillate flowers touching the ground. The population of pollinators (honey bee) is decreased considerably during summer season which affect yield. High temperature favours to induce more number of male flowers and less pistillate flowers. Cultivation of cucurbits under such conditions leads to low yield of poor quality. The existing low productivity of cucurbits in hot arid regions can be increased by adopting innovative approaches like drip irrigation and fertigation, low tunnel technology, spray of micronutrients and integrated crop management (ICM).

Still the vegetables are traditionally grown with little technological inputs, therefore, new technology like tunnel cultivation is to be brought to increase production, productivity and quality of cucurbits. Low tunnel technology can be used to raise off-season crop of cucurbits making cultivation more profitable. The use of low cost protected structures particularly low tunnels with some modification may become a viable option for successful cultivation of cucurbits in arid regions (Choudhary *et al.*, 2015).

Need of Low Tunnels

Presently, river bed cultivation is in practice for production of cucurbitaceous vegetables in off-season in northern parts of India, although area under river bed cultivation is very limited, which cannot be extended further. Besides, glut in the market and low quality produce causes poor economic returns to the farmers. But with the use of protective structures such as row covers or low tunnels vegetable crops like muskmelon, watermelon, longmelon, roundmelon, bitter gourd, bottle gourd, summer squash, *etc*. can be grown very early in the spring or summer season so that the produce to the market can be sent early in the season. It also extends the growing season for selected vegetable crops when large quantities of the crop produce are not available results in higher prices from their off-season produce. For example, crops like longmelon, roundmelon, bottle gourd, bitter gourd, muskmelon, summer squash, *etc*. if grown early in spring or early summer often command a greater price on the market to the growers.

Low tunnels are flexible transparent covering that are installed over the rows or individual beds to enhance plant growth by warming the air around the plants using heat from the sun especially during winter season. Plastic tunnels are transparent which provides required sunshine to the plants, and the plastic also plays a barrier against the cool air in winter.

Normally the economics of protected cultivation directly depends upon the initial cost of fabrication of the protected structure, its running cost and the available market for high quality produce. Therefore, low cost protected structures such as low tunnels, which can generally be fabricated with low cost and the running cost of such structures is also very low. These are highly suitable for off-season cultivation of cucurbits and are also highly economical for peri-urban areas of northern plains of India (Singh, 2014).

What is Low Tunnel?

Low tunnels also known as row covers are low cost miniature structures producing greenhouse like effect. A flexible transparent covering of plastic sheet or non-woven cloth is installed over the rows or individual beds of direct seeded/ transplanted vegetables. It enhances plant growth by warming the air around the plants during winter season and the photosynthetic activities of the plant by increasing the concentration of carbon dioxide inside the tunnels. When the night temperature during winter season goes below 8^{0}C for a period of 30-40 days it warms the soil and protect the plants from hails, cold wind, injury which stimulates germination, early growth and improves the quality of crop. Thus, the plants can be grown successfully during winter or off-season and by this way the crop can be advanced the by 30 to 40 days than the normal season. All together help to increase yield per unit of area and cost invested. Besides being inexpensive, these structures are easy to construct and dismantle.

Plastic low tunnels are miniature form of greenhouses to protect the plants from rains, winds, low temperature, frost and other vagaries of weather. The low tunnels are very simple structures requiring very limited skills to maintain and easy to constructs and offer multiple advantages. For construction of low tunnels, film of 30-50 micron would be sufficient.

Principle of Low Tunnels

In our surrounding atmosphere CO_2 concentration is 0.03% means 300 ppm. Plants use this CO_2 for photosynthesis. In poly house, during night time there is no photosynthesis but CO_2 is given out through respiration. This CO_2 remain accumulated around plants hence the CO_2 concentration is higher inside the polyhouse as compared to outside and this CO_2 is again used by plants growing in ploy house for rapid photosynthesis.

Advantages of Low Tunnels

i. Early and off-season production of cucurbits can be done to get better return.

ii. Provides crop diversification opportunities and supports production of high quality and clean products.

iii. Makes cultivation of vegetables possible in areas where it can't grow in open conditions *viz*., high altitudes and hot arid regions.

iv. It creates optimum microclimate for plant growths thus, increases photosynthetic activities of the plants and thereby yield.

v. Used for raising healthy and early nursery.

vi. Maintains optimum temperature for plant growth.

vii. Enhances nutrients uptake by the plants.

viii. Increases photosynthetic activities of the plants.

ix. Used for cultivation during winter.

x. Provide protection against unfavourable environment like high rainfall, hail, low temperature, frost, wind, insect-pests, *etc*.

xi. Inexpensive, as these structures are easy to construct and dismantle.

xii. It generates better revenue even though total yield may be the same or lower because market prices are higher.

Components of Tunnels

Structural

GI wires (2.0 m long with 4-6 mm diameter) or bamboo sticks.

Covering Material

Transparent and biodegradable polyethylene film of 30-50 microns having 2 meter width.

White coloured non-wooven cloth.

Important Points for Cultivation Under Low Tunnels

i. Prior to start off-season vegetable cultivation in tunnels, the farmer must have practical knowledge about vegetable cultivation.

ii. Soil and water quality should be tested before starting the cultivation.

iii. Recommended package and practices should be followed.

iv. Farmer must have the updated market information to earn high profit.

Field Preparations and Sowing

The land should be prepared to a fine tilth before construction of low tunnels. Application of manures and fertilizers depends upon the crop, variety and soil status. Well decomposed FYM should be to be applied @ 150-200 q/ ha, phosphorus @ 60-80 kg/ ha and potash @ 50-60 kg/ ha at the time of final land preparation. Nitrogen should be applied @ 60-80 kg/ ha in three split doses. One-third of N should be applied basally, about 8-10 cm away from the seeds. Remaining dose of nitrogen should be given at the time of earthing up and initiation of flowering as topdressing. If soil is acidic in nature, it is advisable to apply dolomite/ lime @ 20 q/ ha. Lime should be applied at least 20-30 days before planting. 45-60 cm deep and 45-60 cm wide trenches should be made at a distance of 2.0-2.5 m in east-west direction in December end. Recommended doses of fertilizers (NPK) should be applied in trenches and mixed thoroughly. The excess dose of nitrogenous fertilizers should be avoided as it results in more vegetative growth and less pistillate flowers. For irrigation one lateral (12-16 mm size) in each trench having drippers of 4 litre/ hour discharge spaced at 60 cm distance should be placed.

Before sowing, the seeds should be allowed to soak in water for 3-4 hours (muskmelon, longmelon, cucumber), 6-8 hours (bottle gourd), 10-12 hours (*tinda*, watermelon, ridge gourd) and 24-36 hours (bitter gourd) for quick germination. After soaking, seeds should be treated with Captan or Thiram @ 2 g/ kg, wrapped in gunny bag and kept at warm place (straw) for 2-3 days to facilitate early germination. Daily monitoring of seed in gunny bag should be done to avoid any damage. Sowing of pre-germinated seeds should be done from last week of December to first week of January. Two seeds near each dripper should be done to maintain optimum plant density. The trenches should be irrigated with drip irrigation prior to sowing (Choudhary and Singh, 2016).

Construction of Low Tunnels

Low tunnels are constructed with bamboo or iron rods to form semicircular frames, which are then covered with transparent plastic sheet or white coloured non-wooven cloth. Such structures are temporary and usually no higher than 1.0 m. Just after sowing, the flexible galvanized iron hoops (2 mm thick) are fixed manually at a distance of 3-4 m on trenches. The width of two ends of hoop and height from ground level should be kept 1.0 m. The erected iron hoops should be covered with transparent 30-50 micron biodegradable plastic sheet of 2.0 m width on the day of sowing. Both the perpendicular ends of plastic sheet should be buried in the soil to make low tunnel. It is done manually however; tractor driven machines are also available in advanced countries for this purpose. The prepared tunnel reflects infra-red radiation and keep inside temperature 8-10°C higher than outside. Locally available material such as phalsa twigs can also be used instead of GI wire to reduce the cost.

The increase in temperature inside low tunnel facilitates early germination of seed and crop growth. The plastic sheet serves two purposes: first it traps heat and reduces water loss and second it protects plants from adverse climatic conditions. Good cross ventilation and potential stresses caused by heavy wind, hail or heavy rains must be considered while constructing the structure.

Generally biodegradable plastic film of 30-50 micron thickness is used. It is an eco-friendly and sustainable technology for off season vegetable production. This biodegradable plastic is available according to the requirement of the duration one want to cover the crop or use as mulch in the crop. After that period, the plastic after receiving sufficient sunlight, it becomes brittle. The film eventually breaks down into small flakes and finally completely composted in the soil. The plastic is having vented or silted during the growing season as the temperature increase within the tunnels during peak day time. Vents also help bees to enter inside tunnels for pollination. Generally, 3-4 cm size vents can be made on eastern side of the tunnels just below the top on a distance of 2.5 to 3.0 m. Now a day's non-woven cloth is also being used as covering material instead of plastic which is cheaper than plastic.

Trenches and low tunnels

Crop after removal of plastic

Fertigation, Irrigation and Inter-culture

It is very difficult to obtain full benefit from low tunnels without micro-irrigation and fertigation. Drip irrigation combined with fertigation is essential for growing plants inside the tunnels. It not only applies water in the root zone but also keeps the humidity low leading to less pest and disease problems. Fertilizers are applied through drip irrigation. The water and fertilizer requirement of crops is usually depends upon the crop, variety and soil conditions. Solid water-soluble fertilizers should be used for fertigation as they are 100% water soluble and generally contain two or more major nutrients as well as micro-nutrients. They completely dissolve in water leaving no precipitate therefore; there is no chance of clogging of emitters due to fertigation. Fertigation with water soluble NPK (19:19:19) @ 8-10 kg/ ha should be done at vine development and flowering stages. Spray of 25 ppm boric acid along with 1% urea as the adjuvant 3 times from 8-leaf stage to 45 days after sowing was found beneficial which increase number of pistillate flowers, fruit setting and improve quality of fruits in all cucurbits. Addition of urea as an adjuvant @ 1% to the spray solution improved the absorption of boron by leaves. Optimum soil moisture should be maintained by operating drip system at 5-6 days interval during the first month (*i.e.* January or February) and at an interval of 2-3 days during March-April month for 1-1.5 hour.

Weeding and hoeing should be done along and between the rows. It should be done at the time of topdressing of nitrogenous fertilizer which is generally done before emergence of tendrils. Once the foliage has covered the soil, it is better to stop hoeing since it may damage the roots. Normally two to three hoeing and weedings are required to keep the crop weed free.

Pollination Under Plastic Low Tunnel Crops

Poor pollination is a major problem at high temperature. Inadequate pollination cause drying of ovary, misshapen and undersize fruit. Pollination is an important factor in cucurbits to be taken care of for good fruit setting thereby increasing total yield. The sex form in most of the cucurbits is monoecious or andromonoecious, hence effective pollination is needed. It can be performed by honeybees (*Apis melifera*) which can work in tunnels easily through the vents, made on the plastic. Optimally, one beehive having 30000-50000 workers is sufficient for one-acre area for effective pollination in crops like muskmelon, summer squash, *etc*. It is recommended to keep the beehive box on the north-west side of the field for effective working of the bees. Plastic can also be removed during day time when plants start to produce pistillate flowers to facilitate pollination of the crop by visit of bees. It is very important to ensure pollination when there is complete flowering in the plants inside the tunnels because the yield would be reduced if there is poor pollination.

Hardening of Plants

When plants start to produce pistillate flowers, the plastic should be removed partially during day time and covered during night hours. In second or third week of February when outside temperature rises, the plastic is completely removed from the plants. While removing the plastic care should be taken that it should not be removed suddenly. Hardening of plants is essential to prevent death of plants. Always remove the plastic during morning hours and cover in evening hours. Repeat this process for 2-3 days to harden the plants and avoid shock.

Care of Crop During Summer

With the onset of summer season in hot arid regions the wind velocity increases which increase the soil temperature, collect the vines in rows leading to poor pollination and flower drop. To avoid this situation, *Saccharum* (*sarkanda*) should be put in the space left between rows to facilitate proper vine spread and to avoid the direct contact of vines with warm soil. Proper direction to the growing vines should be given manually so that each and every pistillate flower gets pollinated which ultimately resulted in increased yield.

Management of Insect-pests and Diseases

Monitoring of the crop inside low tunnels should be done to protect the crop from insect-pests and diseases. Hadda beetle, leaf eating caterpillar, leaf miner, mites white fly, aphid and fruit fly are the main insects of cucurbits in hot arid region of Rajasthan. Among the insects fruit fly is one of the most devastating insect of cucurbits. Spray of either Imidacloprid 17.8SL or Thiamethoxam 70 WS @ 0.3-0.5 ml/ litre water control the aphid, white fly and leaf miner. Hadda beetle, leaf eating caterpillar and fruit fly can be controlled by spraying Dimethoate 30EC or Spinosad 45 SC @ 0.5-0.7 ml/ litre water. Mite can be controlled by Spraying propagite @ 2.0 ml/litre of water. Spray should always be done after harvesting of fruits. It is better to adopt integrated pest management (IPM) practices to reduce load of insecticides. Expose the soil to sun by deep ploughing during summer to kill hibernating pupa of insects. Install commercially available Cue-lure traps 7-8 in one hectare area to manage fruit fly. Add 4-5 ml Malathion 50 EC and 250 g jaggery (*gur*) in one litre water. Fill the prepared solution in small 8-10 pots and place them in one hectare field to control fruit fly (Choudhary *et al.*, 2014).

Fungal diseases (downy mildew, *Alternaria* leaf blight, wilt) and viral diseases (cucumber mosaic virus, bud necrosis) adversely affects the yield of cucurbits. Treat the seed with Captan or Thiram or Bavistin @ 2.0 g/ kg seed prior to sowing to protect the crop from fungal diseases. Spraying of Indofil M-45 @ 2.0 g/ litre water is helpful to control downy mildew and *Alternaria* leaf blight. Drenching of crop with Bavisitn @ 2.0 g/ litre water should be made as and when symptoms of wilt appear. The insect vectors (aphid, white fly) of viral diseases can be effectively controlled by spraying the crop with Imidacloprid 17.8 SL @ 0.3-0.5 ml/ litre water.

Cue-lure to manage fruit fly

Harvesting and Crop Advancement

Different cucurbits transplanted from first week of December to first week of February advanced the crop by 30-60 days over their normal season of cultivation. For example, if the muskmelon crop transplanted in first week of February, can be harvested in third week of April and mid January transplanted crop can be harvested in first week of April. Similarly, other cucurbitaceous crops such as bottle gourd, roundmelon, summer squash and bitter gourd can be advanced 40-60 days early than the normal season under low tunnels and on an average the cost benefit ratio of 2.0-3.5 can be obtained.

Inspection of crop

Lack pollination in longmelon

Fruit cracking in watermelon

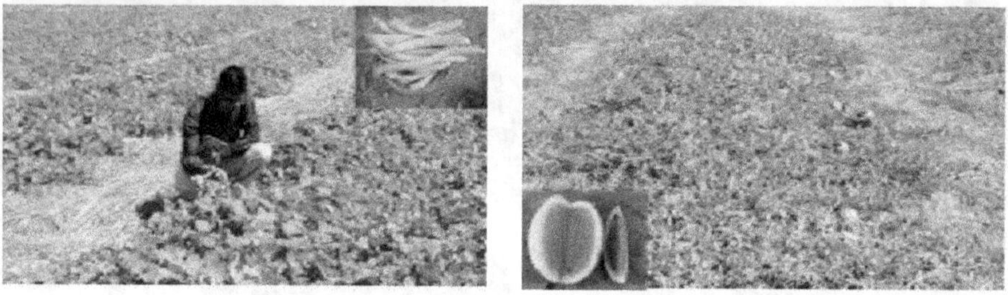

Crop ready for harvesting

Future Scope and Research Need

There are a number of opportunities in various agro-climatic zones in India for low cost protected structure *viz.*, low tunnels. They are highly suitable for growing vegetables in the peri-urban areas of the northern plains of India with location specific modifications (Singh and Sirohi, 2004). The plastic low tunnel technology has great potential for early and off-season cultivation of arid cucurbits during winter. It is highly profitable and can give good economic returns to the farmers. The package of practices including fertigation, need to be worked out for different agro-climatic situations. Use of biodegradable plastics or polymers should be encouraged to combat environment pollution. The further refinement in existing technology will definitely go a long way to harness the full potential of low cost polytunnels in vegetable production in hot arid zone and may be used on a commercial scale by the farmers. Sometimes due to uncertainty of market prices of vegetable produce the expected profit is less. Low tunnel technology is emerging as future horticulture production technology because of its potential to overcome emerging problems of climate variations in arid regions.

The technology has wide scope in arid regions to produce quality yield of cucurbitaceous crops. It is gaining popularity among the farmers however; more research is required on the following points to reduce cost of cultivation:

i. Identification and development of low tunnel responsive varieties.
ii. Technology for raising complete off season crop under tunnels.
iii. Standardization of tunnel technology for cabbage, cauliflower, coriander, tomato, brinjal, *etc*.
iv. Use of non-wooven cloth for tunnel preparation.
v. Research on proper disposal of waste material, especially plastic film.

References

Chandra, P. 1996. Protected cultivation for horticulture development. In: *Strategies for Horticulture Development in India* (Eds. G.L. Kaul and N.K. Dadlani). Hort. Div. DAC, Ministry of Agriculture, GOI. pp. 63-82.

Chandra, P. 2001. Protected cultivation in vegetable crops: current status, problems and future strategies. In: *Emerging Scenario in Vegetable Research and Development* (Eds. G. Kalloo and Kirti Singh). Research Periodicals and Book Publishing House (India). pp. 242-249.

Chandra, P. 2004. Greenhouse technology for improvement of productivity and quality of horticultural crops. In: *First Indian Horticultural Congress*. pp. 79-84.

Choudhary, B.R. and Singh, D. 2016. Tunnel cultivation of cucurbitaceous crops in arid region. In: Good management practices for Horticultural Crops. Edited by M.K. Jatav *et al.* NIPA, New Delhi. pp. 247-250.

Choudhary, B.R., Haldhar, S.M., Maheshwari, S.K., Meena, S.R. and Sharma, S.K. 2014. Maximum profit from cucurbits in hot arid regions under changing conditions. *Phal Phool* (March-April). pp. 25-29.

Choudhary, B.R., Sharma, S.K. and Janakiram, T. 2015. Cultivation of cucurbits in arid Rajasthan fetches more. *Indian Horticulture* (May-June). 60(3):22-24.

More, T.A. (2010). Arid Horticulture-Making greater strides. *Agril. Spectrum*, I(X): 26-29.

Singh, B. and Sirohi, N.P.S. (2004). Protected Cultivation of Vegetables in India: Problems and future prospects. In: *Proc. Int. Symp. Greenhouses, Environmental Controls and In-house Mechanization for Crop Production in the Tropics & Sub-tropics* held at Pahang, Malaysia from June 15-17, 2004. pp. 121-125.

Singh, Balraj. 2014. Protected cultivation of horticultural crops in India: Challenges and opportunities. In: 3rd International Conference on Agriculture and Horticulture held at Hyderabad from 27-29 October, 2014.

19

Diagnosis and Management of Physiological Disorders in Arid Fruit Crops

R.K. Meena, D.K. Sarolia, S.V.R. Reddy, R. Kumar, Chet Ram and Anita Meena

Introduction

India is the second largest producer of fruits crops in the world but productivity of arid fruit crops especially in western Rajasthan is miserably low.The productivity as well as the quality of fruit crops is affected to a greater extent due to various physiological disorders. The extremes of environmental variables like changing environmental conditions such as temperature, moisture, unbalanced soil nutrients, inadequate or excess of certain soil minerals, extremes of soil pH, poor drainage and deficiency of micronutrients as well as macronutrient causes many more disorders (Kumar and Kumar, 2015). At present, the climatic scenario of arid region is also changing and the problems like frost, temperature fluctuation, moisture fluctuation, high evaporation and low precipitation as well as nutrient imbalances are becoming a challenge for cultivation of many arid fruit crops. Reasons of low productivity in hot arid ecosystem of western Rajasthan are primarily physiological or stress related disorders such as styler-end browning in ber, fruit cracking in bael and pomegranate, internal necrosis in aonla and black and white nose in date palm. The world wide approximately 70% of crop yield reduction is the direct effect by abiotic factors (Acquaah 2007). In this chapter, we discussed the problems associated with ber, pomegranate, aonla, date palm & bael especially physiological disorders which will certainly help the farmers in timely overcoming of these maladies for getting satisfactory remunerative prices from arid fruit crops under hot ecosystem of western Rajasthan.

Ber (*Ziziphus mauritiana* L.)

Stylar-end browning

It was identified in ber (*Ziziphus mauritiana* Lamk.) cv. Chhuhara, wherein tip (stylar end) of the ber fruits turns brown. It was found to aggravate with the progress in maturity. The affected fruits contain less soluble solids content, reducing and total sugars and ascorbic acid; besides higher content of secondary metabolites and enzymes responsible for oxidative browning. This disorder was found to occur under high boron accumulation conditions, which was triggered by high temperature, coupled with high illumination and evapo-transpiration (Franck *et al*. 2007, Buts *et a*l. 2016, Kumar and Kumar 2016).

Fruit Cracking

The causes of fruit cracking in arid condition may be due to soil moisture imbalance, low relative humidity and fluctuation in day and night temperature as well as nutrient deficiency at the time of fruit development and ripening stage. This disorder is characterized by cracks developed after a rain on the skin of the fruits, sometimes deep into the flesh, affecting the stem end area, the calyx end and the cheeks of the fruit (side cracks). Cracking in ber can be minimized by spray of Gibberellins @ 20 ppm, 2,4-D, NAA at concentrations 20 ppm or 20 mg/ litre and Boric acid @ 2g/l at the initial stage of aril development reduces the activity of cellulose and reduced cracking. Use of organic manures & irrigated regularly during the entire fruit development stage by conserving soil moisture through mulching, straw, black polythene sheet etc. Erdinc Bektaset *et al.* (2018) sprayed 1% Parka (containing 7.5% stearic acid, 5% cellulose and 1% calcium) and GA_3 (15/mg L^{-1}) were sprayed yielded lower cracking rates.

Fruit Drop

Heavy fruit drop occurs at early stage of fruit development *i.e.* during second fortnight of December. Many reasons such as hormonal imbalance, abortion of embryo and inclement weather have been ascribed to drop immature fruits in ber. The highest fruit drop in cv. Banarasi Pewandi as compared to Thornless and Banarasi Karaka. Single spray of 10-20 ppm Naphthalene acetic acid (NAA) one in mid- October shall be sufficient to control fruit drop (Singh and Bal 2006).

Physiological disorders of ber

Pomegranate (*Punica granatum L.)*

Fruit Cracking

It is a most serious physiological disorder which hinders its cultivation to a great extent. In young fruits it may be due to boron deficiency but fully grown fruit crack due to moisture imbalances or due to extreme variations in day and night temperature (Abd El-Rhman, 2010). Prolonged drought causes hardening of peel and if this is followed by heavy irrigation then the pulp grows and the peel cracks and if harvesting of the mature fruit is delayed for long time or there is attack of insect pests it leads to cracking. This problem can be overcome by maintaining soil moisture and, cultivation of tolerant varieties. The water retention capacity of the plants should be increased by the use of organic manures. The plants should be irrigated regularly during the entire fruit development stage. Early harvesting and spray of calcium hydroxide on leaves and on fruit set reduces the incidence of fruit cracking. Foliar application of boron reduced the extent of fruit cracking in pomegranate. GA is used for improving fruit set and also to control cracking in various fruits including pomegranate. Application of GA_3 @ 40 ppm in pomegranate reduced fruit cracking. Some varieties like Sur-Anar, Francis, Shirvan, Krasnyl *etc*., were reported to be tolerant to cracking (Lal *et al*., 2012).The minimum fruit cracking was observed when $CaSO_4$ 3000 ppm spray was applied followed by $CaSO_4$ 2000 ppm, GA_3 40 ppm and borax 50 ppm Lal *et al.* (2011).

Sun Scald

Sunburn injury is common in pomegranate due to high solar radiation levels and air temperatures, low relative humidity in hot arid ecosystem of western Rajasthan. The incidence and severity of sunburn depends upon climatic factor, cultivars, hormonal, nutritional and soil moisture. Fruit production losses due to sunburn may be 6 to30 per cent depending on seasons and the type of fruit. Grower must follow best management practices to minimize sunburn and grow tolerant cultivars, efficient irrigation, appropriate canopy management, cover or inter-cropping, over tree sprinkler, shade netting, fruit bagging, suppressants

(Kaolin or calcium carbonate) and chemical protectants reported by Lal and Sahu. (2017). Spraying kaolin during the hot summer months is useful in reducing sunscald. First spray of 5% and subsequently 1 or 2 additional spray with kaolin @2.5% at 15 days interval reduce sun scald. Bagging the fruits with butter paper covers is useful in minimizing fruit spoilage due to sun scald. White colour bags are more effective in reflecting sunlight and protecting the growing fruits.

Fruit Cracking & Sun Scald in Pomegranate

Internal Breakdown

It is another serious disorder in pomegranate which leads to 50-60% loss. This disorder is mostly occurs in ambe bahar. Softening of fruits occur 90 days after anthesis and fruits become creamy-browny to dark blackish-brown from outer side and became a severe, if matured fruits retain on the plants for 140 days onwards. So, to avoid this disorder 130-135 days old fruits should be harvested (Sharma, 2005b).

Aril Browning

Aril browning or internal breakdown of arils in pomegranate critically affect fruit quality in some commonly grown cultivars such as, Ganesh and Bhagwa. Aril breakdown or browning is characterized by soft, light creamy, brown, dark blackish or brown and slightly flattened arils which are deformed and possess an unpleasant odour when the fruit is cut open. This disorder is accompanied by desiccation, wrinkling and development of internal spaces in the arils. Experimental studies indicated that the juice and seed content of affected fruits have reduced level of TSS, acidity, ascorbic acid, reducing sugars, calcium, phosphorus, and enzyme catalase and increased level of non-reducing sugars, starch, tannins, nitrogen, potassium, magnesium, boron and enzyme polyphenol oxidase compared to healthy fruits. Studies have shown that the malady is influenced by diverse factors such as genetic background, pruning, growing season, fruit size, harvested date and variety. (Jalikop *et al.* 2010, Sandhu and Gill 2013)

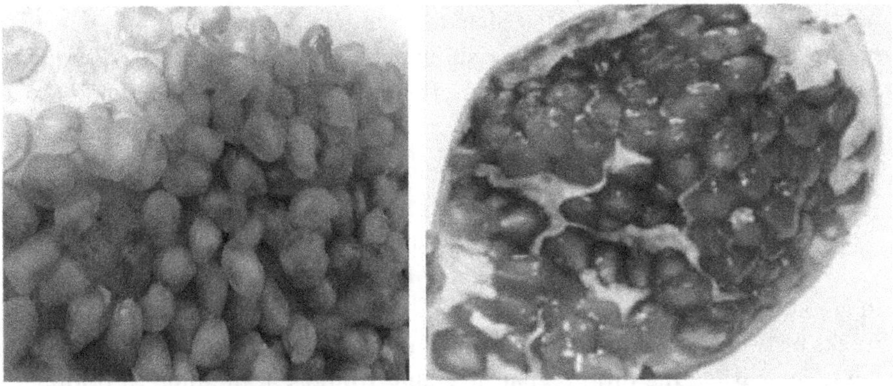

Aril browning & Internal Break down in pomegranate

Aonla (*Emblica officinalis* L.)

Internal Necrosis

Necrosis symptoms starts with the browning of inner most part of the mesocrpic tissues, at the time of endocarp hardening in the second and third week of September. Browning of mesocarp extends towards the epicarp into brownish black areas on the fruit surface in the second and third week of October. Depending upon the severity of the disorder, mesocarp of affected fruit turns black and become corky and gummy pockets develop. Evaluation of aonla cultivars (NA-6, NA-7, NA-10, Kanchan and Chakaiya) at NDUAT, Faizabad found to exhibit higher productivity and fruits are free from necrosis indicating their suitability for fruit processing (Verma, 2015). This disorder is due to the deficiency of boron for which spray of 0.6% borax thrice in the month of September and October at about 10 to 15 days interval are recommended to control the disorder. Combined spray of zinc sulphate (0.4%) + copper sulphate (0.4%) and borax (0.4%) during September-October has been found effective (Sandhu and Gill 2013). Resistant Cv. 'Chakaiya', 'NA-6' & 'NA-7' should be planted in the orchard for avoidance of this disorder.

Off-season Flowering

Off season flowering is also a serious problem in aonla production. An economical loss of 30 to 40% due to off season flowering in aonla was estimated by them. The main reason of this disorder may be global warming or due to some alteration in the tree physiology due to externally governed internal mechanism which leads to absence of distinct phase of dormancy after fruit harvest. Plant comes in partial flowering with predominance of male flower which do not set fruit, exhaust the plant and disturb the physiology and forcing the plant to enter in dormancy by spraying 10% urea after fruit harvest. Restricting soil moisture build up near root zone can lead plant to enter into dormancy.Combined spray of Zink sulphate (0.4%) + copper sulphate (0.4%) and borax (0.4%) during September- October has been found effective reported by Singh and Mishra (2007).

Fruit Drop

It is a serious problem in aonla in western Rajasthan which influences the final yield of crop. There are phases (three waves) of flower and fruit drop, observed in aonla. In case of the first wave, more than 70% flowers drop within three weeks of flowering, which is normally due to lack of pollination, while during the second wave the drop is that of young fruitlets at the time of dormancy break and in last, *i.e.* third wave the droping of fruits is due to embryological and physiological factors and it is spread over the entire period of the fruit development. It starts from later half of August and continues up to harvest (Sandhu and Gill, 2013). Fruit drop results due to many factors such as dry spell, imbalance of hormones, improper nutrition, fluctuation in temperature, cultivars and age of tree, and number of developing fruits. Also delay in harvesting results in heavy fruit drop particularly in Banarasi and Francis cultivars.

Date Palm (*Phoenix dactylifera L.*)

Black Nose

Black nose refers to abnormally shriveled and darkened tip of a date palm plants. Deglet Nour and Hayani cultivars seem to be the most susceptible to this physiological disorder. Black nose results from excessive checking of the epidermis, especially in the form of numerous small, transverse checks or breaks at the stylar end of the fruit. Pronounced shriveling and darkening occur in proportion to the abundance of the checks and are related to humid weather at the Khalaal stage observed by Chao (2007). The checking is induced by high humidity and rainfall; it can be measured to avoid conditions that increase environment humidity. The conditions to be avoided include excessive soil moisture and the presence of intercrops and weeds, especially at the susceptible stage of fruit development. This problem can be managed by bagging of fruits in brown paper. Over thinning can also increase the incidence of checking and subsequent development of black nose.

Crosscuts (V- Cuts)

Crosscuts are a physiological disorder of fruit stalks and fronds. It consists of a slight to deep notch, similar to a cut artificially done by a knife. Fruits borne on strands in line with the break wither and fail to mature properly. Crosscuts result from an anatomical defect in the fruit stalks and fronds involving internal, sterile cavities leading to mechanical breaks during elongation of the stalk or the fronds reported by Yuval Cohen (2010). Crosscuts are commonly found in varieties having crowded leaf bases like Sayer and Khadrawy. Control measures is not use such plants for propagation.

White Nose

White nose in date palm causing due to low RH as the fruit is advanced from the rutab to the tamer stage, especially in the case of Hillawi and Zahidi dates (Ait-Oubahou and Yahia, 1999). It is characterized by the presence of a disco loured ring near the calyx area, which sometimes covers 50% of the fruit. It has been suggested that dry winds for many days during the rutab stage of ripening can cause the basal region of the fruit to ripen more than the rest, causing the ring appearance. However, some studies have shown that this physiological disorder may be due to calcium within the fruit with the basal end containing less calcium than the apical end (Ibrahim, 1995).The disorder is alleviated or reduced in Hillawi dates by soaking in hot water (75°C) for five minutes (Ait-Oubahou and Yahia, 1999).White nose disease is commonly found in Iraq. Dry and prolonged wind in the early Rutab stage causes rapid maturation and desiccation of the fruit resulting in whitish drying at the calyx end of the fruit. The affected fruit becomes very dry, hard and has high sugar content. Hydration may correct this condition in harvested fruits.

Barhee Disorder

It is caused by unbalanced bunches of fruits or insect attack at one side of the shoot bud (Sudhersan and Yousif Al-Shayji, 2007). Barhee disorder is characterized by an unusual bending of the crown of Barhee and Dayri date palm varieties. The disease was first reported in California (USA) and later in Al Basra (Iraq). Affected palms were found to bend mostly to the south and sometimes to the south-west. This phenomenon is severe and bending could reach an angle of about 90. The growers are correcting this situation by fixing a heavy iron bar to the opposite side of the bending; fruit bunches from the opposite side are tied to this bar in order to move the actual weight against the bending side. It seems that within 2 to 3 years, the bending is corrected. Bunch handling is also proposed to correct such an abnormality.

Black Scald

Black scald, different from black nose, is a minor disorder of unknown cause occurring in the United States. It consists of a blackened and sunken area with a definite line of demarcation. The disease usually appears on the tip or the sides of the fruit, and affected tissues have a bitter taste. The exact cause of this disorder is high temperature & humidity (Ait-Oubahou and Yahia, 1999). This physiological disorder occurs particularly in Deglet Noor dates. It can be managed by wrapping the fruits with brown papers.

Frost Damage

The date palm resists large temperature variations (-5 to 50°C) with a growth optimum between 32 and 38°C and a zero of vegetation of about 7°C. The vegetative activity will also decrease above 40°C and ceases around 45°C. When temperature falls below 0°C, it causes serious metabolic disorders with some injury to date palm leaves characterized by a partial or total desiccation. Water of protoplasma freezes after coming out from the cells during defrost, water invaded inter-cellular spaces and affected leaves turn brown and desiccated.

The most practical and available protection for the date growers is to turn on the water and keep the date plantation wet when the temperature begins to get low enough (-5°C and below). A date plantation just irrigated or being irrigated when the temperature falls, has some heat stored, which gives protection.

Lack or Excess of Water

The growth of the date palm is highly affected by variations in water availability and the water content of the soil. A decrease in yield or complete failure in fruit production could result from these water variations. To compensate for high evapotranspiration, the date palm requires a quantity of water from 1,500 to 2,800 mm/year. Prolonged water stress will significantly decrease growth and yield, and if the drought continues for several years, date palm can dry up and die.

On the other hand, when the water table is high and drainage is inadequate and/or the leaching and transport of soluble salts is not complete, high evaporation rates tend to increase the concentration of salts in soil and in surface water. However, there are limits of salt tolerance and the date palm will not grow when soluble salt of the soil is above 6 percent. Although date palms are resistant to flooding, healthy growth of palms requires a well-drained soil, and it is clear that irrigation must always go hand in hand with drainage.

Bale (*Aegle marmelos L.*)

Fruit cracking

Fruit cracking is an important physiological disorder found in some commercial cultivars throughout the growth and development period of fruit. Spray of growth regulators also help to reduce the ill effect of hormonal imbalances (Sandhu and Gill, 2013). From the review, it is clear that physiological disorders have become menace in many fruit crops resulting in huge losses to growers. There is a need for long-term quantitative documentation of tree phonological patterns in diverse climatic zones of India. Recent advances in physiology and genetics may help solve problems of perennial fruit tree production. Though molecular biology have greatly improved our understanding of plant responses to stresses in many important commercial horticultural crops, a lack of transcriptomic and genomic information hinders our understanding of the molecular mechanisms underlying fruit-set and fruit development. There is need for exhaustive studies to know the precise physiological significance of radiation effect in climatic fastidious fruit crops. Intelligent anticipatory management strategies and adaptation will be the critical components for successful and sustainable quality fruit production. Management aspects may include biophysical treatments including reclamation of deficiency and excess of nutrient elements by proper fertility status maintenance, timely agronomical operations and input application. Location specific management strategies for important perennial fruit crops will require focused attention through multidisciplinary approach including adaptation and management.

Cracking and fruit drop in Bael

Fruit Drop

Fruit drop in Bael occurs due to embryo abortion, physiological imbalances, fruit borer attack, fruit rotting and fruit cracking. Fruit drop was recorded high in month of July in all varieties which continued till harvest and again showed the peak in the month February. Although, the crossing between Pant Shivani x Pant Sujata and spray of NAA 20 ppm reduced the fruit drop percentage but the spray of Borax @ 1.0% was found most effective in reducing the fruit drop and increasing the physicochemical characters of fruits.Fruit drop in bael can be minimized by spray of NAA (10, 20 and 30 ppm), 2, 4-D (5, 10 and 20 ppm), GA3 (25, 50 and 50 ppm) and ethrel (50,100 and 150 ppm) in bael cv. Pant Shivanireported by Uniyal and Mishra (2015).

Conclusion

Fruit drop is complex phenomena probably due to environmental changes coupled with nutritional deficiencies as well as endogenous hormonal manipulation. Therefore, growers must gain knowledge to recognize the various physiological disorders that occur in their climatic regions and manipulation with available resources like timely application of cultural practices, manuring & fertilization of fruit crops with organic and inorganic fertilizers to mitigate nutritional deficiencies. Evaluation of cultivars adaptable to particular region coupled with sound horticultural practices such as timely planting, harvesting, adequate fertilization and irrigation will mitigate the unavoidable effects of physiological disorders of fruit crops in western Rajasthan. Further work is necessary pertaining to breeding and agronomical practices of fruit crops in agro ecological zones of Rajasthan with the goal of developing potential cultivars withstand against abiotic stress.

References

Abd El-Rhman 2010. Physiological studies on cracking phenomena of Pomegranates. *J. Appl. Sci.*, 6(6): 696-703.

Acquaah G. 2007. Principles of Plant Genetics and Breeding. Blackwell, Oxford, UK.

Ait-Oubahou A and Yahia EM (1999), Postharvest handling of dates, Postharv News In/o, 10(6), 67N-74N. Buts K, Hertog M L, Ho Q T, America A H, Cordewener J, Vercammen J, Carpentier S C and Nicolai B.

2016. Influence of pre-harvest calcium, potassium and triazole application on the proteome of apple at harvest. Journal of Science of Food and Agriculture doi: 10.1002/jsfa.7664

ChihCheng T. Chao. 2007. The Date Palm (Phoenix dactylifera L.): Overview of Biology, Uses, and Cultivation, Hortscience vol. 42(5), pp 1077-1082.

ErdincBektas, ErdalAglar, OrhanKarakaya and Sefa Gun. 2018. Cracking and quality attributes of jujube fruits as affected by covering and pre-harvest Parka and GA_3 treatments, *Scientia horticulture Volume* 240, pp 65-71

Franck C, Lammertyn J, Ho Q T, Verboven P, Verlinden B and Nicolaï B M. 2007. Browning disorders in pear fruit. *Postharvest Biology and Technology* 43: 1–13.

Gorakh Singh Rajneesh Mishra. 2007. Aonla Rejuvenation. Extension bulletin, CISH, Lucknow, pp 13

Hoda A.K. and Hoda, S.H.A. 2013. Cracking and fruit quality of pomegranate (*Punicagranatum* L.) as affected by preharvest sprays of some growth regulators and mineral nutrients. *J. Horticult. Sci. Ornam. Plants,* 5(2): 71-76.

Ibrahim AO (1995), The physiological relationship between growth regulators and characteristics of date palm (Phoenix dactyli/era L.) cv. Hillawi, PhD thesis, BasrahUniv, Basrah, Iraq (In Arabic with English summary), 97 p

Jalikop S H, Venugopalan R and Kumar R. 2010. Association of fruit traits and aril browning inpomegranate (*Punicagranatum* L.). *Euphytica* 174: 137–41

Kumar R and Kumar. V. 2016. Physiological disorders in perennial woody tropical and subtropical fruit crops: A review. *Indian Journal of Agricultural Sciences* 86(6): 703–17.

Lal, S., Ahmed, N. and Mir, J.I, 2012. Effect of different chemicals on fruit cracking in pomegranate under karewa condition of Kashmir Valley. *Indian J. Plant Physiol.*, 16(3&4): 326-330.

Lekhram Verma. 2013. Improved cultivation technique of Aonla, Rashtriya Krishi Volume 8(2) December, pp 7-10

Narayan Lal and Nisha Sahu. 2017. Management Strategies of Sun Burn in Fruit Crops-A Review, *Int. J. Curr. Microbiol. App. Sci.* 6 (6): 1126-1138

Pathak R K, Pandey D, Haseeb M and Tandon D K. 2003. *The Aonla. Bulletin, CISH, Lucknow.*

Rajesh Kumar and Vinod Kumar. 2015. Physiological disorders in perennial woody tropical and subtropical fruit crops: A review. *Indian Journal of Agricultural Sciences,* 86 (6): 703–17.

S. Lal, N. Ahmedand J.I. Mir. 2011. Effect of different chemicals on fruit cracking in pomegranate under karewa condition of Kashmir valley.Indian J. Plant Physiol., Vol. 16, pp. 326-330

Sandhu S and Gill B S. 2013. Physiological Disorders of FruitCrops, p 189. New India Publishing Agency, New Delhi.

Savreet Sandhu, Bikramjit Singh Gill. 2013. Physiological Disorders of Fruit Crops. New India Pub. Agency

Sepahi, A. 1986. GA3 concentration for controlling fruit cracking in pomegranates. *Iran Agric. Res.*, 5: 93-99.

Singh, D.B., Kingsly, A.R.P. and Jain, R.K. 2006. Controlling fruit cracking in pomegranate. Indian Hort. 51(1): 14.

Singh, D.B., Sharma, B.D. and Bhargava, R. 2003. Effect of boron and GA3 to control fruit cracking in pomegranate (*Punicagranatom* L.). *Current Agri.*, 27 (1/2): 125-127

Sudhersan, C. and Y. Al-Shayji, 2007. A study on physiological disorders of tissue cultured date palms in Kuwait. Technical Report, KISR No. 8800, KISR, Kuwait.

Sunita Uniyal and K.K. Mishra 2015.Effect of plant growth regulators on fruit drop and quality of Bael under Tarai conditions of Uttarakhand, *Indian Journal of Horticulture*, issue (1) pp 126-129.

Yuval Cohen, Stanley Freeman and Aida Zveibil, Rachel Ben Zvi, Yaakov Nakache, Shimon Biton and Victoria Soroker. 2010. Reevaluation of Factors Affecting Bunch Drop in Date Palm, *Hort science* 45(6) pp 887–893.

20

Management of Abiotic Stresses in Horticulture Crops

Narendra Pratap Singh, D D Nangare and Yogeshwar Singh

Introduction

India is a second largest producer of fruits after China contributing about 11 per cent to the global production. India leads the world in productivity of papaya, grapes and banana but average productivity of country is much below as compared to other leading countries, which is merely 11.6 t/ha. Abiotic stress is the primary cause of crop loss worldwide, reducing average yields for most major crop plants by more than 50%. Drought and salinity are becoming particularly widespread in many regions, and may cause serious salinization of more than 50% of all arable lands by the year 2050 (Wang, 2003). This leads to a series of morphological, physiological, biochemical and molecular changes that adversely affect plant growth and productivity. There are various factors, which contributes to low productivity like limited availability of quality planting materials, old and senile orchards, poor orchard management practices, insect pest and diseases, poor post-harvest management practices *etc*. There is a much scope in expansion of horticultural crops in arid region and it has vast potential for changing scenario of horticulture of the country. Land resource, surplus family labors, increasing canal irrigated area, developing infrastructural facilities, plenty of solar and wind energy, etc. are the strength in the region for research and development of arid horticulture. Further, low incidence of diseases/insects in the region provides good scope for production of quality seed and planting material of horticultural crops (Sharma *et al*., 2013).

Gaps and Major Problems in Prevailing System in Growing Horticulture Crops

- Water is most precious input in hot arid region for the horticulture crops of the country therefore, adoption of micro-irrigation system is desirable to save water and enhance productivity.
- For arid climatic conditions, the variety is needed which is resistant to biotic and abiotic stresses for sustainable production.
- In some parts of arid region, occurrence of frost is also a common feature during winter season, which affects vegetative growth of plants as well as productivity, quality of fruits especially in ber, lasoda and aonla.
- There is no heat tolerant variety among arid horticultural crops, which should be developed to achieve higher production.
- One of the major problems in development of horticulture scenario in arid parts of the country is lack of sufficient quality seed and planting material.
- Seed is a precious input to increase quality production of vegetable, spices and flowers as well as some fruit crops. Thus, production of quality seed material and their supply to farmers will boost up the production of arid horticultural crops.
- The post-harvest management is essential to overcome the losses at different stages of grading, packing, storage, transport and finally marketing of both fresh and processed products.
- The weak processing infrastructure, as it exists today, has been one of the contributing factors for ineffective utilization of the raw materials resulting in huge post-harvest losses.
- Lack of sufficient processing units for production of quality output is a major bottleneck for the arid fruit crops.
- Marketing of horticultural produce is a major constraint in the production and disposal system and has a major role to play in making the industry viable.
- The high capital cost involved in establishing orchards, or rejuvenation of existing old unproductive plantation poses serious constraint in area expansion.
- The long gestation period of the perennial horticultural crops like mango, sapota, citrus and date palm coming to the economic bearing age is also a constraint for early returns.

- Lack of proper storage facility and knowledge and equipment for grading and packaging of fruits and vegetables is also constraint for the growers of hot arid region.
- Lack of proper marketing of produce to get better returns. The development of storage facilities and agro-based processing units in the region is also needed for development of horticulture (More *et al.*, 2012).

Good agricultural practices (GAP) are Practices that which addresses environmental, economic and social sustainability for on-farm processes and result in safe and quality food and non-food agricultural products. (FAO COAG 2003 GAP paper). In simple language, GAP stands on four pillars (economic viability, environmental sustainability, social acceptability and food safety and quality).In recent years, the concept of GAP has evolved to address the concerns of different stakeholders about food production and security, food safety and quality, and the environmental sustainability of agriculture. These stakeholders include governments, food retailing industries, farmers and consumers who seek to meet specific objectives of food safety, food production, production efficiency, livelihood and environmental benefits. GAP offers means to help reach those objectives.

Advance Approaches for Fruits and Vegetables in Arid Climate

New Traits, Varieties & Crops

Increasing agricultural productivity requires technological advances in crop yields. In addition to increasing productivity generally, several new varieties and traits offer farmers greater flexibility in adapting to climate change, including traits that confer tolerance to drought and heat, tolerance to salinity (e.g., due to rising sea levels in coastal areas), and early maturation in order to shorten the growing season and reduce farmers' exposure to risk of extreme weather events. These promising new traits and varieties, which are mostly still in development, can emerge from traditional breeding techniques that leverage existing varieties that are well suited to vagaries of the local production environment as well as from more advanced biotechnology techniques such as marker assisted selection and genetic modification. Climate change will also lead to new pest and disease pressures. The nuances of temperature changes e.g. higher low temperatures and fewer freezes – could shorten dormant periods, speed pest and disease growth and change the dynamics of these populations and their resistance. Crops, varieties and traits that are resistant to pests and diseases will improve producers' ability to adapt to climate change. To the extent that these varieties reduce the need for pesticides, they also reduce carbon emissions by decreasing pesticide demand as well as the number of in-field applications. The mitigation potential of new crops and varieties extends to direct carbon sequestration and perhaps to second generation bio-fuel crops. The damage to fruit trees was relatively more in low-lying areas where cold air settled and remained for a longer time on ground. But temperate fruits such as apple, peach, plum and cherry gave higher yield due to extended chilling. One of the best way to reduce the impact of the cold wave is to go in for proper selection of fruit species and varieties as per the site conditions. Apart from that, wind breaks or shelter belts should be provided. There should be frequent irrigation, smoking, covering young fruit plants with thatches or plastic sheets.

Adoption of Soil and Moisture Conservation Techniques

Some of the basic methods to collect and conserve water are; harvesting of rain water, harvesting of water from snow melting, development of catchment area and storing runoff water for recycling, check dams and construction of waterways. Contour cultivation, contour trip cropping, mixed cropping, tillage, mulching, zero tillage, are some of the agronomical measures for the *in-situ* soil moisture conservation. Mechanical measures like contour bunding, graded bunding, bench terracing, vertical mulching etc. also need to be followed for effective soil and moisture conservation in dry lands. Another technology for efficient utilization of runoff is water harvesting recycling. Rainwater harvesting includes collecting runoff water into dug out ponds or tanks in small depressions, gullies and into storage dams of earth or masonry structures. Rain water harvesting is possible in areas having rainfall as little as 500 to 800 mm. Depending on the rainfall and soil characteristics, 10-50 % of the runoff can be collected in farm pond. Surface run off thus collected in a farm pond can be used to provide protective irrigation in the period of prolonged dry spell. Trench planting (0.5-1.0 m) is recommended for ber, *aonla* and custard apple to conserve moisture. The trenches collect rain water along with silt and organic matter and thus promote tree growth. Planting in trenches is common for pineapple under dry conditions. contouring, contour bunds, sloping area land technology (SALT), contour drains, graded bunding, terraces, broad bed furrow system for raising crop and conservation of water, contour Furrow, stabilization structures, curved land ploughing for intensive land preparation; soil conservation and water retention through use of vegetation, live check of bamboo pieces and loose boulders are some of the other agronomical measures used to conserve moisture.

Water Management & Irrigation

In the midst of increasing urban and environmental demands on water, agriculture must improve water use efficiency generally. Adding climate change to this mix only intensifies the demands on water use in agriculture. With hotter temperatures and changing precipitation patterns, controlling water supplies and improving irrigation access and efficiency will become increasingly important. Climate changes will burden currently irrigated areas and may even outstrip current irrigation capacity due to general water shortages, but farmers with no access to irrigation are clearly most vulnerable to precipitation volatility.

In places with limited access to irrigation, well timed 'deficit irrigation' can make a substantial difference in productivity. With dwindling water supplies, such deficit irrigation techniques will become increasingly important. In non-irrigated areas, water conservation and water harvesting techniques may be farmers' only alternative to abandoning cultivation agriculture all together. Drip irrigation has proved its superiority over other conventional method of irrigation, in horticulture due to precise and direct application of water in root zone. A considerable saving in water, increased growth, development and yields of fruits and vegetables and control of weeds, saving in labour under drip irrigation are the added advantages. Drip irrigation can be adopted in fruit crops and also to all vegetable crops including closed spaced crops like onions and beans. The saving in water is to the tune of 30-50% depending on the crop and season.

The following methods may be adopted under limited water conditions to save water:

(a) **Water saving irrigation method**: Under limited water situations, water-saving irrigation methods like alternate furrow irrigation or widely spaced furrow irrigation and drip irrigation systems can be adopted. Studies conducted on methods of irrigation in capsicum, tomato, okra and cauliflower indicated that adopting alternate-furrow irrigation and widely-spaced furrow irrigation saved 35 to 40 per cent of irrigation water without adversely affecting yield.

(b) **Mulching practices in vegetable production** - The technique of covering the soil with natural crop residues or plastic films for soil and water conservation is called mulching (Kumar *et al.*, 2012). Mulching can be practiced in fruits and vegetable crops using crop residues and other organic material available in the farm. Recently plastic mulches have come into use due to the inherent advantages of efficient moisture conservation, weed suppression and maintenance of soil structure. Wide variety of vegetables can be successfully grown using mulches. In addition to soil and water conservation , improved yield and quality, suppression of weed growth , mulches can improve the use efficiency of applied fertilizer nutrients and also use of reflective mulches are likely to minimize the incidence of virus diseases. For vegetable production generally polyethylene mulch film of 30 micron thick and 1 to 1.2 m width is used. Generally raised bed with drip irrigation system is followed while laying the mulch film.

Application of Foliar Nutrition

The foliar application of nutrients during water stress conditions helps in the better growth by quick absorption of nutrients. The spraying of K and Ca induces drought tolerance in vegetable crops. Spraying of micronutrients and secondary nutrients improves crop yields and quality. Anti-transpiration coatings have been shown to be effective for maintaining quality through control of water loss. Anti-transpirants are chemicals which when sprayed on plants form a film which increases the diffusion resistance of water from stomata and thus reduces transpiration losses of water. Several chemicals have been successfully used like Acropyl in grapes, polycot in banana and kaolinite (3-8%) in different fruit plants. The energy input can be reduced by increasing plant reflectivity by using effective chemicals like zinc oxide, kaolinite, chalk etc alone or in combination with other anti-transpirants. These chemicals are used to reduce temperature on plant parts. Film forming compounds like wiltpruf, mobileaf, clear spray, vapourgard and folicoat can be used to reduce transpiration and water loss. Water loss can also be restricted by applying chemicals which facilitate stomatal closure. Some of these are; phenyl mercuric acetate (PMA), Decenyl succinic acid (DSA), Atrazine and Sodium azide.

Conservation Horticulture

Production techniques may be as important as production technologies in climate change adaptation and mitigation. One such technique stands out in particular: ***conservation or reduced tillage agriculture***. This technique aims to build up organic matter in soils and create a healthy soil ecosystem by not tilling the soil before each planting. Seeds are planted using seed drills that insert seeds to a precise depth without otherwise disturbing the soil structure. By increasing the organic matter in soils, conservation agriculture improves the moisture capacity of the soil and thereby increases water use efficiency. The practice also reduces carbon emissions by reducing tilling, although it also requires more sophisticated pest and disease control because the system is not 're-booted' at each planting.

The basic principles of conservation agriculture are soil cover, particularly through retention of crop residues on the soil surface as mulch or cover crops; a minimum level of soil disturbance, e.g., reduced or zero tillage practices; and sensible, profitable crop rotations. Soil moisture can be conserved through mulching either with black polythene or locally available mulches, growing cover crops or inter-culturing in the orchards to check soil erosion and runoff rain water. If area around tree basin remains covered with mulches, like grass mulch (10-12 kg/basin) and black polythene mulch (100 micron) throughout the growth period, it helps in conserving the soil moisture, saving water for more critical stages during summer and reduced the weed population by 60 % with grass mulch and

100% with black polythene mulch. The polythene mulch maintained 29% more soil moisture compared to un-mulched trees on soil available water content basis. Surface-applied mulches provide several benefits to crop production by controlling evaporation from the soil surface, heat energy and nutrient status in soil, buffering drastic changes in soil temperature. Soil temperature, especially at the surface layer is reported to be important for the translocation of photo-assimilates. Straw mulch ameliorates environmental stresses and improves the product quality and safety.

Wind breaks, Hedges and Intercropping

To overcome the adverse effect of high temperature and dry winds, tall growing trees need to be planted all along the boundary of the farm. Windbreaks help to deflect and to filter through the wind current thereby reducing the velocity of wind resulting lower displacement of wind around tree and cause reduction in transpiration and evaporation. The orientation of windbreak should more or less be at right angles to the prevailing winds. It is believed that the windbreak is effective for a distance equivalent to 3 to 4 times of tree height. *Acacia tortalis, Cassia siamea* and *Prosopisjulifera* are some of the useful tree species suitable as wind break under less water conditions. Inter cropping of vegetable crops of the area can be practiced in orchards during summer months. Maize/ Sorghum can be grown all along the border of the plot to mitigate the effect of desiccating winds. Crop rotation, cover crops, strip cropping, mulching (Kumar *et al.* 2011), contour hedge row intercropping system are used to conserve moisture.

Use of Plant Growth Regulators

Foliar sprays of 50 mM IAA, GA_3, or benzyl amino purine (BAP) partially counteracted the effect of water deficit on photosynthesis and transpiration. Growth regulator applications can also potentially enhance stress resistance, particularly for fruits and vegetables which are prone to show accelerated ripening or senescence in response to drought stress. Accelerated ripening or senescence is most often mediated by ethylene production in response to the stress. As consequence anti-ethylene products such as aminovinylglycine (AVG) and 1-methylcyclopropene (1-MCP) could be used to mitigate drought stress. While foliar cytokinin (CK) application can prevent ABA-induced photosynthetic limitation, the effects can be transient and of little consequence in the long term. Paclobutrazole (10 mg/ lit) is used to avoid moisture stress in mango. Accelerated ripening or senescence is most often mediated by ethylene production in response to the stress. As consequence anti-ethylene products such as aminovinylglycine (AVG) and 1-methylcyclopropene (1-MCP) could be used to mitigate drought stress. Other growth hormones, such as methyl jasmonate (which promotes leaf senescence), can enhance chilling resistance in avocado, grapefruit, bell peppers and zucchini squash. Abscisic acid has been demonstrated to reduce chilling-induced injury in some crops (Wang, 1993).

Beneficial Microbes

Microbial activity in soils is stimulated by the release of carbon-rich material in the form of root border cells, and/or the selective exudation of specific sugars, carboxylic acids, or amino acids that encourage the development of cultivar-specific, plant-beneficial, microbial communities. Some of these microorganisms benefiting from the availability of fresh energy are also capable of enhancing plant growth and development, which are generally called as Plant growth promoting (PGP) rhizosphere microorganisms. Plant growth-promoting rhizobacteria (PGPR) and fungi (mycorrhizae) can facilitate plant growth directly by facilitating the uptake of nutrients from the environment, by influencing phytohormone production (e.g. auxin, cytokinin, or giberallin), and/or by enzymatic lowering of plant ethylene levels. In addition to facilitating the growth of plant, these microorganisms can protect plants from the deleterious effects of flooding and drought.

The beneficial plant-microbe interactions in the rhizosphere are the primary determinants of plant health and soil fertility. Rhizobacteria include mycorrhization helper bacteria (MHB) and plant growth promoting rhizobacteria (PGPR), which assist AMF to colonize the plant roots. Synergistic positive interactions have been reported between AMF and plant growth-promoting bacteria (PGPB) such as nitrogen fixers, fluorescent *pseudomonad's*and sporulating *bacilli*. The most common bacteria in the mycorrhizosphere are *Pseudomonas*. Some extremely mycorrhizal dependent plants, including grapes, citrus, melons, oaks and pines may quite literally starve to death in soils that lack this helpful fungi. These benefits of mycorrhizal symbioses in vegetables, fruits and tree species, both agronomically by increased growth and yield as well as ecologically by improved fitness, indicate that mycorrhizal plants are often more competitive and better able to tolerate environmental stresses.

Hail Control Mechanisms

Artificial hail control is an important measure in disaster prevention and mitigation. Cloud seeding for hail suppression is based on the cloud microphysical concept in which seeding is postulated to reduce hail severity. Another concept is to create shock waves which can prevent the formation and growth of hail by melting altogether. Shockwaves are produced using hail guns/cannons. The super-cooled water situated on the external layer of hailstone is transformed from liquid state to solid state. Therefore the hail nuclei are not able to melt anymore and remain at a small size which thus minimize the damage when they hit the ground. Nowadays, acetylene or butane gas is used to generate hail disruptive shockwaves this allows the emission of a more powerful shockwaves.

Protective screens termed as anti-hail nets above the crop can be appropriately utilised especially for high value crops. Tree shelterbelts can markedly reduce hail damage in their immediate vicinity since hails are usually associated with strong winds. Some hail is intercepted directly by the trees protecting crops immediately downwind. The trees also create a change in the air flow so that the area in the lee of them is particularly sheltered with hail deflected laterally. Wind speeds will also be less in the lee of the shelter so the total hail kinetic energy, which results both from the vertical fall speed of the hail and the wind speed will be less.

Wind machines can raise air temperatures around plants by about a third to half the temperature inversion difference. One protection method is to use wind machines in orchards of high value crops. Wind machines are tall, fixed-in-place, engine-driven fans that pull warm air down from at least 15 m above ground during strong temperature inversions, blowing it down and out, pushing away and replacing cold air near target crops. This raises air temperatures around cold-sensitive perennial crops such as grapes.

Nets hold heat radiating from the ground and raise the temperature in the orchard. Nets also block much of the cold air flow associated with advection frosts. Frost protection is really only effective for radiant frosts when winds are light or calm, and in low-lying, frost-prone areas. Protection efforts need to focus on the most frost-prone blocks and in response to forecasts of the critical temperatures needed for substantial crop damage. Protection has three basic objectives.

1. reduce heat loss from the surface
2. stir the air and break up the temperature inversion layer
3. add heat to maintain the temperature above the danger level.

Marketing & Supply Chains

Post harvest losses represent one of the single greatest sources of inefficiencies in food production worldwide and therefore one of the best opportunities for effectively improving crop productivity. These losses – which are due to poorly timed or executed harvesting, exposure to rain, humidity and heat, contamination by microorganisms, and a host of other sources of damage and deterioration – often get far less attention than they deserve. Half or more of the total harvest of some crops can be lost post harvest. Investments in improved harvesting, processing, storage, distribution, and logistics technology and necessary training investments can pay off as well as improved crop yields in terms of gains to consumers and the climate. As climates become hotter and precipitation more erratic, the potential for postharvest losses may increase and thus improved transport and storage become even more important.

Innovative and Improved Postharvest Management Strategies

Simple modifications to postharvest handling systems can sometimes result in significant reduction in stress exposure and consequently result in storage and/ or shelf life extension. One of the most successful strategies is the application of plastic film packaging or wraps to prevent desiccation, resulting in significant improvements and shelf life and quality of many fruits and vegetables. In many cases, modified atmosphere packaging is considered to largely control humidity around product and thus prevent moisture loss of fresh-cut and whole fruits and vegetables. Also, anti-transpiration coatings have been shown to be effective for maintaining quality through control of water loss. In regards to maintaining water content on the retail shelf, the application of misting systems can 'recharge' the vegetable and in so doing maintain quality over longer durations at less than ideal storage temperatures.

Potential Benefits of GAP

- Appropriate promotion and adoption of GAP from farm to fork will help improve the safety and quality of food and agricultural products.
- Producers and consumers will benefit from global markets and improve their livelihoods and the national economy as a whole.
- Adoption of GAP will help promote sustainable agriculture and contribute to meeting national and international environmental and social development objectives.
- Adherence to food quality and safety will protect people's health – an important factor in national development.

Future Scope and Research Needs

- The best policy and institutional responses will enhance information flows, incentives and flexibility
- Policies and institutions that promote economic development and reduce poverty will often improve agricultural adaptation and may also pave the way for more effective climate change mitigation through agriculture.

- Existing technology options must be made more available and accessible without overlooking complementary capacity and investments
- Adaptation and mitigation in agriculture will require local responses, but effective policy responses must also reflect global impacts and inter-linkages.

Conclusion

As climate change affects input availability, especially water in many places, input use efficiency must increase with these productivity demands. Carbon emission polices may simultaneously encourage or force producers to recognize GHG emissions as an important and costly "input" in production processes and open new opportunities and incentives for on farm GHG mitigation. Innovations in agriculture have always been important and will be even more vital in the context of climate change. Thoughtful policy responses that encourage the development and diffusion of appropriate agricultural technologies will be crucial to enabling an effective technological response. A careful balance of institutional change and wise investments is required to deal with both the demands of climate change and the demands of improving lives of the poor. As we consider implications of and responses to climate change, continuing concerns for improvements in nutrition, food security, food safety, local environments, and rural communities must not be neglected. Given the reliance of the poor on agriculture and the sensitivity of agriculture to climate change, impending climate changes will almost certainly hit (currently) developing countries and vulnerable populations within these countries hardest. While this reality seems to make the development process more complex, it should also stimulate greater urgency in addressing rural poverty and vulnerability. These twin imperatives of climate change – greater complexity and greater urgency – are important to keep in mind when formulating policies and institutions aimed at improving climate change adaptation and mitigation in agriculture.

References

Baldwin EA 2003. Coating and Other Supplemental Treatments to Maintain Vegetable Quality. In: Bartz JA, Brecht JK (ed) Postharvest Physiology and Pathology of Vegetables, 2nd Edn. Marcel Dekker, Inc, New York, p 413-456

Beck EH, Fetitig S, Knake C *et al* 2007a Specific and unspecific responses of plants to cold and drought stress. *J Bio Sci* 32(3): 501-510

Cline, W.R. 2007.*Global Warming and Agriculture : Impact Estimates by Country*. Washington, DC: Center for Global Development : Peterson Institute for International Economics.

Goyary D 2009. Transgenic Crops, and their scope for abiotic stress environment of high altitude: Biochemical and Physiological Perspectives. DRDO Science Spectrum, March 2009, p 195-201

GRDC 2014. Frost risk on the rise despite warmer climate. Ground Cover supplement: Turning up the heat on frost in cereals. Issue 109 March-April 2014. http://grdc.com.au/ Media-Centre/ Ground-Cover-Supplements/ GCS109

IPCC, 2013: Summary for Policymakers. In: Climate Change 2013: The Physical Science Basis. Contribution of Working Group I to the Fifth Assessment Report of the Intergovernmental Panel on Climate Change [Stocker, T.F., D. Qin, G.-K. Plattner, M. Tignor, S. K. Allen, J. Boschung, A. Nauels, Y. Xia, V. Bex and P.M. Midgley (eds.)]. Cambridge University Press, Cambridge, United Kingdom and New York, NY, USA.

IPCC. 2007. Climate Change 2007: Synthesis Report. Contributions of Working Groups I, Ii, and Iii to the Fourth Assessment Report of the Intergovernmental Panel on Climate Change. Geneva: IPCC.

Kumar PS, Minhas PS, Govindasamy, V, Choudhary RL (2014). Influence of Moisture Stress on Growth, Development, Physiological Process and Quality of Fruits and Vegetables and Its Management Strategies. In.(Eds. R.K. Gaur and Pradeep Sharma) Approaches to Plant Stress and their Management, Springer publications., New Delhi, pp.125-148

Lehmann, J.J. Gaunt, and M. Rondon. 2006. "Bio-Char Sequestration in Terrestrial Ecosystems– a Review." *Mitigation and adaptation strategies for global change* 11(2):395-419.

Suresh Kumar P, Choudhary VK, Bhagawati R (2012) Influence of mulching and irrigation level on water-use efficiency, plant growth and quality of strawberry (Fragariax ananassa). *Ind J Agric Sci* 82(2):127-133

Wang CY 1993. Approaches to Reducing Chilling Injury of Fruits and Vegetables.Hortl Rev 15:63-95

Section 02: Economics, Management and Extension of Arid Horticultural Crops

21

Concept and Strategies for Doubling Farmers Income: Focus on Horticulture

P. L. Saroj and Mukesh K. Berwal

India's economic security continues to be predicated upon the agriculture sector, and the situation is not likely to change in the foreseeable future. Even now, agriculture supports 58% of the Indian population, as against about 75% at the time of independence. In the identical period, the contribution of agriculture and allied sector to the Gross Domestic Product (GDP) has fallen from 61 to 13%. As of today, India supports 16.8% of world's population on 4.2% of world's water resources and 2.3% of global land resources. Also per caput availability of resources is about 4 to 6 times less than that of world average. This will further decrease due to increasing demographic pressure and consequent diversion of the land for non-agricultural uses. Around 51% of India's geographical area is already under cultivation as compared to 11% of the world average. The present cropping intensity is 136% and it has been registered an increase of only 25% since independence. Further, rainfed and dry-lands area constitute approximately 65% of the total net sown area. There is also an unprecedented degradation of land (107 million ha) and groundwater resource, and also fall in the rate of growth of total factor productivity. This deceleration needs to be arrested and agricultural productivity has to be doubled to meet growing demands of the population by 2050. Efficiency-mediated improvement in productivity is the most viable option to raise production of various commodities including horticulture.

The country has made tremendous achievements in agriculture during five decades since the onset of green revolution in late sixties. This enabled the country to overcome prevalent hunger and starvation; achieve self-sufficiency in food; reduce poverty and bring economic transformation in millions of agrarian rural families. The situation, however, started turning adverse for the sector around mid-nineties, with slowdown in growth rate of output, which then resulted in stagnation or even decline in farmers' income leading to agrarian distress, which is spreading and turning more and more serious. Natural resources are base for agriculture, which provides for sustainable production, is shrinking and degrading and are adversely affecting production capacity of the ecosystem. However, demand for agriculture is rising speedily with increase in population and per caput income and growing demand from industry sector. The share of agriculture in national GDP is not improving (Fig. 1). There is, thus, an urgent need to identify severity of problem confronting agriculture sector to restore its vitality and put it back on higher growth trajectory. The problems, however, are surmountable, particularly when new tools of science and technology have started offering tremendous opportunities for application in agriculture.

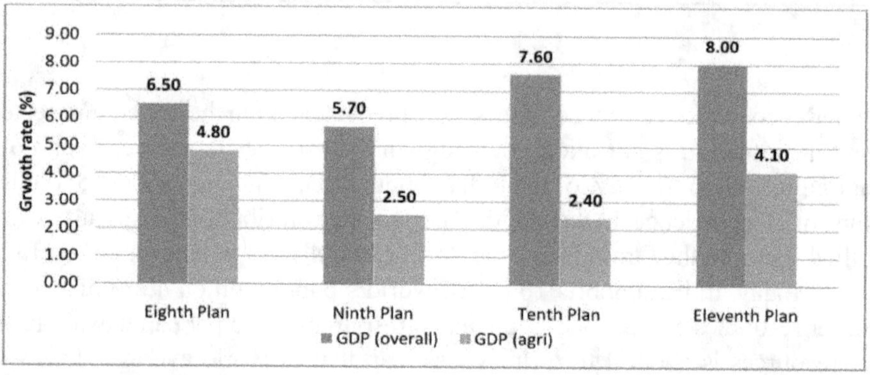

Fig. 1: Agricultural growth rate during different plan periods (*Source*: Pattnayak, 2016)

India is primarily an agrarian economy, but still Indian farmers continue to be among the poorest in the world due to continuously reduction in operational farm holdings and holding size (Figure 2). With the diverse climatic conditions found and large areas of rich fertile land, India has a lot of potential to become one of the leading producers of various agriculture products. With presence of abundant raw materials in the form of natural resources, India lags far behind in ensuring nutritional security for its own citizens. This state is a combined consequence of various factors, such as lack of advanced scientific methods in agriculture, unavailability of modern machinery to the farmers, lack of electricity in villages and lack of financial support. Moreover, storage and maintenance inefficiency leads to tones of food grains in the country being wasted as they rot in the storage godowns. Similarly, traditional manner of farming by the farmers also lead to low yield. Farmers across the country depend more on the practices learnt by their ancestors and take various decisions based on myths. Being uneducated, they had a belief that anything for improving soil fertility must have a direct response and the more they add it, the better things should become. Same is their concept regarding fertilizer use like what the soil lacks and what kind of fertilizers should be used is not known to them many a times. But some good practices like crop rotation are also followed by them since generations.

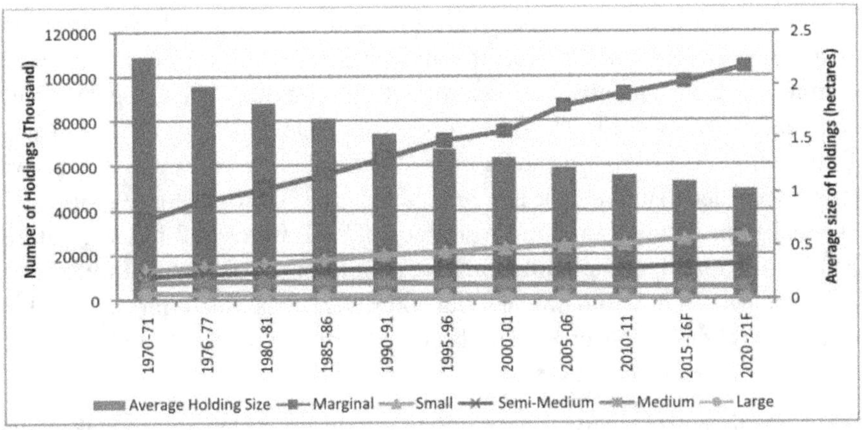

Fig. 2: Number of operational holdings and average holding size in India (Source: Fourth Semi-Annual Medium Term Agricultural Outlook Report, September, 2015, NCAER.)

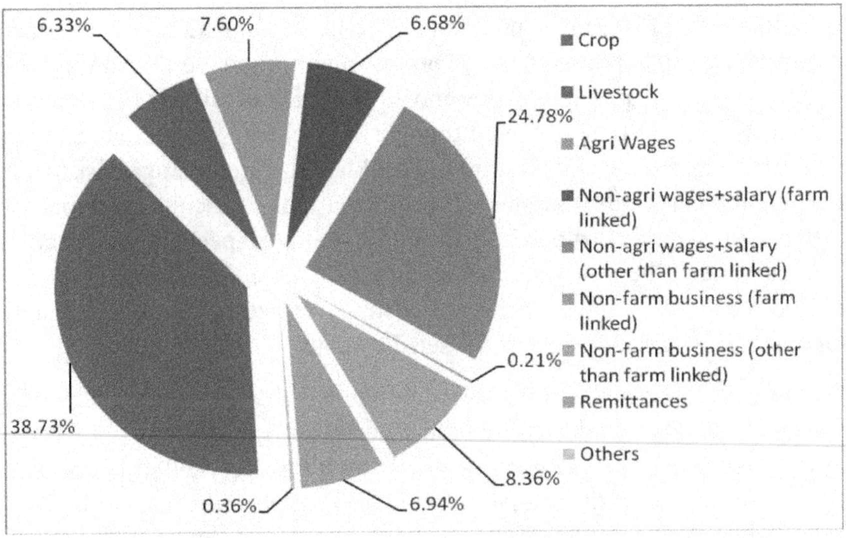

Fig. 3: Share of different components in total income of Agricultural households 2011-12

The agriculture and allied sector continues to be pivotal to the sustainable growth and development of the Indian economy. It does not only meet the food and nutritional requirements of 1.3 billion Indians but also contributes significantly to the employment generation and maintain ecological system. Moreover, the role of the agricultural sector in alleviating poverty and in ensuring the sustainable development of the economy is well established. The sector is, however, currently facing a dilemma. While it has made large strides in achieving the agricultural development goals of food security, availability and accessibility, it is still being challenged by a formidable agrarian crisis. This situation has recently led to fresh thinking on the developmental approach in the agriculture sector. The need for focusing on the welfare and prosperity of farmers has gained prominence. Consequently, the Department of Agriculture and Cooperation was renamed, by the Hon'ble PM on Independence Day 2015, as the Department of Agriculture, Cooperation and Farmers Welfare. In this fresh approach, priority is to be accorded to making the agriculture and allied sector not only ecologically sustainable in its use of natural resources of soil, water and forests, but also socio-economically sustainable to farmers in terms of prosperity, welfare and social security. Innovating managerial solutions to maximize farmers' welfare rather than relying solely on modern farming to raise productivity and production is the clarion call of the day.

Agricultural sector received continuous attention of the policy makers and stakeholders. A number of initiative were undertaken to improve the performance of this sector. For the first time in our history, Hon'ble Prime Minister of India exhorted to "Double the Farmers' Income" (DFI) by 2021-22 and helped in channelizing the efforts in the unified direction, and a holistic approach is being followed from top to bottom in an integrated manner and the slogan is catching the momentum and attention of one and all. Along with this, DFI goal was also coupled with many new and well-thought out schemes like Pradhan Mantri Fasal Bima Yojana, e-National Agricultural Market, Paramparagat Krishi Vikas Yojana, Pradhan Mantri Krishi Sichai Yojana etc. Consequently Honorable Finance Minister Mr. Arun Jaitley during his budget speech on February 29, 2016 announced the government's resolve to double the income of farmers. Perhaps, for the first time in the history of Indian agriculture, a goal of doubling of income of farmers in six years is set. To quote:

"We are grateful to our farmers for being the backbone of the country's food security. We need to think beyond food security and give back to our farmers a sense of income security. Government will, therefore, reorient its interventions in the farm and non-farm sectors to double the income of the farmers by 2022."

—Finance Minister Arun Jaitley, Budget Speech, February 29, 2016.

Based on the recommendations of National Commission on Farmers, Government announced measuring agricultural progress by real income of farmers and not by gross production of agricultural commodities. A Task Group has been formed by the Ministry of Agriculture to come out with a detailed plans and its operationalization. In this context, Indian Council of Food and Agriculture organized a national consultation on 30 April, 2016 in New Delhi, involving top experts from the Government, industry, research and academic bodies, farmers organizations, parliamentarians and NGOs. The meet was chaired by Prof. MS Swaminathan and co-chaired by the Secretary – Agriculture, Government of India.

The average monthly income per capita from farming increased from Rs 1,060 in 2003 to Rs 3,844 in 2013, according to the report Situational Assessment of Agricultural Households by the NSSO with a compounded annual income growth rate of 13.7%. To double the income of farmers by 2022, in nominal (numerical) terms, which do not take inflation into account, would require a 15% compounded income growth rate, which is a marginal increase over the achieved increase from 2003 to 2013. However to increase the income in real terms would imply restructuring agriculture processes and policy interventions. There is however, almost unanimity that the net income of farmers can surely be doubled well within the period of six years.

Honorable Prime Minister has outlined 7 strategies to help for doubling the incomes of farmers. They are:

1. Big focus on irrigation with large budgets, with the aim of "per drop more crop".
2. Provision of quality seeds and nutrients based on soil health of each field.
3. Large investments in warehousing and cold chains to prevent post-harvest crop losses.
4. Promotion of value addition through food processing.
5. Creation of a national farm market and removing distortions.
6. Introduction of a new crop insurance scheme to mitigate risks at affordable cost and,
7. Promotion of ancillary activities like poultry, beekeeping and fisheries.

Of these, strategies at S. No.1 and 2 addresses the farm level production related issues and 7th addresses the issue of augmenting income through off-farm and non-farm activities. Other strategies are more useful for better price realization and post-harvest management. Hence, they can help partially. Further, strategies need to be built around natural resource management, social sector policies such as health and education. For example, farmers' expenditure on health and education is substantial enough to topple his balance sheet.

This national consultation gave a total of 40 recommendations under five broad heads for increasing incomes of farmers (Source: Indian Council of Food and Agriculture-2016):

A. Increasing incomes by improving productivity

- Biotechnological intervention
- Improving agricultural productivity in rain fed regions of India,
- Bridging yield gaps among the States,

B. Water and Agri-Input policies

- Fertilizer Subsidy and Rationalizing the NPK pricing
- Minimizing Crop losses.
- Farm Mechanization.
- Integrated water use policy.

C. Integrated Farming System

- Promotion of integrated farming system approaches.
- Integration of dairy husbandry with field crops.
- Integration of horticultural crops.

D. Better market price realization

- Revision of the agricultural produce market committee (APMC) Act and monitoring.
- The launch of NAM requires easing of norms of licensing to enable seamless participation of buyers from across the country, movements of goods without restriction.
- Reforms to the APMC Acts to permit pan-India trades.
- Agri-infrastructure development.
- Minimizing post harvest losses by strengthening grain storage infrastructure

E. Special Policy Measure

- Structural reforms in agriculture pertaining to land leasing and market restrictions need to be addressed.
- Through a nationwide crops competitiveness study.
- Agriculture to be brought to the Concurrent List.
- Implementing ambitious Agribusiness Hubs Model.
- Multi-functional Agribusiness hubs in all the Gram Panchayats of the country.

- ICT-based agricultural extension. (Information and Communication Technologies)
- Diversification of agriculture in the First Green Revolution areas.
- Integration of all central and state government subsidies.
- Strengthening organic food production.
- Establishing Special Agriculture Zones (SAZ) by selecting export oriented and industrial use crops.
- Promoting scientific agriculture micro-irrigation on a very large scale.

Strategies to Double Farmer's Income Through Horticulture

Horticulture, which includes fruits, vegetables, flower and ornamental plants, tuber crops, species, medicinal and aromatic plants, has become a priority area in the diversification of Indian agriculture and is improving economic conditions of farmers and entrepreneurs as also contributing significantly to the national economy. Indian horticulture has achieved a growth rate of 7% during 2004-05 to 2014-15 with an increase in area by 2.7%, productivity by 37% that has resulted in daily per capita availability of about 200g of fruits and 400g of vegetables. With an annual production is over 280 MT and contributing a tune of 33% in Agriculture GDP. At present India is the second largest producer of horticulture produce. India has marched ahead from food sufficiency to nutritional security, which is visible with rise in the production, productivity and availability of an array of horticulture produce / products round the year (Fig. 5). This is a welcome sign as it addresses the burning agrarian issues like malnutrition, greening of marginal and arid regions, dwindling rural employment, empowering women, availability of safe foods, post harvest management and value addition, enhancing export opportunities *etc.*

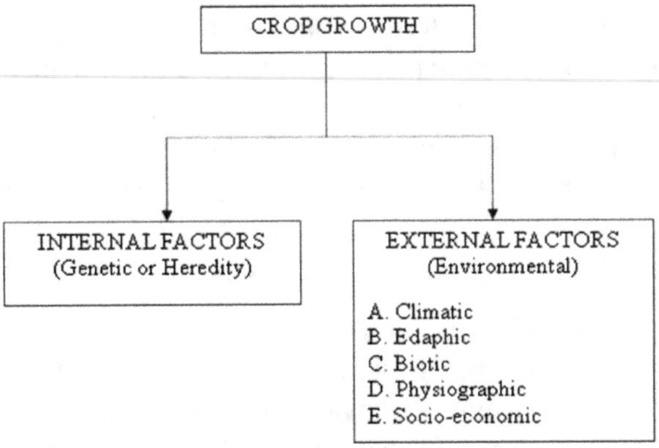

Fig. 4: Factors affecting crop production

Despite the impressive achievements in Horticulture sector, still low productivity is a challenge and matter of concerned. The phenomenon of climate change and aberrant weather conditions are the major issues to be tackled with new innovations in crop improvement and production system (Figure 4). Here we are discussing some challenging but very vital strategies to overcome these barriers in production system and reach to the goal of doubling farmers income by 2022 through horticulture.

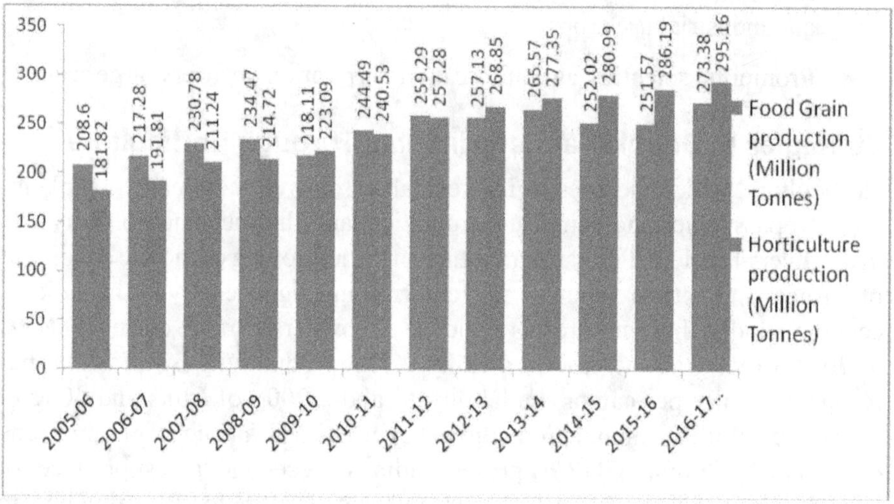

Fig. 5: Trend in foodgrains and horticulture production since 2004-5 to 2014-15. (*Source*: Ministry of Agriculture and Farmers Welfare)

1. Policy initiatives for Doubling Farmers' Income: Doubling farmers' income or raising it to a significant degree requires re-orientation in the agriculture policy stance. it includes (i) Identification of viable strategies and their proper implementations. (ii) Identify who within agricultural sector have low incomes and/or are disadvantaged in terms of access to technology, markets, credit, information and infrastructure (iii) Identify where the low-income households are located, and the location characteristics in terms of infrastructure, information investment, technology penetration. It will provide a menu of policy choices, and pathways to enhance farmers' income. Efforts has been made to enhance the farmers income by Union government since 2014-15 through budgetary allocation toward different farmer friendly schemes (Figure 6).

2. Improving availability and quality of panting materials and seeds: Seed or planting material, comprising all forms of propagules, acts as a catalyst and a starting point for realising potential of all other inputs in agriculture. All forms of propagules are used as seed specially in case of vegetables, ornamentals and

Budget 2014-15	Budget 2015-16	Budget 2016-17
➢ Established National Adaptation Fund for Climate change ➢ Set up Long-term Rural Credit Fund ➢ Bhoomi Heen Kisan- Providing financial assistance to 5 lakh joint farming groups of Bhoomi Heen kisan through NABARD	➢ Emphasis on Rural Infrastructure Development ➢ Long Term Rural Credit Fund ➢ Short term Cooperative Rural Credit Refinance Fund	➢ Doubling Farmer's Income by 2022 ➢ Long-Term Irrigation Fund ➢ Expansion of Soil Health Card Scheme
➢ Launching of Soil Health Cards Set up-100 mobile Soil Testing Laboratories	➢ Parampragat Krishi Vikas **Yojana**	➢ Continuation of Parampragat Krishi Vikas ➢ Organic Value Chain Development in North-East
➢ Pradhan Mantri Krishi Sinchayee Yojana		
➢ Agri-Tech Infrastructure Fund		

Budget 2017-18	
Agricultural Credit	➢ Ensure adequate flow of credit to under serviced areas (Rs. 10 lakh crores allocated)
Pradhan Mantri Fasal Bima Yojana	➢ Coverage of the scheme to be increased from 40% in 2016-17 to 50% 2018-19 (Rs. 9000 carore allocated)
Soil Health	➢ Establishment of mini labs in Krishi Vigyan Kendras (KVKs) and ensure 100% coverage of all 648 KVK
National Agriculture Market	➢ Coverage of National Agriculture Market (e-NAM) is expanded from 250 to 585.
Irrigation	➢ Corpus of long Term irrigation fund has been doubled (20000 carore allocated). ➢ A dedicated micro-irrigation fund will be setup in NABARD to achieve the goal "Per Drop More Crop" (Rs. 5000 carores allocated)
Agro Processing Unit	➢ Emphasis on contract farming for strengthning linkage between fruits & vegetables cultivation and agro processing Units
Dairy	➢ A dairy processing and infrastructure development fund would be set up in NABARD for moderanizing milk processing unit set up under operation flood programme (Rs. 2000 carore allocated)

Fig. 6: Budgetary allocations towards farmers friendly schemes since 2014-15
(*Source: Chadha et al. 2017*)

horticultural plants. The use of virus-free mini-tubers and micro-tubers of potato has started, and is increasing day-by-day including export of these. The Indian Council of Agricultural Research is engaged in harnessing modern scientific knowledge to develop high-yielding, open-pollinated varieties and hybrids that are resistant/ tolerant to major biotic and abiotic stress factors, efficient in input use, and adapted to different agro-climates, suited for irrigated and rainfed areas. Varieties and hybrids have also been tailored with ideal maturity and plant architecture for intensification and diversification of cropping systems.

3. Crop management through canopy management and HDP: Canopy architecture and management is the manipulation of tree canopy to optimize production of quality fruits. It is not only involve training and pruning, but also regulate flowering and fruit growth. It is well established that small trees utilize sunlight in to better fruit yield than that of large trees. With stagnation in productivity and competitiveness in international market, the demand of high quality fruits by consumers and to enhance production efficiency, fruit growers (specially temperate fruits) must establish high density orchards (HDP) on size controlling rootstocks wherein the potential returns can be increased by 3-4 fold as compared to low density plantation particularly during the first ten years of establishment (Ahmed *et al*. 2016). In a nutshell, the architecture design for any fruit species depends on the system of planting and space provided for each plant. Further the concept of mechanization and ease of orchard operation is kept in mind while designing the canopy skeleton. The major principal followed in designing canopy are distribution of branches of arms and branches, fruiting areas, space for aeration and light penetration. High density specially ultra high density plantation of tree crops is capable to reduce gestation period to 1-3 years depending upon crop species with ability to increase productivity by two to three fold (Krishna *et al*., 2016).

4. Nutrient Management: Site-specific integrated nutrient management, envisaging precise use of chemical fertilizers including secondary- and micro-nutrients, organic manures, composts/vermicomposts, biofertilizers and green manures, is ideal. This will maintain soil health, enhance nutrient-use efficiency, provide sustained yields, ensuring better economic returns to farmers. The system is also benign environmentally in containing the emissions of green house gases and pollution of surface and groundwater resources. The ICAR has generated site-specific nutrient management packages that would enable additional yield of 50-60 million tonnes of foodgrains and this is also true for horticultural crops as well.

5. Irrigation management "Per Drop - More Crop": The agricultural sector is going to face grim competition for supplies of fresh water, with its share dropping to 75% from the present 83% in the near future, in the wake of growing industrial and domestic sectors. The judicious management of water resources is, therefore, going to be crucial to sustain agricultural growth in the country. Currently agriculture consumes 70 percent or more of the world's developed freshwater supplies. Water is one of the crucial inputs in agricultural production accounting for about 80 percent of the water withdrawals in India. Alarming rates of ground water depletion and serious environmental and social problems of some of the major irrigation projects on one hand, and the multiple benefits of irrigation water in enhancing production and productivity, food security and poverty alleviation highlights the importance of proper irrigation system for agriculture sector. India will be required to produce more and more from less and less land and water resources. The fact that 2000 to 5000 liters of water is required to produce enough food to feed one person each day, underscore the importance of increasing the productivity of water in agriculture. In India only 40 per cent of the agricultural land is under irrigation and the rest sixty percent area is rain fed. The excess use of canal water in south-western Punjab, Haryana and Rajasthan is leading to water-logging and development of secondary salinity. Large volumes of waste water (18.4 million m^3/day) need to be utilized for irrigation after their proper treatment, especially in peri-urban areas. The micro-irrigation and resource conservation technologies (RCTs), economizing on water and nutrients require to be promoted in a big way.

6. Management of emerging insect pests and diseases: Several constraints hinder the horticulture sector from realizing its full potential, of which insect pest and diseases cause a considerable extent of damage of the crops. It cause nearly accounts 25-30% losses in fruits and vegetable produce annually. Without shifting the focus on the existing panoply of insect pest and pathogens, the looming threat of new biotic constraints needs to be addressed to enhance the farmers income. The adoption and promotion of IPM and IDM in fruit and vegetable crops still remain a challenge in view of several constraints faced by farmers because of small holdings, their raw consumption, harvest at short intervals and less time available for reaching from harvest to table. Wider area adoption of IPM and IDM is further challenging because vegetable cultivation is often fragmented in terms of area and space and IPM Being knowledge intensive, dependent on quality critical inputs like bio-pesticides and often require community approaches, thereby making adoption in larger areas extremely difficult task. Extreme climatic conditions, regional variations in crops and pest problem warrants frequent orientation in IPM format for development in season and location specific IPM practices. Establishment of national IPM network through using ICT technology for major pest across crops and agro-ecological regions for real time pests scenario building and the implementation of IPM practices is the need of time. Effectiveness of IPM modules should be made obvious to reach managers/policy makers and state Govt. agencies to support IPM programmed for quality produce (Ahuja and Sardana, 2016). For effective IPM measures against pest and diseases in horticultural crops, real time weather data based online advisory system can also play an important role.

7. Ensuring higher price of farm produce and product: Farmer's income can be increased by decreasing the profit made by middle men through self management of activities like packaging, transportation, sale and distribution of horticultural produce. Market driven quality production with grading and packing can help in fetch higher price in market. Horticultural produce prices tend to be more volatile owing to highly perishable nature, seasonality, inelastic demand and high risk of biotic and abiotic stresses. Farmers go for distress sale to avert the price risk which jeopardizing higher return. Therefore, accurate and on time market intelligence of horticultural produce is required to ensure higher prices. Application of advanced statistical methods and software to understand huge market data and information is essential. Effective market intelligence would enable efficient cropping pattern and ensure higher prices.

8. Technology-led organic horticulture and safe food: Organic horticulture has great potential to provide benefits in terms of environmental production, conservation of non-renewable resources and improved food quality. According to the FiBL –IFOAM, 37 million hectare area is under organic agriculture worldwide and India having 0.8 million hectare under organic agriculture. In arid region some of the fruits and vegetable like cucurbitaceous vegetables, sangri, kair, karonda and cluster bean are produced organically due traditional farming system and non-availability of chemical inputs in remote areas or economical availability of organic input. Food produced through conventional horticultural practices by using of agrochemicals is unsafe or harmful human health and also toxic to environment. Organic food production in horticultural crop is highly remunerative than the conventional production system. Further safety of horticulture produce can be ensured using new generation organic inputs like organic fertilizers and pesticides and bio-agents.

9. Value addition and reduction in post-harvest losses: Value addition in horticulture is an integral part in increasing farmer's income by reducing post harvest losses and transforming horticultural produce to nutritionally rich value added quality product. Low cost drying technology of vegetables offers a highly effective and practical means of preserving vegetables in large quantity with extended shelf life of 6-8 months at ambient storage condition. Horticultural crops are rich in phyto-chemicals like anthocynins, carotene, phenols, flavonoids which act as antioxidants, anti-diabetics, anti-inflammatory and also improve digestion. Many horticultural produce like Bael, Aonla, Jamun, Moringa, Kair, and Tumba have been recognized for their health promoting and disease preventing properties. Extraction and utilization of these phytochemicals by highlighting their health promoting and disease preventing properties can add value to the produce. Storage temperature is most important factor for maintaining quality of horticultural produce after harvest. Storage in refrigerated vehicles and cool chain facility retards the aging due to ripening, softening and colour change.

10. Mechanization in horticulture: The major field operations for horticultural crops includes nursery preparation, planting, intercultural operation, irrigation, plant protection, harvesting, grading, packaging and transport. All these operations are predominantly depends on human labour owing to shortage of appropriate economical machinery and lack of commercial cultivation. Horticultural crops are highly suitable to mechanization from nursery to post harvest handling. Now a day's various tools and machinery are available for nursery seedling production, transplanting, spraying of chemicals etc. Drip irrigation not only save water but also helps in efficient utilization of chemicals and fertilizers. Varieties suitable for mechanical harvesting and mechanical harvesters are developed for various horticultural crops which ensure timely harvest with minimum crop losses.

11. Sustainable management of natural resources: Continued degradation of natural resources under intensive agriculture to attain goals of food sufficiency is one of the reasons for the declining factor productivity and stagnation in food grains production in the country. The health of our soils has been impaired due to emergence of multi-nutrient deficiencies and falling of organic carbon levels. The soils are, generally, not replenished adequately with nutrients removed by crops, particularly micro and secondary nutrients. The wider fertilizer consumption ratios for many states corroborate nutrient imbalance in soils. The soils are, presently, operating on a negative nutrient balance of about 10 million tonnes per annum. Limiting nutrients, not allowing full expression of other nutrients, obviously, lower overall fertilizer-use efficiency and crop productivity. Farmers are resorting to addition of more and more fertilizers to obtain yields similar to previous years, specially in rice-wheat belt of Indo-Gangetic plains and the increased inputs are thus adding to the production cost; marginalizing net returns to farmers.

12. Climate smart horticulture: Climate Smart Horticulture is resilient to climate change, reduces harmful emissions and improves carbon sequestration, maintains healthy ecosystem and biodiversity with sustainable production. As agriculture accounts for up to 30 per cent of global greenhouse gas emissions. Therefore Climate Smart Horticulture is required to achieve nutritional security and climate change goals. Climate Smart Horticulture play important role in addressing the food and nutritional security along with increased productivity and reduced emissions. The overall efficiency horticultural production system can be enhanced through improving its various components such as genetic resources, soil and nutrient management, irrigation management, biotic and abiotic stress management, and harvesting, value addition and cool chain managements. Studying the effect of different abiotic stresses on growth and development and their mechanisms (stress response, acclimation and adaptation) can facilitates the identification of environmental changes such as lack of light, high temperatures or water deficit (Restrepo-Díaz, 2010).

Conclusion

Doubling of the income of farmers in nominal terms has already been happening in recent periods and it is not a challenge. Doubling the income in six years, in real terms, however, is a formidable challenge and needs large scale revamping, reorientation and innovation in the initiatives. Farmer's income can increase through increasing total output and their prices, reducing production costs through lowering input use and/or reducing input prices, diversifying production mix towards more remunerative enterprises and providing earning opportunities in non-farm sector. Apart from the traditionally known risks to farmers, climate change is an additional risk factor that can cause loss of farm income. Apart from this, access to good physical, economic/financial, social infrastructure such as marketing and processing facilities, godowns and cold storage capacity, banking network that can provide much needed capital, educational, medical facilities and training facilities for imparting skills that market demands is important.

References

Ahmed, N., Pandit A.H. and Mir, J.I. 2016. Canopy architecture and HDP in temperate fruits: An innovative approach for doubling farmer's income. In Proc. 7th Indian Horticulture Congress-2016 on Doubling Farmers Income through Horticulture, being held at IARI, New Delhi during November 15-18, 2016. Pp. 50-51.

Ahuja D.B. and Sardana H.R. 2016. Role of IPM protocol in horticulture crops for improving profitability. In Proc. 7th Indian Horticulture Congress-2016 on Doubling Farmers Income through Horticulture, being held at IARI, New Delhi during November 15-18, 2016. Pp. 96-97.

Birthal, P.S., Negi, D.S. and Roy D. 2017. Enhancing farmers' income: Who to target and How? Policy Paper - National Centre for Agricultural Economics and Policy Research No.30 pp.xiv + 38 pp.

Chadha, R., Vashishtha, P.S., Pratap, D., Elumalai, K., Bathla, S., Pohit, S., Tandon, A., Jain, R., Ruhil, N., Das, G. K., and Rathore, Y.S. 2017. Doubling farmers' income in India Backdrops and challenges. National Council of Applied Economic research. 1956-2017 Quality, Relevance, Impact. PP 1-25.

Gulati, A., Jain, S and Satija, N. 2013. "Rising Farm Wages in India: The 'Pull' and 'Push' Factors", Discussion Paper No.5, Commission for Agricultural Costs and Prices. Department of Agriculture & Cooperation, Ministry of Agriculture, Government of India. pp. 1-29

Horticultural statistics at a Glance 2017. Horticulture Statistics Division Department of Agriculture, Cooperation & Farmers Welfare Ministry of Agriculture & Farmers Welfare Government of India. pp. 1-481.

Indian Council of Food and Agriculture. 2016. Report on Doubling Farmer's Income by 2022 Farm Crisis and Farmers' Distress. pp. 1-10

Krishna B., Chaudhary A.U., Dhake A.V. and Balasubramaniam V.R. 2016. High density planting in tree crops for higher productivity and profitability. In Proc. 7th Indian Horticulture Congress-2016 on Doubling Farmers Income through Horticulture, held at IARI, New Delhi during November 15-18, 2016. Pp. 59.

Pattnayal, S.K. 2016. State of Indian Agriculture 2015-16. Government of India Ministry of Agriculture & Farmers Welfare Department of Agriculture, Cooperation & Farmers Welfare, Directorate of Economics and Statistics, New Delhi, pp. 1-252.

Restrepo-Díaz, H. Melgar, J. and Lombardini, L. 2010. Ecophysiology of horticultural crops: an overview. *Agronomía Colombiana*, 28(1), 71-79.

22

Improving Yield and Income of Arid Vegetables

D. K. Samadia

Introduction

The Indian *Thar Desert* is a marginal zone for agricultural production. Scanty and un-even rain, extremes of high (March–October) and low (December–January) temperature condition for the prolong duration and in-together these factors creates an associated abiotic stressed environment that limiting choice of crops, quality of produce, productivity and production under hot arid agro-climate. Out of 9–12 mean rainy days/annum in the hot arid region of north-western parts of Rajasthan, much of the rain often comes in only a few (1–3) heavy rainfall events during the monsoon season (June–September), which can be erratic and unpredictable. In kharif season, higher day temperature range (35–42°C) and prolonged drought periods in combination with sandy soil creates an environment where few crops can survive under rainfed situations. Here, the native crop-plants both annual and perennial can play an equally vital role for food and fodder especially when the livestock population is an important element of farming system. Further, horticultural exploitation with native crop-plant species in general is found to be potential and stable in the arid and semi-arid areas which receive about 150–650 mm of rains.

Horticultural significant crop-plants produce and their product are rich source of carbohydrate, nutrients, vitamins, anti-oxidant and fibre and are thus essential part of nutritious diet. In India, about 175 types of crop-plants are being used for vegetable purpose. The vegetable production in the country is largely uneven and concentrated mostly into the few states and areas where the agro-climate is much favourable and mild for their production. In addition, emphasis is only with few numbers of vegetables and that too are grown under high input and assured irrigation facilities. To meet-out ever increasing vegetable demand and utilization of indigenous resources, now, there is utmost need for massive and integrated efforts, and it should be firstly by developing high yielding crop-genotypes suited to the prevailing situations of defined agro-climatic zone and resource based production technology under the climatic variability in region. Second is through exploitation of native resources *i.e.* vast, varied and un-tapped land-scape and under-exploited horticultural species for their vegetable potentialities. In this way, the desert, arid, semi-arid and tribal land area of Rajasthan state is a vast un-tapped resource where several popular and native crop-plants having horticultural significance can be exploited for diverse and organized vegetable culture (Samadia *et al.*, 2004 and Samadia, 2007) adopting production site management approaches (Samadia, 2016).

Arid Agro-climate – Need Special Attention for Vegetable Culture

Production and marketing of high value and diverse form of vegetable is now seen as key to generate more income per unit area, time and round the year and also reducing the poverty amongst small farmers in the developing countries. India is the second largest producer of vegetables and share is about 12% of the world production. On the other hand, its distance from the world's number one (China) is quite large and this wide-gap arises several questions related to the productivity of vegetable crops in Indian fields, types and quality of genotypes and seed/planting material being used, adoption level of technologies, management and utilization of resources at the production sites (Samadia, 2009). In recent years, there is a significant change in consumption of vegetables in the country. The increase vegetable consumption is further expected to enhance, and it is due to rising income levels, changing pattern of dietary mix, nutrition and health awareness, ease and round the year availability. On the other hand, present productivity level is lower than the world average and, accordingly it cannot meet-out the increasing demand without integration, diversification and management of production sites for vegetable culture in the variable climatic regions (Samadia, 2008 and More and Samadia, 2009).

In the hot arid region, mixed cropping of bajra with legumes such as cluster bean, moth bean or cowpea dominates the traditional farming system. The indigenous cucurbit vegetables such as kachri, snap melon, mateera and tinda are the rescue crops and provide subsistence income under rainfed conditions. In addition, native species such as khejri, ker, lasora, kumat, khimp, phog, guarpatha, tumba and khumbi have horticultural significance and ensuring nutrition rich vegetables to the desert dwellers. Here, systematic vegetable culture is negligible and productivity is very low, and it is primarily owing to unavailability of desirable crop-plant genotypes suited to environmentally stressed production sites, unavailability of quality seed / planting material of recommended genotypes for growing with varying situation and lack of crop-genotype-climatic matching techniques. Therefore, there are essentially two complementary requirements for obtaining high quality vegetable production and productivity *i.e.* improvement in the genetic make-up of prioritized crops and development of favourable micro-climate at the production sites (Samadia, 2008).

Table 1: Prioritized crop-plants for vegetable culture under hot arid agro-climate

Group	Crop-plants
Cucurbitaceous	Kachri (*Cucumis melo* var. *callosus/agrestis*), snap melon (*Cucumis melo* var. *momordica*),kakri (*Cucumis melo* var. common / *utilissimus / fluxuosus / acidulus)*, muskmelon (*Cucumis melo*), mateera / watermelon (*Citrullus lanatus*), round melon / tinda (*Praecitrullus fistulosus*), bottle gourd (*Lagenaria siceraria*), ridge gourd (*Luffa acutangula*), sponge gourd (*Luffa cylindrica*), ivy gourd / kundru (*Coccinia indica*), bitter gourd (*Momordica charantia*) and spine gourd / kankoda (*Momordica dioica*).
Leguminous	Cluster bean / guar (*Cyamopsis tetragonoloba*), cowpea (*Vigna unguiculata*), Dolichos / Indian bean (*Lablab purpureus*), sword bean (*Canvalia gladiata*), pea (*Pisum sativum*)
Solanaceous	Chilli (*Capsicum annuum*), brinjal (*Solanum melongena*), tomato (*Lycopersicon esculentum*)
Cole crops	Cabbage (*Brassica oleracea* var *capitata*), cauliflower (*Brassica oleracea* var. *botrytis*), knol-khol (*Brassica oleracea* var. *caulorapa*)
Bulbous	Onion (*Allium cepa*), garlic (*Allium sativum*)
Root crops	Radish (*Raphanus sativus*), carrot (*Daucus carota*)
Okra	Okra (*Abelmoschus esculentus*)
Leafy	Methi (*Trigonella foenum graecum*), palak / spinach (*Beta vulgaris / Spinacia oleracea*), coriander (*Coriandum sativum*), chaulai (*Amaranthus* spp), bathua (*Chenopodium* spp.), sowa (*Anethum sowa*)
Perennials	Khejri /*sangri* (*Prosopis cineraria*), drumstick / *sehjna* (*Moringa oleifera*), Indian aloe / *guarpatha* (*Aloe barbadensis*), phog (*Callgonum polygonoides*), khimp (*Laptadenia pyrotechnica*), ker (*Capparis decidua*), lasora (*Cordia myxa*), kumat (*Acacia senagal*), khumbi (desert mushroom)

(A) Arid Zone Vegetable Culture

The Indian arid zone can grow a large number of vegetable crops compared to any hot desert of the world (Samadia *et al.*, 2004). Some of well recognized and potential vegetable crop-plants have been identified for diversification of dry-land horticulture and are given in the table. Based on region specific surveys in the hot arid, semi-arid and tribal areas of Rajasthan and Gujarat and field research work over 25 years by me, the extent of restrictions and opportunities have been assessed in respect to the horticultural development and recommendations have been carved out for vegetable production. In addition, innovative and resource based conceptual production approaches have been suggested for cultivation of vegetable crops under rainfed, limited or assured irrigated conditions with three distinct agro-climate of the Rajasthan state. The importance of eco-significance native trees/shrubs species having the potentialities of regional, traditional and minor vegetables in fresh, dehydrated or processed forms have also been highlighted for promoting resource based native or intensive crop production model under the climatic adversity. By incorporating the native and unexploited indigenous floras at the production sites as component of wind-break, boundary-side, bio-fence, hedge-row, field-divider, strip-block or planting model would not only help in creation of favourable micro-climate for cropping system but also ensure regular income from the socio-welfare species as monsoon supported harvest.

a) Environmental Factor Analysis for Vegetable Culture

Although, hot arid zone can grow a large number of vegetables but it possesses environmental and bio-physical constraints and thus limiting the extent and type of diversification and productivity levels. The major restrictions are related to the extremes of temperatures, scanty rains and scarcity of water, excessive evaporation and soil conditions. In the north-western part of hot arid region, March to October months are very hot, dry and bright with continuous higher day temperature range 35-38 °C for the prolonged period of time. The mean maximum air temperature in May and June varies from 42 °C in the west to 40 °C in the east but it is very common to experience high temperature range from 48-50 °C for number of days. In hot arid region, winter is dry, severe and for prolonged duration i. e. November to February. In December–January, mean minimum air temperature ranged from 4 to 10 °C and it is foggy. Most of the years, here, low temperature dips-down below freezing point (-1 to -6 °C) and occurrence of frost is quite frequent.

The region encounters strong winds during the hot summer months, which blow away sand and other finer particles that are deposited at other localities. In such areas the fertile top-soil is continuing blow away leading behind a coarse substrate poor in nutrients, thus, affecting production potentials of the farm-land. At the sites where deposition takes place the coarse sand grains veneer the surface and buried the productive soil. The problem becomes more serious in the vegetable fields which have to be ploughed before or during summer months and thereafter remain without any protective plant cover till the onset of monsoon. Vegetable crops also get damaged from the strong winds through sand blasting on seedlings, leaves, stem, buds, flowers and tender fruits during the sand-storms (ANDHI) in the hot months. The smaller seedlings or plants get buried under sand and vine plants disturbed badly. The crop-plants thereby get mutilated and cannot make normal growth leading to failure in flowering or reduced productiveness. The extreme high temperature and aridity condition from March to October month resulting to excessive evapo-transpiration from the crop-plants and soil surface. In addition, dry, hot and dusty winds (LOO and ANDHI) drastically affecting vegetable crops and quality of produce.

By and large desert sand-dunes field soils are sandy and coarse texture, and poor in fertility and water holding capacity. The erratic and low rains, acute water stress, extremes of low / high temperatures, high evapo-transpiration, strong cold / hot wind currents and poor soil conditions are the most important abiotic factors imposed by nature, and these all together distracting crop culture in hot arid areas, and thus micro-level and strategic planning is required for success in vegetable production and also for better resource utilization. Therefore, it is essential to select proper land-scape, crop-fields, perennial component and planting models for development of production sites and minimizing the ill-effects from the environmental factors (Samadia, 2004).

b) Crop Production Factor Analysis for Vegetable Culture

Owing to variable and to some extent manageable agro-climatic conditions, availability of limited irrigation water facilities and potentials for different types of production systems (rainfed, limited or assured irrigation) in the vast spread arid farm-land, a careful selection of crops and their genotypes for specific micro-zone could be a gainful boon for success in vegetable culture. The peculiar arid agro-climatic conditions impart unique quality in kachri, snap melon, mateera (watermelon), bottle gourd, muskmelon, chilli and seed spices. An enormous scope to promote large-scale seed production in cucurbits, tomato, chilli, brinjal, okra, early cauliflower, pea, cluster bean, cowpea, Indian bean, palak, methi, coriander, cumin, and truck gardening for distance market in particular to the cole crops, onion, carrot, melons and chillies is do exists with irrigated farm-land. The indigenous products from khejri, ker and kachri can be exploited for preparation of value added products. Dry vegetables products have lot of market potential as concentrated vegetables. A condition of high temperature and low humidity help in solar drying of number of vegetables, and the practice is already common for drying of pods khejri (*sangri*) and cluster bean and fruits of kachri, snap melon, round melon and chillies (Samadia, *et. al*, 2004, Samadia, 2007 and More and Samadia, 2007).

In hot arid region, existing low productivity in vegetables can be enhanced by improving the genetic make-up, adopting production technology and innovative approaches under integrated concept as crop-genotype-environment accepting. Here, special attention is needed to exploit native crop-plant genetic resources of horticultural significance species for developing systematic and viable vegetable culture. For this, systematic breeding approaches for the targeted crops should be required on high priority for developing better genotypes with multiple-use attributes including processing qualities, and wider adaptability under abiotic stresses. In addition, vegetable production, quality and productivity levels would also be depend on the availability and management of limited rain and irrigation water. The marketable yield of recommended crop-genotypes and productivity levels can be further enhanced manifold when due consideration are given to *in situ* rain water harvesting, moisture conservation and adoption of improved practices and techniques in a strategic manner under the innovative concept- Horticulture Based Crop Production Site Management Approaches (HBCPSMA, Samadia 2004).

c) Rainfed Vegetable Culture

The potentialities of growing vegetables under the rainfed situations is recognized in the Indian desert in particular since the conditions are not as extremes as in the desert of Sahara and Arabia. The arid agro-climate is conducive for the prioritized vegetables and quality production under conventional systems is much successful provided the appropriate crop-genotypes and good management practices. Systematic rainfed vegetable culture is yet not exploited or popularized adopting limited rain or water and khejri based production site management approaches.

In hot arid region, rainfall is low, erratic and restricting to a few short spells causing to soil and atmospheric water-deficit, acute moisture stress and frequent droughts. Most of rains occur during the monsoon season (July-September) and distribution is unpredictable and there is wide variation between the years. In north-western part, about 54% of the area receives less than 250 mm, 26% area 250-300 mm and 17% area 350-600 mm rains, and it can supports profitable native horticulture. Thus, rainfed vegetable culture is possible and it could be profitable only with the adoption of techniques for water harvesting, moisture conservation and management of production site.

The occurrence of summer rainfall has thunderstorm pattern resulting in run-off generation, and winter rains due to the western disturbances (*mavath*) is also a good source of supplement watering. In western Rajasthan, it is demonstrated that even under 250-300 mm rainfall, the runoff is about 25% and it is about 40 % in the years with 400 mm rain. The runoff quantity can be increased by treatment of catchments area and its size, slope and length of run also influence the runoff yield, and thus run-off offers possibilities of *in situ* water harvesting and life-saving irrigation for vegetable culture.

In arid region, the high evaporation rates causing to the problem of water deficit condition, and after completion of the monsoon season such conditions resulting into high soil moisture deficit and atmospheric water stress. The vapour pressure deficit becomes more than 20 mb in the hot months and exceeds 30 mb during May-June. Therefore, the best suited vegetables for cultivation in the arid zone are those having the maximum growth, flowering and fruiting phase during the period of maximum water availability in the soil and low vapour pressure deficit in the atmosphere. The indigenous crops which are found to be potential for commercial exploitation are kachri, snapmelon, mateera, roundmelon, kakri, clusterbean, cowpea, Indian bean, sword bean and these can be sown with the onset of monsoon rains from end of June to July so that flowering and fruiting taken place between August to November. The crop like brinjal can be taken in dry-land by using runoff concentration techniques provided the seedlings are kept ready in nursery for transplanting with the onset of monsoon in early July. The potentialities of perennial vegetables such as khejri, moringa, curry-leaf, ivy gourd, spiny gourd, Indian aloe and tumba is much viable adopting khejri based planting models and production site management approaches.

At present, few native cucurbits and clusterbean are grown under rainfed conditions. Among prioritized crops, the hardiest type belongs to cucurbitaceous where as clusterbean, cowpea, Indian bean, sword bean, brinjal and okra are the drought and stress tolerating. Here, early maturing and short duration cultivars which makes maximum use of residual water from the monsoon rains are found to be suitable and thus, there is need to develop trait-specific genotypes for the rainfed vegetable culture. There is need to promote khejri based cropping models incorporating above said ground storey and khejri, ker, kumat, lasora,drumstick, khimp, phog and Indian aloe as perennial component crop-plants and which are well-liked for their vegetable products.

d) Irrigated Vegetable Culture

In hot arid region, main sources of irrigation are open-well and tube-wells in Jodhpur, Barmer, Churu, Sikar, Jhnunjhunu, Pali, Jaipur, Nagour, Jalore, Bikaner and Jaisalmer districts of Rajasthan as well in a considerable areas of Gujarat and Haryana, and canals in Ganganagar, Hanumangarh, Bikaner and Jaisalmer districts in Rajasthan as well in a considerable areas of northern Gujarat, Punjab and Haryana. About 30 % irrigated area of the arid region is irrigated by wells or tube-wells and 60 % area is under canal command, and it is primarily with agronomic crops. At present, numbers of vegetables of arid areas are coming from small pockets having assured irrigation facilities, and it is unorganized and around the towns. It is a matter of concern looking to the vast area of the state under arid and semi-arid conditions where cultivation of vegetable crops is very negligible.

Arid and semi-arid districts of Rajasthan have nearly 65 lakh hectare area under irrigation and out of which about 20 lakh ha area is canal irrigated and remaining with other sources such as wells, tube-wells and pond. Field crops such as wheat, mustard and gram in rabi and cotton, groundnut, guar and bajra during kharif are grown. Traditionally, irrigation is given through flood method causing great loss of valuable water resource, and now this loss can be minimized greatly through improved methods (sprinkler and drip) and thereby increasing the irrigation water availability. Adoption of pressurized irrigation methods can bring additional area under irrigation with replacing the field crops, and part of that area can be brought under vegetable culture.

There is considerable increase in the number of sprinklers installed in the arid and semi-arid districts of Rajasthan state in recent years. Similarly, construction of diggi (water reservoir) is also increasing in command areas which required for sprinklers and drip irrigation with the liberal help of government agencies. As a result, 10%additional area (5.0 lakh ha) can be brought under irrigation which is presently under the fallow and rainfed culture and here, vegetable crops can be promoted adopting production site management approaches for distant market, truck gardening, seed or processing industry. Owing to the agro-climatic variability within sub-zones, availability of limited irrigation water and systems of diversified crop cultivation with limited or assured irrigation facilities at the production sites, it is therefore, a careful selection of the crop and commodity combinations for each defined location (specific production sites) would be primary obligation for success in vegetable culture under abiotic stressed arid environment.

Most of the vegetable crops are shallow roots and a short period (3-7 days) of moisture stress can affect growth, flowering, fruit-set, quality and yield potential under the hot arid environment. The common method of irrigation is by flow (check basin, furrow or channel) since it is simple and low cost but is very inefficient causing water conveyance and percolation losses. Among the controlled method, bubbler or hose-pipe up to the crop-fields for channel or deep furrow system and pressurized (drip and sprinklers) technology are gaining popularity for vegetable cultivation. To standardize the production technology for irrigated vegetable culture in the sandy soils, a series of experimental trials were conducted at CIAH, Bikaner (1997-2016) adopting water management techniques and irrigation methods (check basin, channel, drip and sprinklers). For controlled pipe-line and pressurized irrigation, diggi based technological advancement have been demonstrated by me at CIAH, Bikaner to promote vegetable culture adopting channel and drip method of crop cultivation and it is described "No–pole electricity–water resource utilization concept (Samadia, 2016).

(B) Technological Interventions for Arid Vegetable Culture

Now, it is well recognized that the majority of adaptive vegetables can be grown successfully in the Rajasthan state through proper selection and / or slight modifications in the genotypes and adopting desirable production techniques and management practices. Hi–tech production does not mean the cultivation only under the greenhouses or controlled conditions. In my opinion, definition of hi-tech crop production is very wide and still there is ample scope to recast it in relation to the crops and commodity produce to be desired, technology required to be modified for improving the productivity of available resources and minimizing the ill-effects of environmental factors at the production sites. Thus, hi-tech production can be summarized as interventions in improved, modern and innovative technologies for enhancing the productivity of resources in obtaining the higher crop yields under environmentally stressed areas.

The technologies required to be improved on priority for arid zone vegetable culture consisted are of better genotypes for high quality marketable yield under the defined production system of sub-zone; modification in the time and techniques of sowing / planting and plant population/unit area; integration of crop management and protection practices and on-farm value addition and post-harvest management. The modern ways needs to be understand are of water utilization through drip / mini-sprinkler systems for vegetables; crop-plant architecture, use of bio-regulators, crop mechanization and trellis for vertical harvest; and protective techniques for nursery-raising; surface covering, zero-energy and low cost structures for off-season vegetable raising and to escape ill-effects of winter temperature conditions. The innovative concepts which need higher attention are of development and management of khejri based production sites (HBCPSMA) for promotion of vegetable culture and to minimize the ill-effects from abiotic and biotic factors and inter-linking of crop-commodity production sites together with stable seed/planting material supply agencies, and storage, processing and marketing chain.

(1) Varietal Development for Arid Vegetable Culture

The concerted germplasm utilization and improvement programmes in the country resulted to development of good number of high yielding varieties and hybrids in the potential vegetable crops suited to the different agro-climatic zone. To meet-out the increasing vegetable demand with climatic resilience, the foremost need is massive efforts for developing high yielding varieties / hybrids / trait specific value added genotypes and then seed availability of recommended crop-genotypes suited to the production sites of the defined region of the state, because this is one of the most efficient and rapid means used for the increased production world-wide in native and adaptive vegetables. Despite of large number of varietal development in vegetable crops under the national net-work and their zonal recommendations, the requisite productivity and quality of produce is yet not improved significantly. There are very few numbers of crop-genotypes with wider adaptability for marketable and quality yield levels with varying environment, and doing well only under favourable sub-tropic agro-climate of the state. However, much research work on developing varieties and value added genotype is needed for native and unexploited crop-plants of the regional significance, and also on adaptive and potential vegetables for cultivation in non-conventional areas with arid to semi arid agro-climate of the state.

While assessing the arid agro-climate during 1994–1997, it was concluded that environmental constrains *viz.*, low and erratic rainfall, extremes of high and low temperature and aridity conditions, and un-availability of appropriate crop–genotypes and production technologies are the limiting factors for success in vegetable culture. With the establishment of national centre at Bikaner in 1994, initial research efforts were for germplasm conservation and utilization in the native crop–plant species of the horticultural significance. Traditionally, kachri (*Cucumis melo* var. *callosus* / *agrestis*), kakadia / snap melon (*Cucumis melo* var. *momordica*), mateera–a drought hardy watermelon (*Citrullus lanatus*), tinda (*Praecitrullus fistulosus*) and guar (*Cyamopsis tetragonoloba*) is grown with mixed cropping under rainfed conditions. Besides, khejri (*Prosopis cineraria*) is a perennial tree species and playing vital role in long-established farming systems, and its tender pods are used as vegetable. The crop evaluation and performance studies on 45 vegetables and their potential genotypes from 1995–2005 at CIAH, Bikaner clearly demonstrated that there is tremendous scope for cucurbits, chilli, brinjal, tomato, pea, beans and cluster bean, onion, garlic, radish, carrot, cole and leafy crops with irrigation facilities and adopting the production site management approach as a conceptual tool for obtaining high quality marketable yield, provided better and trait specific genotype in the prioritized arid vegetables.

a) Conservation and Maintenance of Arid Vegetable Germplasm

The landraces and local types / cultivars are source of genes for stresses, adaptability, quality and yield and therefore, collection and evaluation is the pre-requisite for the potentially utilization of genetic resources. Realizing the importance of vegetable crops in the hot arid region of northwestern parts of India, the systematic collection of germplasm was started since 1994 at Bikaner under mission mode of the NRCAH, NATP on plant bio-diversity and institute programme of CIAH. During last two decades, several crop-specific and multi-crop explorations were conducted for survey and collection of vegetable germplasm from parts of arid, semi-arid and tribal areas of Rajasthan and Gujarat, and also augmentation of germplasm in targeted crops was done for varieties and lines from national net-work including AICRP (VC) and NBPGR.

A data base was developed for compilation of R & D work on genetic resource management in vegetables at the institute and up to 2008, a total 1725 accessions were collected and evaluated and out of them 1060 germplasm were deposited in NGB at NBPGR for long-term conservation. The vegetable germplasm either collected by CIAH or assembled from NBPGR under NATP (PBD) or national net-work was evaluated for use in the crop improvement programme at Bikaner. On characterization and evaluation under the extremes of arid agro-climatic conditions, vegetable crop germplasm exhibited wide range of variability for agro-morphological traits, flowering and fruiting behavior, maturity, yield and quality contributing characters, besides resistant or tolerant to abiotic and biotic stresses.

Based on intensive field-crop performance studies over the seasons and years (1994–2016), the potential arid vegetable crop germplasm were identified and developed genetic material from the base germplasm is being maintained as active breeding lines (500) for use in the improvement programmes. The maintained material is comprised of kachri (68), snap melon (65), mateera (65), muskmelon (60), round melon (10), kakri (18), bottle gourd (20), ridge gourd (20), sponge gourd (15), bitter gourd (4), chillies (45), brinjal (30), tomato (14), khejri (15), India bean (30), sword bean (01), cluster bean (02) and others (15) including perennial vegetables.

b) Utilization of Arid Vegetable Germplasm

To promote profitable vegetable culture under arid agro-climate, systematic crop improvement work plans were formulated first time by me as PI of the project and stated in 1994. During the first five years (1994–1999), the prime emphasis was on the native and drought hardy vegetable crops such as kachri, snap melon, kakri, mateera, tinda and cluster bean. The chilli work for Mathania type was incorporated in 1998 and then intensive work on potential crops such as muskmelon, gourds, brinjal, beans and moringa was started from the year 2000 considering the requirement of superior genotypes for cultivation under the extremes of climatic variability. A mission mode approach for improvement in the khejri was started in 2001. Trait specific breeding for heat tolerant in round melon, bottle gourd, muskmelon, tomato and brinjal was started from 2007. The germplasm characterization and utilization at CIAH resulted to the purification of promising and trait specific lines for evaluation, and followed by selection and combination breeding over the period (1996–2016) given out about 725 lines in the prioritized vegetables and were used for varietal development under hot arid agro-climate.

The result of systematic purification followed by selection and hybridization breeding at CIAH, some high yielding and better quality producing genotypes have been developed in arid vegetables and recommended for commercial cultivation, large-scale testing at farmers fields / national net-work or use in breeding programmes. The varieties / selections recommended for the cultivation in arid region are in mateera (AHW-19, AHW-65 and Thar Manak), kachri (AHK-119 and AHK-200), snap melon (AHS-10 and AHS-82), kakri (AHC-2 and AHC-13), bottle gourd (Thar Samridhi), cluster bean (Thar Bhadavi), Indian bean (Thar Kartiki and Thar Maghi), sword bean (Thar Mahi), khejri (Thar Shobha) and brinjal (Thar Rachit). The developed lines of bottle gourd (AHLS-24), ridge gourd (AHRG-1 and AHRG-8), sponge gourd (AHSG-4 and AHSG-5), round melon (AHRM-1 and AHRM-2), muskmelon (CIAH-1), chilli (HRM-1), brinjal (CIAH-22), ivy gourd (AHIG-1), palak (AHLP-1) and moringa (AHMO-1-4s) were used in combination / selection breeding from 2001-2010 and advanced material is under testing as value added genotype.

In addition, some new breeding material is developed in the targeted vegetable crops at CIAH, Bikaner and evaluated over the seasons and years (2005-2015) for their performance studies under high temperature and abiotic stressed arid environment. These lines consisted of kachri (AHK-411, AHK-564 and AHK-572), snap melon (DKS/AHS-2011/2), mateera (AHW RSS-1 and AHW BSM-1), ridge gourd (AHRG-15-4-1), sponge gourd (AHSG/2015/F5/1), bottle gourd (AHLS Oblong-1 and AHLS-Long/2015/F6/1), bitter gourd (AHBT-2), tomato (AHSL-1 and AHSL-2), cowpea (AHCP-1), Indian bean (KSB-66), cluster bean (AHG-20), velvet bean (AHVB-1), methi (AHLM-1), bathua (AHLB-1), carrot (AHDC-1) and khejri (CIAH Selection-2).

c) Priorities for Germplasm Utilization and Focus on Future T hrust

Primarily I am working on the native crop-plants to promote horticultural production under hot arid environmental restrictions. Therefore, the foremost concern is on safe conservation and effective utilization of germplasm for varietal development and also systematic arid vegetable culture in particular with kachri, snap melon, mateera, tinda, bottle gourd, cluster bean and khejri at the resource constraints production sites. In additions, developing of appropriate breeding material and trait specific lines through germplasm utilization in targeted vegetable crops is also in the top priority for promotion of their cultivation under the abiotic stressed conditions such as high and low temperature, drought, limited rains or irrigation water, and suitability for off-season or prolonged period of harvesting.

Breeding for improvement in yield, quality and productivity is a continuous nature of work, and in addition, varietal requirements for the three environmentally distinct agro-climatic zones of the state and continual increasing demand of vegetables inviting timely up-gradation and prioritization in the research programmes. Realizing the need of hour, the vegetable breeding work should be focused on prioritized crops and their objectives through germplasm utilization for developing varieties and value added genotypes in particular to the defined zones of the state for which crops are targeted. Beside, the potential and native crop objective of commercial exploitation should be defined in respect to national and regional perspective. Based on research work progress and gap analysis in germplasm utilization and varietal development in last 25 years, crop and trait specific prioritized needs have been identified for the further in-depth analysis and research strengthening to achieve targets on the issues highlighted as constraints and work-plan given below.

(i) Research Constraints

1. Insufficient germplasm in most of vegetable crop and ignorance of old lines / varieties in breeding.
2. Low priority on germplasm utilization to solve the problems through combination breeding.
3. Low priority to native and uncommon vegetables for region and resource specific research.
4. Lack of new high yielding genotypes with in-built resistance to abiotic and biotic stresses.
5. Absence of basic research and promotion facilities to work on wild and related species.
6. Low level of scientific interest to carry-out long-term research programme on native vegetables.
7. Low interest to solve long-standing crop problems with advancement of science and technology.

(ii) Research Work-plan

a) Utilization of national genetic resources for breeding high temperature and abiotic stresses tolerant genotypes in round melon, bottle gourd, mateera, muskmelon, ridge gourd, sponge gourd, bitter gourd and tomato with better marketable fruit quality yield and cultivation under varying agro-climate of the state,

b) Utilization of native genetic resources for developing high yielding and better marketable quality varieties or value added genotypes under abiotic stresses of the respective zones in kachri, snap melon, spiny gourd, ivy gourd, cucumber (chikan-kakri of tribal area), arya-kakri, tur-kakri, palak, methi, khejri, sehjna, curry-leaf, guarpatha and other regional adaptive vegetables including leafy, roots, tubers (yam, aerial yam and sweet potato).

c) Utilization of national genetic material and large-scale varietal trials for identification of region and trait specific genotypes in commercially viable (chilli, ginger, turmeric, garlic, onion) and seed spices crops in respect to hot arid to semi-arid and/or sub-humid and tribal zones.

d) Distant hybridization and breeding for developing virus free genotypes in watermelon and chillies for cultivation under abiotic and biotic stresses,

e) Utilization of native germplasm for developing suitable brinjal genotypes having good quality fruit characters with the regional preferences and cultivation under high temperature conditions, and tolerant to FSB.

f) Utilization of national genetic material and breeding for developing high and better quality pod yielding genotypes in potential vegetable legumes such as peas, cluster bean, cowpea, Indian bean, sword bean and velvet bean for production under variable resources and stressed conditions.

g) Heterosis or combination breeding for developing genotypes resistant to fruit-fly infestation and virus free in bitter gourd, ridge gourd, sponge gourd and pumpkin for cultivation under semi-arid and sub-humid zones.

h) Utilization of national genetic material for breeding better quality and high yielding genotypes for low-cost protected cultivation and vertical harvesting in the prioritized vegetable crops under hot arid agro-climate.

(2) Agro-techniques Developed for Arid Vegetable Culture

To standardize the production techniques and good management practices for vegetable culture in the sandy soils of hot arid agro-climate, a series of crop specific and combination experimental trials were conducted adopting check basin, channel, drip and sprinkle method of cultivation either under rainfed or with limited irrigations at CIAH, Bikaner (1997-2016). The experimental conclusions and recommendations are promoted as the practical approaches of crop management and agro-techniques for vegetable production. For agro-technique standardization, potential and native vegetables such as cucurbits, brinjal, tomato, pea, cowpea, bean and cluster bean, cole crops, palak, seed spices, sehjan and khejri were studied with HBCPSMA for developing packages of practices.

a) Khejri based crop production site management approach for vegetable culture

Khejri is versatile tree species of the *Thar* Desert and it is most potential perennial component compatible to almost any companion crop-plant in the long-established farming systems of the hot arid and semi-arid areas of north-western parts of India. The budded khejri tree grows luxuriantly under the extremes hot and dry conditions. This eco-restoring plant not only helping in soil fertility build-up but also minimizing the ill-effects from high solar radiation, high temperature and hot wind at developed production site. Systematic plantation of khejri variety Thar Shobha both with high density and paired-rows at wider-space exhibited significant effect in creating favourable micro-climate at the developed production site. Khejri based approach is also helping in promotion of pollinating agents such as ants and bees at production site. Therefore, this innovative concept offers incredible scope for exploiting the high-value vegetable crop sequence / combination models adopting khejri based production site management approaches (Samadia, 2015).

Under the climatic resilience and changed scenario, now, the horticultural focus is shifted from sustenance to measurable economic farming but the mono-cropping is much risky for developing arid zone vegetable culture. Therefore, the traditional systems pre-dominantly mixed cropping needs multi-dimensional interventions through prioritized vegetables, native crop-plants and newer technological advancements and now, it can be performed with the HBCPSMA concept and adopting khejri based crop production site management approach as innovative tool for doubling income (Samadia, 2016).

The region and crop specific long-term research findings under the HBCPSMA concept (1994 to 2016) enables to conclude it with recommendations and based on that some principles have been suggested for development and management of khejri based crop production sites. The production sites should be developed in accordance to the topography of sand-dune landscape and soil conditions. Fencing of production site and development of multi-tier rows of seedling plantation with desert tree species (khejri, lasora, rohida, bordi, ker, kumat) and shrubs (jharber, phog) all-around the field block is the prime consideration for creation of favourable micro-climate and it is must for protection of crop production sites from the extremes of hot, cold and high wind conditions. The four-side field area of production site is developed as multi-tier rows of seedling plantation (single or paired, 4mx4m) with the directions for khejri and lasora or rohida, kumat (south), khejri and lasora or ker, rohida (west), khejri and kumat or rohida, bordi (north) and khejri and ker or rohida (east).

The innovative practices / technologies under HBCPSMA includes selection and development of khejri planting model based on available resources at production site, preparation and maintenance of the production sites for *in–situ* rain water harvesting and soil moisture conservation, and crop cultivation. Adoption of pre–monsoon deep ploughing in June prior to crop sowing and post–monsoon ploughing after crop harvesting November month as techniques resulted into more *in–situ* rain-water harvesting, moisture conservation and weed free fields. Keeping of crop field fallow for 1–2 months either April–June or October–November, and is an essential practice for the soil health security and fertility build-up. Besides, seed/plant selection, sowing/planting time and techniques, maintenance of crop plant population and prophylactic protective measures are good management practices essentially required for improving yield and productivity of the resources, and this is only worthy when the best suited genotype is identified for vegetable cultivation at the production site.

Based on the crop potentialities and critical analysis over the 15 years, so far, three horticultural combinations are suggested for the developing khejri based crop production sites. The first two are primarily with rainfed or limited irrigations, and namely are organic *Panchkuta* production technology (khejri, ker, lasora, kumat and kachri) and native crop production technology (khejri + native vegetables) where khejri planting model KM–1 (4m x 4m) or KM–3 (8m x 8m) is practiced, respectively with hectare block lay-out. For prioritized vegetable cropping sequence and irrigated cultivation, wide spacing planting models (KM–9 or KM–11) are most appropriate and here 4mx4m paired row khejri plantation is done at 24 m or 44m distances, respectively with four hectare block layout and it is intensive crop production technology. For systematic block planting, single or paired rows of khejri var. Thar Shobha should be laid-out in the east-west direction and wider area is the developed fields for vegetable crop cultivation.

b) Channel Technology of Vegetable Cultivation

The channel system of cultivation as standardized at CIAH, Bikaner under the sandy soils of arid region is found to be the most practicable for vegetable culture such as cucurbits, brinjal, tomato, chilli and beans. This technology of crop cultivation save valuable inputs such as irrigation water (30–50%), manure and fertilizes (30–35%) and labour cost (25–30%). The channel cultivation helps in better care of young seedling, ease in inter-culture, irrigation, sprays and harvesting and permitting mechanization for weed management. This technology also reduces the pre-harvest losses of fruits by restricting water-flow only in the channels. Under channel system, total irrigation water requirement and water losses are lesser than the check basin. The water distribution in root zone of the crop is more uniform. Labour and time requirement for irrigation is also less. This system of vegetable cultivation also offers an additional opportunity to harvest maximum rain water within the channels, and it is even with the scanty rainy day (10–15 mm), which is directly available nearby root-zone of the crop-plants.

After selecting proper field at production site, the marked area should be cross-ploughed 2–3 times followed by planking, and it should be started prior to 10–15 days of crop sowing. About 50–60 cm wide and 20–25 cm deep furrows (channels) are made at 1.5–2.5 m distances depending upon the crop, cultivar and season for cucurbitaceous vegetables. For solanaceous crops, channels of about 50–60 cm wide are made in continuous manner. With trellis system of beans and gourds production, the distances between the channels should be about 2.0 meters. For uniform irrigation water distribution, the maximum length of the crop-sown channels should not be more than 25 m on one side. Field plot of 50 x 50m size should be developed keeping one meter wide water delivery channel in centre area. Application of basal doses of manures and fertilizers is done only in the channels. The seeds are sown at 50 cm distances only on down-side of the northern slope of each channel for ground storey cucurbitaceous crops and also for trellis system of cucurbits and beans. At each sowing point, 1–3 seeds are dibbed. When the seedlings are 8–10 cm in height, thinning should be done keeping one or two plants at each point. In solanaceous crops, seedlings should be transplanted at 50 cm distances in the centre area of the channels.

In cucurbitaceous crops, less number of irrigations is required for kachri, snap melon and mateera as compared with watermelon, kakri, muskmelon, tinda and gourds. With channel system of cultivation, these crop required irrigation at an interval of 7–10 days depending upon stage of growth, fruiting behavior, season and weather conditions. During very early stage of crop growth, irrigation should be done more frequently. A normal frequency of irrigation at 7–8 days is the ideal for brinjal, tomato and beans under channel system of crop cultivation. Studies conducted in sandy soils at CIAH, Bikaner revealed that even in the conventional flood method of irrigation 30–50 per cent water saving can be done when channel system of vegetable cultivation is practiced over check basin of 5x5 m plots.

c) Drip Technology of Vegetable Cultivation

Irrigation water management is important to ensure that crop-plant needs can be meet-out without under or over watering. The increasing crop-yield level obtained with drip irrigation can be attributed to the several factors such as high water use efficiency because of the precise application directly to the root-zone of plant and less water losses due to the reduced evaporation, run-off and deep percolation, and less fluctuations in the soil-water content resulting into avoidance of water-stress. The principle of drip irrigation is to supply water at the very low-rate and in the region of maximum root activity and thus, there is considerable saving in the water consumption. Surface or sub-surface drip irrigations do not wet plant foliage and thereby reducing disease spread significantly. Drip irrigation maintains crop rhizosphere almost at field capacity so that the plants do not experience moisture stress conditions. Therefore, drip technology is the most efficient method of irrigation when it comes to the water saving and its productivity. For technological advancement adopting pressurized irrigation and crop production, a series of experimental trials were conducted for water management studies under sandy soils of arid agro-climate from 2002–2012 at CIAH, Bikaner and it was in particular to assess the feasibility of different combinations of drip and micro-sprinkler system and also find out practical recommendations for vegetable cultivation with limited irrigation water.

The R & D studies with drip technology led to conclude that single lateral lines (14–16 mm) at 2.0 m distances with in-line drippers (4 lph) at 50 cm distances is found to be the most appropriate for cucurbitaceous crops such as kachri, snap melon, round melon, mateera, muskmelon, bottle gourd, sponge gourd and ridge gourd under limited irrigation water in sandy soils of hot arid agro-climate. Similarly, with trellis system of crop cultivation, the laterals at 2.0 m and drippers at 50 cm distances is also found to be the optimum in Indian bean, sword bean, ivy gourd and bitter gourd. Drip technology has also given excellent results in brinjal, tomato, cowpea and cabbage (laterals at 1.0 m apart and drippers at 50 cm) at CIAH, Bikaner. In conclusion, with limited water these crops should be irrigated for 1–2 hours at 3–4 days intervals by assessing the age of crop-plants, growth phases and seasonal weather conditions. About 15–50 per cent higher yield with better marketable quality was obtained in the above said crops with limited water under drip technology in comparison to the channel system of production. For better water distribution and crop management, one side length of lateral lines should not exceed 25 m and unit plot with control valves should be of 50x50 m size, and 4 ha block area can be controlled with a drip system as unit.

With drip technology of crop cultivation, the increased yield was because of higher number of fruit-set and marketable fruits/plant (snap melon, kachri, round melon, gourds, tomato, brinjal and beans), more number of high quality fruits (muskmelon, watermelon, bitter gourd, bottle gourd, ivy gourd, tomato and brinjal) and better plant growth (muskmelon, round melon, gourds, brinjal, tomato, cowpea and beans) under high temperature and abiotic stress conditions. Based on crop potentialities, seasonal weather fluctuations and available resources, hi-tech open field vegetable production adopting drip technology has been standardized and recommended for commercial cultivation of the arid zone vegetables under limited irrigation with diggi based assured water supply system at the production site. The efficiency of the drip technology in the Indian arid zone would only be at the highest level if it is planned, layout and maintained after understanding the prevailing climatic situations, crops under cultivation and competency in the resource management (Samadia and More, 2008).

d) Tent and Channel Covering Technology for Protected Vegetable Cultivation

On the basis of environmental and vegetable production constraints analysis in the arid zone of Rajasthan, two prime limiting factors have been identified in particular to protected cultivation, and these are strong wind velocity with sand-storms and continuous high temperature conditions (March–October). Therefore, prior to recommendations on protected technology for the production site of hot arid areas, complete analysis is to be done on structures type (low-tunnels, poly-houses, shade-houses or controlled green-houses), crop-genotype and cultivation season (Samadia, 2004). Based on this preliminary work, systematic studies on low cost protected crop cultivation (tomato, chilli, capsicum and summer squash) was undertaken at CIAH, Bikaner from 2007–2010.The plant growth, fruit set, quality and yield potential was much inferior in tomato, capsicum and chilli under green shade-net (50 and 75%) and insect's proof shade-net covering structures during summer and rainy season. Similarly, extremes of high temperature and high wind conditions affecting poly-tunnel cultivation of cucurbits as complete crop from sowing to harvesting. Therefore, the said experimental results on protected vegetable cultivation here discouraged me over the open field hi-tech crop production practices.

However, movable and folding type structure (modifications in low tunnel to tent type) with polythene as covering material was found to be the best for early cucurbits sowing under protection and to save them from the low temperature / frost conditions during the peak winters (December and January), and followed by removal of polythene covering between the last week of January to first week of February, and allowed it as early spring summer season crop of open field, and this can be popularized. Tent and channel covering protected cucurbits cultivation as standardized at CIAH is a technological advancement for early harvesting, and bottle gourd was test crop for the experiments.

The adversity of extremes of low and high temperature conditions in the north–western parts of arid region restricting the quality fruit yield, period of crop harvesting and productivity in bottle gourd. In this part, the winter season is very severe and starts from November, and continues to the end of January and sometimes it extends up to the second week of February, and therefore, sowing of bottle gourd under open field is performed in the middle of February for spring-summer season crop. In this part, high temperature picks up very fast with the end of March and extremes of high temperature conditions is continuing from April–June. The high temperature and aridity situations in May and June month severely affecting the development of female flower, fruit set and marketable quality fruits in bottle gourd and thereby very low productivity and benefits to the growers. Under this situation it is the only possibility to sow the crop at the earliest under protection from the severe winter in January and then allowing the crop under open field condition after risk dates or with the rising temperature from first week of February for development of plants and fruiting.

To standardize this technique, a series of experiments on low cost and zero energy protected methods along with drip technology of cucurbits cultivation were conducted from 2007–2010 at CIAH, Bikaner for winter–summer season crop. Different types of structures were designed considering the environmental situations, crop feasibility and polythene sheet as covering material was used to protect the early sown crop, and with tent type structure sowing was done on 1st and 15th January. With the channel technology, sowing was done on 15th January, 25th January and 5th February and surface was covered with polythene sheet also. Similarly, crop was also sown under open channels on the above said dates. Beside, normal crop was sown on 15th February under the open field conditions. Tent type structures covered with polythene sheet and where crop sown on 1st January exhibited earliness of about 30 days for first marketable harvesting (15 March) in comparison to the normal open crop harvesting (16 April). The tent technology for early crop production exhibited improvement for high yield potential of about 30–35% with better fruit quality in bottle gourd. Besides, channel technology with polythene as surface covering material is also beneficial for cucurbits sowing from second week of January, and it is very simple and low cost.

e) Trellis Technology of Vertical Vegetable Cultivation

Trellis system of vertical harvesting and vegetable cultivation with practices of drip / channel method, provision of tent-type structure or surface covering with polythene at sowing to initial plant growth stages for protection from the extremes weather conditions, and thereby crop management can be a advantageous approach for early harvesting, higher yield, better quality and doubling income under resource poor arid environment (Samadia, 2016). For developing trellis technology, paired-row seedling plantation of khejri in arid and semi-arid areas and sandheda (*Painshyana ellata*) through limb planting in sub-humid tribal dominating area should be promoted alongwith use of same plants twigs or phalsa or bamboo sticks for creation of vine spreading system at prepared field for crop cultivation. Otherwise, steel iron angle, post or poles of 6–7 feet height cab be fixed at appropriate distance and wire-net trellis system is developed. In between the developed trellis rows, deep furrows or channels are prepared as seed-bed for sowing / planting of prioritized vegetable crops, and cultivation of kachri, ridge gourd, sponge gourd, bottle gourd, bitter gourd, spiny gourd, ivy gourd, aerial yam. Indian bean, sword bean and velvet bean can be promoted.

f) Innovative Vegetable Type Cluster Bean Production Technology

For improving tender quality pod yield in vegetable type cluster bean var. Thar Bhadavi under rainfed or crop saving irrigations and also enhancing productivity of resources in the hot arid agro-climate, an innovative crop production technology has been standardized with HBCPSMA concept and recommended. Here, the technological advancement includes selection and development of khejri var. Thar Shobha based production site adopting planting model KM-1, KM-3, KM-9 or KM-11 to create favourable micro-site for cluster bean cultivation. Besides, field preparation, sowing techniques and plant population maintenance, crop protection measures and pod harvesting are management practices.

The experimental results in cluster bean adopting the above said technological interventions at CIAH, Bikaner demonstrated that variety Thar Bhadavi can be grown successfully under absolute rainfed conditions with evenly distributed monsoon rains (7–12 days from end of June to September & 250–325 mm) in the kharif season. With irregular rains, however, better yield can be obtained with one or two supplemental irrigations during long dry-spells in the normal monsoon season. Field preparation was done following the weather forecast of the region and it was only from end of June in accordance to the pre-monsoon rains during 2005–2015 and sowing was done in July after good rains. Sowing was done within 24 hours of rains and about 12 kg seed is enough for hectare under broadcasting.

Pre-monsoon field ploughing and cross-harrowing in the June prior to crop sowing and post monsoon field ploughing in the month of November after crop harvest resulted in to more *in situ* rain water harvesting, moisture conservation and weed free field at production site. A complete cultivation packages is developed for field operations such as weeding and inter-culture (25–30 and 45–50 DAS), top-dressing of urea on rainy days (50 kg/ha) between 25–45 DAS, maintenance of weather forecast records on occurrences of rains, its quantity and intervals for crop performance and scheduling of life saving irrigation at the production site.

Out of eleven years, nine good crops were harvested as tender pods yield (42–85 q/ha) under absolute rainfed or with 1–2 life saving sprinkler irrigations. The crop was irrigated for 2.0–2.5 hours which delivers water equivalent to 20–25 mm of rains in each irrigation. Under sprinkler irrigation, an interval of 18–21 days from the preceding rainy-day was found to be the best in normal year (3–4 good rainy-days within 35 days from sowing) to save the crop due to irregular monsoon rains during August or September, and considered that it was the critical stage for life saving irrigation and to have good economic harvest in cluster bean. Regular harvesting of tender pods and marketing, and also on-farm value addition for dehydrated tender pods and seed production techniques for maintenance of variety are the technological considerations for better economic returns.

References

Anonymous 1994–2016. Annual reports. National Research Centre for Arid Horticulture / Central Institute for Arid Horticulture (ICAR), Bikaner (Rajasthan) India.

Anonymous 2005. Final report - NATP on sustainable management of plant bio-diversity - Collection, evaluation and maintenance of arid horticultural crops. Co-operating centre of NBPGR, Central Institute for Arid Horticulture (ICAR), Bikaner (Rajasthan) India.

Khan, H. and D.K. Samadia 2012. Variability and association studies in tomato germplasm under high temperature arid region. *Journal of Horticulture Sciences*, 7 (2): 194 – 198.

More, T.A. and D.K. Samadia 2008. Prospect of horticulture in arid zone. In: Dryland Ecosystem: Indian Perspective. (ed. K.P.R.Vittal, R.L. Shrivastava, N.L. Joshi, Amal Kar, V.P. Tewari and S. Kathju), CAZRI and AFRI, Jodhpur, pp 149-167.

More, T.A. and D.K. Samadia (2008). Breeding strategies for vegetable crops with special reference to arid region. In: Souvenir and proceedings of national seminar on opportunities and challenges of arid horticulture for nutrition and livelihood, CIAH, Bikaner. pp. 1-10.

Pareek, O.P., B.B. Vashishtha, Vishl Nath, D.K.,Samadia and R.S. Singh 1999. Horticultural resources in *Thar* dessert. In: Natural resources in Rajasthan (eds. AnilKumar, Rajesh Sharma and Madhu Sharma), SKRAU, Bikaner. pp 4-9.

Samadia, D.K. 2003. Genetic resource management for sustainable horticultural development in arid regions of India. In: Sustainable Resource Management (eds. B.B.S. Kapoor and Ali Ahemad), Madhu Publications, Bikaner (India). pp 163-173.

Samadia, D.K. 2004. Protected cultivation technology for arid regions of India. In: Advances in Arid Horticulture, Saroj, P.L., B.B. Vashishtha and D.G. Dhandar (ed.), Vol. I. International, Book distributing Co., Lucknow (India). pp. 267-289.

Samadia, D.K., B.B. Vashishtha and G.B. Raturi 2004. Prospects of vegetable production in the arid zone of India. In: Impact of vegetable research in India, Sant Kumar, P.K. Joshi. and Suresh Pal (ed.), National Centre for Agricultural Economics and Policy Research (ICAR), New Delhi. pp. 101–114.

Samadia, D.K. and O.P. Pareek and B.B. Vashishtha 2005. AHS 10 and AHS 82: New snap melons. *Indian Horticulture*, 50 (2): 10-11.

Samadia, D.K. 2006. Technology for cucurbitaceous crop cultivation in arid regions. *Intensive Agriculture*, 44 (1-2): 3-7.

Samadia, D.K. 2006. Performance of muskmelon genotypes under arid conditions. *Indian Journal of Arid Horticulture*, 1 (2): 24-27.

Samadia, D.K. 2007. Arid Vegetables. In: Under utilized and under exploited horticultural crops.Peter, K.V. (ed.), Vol. 2 New India Publishing Agency, New Delhi (India), pp.153-174.

Samadia, D.K. 2007. Genetic variability studies in chilli germplasm under hot arid eco-system. *Indian Journal of Horticulture*, 64 (4):477-478.

Samadia, D.K. 2007. Studies on genetic variability and scope of improvement in round melon under hot arid conditions. *Indian Journal of Horticulture*, 64 (1): 58-62.

Samadia, D.K. 2007. Genetic variability and correlation studies in brinjal. Indian Journal of Arid Horticulture, 2 (2): 18-21.

Samadia, D.K. 2008. Evaluation of Indian bean genotypes under hot arid environment. *Indian Journal of Arid Horticulture*, 3 (2): 22-25.

Samadia, D.K. 2008. Hi-tech approaches for vegetable and seed production under hot arid region of India. Hi-tech production of arid horticulture, More *et al.* (ed.), CIAH, Bikaner, pp. 241-248.

Samadia, D.K. 2008. Improvement of vegetable crops for hi-tech production. Hi-tech production of arid horticulture, More *etal.* (ed.), CIAH, Bikaner, pp. 391-398.

Samadia, D.K. (2008). Hi-tech approaches for vegetable and seed production under hot arid region of India. Hi-tech production of arid horticulture, More *et al.*(ed.), CIAH, Bikaner, pp. 241-248.

Samadia, D.K. and T.A. More 2008. New bottle gourd: *Thar Samridhi. Indian Horticulture*, 53 (2): 21.

Samadia, D.K. and T.A. More 2008. Drip irrigation technology for cucurbit production under hot arid environment. Abstract of paper In: New R & D Initiatives in Horticulture for Accelerated Growth and Prosperity. 3rd Indian Horticulture Congress-2008. November 06-09, 2008, Bhubaneswar, (Orissa). 1.49: 38-39 pp.

Samadia, D.K. and T.A. More 2009. New Indian beans for more protein-rich foods. *Indian Horticulture*, 54 (6): 20-21.

Samadia, D.K. and T.A. More 2009. AHSB-1 for protein-rich diet in arid region. *Indian Horticulture*, 54 (6): 24-25.

Samadia, D.K. and T.A. More 2010. Tent technology: An innovative technological advancement to harvest early marketable fruit yield in bottle gourd under hot arid environment. Abstract of paper In: Book of abstracts of 4th Indian Horticulture Congress-2010. November 18–21, 2010. New Delhi. III-33: 184 p.

Samadia, D.K. 2010. Genetic variability studies and scope of improvement in sponge gourd under hot arid agro-climate. *Indian Journal of Arid Horticulture*, 5 (1-2): 37 – 39.

Samadia, D.K. 2011. Genetic variability studies for improvement in bottle gourd under hot arid agro-climate. *Indian Journal of Arid Horticulture*, 6 (1 – 2): 15 – 18.

Samadia, D.K. 2011. Genetic variability studies in ridge gourd under arid environment. *Indian Journal of Horticulture*, 68 (2): 275 – 277.

Samadia, D.K. and T.A. More 2011. Thar Manak to quench thirst in desert. Indian Horticulture, 56 (5): 14 – 15.

Samadia, D.K. 2012. Thar Bhadavi – new cluster bean. *Indian Horticulture*, 57 (4): 15 – 16.

Samadia, D.K. 2014. Innovative concepts, techniques and practices for enhancing productivity of vegetable crops under hot arid agro-climate. Chapter in compendium of short course: Advances in water management and fertigation in fruit and vegetable crops of hot arid region in India (eds. Sharma, *et al.*), CIAH: Bikaner. pp 47–53.

Samadia, D.K. 2014. Utilization of genetic resources of vegetable crops for varietal development under hot arid agro-climate. Chapter in compendium of short course: Advances in water management and fertigation in fruit and vegetable crops of hot arid region in India (eds. Sharma *et al.*), CIAH: Bikaner. pp 189–193.

Samadia, D.K. 2016. Horticulture based crop production site management approach (HBCPSMA) - an innovative concept for doubling farm income under dry-lands. *Journal of Agriculture and Ecology* (http://saaer.org.in), 2016-1: 1–9.

Samadia, D.K. 2016. Innovative concepts, techniques and practices for managing vegetable culture under environmental stresses of hot sub-tropic climate. Book chapter In: Good management practices for horticultural crops. New India Publishing Agency. New Delhi, pp199–216.

Samadia, D.K. 2016. Breeding strategies for enhancing vegetable production under environmental stresses of hot agro-climate. Book chapter In: Good management practices for horticultural crops. New India Publishing Agency, New Delhi, pp217–226.

23

High Value Seed Spices for Better Profitability-Doubling Farmers Income

Gopal Lal

Introduction

Arid and semi arid parts of the India are known as Seed Spices Bowl (Rajasthan and Gujarat) and contribute more than 80% of total seed spices production. Total 109 spices are listed by ISO and 63 spices are grown in India and out of which 20 are being classified as seed spices. Out of 20 seed spices, ICAR-NRCSS working on 10 most important seed spices namely cumin, coriander, fennel, fenugreek, ajwain, anise, caraway, celery, dill and nigella (NRCSS Vision, 2050). Seed spices also grown in some other states on considerable areas such as Uttar Pradesh, Madhya Pradesh, Bihar, West Bengal, Orissa, Punjab, Karnataka and Tamil Nadu. Since there is a large scope of seed spices by introducing them in new areas, the higher yields can also be achieved effortlessly by implementing new technologies. Among seed spices, maximum area is under cumin followed by coriander, fennel fenugreek, ajwain, celery, nigella, dill, anise *etc*. Being seasonal crops, these are grown extensively in rotation with food crops and also as inter/mixed crops under rainfed/irrigated conditions. All these crops mostly grown in rabi season. India is the largest producer, consumer and exporter of seed spices in the world. Out of total area and production in spices, seed spices contribute about 51.79% of total area and 19.06% of production of spices in the country. The area under seed spices is about 1.74 million hectare and 1.45 million tonnes. India is consistent source of seed spices for importing countries worldwide. There has been ever increasing demand of seed spices and importing countries look at India for quality produce of seed spices.

India is exporting only 10 percent of its production annually and full fill the 50-60 percent of world demand. Seed spices play a significant role in our national economy because of its large domestic consumption and growing demand for export. The total export of seed spices is worth of more than Rs 2800 crore annually out of which cumin alone contributed more than Rs 1900 crore (DASD, 2016-17). The seed spices are grown in different part of the world covering mainly Mediterranean region, South Europe and Asia. Morocco, Russia and Bulgaria are the major producing countries for coriander, Turkey, Iran and Egypt for Cumin; Egypt, China, Romania and Russia for fennel; Morocco and Bulgaria for fenugreek; Iran and Egypt for Ajwan; Germany and Hungry for dill; southern France and china for celery; Bulgaria, Cyprus, Germany, and Russia for anise; Pakistan, Sri Lanka and Egypt for Nigella. Among these countries, India is reported to be largest producer of seed spices.

A "New India" is the latest national agenda. Prime Minister, in his fourth Independence Day speech, made an appeal with his trademark gesture of both hands pointing towards the gathering: "A new India that would fulfil the dreams of the young and women, and see the income of farmers double." Doubling of farmers' income by 2022. The seed spices one of the agriculture commodities that certainly double the farmer's income. The list of seed spices crops grown in India given below:

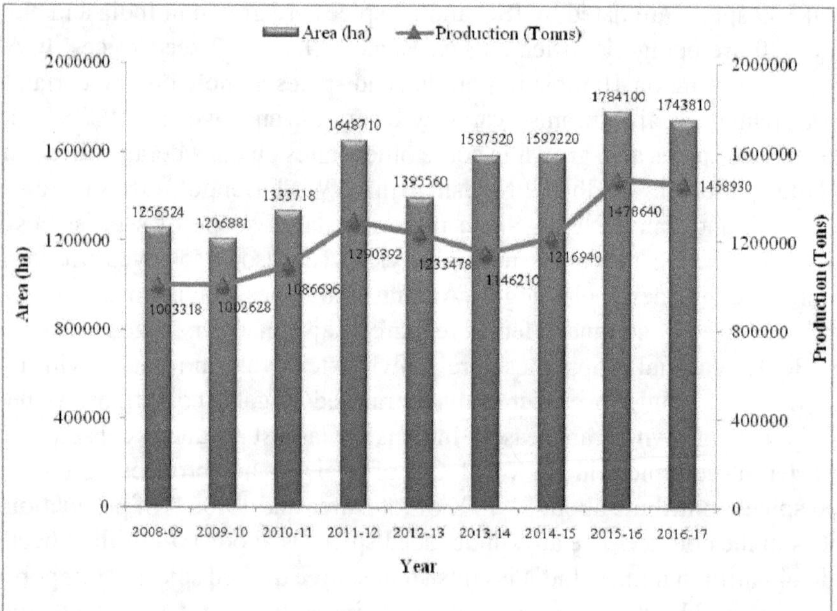

Fig. 1: Status of area and production of seed spices in India (DASD, 2016-17)

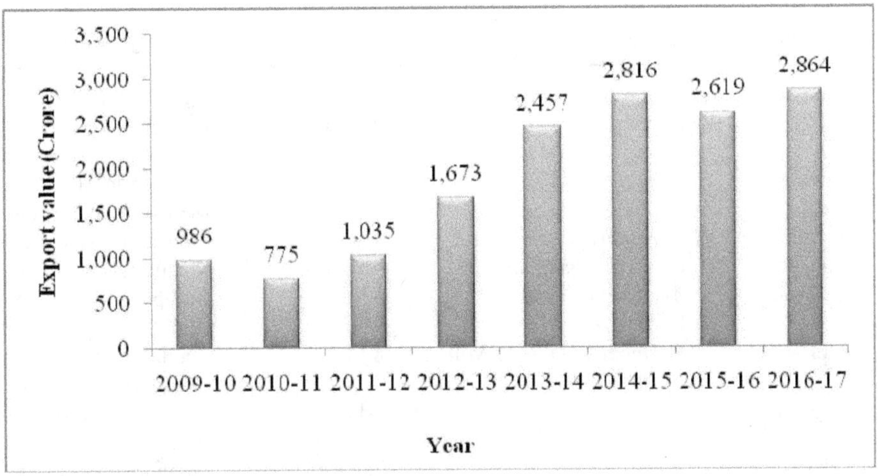

Fig. 2: Export of seed spices from India (DASD, 2016-17)

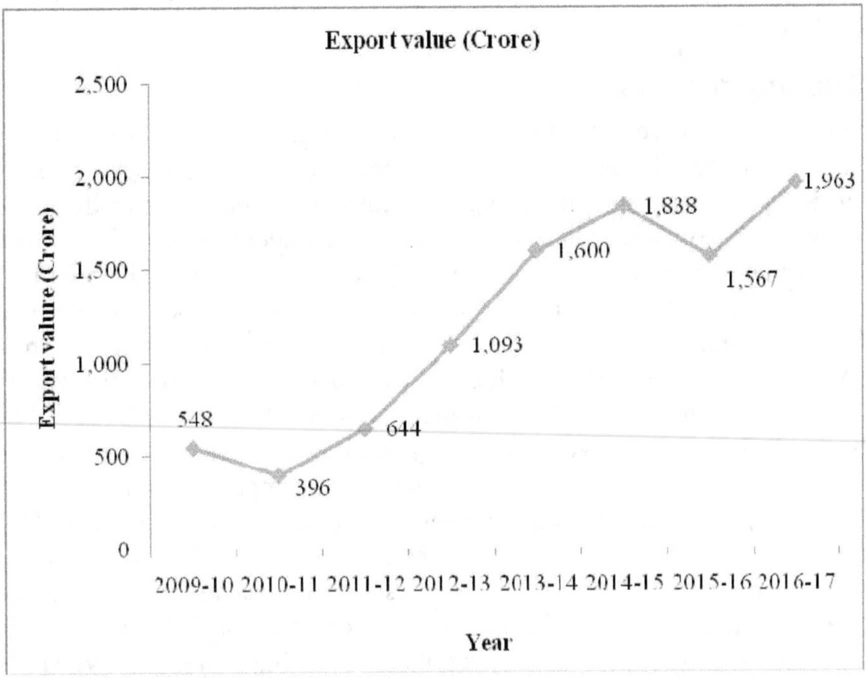

Fig. 3: Export value of cumin (DASD, 2016-17)

Table 1: Important seed spices grown in India

Name of Crops	Scientific Name	Family	Centre of origin
Ajwain	*Trachyspermum ammi*	Apiaceae	Egypt and India
Aniseed	*Pimpinella anisum* L.	Apiaceae	Eastern Mediterranean region
Caraway	*Carum carvi* L.	Apiaceae	Mediterranean
Celery	*Apium graveolens* L.	Apiaceae	Mediterranean region
Coriander	*Coriandrum sativum* L.	Apiaceae	Mediterranean region
Cumin	*Cuminum cyminum* L.	Apiaceae	Mediterranean region
Dill	*Anethum graveolens* L. *A. sowa*	Apiaceae	Europe, Africa and Asia (India)
Fennel	*Foeniculum vulgare* Mill.	Apiaceae	South Europe and Mediterranean region
Fenugreek	*Trigonella foenumgraecum* L., *T. carniculata*	Fabaceae	South East Europe and West Asia
Nigella	*Nigella sativa* L.	Ranunculaceae	South and Southwest Asia

Source: NRCSS Vision 2050

Technology Interventions to Double the Farmer's Income Through Seed Spices

Crop Improvement

Systematic improvement in seed spices was started after the initiation of All India Coordinated Spices and Cashew Improvement Project (AICSCIP) in 1971. The breakthrough in the improvement of seed spices was made in 1975 when cumin, coriander, fennel and fenugreek were added in the AICSCI project by starting four new seed spices research centres, *viz*. at Coimbatore (Tamil Nadu), Guntur (Andhra Pradesh), Jagudan (Gujarat) and Jobner (Rajasthan). Exclusive start of All India Coordinated Research Programme on Spices (AICRPS) at Calicut in 1986 made significant contribution in seed spice research (Singh and Solanki, 2015). Establishment of National Research Centre on Seed Spices (NRCSS) in the year 2000 at Ajmer, Rajasthan has made outstanding impact on seed spice breeding programme. Genetic improvement in seed spices was attempted mostly through selection, hybridization and mutation breeding. Classical breeding in seed spices has its own limitations including restricted gene pool, small flower size that hampers hand emasculation and pollination (Verma *et al*., 2017a&b). In the crop improvement programme the prime necessity is to know the extent of variability available in hand. The variability may be from the natural variants available in nature or it may be created. Since the variability naturally present in these crops is very low, so, crop improvement in these crops in not so easy. Many varieties in seed spices developed through selection from the exotic materials and many selected from the available germplasm. There is not much more literature available on artificial hybridization in seed spices but the sib mating by natural hybridization is common in almost all seed spices except fenugreek. Seed spices are lacking behind with other agricultural crops in terms of candidate genes identified for biotic and abiotic resistance and desired quality traits. Wild species in seed spices are not much available and only traditional varieties/ improved varieties as well as exotic collections are available for breeding purpose. The most of the seed spices are grown in the areas where the scarcity of natural resources as well as the condition of soil is poor, so development of varieties which are suitable for these area is also important. The improved varieties developed so far in seed spices are given below:

Table 2: Improved varieties of seed spices developed by different Institute/University in India

S.No.	Crop	Variety
1.	Cumin	RZ – 19, RZ – 209, RZ – 223, RZ-345, GC- 1, GC – 2, GC – 3, GC – 4
2.	Coriander	Ajmer Coriander -1 (ACr 1), Ajmer Coriander -2 (ACr-2), RCr – 41, RCr – 20, RCr – 435, RCr – 436, RCr – 684, RCr – 446, RCr-728, GCr – 1, GCr – 2 , CO- 1, CO- 2, CO- 3, CS- 287, Rajendra Swathi, Rajendra Sonia, Sadhna, Swathi (CS – 6), Sindhu, GCr – 1, GCr – 2, CO- 1
3.	Fennel	Ajmer Fennel-1 (AF – 1), Ajmer Fennel-2 (AF-2), RF – 101, RF – 125, RF – 143, RF-205, PF – 35, GF – 1 (Gujarat Fennel – 1), GF – 2, GF – 11, Guj. F.-12, Co – 1, Hisar Swarup
4.	Fenugreek	Ajmer Fenugreek 1, Ajmer fenugreek 2 , Ajmer Fenugreek 3, Ajmer Fenugreek 4, Ajmer Fenugreek- 5, RMt – 1, RMt – 143, RMt – 305, RMt-361, GM – 1, CO – 1, Rajandra Kranti, Andhra Pradesh, Lam Selection–1, APHU Methi-1, Hisar Sonali, Hisar Suvarna, Hisar Mukta, Hisar Madhavi (HM – 350), HM-219, Pant Ragini, Pusa Early Bunching, Pusa Kasuri
5.	Dill (Sowa)	Ajmer Dill 1 (AD–1), Ajmer Dill 2 (AD–2), GD-1, GD-2, RSP-11
6.	Ajwain	Ajmer Ajwain 1 (AA-1), Ajmer Ajwain 2 (AA–2), Ajmer Ajwain - 93 (AA-93), Pratap Ajwain 1, GA – 1, Lam Selelction–1, Lam Selection–2, Pant Ruchika, RPA-68,
7.	Nigella	Ajmer Nigella 1 (AN – 1), Ajmer Nigella 20 (AN-20), Azad Kalaungi (1998), Pant Krishna (2001), Rajendra Shyama, Rajendra Mani
8.	Anise	Ajmer Anise 1 (AAni – 1)
9.	Celery	Ajmer Celery-1, RRL-85-1,

Source: www.seednet.gov.in; http://www.aicrps.res.in/index.php/download-it/varieties-released

Production of Quality Seed of Seed Spices

The seed replacement rate in seed spices is very low (15%) which is a major challenge for enhancing production and productivity of seed spices. Providing good quality seeds is one of the most important and easiest means to accelerate the productivity of seed spices in the country. Importance of quality seed includes seed is a vital input in crop production, it is the cheapest input in crop production and key to agriculture progress, crop status largely depends on the seed materials used for sowing, response of other inputs in crop production depends on seed material used, the seed required for raising crop is quite small and its cost is so less compared to other inputs, this emphasis the need for increasing the areas under quality seed production and it is estimated that good quality seeds to improved varieties can contribute about 20-25% increase in yield.

Production Technologies

Suitable package of practices including, efficient water, optimum sowing time, seed rate, nursery management, transplanting, spacing, intercultural practices, weed management, cropping system, intercropping and organic farming practices has been standardized for different seed spices crops for realizing the optimum potential of improved varieties in the different agro-climatic conditions. The production technology for off season cultivation has also been worked out for leafy coriander and fenugreek production. In order to boost up production it is important to popularize improved and innovative production technologies for adoption by the farmers. In coriander, plastic covered walk-in-tunnel and shade net covered walk-in-tunnel with low pressure drip system was found highly suitable for off-season cultivation for green leaves. Raised bed technology with drip fertigation and plastic mulching has been found effective for fennel, dill, celery and ajwain. The production potential of seed spices is much larger than that has been achieved so far. In the endeavour to meet the increasing demand of seed spices crops it is now essential to work out complete package of practices varying across seasons, zones and production systems.

Seed spices are basically low nutrient and water requiring crops and seed spices cultivation is confined to the areas where limited water is available for irrigation. Irrigation efficiency is only to the tune of 30-35%, therefore only means to provide the water is effective management and enhancing the water use efficiency to the level of 80% which is possible with the use of micro irrigation. Lots of initiatives have been taken to apply water in such a manner which can provide maximum output. When, where and how water should be applied has been worked out. Precise irrigation management practices have been developed for seed spices at NRCSS (Lal *et al.*, 2011). Normally, all these crops are grown in flat bed, practice of raised bed sowing may be very effective in saving irrigation water as well as maintaining favourable aerobic condition in the root zone throughout the growing season of the crop resulting in better yields. Among various methods tried drip irrigation has proved successful in exhibiting high water productivity by saving irrigation water from 25-60% along with 10-60% increase in yield as compared to conventional method of irrigation.

Different cropping system has been helpful for crop diversification, efficient utilization of resources and making intensive use of inputs for sustainable income to marginal farmers. The diverse favourable agro-climatic conditions of the country provides opportunity to cultivate variety of horticultural crops such as fruits, vegetables including root and tuber crops, flowers, plantation and spices crops and medicinal and aromatic crops. These crops provide additional income to farmers. Intercropping of fennel or with knolkhol in 1:2 ratio followed by fennel/ dill with cabbage in 1:1 intercropping ratio is very effective for realising higher yield and benefit. Cluster bean- cumin- summer moong followed by pearl millet- cumin- summer moong cropping sequence is very effective cropping sequence for cumin growing farmers. Intercropping of seed spices in arid fruit orchards has also been assessed and found economical.

However, most of the seed spices are low nutrient requiring crops but fertilizers alone account for 20-30% of the total cost of production in spices crops. Fertilizer efficiency studies using isotope labelled fertilizers have shown nutrient derived from fertilizer (NDFF) values of 0.5 to 30%. Monitoring soil health and leaf analysis will be useful to ensure a balanced nutrient management and safeguard against hidden hunger. The phosphate rich organic manure (PROM); a value added product by co-composting different organic wastes with high grade rich phosphate in fine size can be useful in many spices crops and is effective alternative to costly chemical fertilizers. Foliar nutrition method for some of the spices crops has been worked out but there is need to cover more seed spices for high nutrient use efficiency. Besides, use of liquid form of fertilizers through fertigation appears to be promising for deeper application with sizeable input saving. The use of bio-fertilizers, VAM fungi, biological nitrogen fixers and other beneficial microbial agents also need to be optimally used to attain an efficient management.

Seed Spices Based Farming System in Arid and Semi-Arid Agro-Ecosystems of India

The farming system is a profitable and sustainable production system which allows the production of diversified agricultural produce. The systems approach recognizes the importance of interconnections and functional relationships between different components of the farming system (plants, soils, insects, fungi, animals, water). In cropping system we can ensure better management of land, labor and capital resources. The crop productivity maintain despite of continuous cropping by applying a high rate of organic fertilizer in every cropping season. The production cost of farm can be reduced by applying organic farming and using biological and cultural methods in controlling plant pests and diseases. The income of farmers sustaining this system and the dream for doubling of farmer's income can become reality. Mixed farming with seed spices could be possible even in fully grown orchards. A combination of one or more enterprises with cropping when carefully chosen, planned and executed gives greater dividends than a single enterprise, especially for small and marginal farmers. Farm as a unit is to be considered and planned for effective integration of the enterprises to be combined with crop production activity. Therefore, to mitigate the risk and ambiguity of the income from traditional cropping system, it is crucial to combine the seed spice crops with different agricultural enterprises in the production programme that yield continuous and evenly distributed income, cater diverse needs of the farmer's family along with imparting sustainability through conservation and improvement of natural resources in fragile arid and semi- arid ecosystem.

Farming system is a mix of farm enterprises such as crop, livestock, aquaculture, agro forestry and fruit crops to which farm family allocates its resources in order to efficiently manage the existing environment for the attainment of the family goal (Pandey *et al.*, 1992). Appropriate combination of different available farm components certainly enhances the profitability of farmer's farm. It interacts adequately with environment without dislocating the ecological and socio-economic balance on one hand and attempt to meet the national goal on the other. In its real sense it will help in lifting the economy of agriculture and living standard of the farmers of the country as a whole.

Seed Spices Based Farming System

Seed spices are an important horticultural crops and these area mainly concentrated in arid and semi-arid parts of India. Seed spices can be accommodated in other crops in different cropping and/or farming system (Malhotra, 2004). The fenugreek can be grown as a fodder crop which is rich in protein. The straw from different spices can be used as a feed for the animal by blending it in different ratio. Seasonal vegetables and flowers can be rotated intercrop with seed spices. Following cropping system can be adopted by farmers to enhance the farm income:

Farming Systems in Arid and Semi-Arid Regions of India

Gujarat: Cumin, coriander, fennel

Karnataka and Andhra Pradesh: Fennel, coriander, ajwain

Madhya Pradesh: Coriander, fenugreek and ajwain

West Bengal, Bihar, Estern U.P. and Bundelkhand: Fennel, nigella, and coriander

Other predominant crops of semi-arid regions: Wheat, rice, sugarcane, vegetable, moong bean, urdbean, pea, pigiopea,

Live stock components of semi-arid regions: Buffalo, cow, goat, sheep, fisheries, poultry

Protection Technology in Seed Spices

Insects Control

Seed spices crops attacked large number of predators and parasitoides due presence of its specific aroma, nectar and pollen on these crops. The population of parasitoides found increased with the increase of temperature in Rajasthan and Gujarat condition. The parasites *Aphaleeinus kurdijmovi* occurs regularly, reaches to its maximum (98 per cent) which was observed during November-December. On artificial release, the parasitism increases gradually and reaches up to a maximum in 30 days (Kumaresan *et al.* 1988). The noctuids moth larvae *Spodoptera* and *Helicoverpa* attracted large number of parasitoides in field conditions. The common parasitoides are *Sturamia inconspicuoides, Actia monticola, Euplectrus gopimohani*. For managing noctuids moth larvae at field conditions 4-6 inaculative releases egg parasitoids 150,000 like *Trichogramma chilonus*. *T. brasielensis* @ 150000 parasitoides/ha or starting at first appearance of the moths at 1-15 days interval is found useful., release of larval parasitoids such as *Chilonus blackburnii*, *Bracon brevicornis, ,Telenoums heliothinae*, *Carcelia illota ,Coteria kazat* or *Campoletiscloridae*, take care of the already hatched larvae. Coccienellied consist of major predator found feeding on various sucking pests of seed spices. Major coccienellid found predating on seed spices crops are, *Coccinella septumpunctata* L. *Bromoides suturalis* F. *Menoochilus sesxmacalatus* and *Adonia* sp. Predatory bird myna (*Acridotheres tristis*) was also found feeding on the aphid. The other common predators of aphids are Chrysoperla carnea, Episyrphus balteatus and Ischiodon scutellaris (Singh, 2007; Mittal and Bhutani, 1994).

Application of neem based commercial formulation like Neemarin at 1% and seed extract of neem (*Azadirachata indica*), karanj (*Pongamia* sp.), buken (*Melia* sp.) and pride of India (*Lagerstroemia indica*) reduce the aphids population by 50 per cent within 7 days of application (Singh, 2007). It was observed that 80 per cent reduction of aphid *H. coriandri* population on coriander crop through spraying of 1.0% neem oil and achieve highest grain yield 983 kg/ha. In cumin crop aphid *M. Persicae* population was reduced more than 50% for 15 days by application Neem Seed Kernal Extract (NSKE) at 5%. Applications of neem oils and karanj oils successfully controlled early instars larvae of *H. armigera* and *S. litura* (Sharma *et al.*, 2007)

In cumin crop spraying of phosphomidon-0.03%, monocrotophos -0.03% at 2 week intervals after first appearance of aphids give effective protection. The seed yield o treated plot was 4.7 q/ha in compared to 3.0 qt/ha on untreated plot (Gupta and Yadava, 1986). Sharma *et al.* (2007) found two application of dimethoat 0.03%, thiomethoxam 0.025%, Imidachlorprid 0.005% and acephate 0.037% reduce more than 90 % of aphid (*Myzus persicae*) in cumin in 3 days of treatment and give 40 % more yield in comparison to untreated plot. The yield was also at par in both the treatments. Two spray of carbosulfan 25 EC at 1250 ml./ha was found an optimum dose for effectively control of sucking pest population in cumin especially aphids and white fly. Corbosulfan did not showe any adverse effect on natural enemy population and also leave no phytotoxic effect on plant (Ameta and Gupta, 2007).

Uses of aldicarb side dressing at 1.0 kg/ha, corbofuran 3 G at 0.5 kg ai /ha., phorate 10 G at 1.0 kg /ha., endosulfan, 0.03%, phosphomidon 0.03%, monocrotophos 0.03% effectively reduce the aphid population on coriander crop and increased yield. (Jain and Yadava, 1989; Meena *et al.*, 2003) .Thrips (*Thrips tabaci* and *Scirtothrips dorsalis*) and mite on coriander was controlled by use of monocrotophos at 0.5 kg/ha and chlorpyriphos at 0.05- 1.0%.

Insecticidal control of aphid *Hyadaphis coriandri* on fennel was achieved by application of foliar spray of 0.05% quinolphos, 0.07% endosulfan, 0.009% cypermethrin, 0.05% dichlorvos, 0.025% methyl parathion 0.03% phosphomidon, 0.03% dimethoate and 0.01% fluvalinate (Mittal, and Bhutani, 1994). At flowering stage application of methyl parathion 0.025% and endosulfan 0.07% minimizes infestation of seed midge in fennel crop during storage. Two application of metasystox at 0.05% significntely reduce thrips population on fennel crop (Sagar, 1987). Application of dimethoate at 0.045% and metasystox at 0.03% gave maximum aphid control in fenugreek crop and give highest yield (6.87 and 6.27 qt./ha) (Kumawat *et. al.*, 2001). Effective management of *Spodopter litura* and *Helicoverpa armigera* on cumin, coriander, fennel fenugreek and nigella was achieved by application of 0.05% monocrotophos fallowed by 0.2% aldicarb after 10 days, fenvalerate 50 g ai /ha or endosulfan at 700 g aµ / ha.

Disease Control

The disease infestation in seed spices lower the production and productivity. Hence management of disease one of the important practices which can enhance the productivity of seed spices. Diseases like wilt, blight and powdery mildew in cumin; wilt, powdery mildew and stem gall in coriander; blight and gummosis in fennel, and powdery mildew, downy mildew and mycoplasma-like organisms (MLOs) in fenugreek frequently attack these crops causing heavy loss of yield and deteriorate the quality of the produce. Though the package of practices for the control/ management of each of these diseases have been formulated but still diseases like wilt (cumin and coriander), root rot (fenugreek) and gummosis (fennel) are difficult to be controlled completely. Concerted efforts are needed to evolve appropriate technology for their effective control or varieties which can tolerate or resist this disease. Methodology developed and recommended by NRCSS, Ajmer for control of important diseases are as under.

- Application of mustard residue (2.5 t/ha) + mustard cake (0.5 t/ha) + neem cake (0.5 t/ha) as soil amendment with one irrigation during hot summer.
- Application of *T. viride* or *A. versicolar* as seed treatment @ 10g/kg seed and soil application @ 2.5 kg/ha mixed with 50 kg of FYM at the time of sowing
- ST with *Trichoderma,* SA with cator cake and FS with chlorothalonil followed by two sprays of difenoconazole and azoxystrobin was found most effective for the management of *Alternaria* blight of cumin (Lal *et al*., 2012).
- Soil Solarization in combination with oil cakes has been found highly effective
- Two sprays of neem oil (2%) at 10 days intervals during flowering in evening hours
- Use of chemicals

Organic Seed Spice Production Technology

The pesticides residue risk is major concern to consumers in developed countries and peoples are looking to have organic produce. The production of seed spices in arid region is organic by default and it is rather easier to shift towards certified organic farming of seed spices. Due to low inputs requirements seed spices can easily be grown as organic. Moreover there is lack of specific varieties for organic cultivation and responsiveness to organic nutrient sources. Seed spice crops such as fenugreek, anise, coriander, ajowan, fennel, dill can be grown in organic farming module, while cumin is very risky crop and organic cultivation of this crop is quite difficult. To increase the soil fertility, green manuring with sunhemp, dhaincha, legumes are beneficial. In a study at NRCSS, application of organic Module-1 comprising of soil application of vermin-compost, *Trichoderma*and neem cake as well as seed treatment of *Azotobacter,* PSB and *Trichoderma* with the foliar spray of garlic extract and neem oil is better for realizing higher, growth and yield parameters as wells as yield and net return. Application of 7.5 t/ha sheep manure followed by 4.0 t/ha vermi-compost with bio-fertilizer *Azotobactor@* 250g/ ha seed in coriander is better for realizing higher yield in cumin and coriander. Application of 75 % N & P through vermin-compost and enriched vermi compost with seed inoculation by *Azotobactor and PSB* is economical organic nutrient management strategy in coriander and cumin. Seed treatment along with soil application of *Trichoderma viridae @* 2.5 kg/ha and soil application of neem cake @150 kg/ha and foliar spray of karanj oil for is an integrated approach for disease management in coriander under organic coriander production. Insecticidal soap developed by NRCSS (neem oil based- 1%). In coriander application of PSB + *Azotobactor* along with FYM. In fenugreek *Rhizobium* treatment along with FYM.

All over the world, people are becoming more conscious of health problems due to consumption of agricultural produce contaminated with pesticide residue and the adverse impact of modern agriculture on the environment. Major advantages of organic farming are that organic seed spices are highly remunerative because of export demand and moreover the organic produce is of high quality, safe, nutritive and eco-friendly. The related benefits accrued are that long term fertility of soils is protected as it is based on the principle of sustainable farming and the insoluble nutrient sources can be utilized through

Postharvest Management and Processing of Seed Spices

The postharvest handling activities start after the harvesting of crop, first with drying of freshly harvested crop, threshing, seed drying, cleaning, grading, packing and transportation of whole seed to market, processing plant or storage place. On farm processing is an important aspect to prevent deterioration of the produce and maintain the quality. Improvements in hygiene, packing and storage facilities help in keeping quality of spice flavours and reduce chances for afla-toxin and salmonella contamination of seed spices. Many products have been developed by NRCSS in seed spices such as essential oil, oleoresins, cryogrind powder of seed spices, coriander, fennel, dill squashes, RTS, biscuits *etc.*

Transfer of Technologies

Dissemination of developed technologies through institutional projects, TSP/ NEH and MIDH for expansion of seed spices in new areas (Lal *et al.*, 2015b&c). ICAR-NRCSS disseminating the developed technologies to stakeholder through Front Line Demonstration (FLDs), Fields Days, OFT (On Farm Trials), Farmers Trainings, Krishi Vigyan Melas, Exhibition, District and State Level Seminars, National Seminars/Symposium, Farmers Field Visits and extension bulletins, folders, leafleted, books, manuals etc.ICAR-NRCSS conducted FLDs of seed spices in different district of major seed growing areas of Rajasthan, Gujarat, M.P., Andhra Pradesh, J&K and Chhattisgarh. ICAR-NRCSS have also conducted several transfer of technology (ToT) activities such as farmers trainings and FLDs in Meghalaya, Assam, Arunachal Pradesh, and Sikkim for expansion of areas under seed spices in different agro climatic Zones and Sub Zones as well as new areas in the country (Lal , 2014; Lal *et al.*, 2013; Lal *et al.*, 2015a).

References

Ameta, O.P. and Gupta, H.C.L. 2007. Studies on bioefficacy , biosafety and dissipation of carbosulfan 25 EC in cumin, for the management of sucking insect pests. Paper presented in National Seminar on Production, Development, Quality and Export of seed spices. Feb, 2-3, 2007. NRC Seed Spices, Tabiji, Ajmer, Rajasthan- India. Pp – 357-364.

Kumaresan, D.; Regupathy, A and Baskaran, P. 1988. Pest of Seed Spices. Rajalakshmi Publication. pp. 1-241.

Kumawat, K.C. and Singh, S.P. 2001. Bio-efficacy of some insecticide against aphids, *Acyrthosiphon pisum* on fenugreek. *Annals of Plant Protection Science*. 9: 2, 320-322.

Lal, G. (2014). Impact of adoption of scientific interventionsin fenugreek on grain yield and farmers income:An assessment by FLD's in Arid Zone of Rajasthan,India. *African J. Agril. Res.*,9(31): 2391-2396.

Lal, G., Mehta, R. S., Chand, P., Godara, A. S. andH. Cherian. 2015a. Performance of improvedvarieties and technological interventions at farmers'fields for cumin cultivation.*J. Spices and AromaticCrops,* 24(2): 145-148.

Lal, G., Meena, S. S., Maheria, S. P., Mehta, R. S. and Anwer, M. M. 2011. Growth and yield of cumin(*Cuminum cyminum*L.) as influenced by irrigationmethods and protected cultivation practices.*Int. J. Seed Spices,* 1(1): 13-17.

Lal, G., Mehta, R. S., Singh, D. and Choudhary, M. K. 2013. Effect of technological interventions oncumin yield at farmers' field. *Int. J. Seed Spices,*3(2): 65-69.

Lal, G., Sharma, Y.K., Meena, R.S., Meena, N.K. and Maheria, S. P. 2015b. Performance of front linedemonstrations of fenugreek on yield and farmers'returns in Kukanwali area of Rajasthan.*Int. J. SeedSpices,* 5(2):38-42.

Lal, G., Singh, B., Singh, R., Mehta, R. S., Sharma,Y. K., Meena, R. S. and Maheria, S. P. 2015c. Assessment of varietal performance, technologicalgap and productivity gain through front linedemonstrations in fenugreek. *Legume Res.* 38 (5):616-620.

Malhotra, S.K. 2004. Potential of seed spices towards crop diversification. (In) Compendium of winter school lectures pp. 55-61.

Meena, P.C.; Sharma, J.K. and Noor 2003. Effect of a biotic factors on the occurrence of aphid, *Hyadaphis coriandri* (Das) and *Coccinella* predator on coriander varieties. *Annal of Agri Bio Research.* 7: 2, 181-186.

Mittal, V. P. And Butani, P.G. 1994. Pests of Seed Spices. In Advances in Horticulture Volume-10Plantation and Spice Crop part-2. (Chadha, K.L. and Rothinan, P.eds.) .Malhotra publishing House, New Delhi 110064. Pp. 825-855.

NRCSS Annual Report, 2015-16

NRCSS Vision 2050. NRCSS, Ajmer, India

Sagar, P. 1987. Efficacy of some insecticide against thrips, *Thrips flavus* Schrank, a pest of fennel in the Punjab. *International Pest control*. 29: 6, 130.

Sharma, K.L.; Kumawat, K.C. and Yadav, S.R. 2007. Relative bio efficacy of different insecticide against aphids *Myzus persicae* Sulzer on cumin. Paper presented in National Seminar on Production, Development, Quality and Export of seed spices. Feb, 2-3, 2007. NRC Seed Spices, Tabiji, Ajmer, Rajasthan- India. Pp – 336-341.

Singh, B. and Solanki, R.K. 2015. Status of seed spices research and development in India. *Indian Journal of Agricultural Sciences,* 85(2): 151–6.

Singh, M.P. 2007. Integrated pest management (IPM) in seed spice. Paper presented in National Seminar on Production, Development, Quality and Export of seed spices. Feb, 2-3, 2007. NRC Seed Spices, Tabiji, Ajmer, Rajasthan- India. pp – 139-148.

Verma, A.K., Kakani, R.K., Solanki, R.K. and Meena, R.D. 2017a. Improvement in yield attributing traits of cumin (*Cuminum cyminum*) through acute exposure of gamma ray. *International Journal of Pure & Applied Bioscience*, 5(2): 312-318.

Verma, A.K., Sharma, S., Kakani, R.K., Meena, R.D. and Choudhary, S. 2017b. Gamma radiation effects seed germination, plant growth and yield attributing characters of fennel (*Foeniculum vulgare* Mill.). *International Journal of Current Microbiology and Applied Sciences*, 6(5): 2448-2458.

24

Nutritional Security of Semi Arid Farmers Through Underutilized Fruit

Sanjay Singh, A.K. Singh, D.S. Mishra and V.V. Appa Rao

Introduction

India is emerging as a major stake holder in the global horticulture scenario. More emphasis is given to underutilized fruits due to their high nutritive and medicinal value in addition to being resistant/tolerant to many biotic / abiotic stresses. Jamun (*Syzygium cuminii* Skeels), fruits are used as an effective medicine against diabetes, heart and liver trouble (Singh and Singh, 2012). The powder of seeds has high value being useful in the treatment of diabetes. Therefore, the *jamun* fruits are having high value in terms of therapeutic and nutrition. Fruits and seeds of tamarind have tremendous potential for industrial use. Mahua (*Bassia latifolia*) is a boon for tribal farmers of Gujarat. The corolla commonly called as *mahua* flowers is a rich source of sugar containing appreciable amount of vitamins and minerals. Flowers (fresh and dried) and spent flowers (after fermentation) are also used as feed for livestock. Fruits are eaten as raw or cooked. Seeds are good source of oil. The oil obtained from Mahua kernel is used for edible purpose and permitted for preparation of vegetable oil. Mahua oil is used in manufacture of soap, lubricating grease, fatty alcohols and candles. Chironji (*Buchanania lanzan* Spreng) is excellent tree of agro forestry and social forestry. The flesh of ripe fruit is very palatable and the oily kernels are good source of protein (28-30%) and are used in preparation of sweets. Kernel oil can be used to substitute olive and almond oil. Khirni (*Manilkara hexandra*) thrives well on rocky, gravelly, saline and sodic soils and is good source of sugars and antioxidants. It is being used commercially as rootstock for sapota. Tamarind has tremendous potential under semi arid conditions (Singh and Singh, 2013). The medicinal properties of the bael (*Aegle marmelos* Correa) fruit are due to high marmelosin content. The pulp of mature bael fruit is very effective to cure dysentery and diarrhoea Varieties on jamun (Goma Priyanka and Thar Kranti), bael (Goma Yashi, Thar Divya and Thar Neelkanth), tamarind (Goma Prateek) and chironji (Thar Priya) have been developed. These varieties are precocious bearer with good yield and quality attributes and suitable for high density planting. These minor fruits have potential in secondary agriculture through value addition. Minor fruits have been conserved at the station and it has great potential under semi-arid regions of western India.

Present Scenario

At present, a sizable area (171 m ha) in India is categorized as degraded and marginal lands which are subjected to problems of water and wind erosion, salinity, alkalinity, acidity, water logging, gullied and ravine lands, shifting cultivation, etc. are either lying vacant or underutilized. Out of 143 million hectares of arable lands in the country, nearly 74% is rain-fed, which contributes about 42% of the total food production. Productivity of the crops is low due to prevailing adverse agro climatic conditions and poor knowledge of farmers regarding technological advancements and most of farmers are living below the poverty line. Therefore, this trend needs to be discouraged and minor fruit crop based models needs to be adopted for higher production and income generation as they are, known to withstand vagaries of monsoon better than sole agricultural crops. Although, India is the second largest producer of fruits in the world, however, availability of fruits to the resource poor people of semi-arid region is very low. Considering the present trend of population growth and also to meet the minimum requirement of the population, there is a need to increase production of fruits. A good deal of germplasm of semiarid fruits is available in different parts of the country in the form of land races and primitive cultivars. Many of them are mainly propagated by seed over the years generating a lot of variability, e.g., ber, bael, aonla, jamun, karonda, phalsa, wood apple, manila tamarind, charoli, custard apple etc. Therefore, seedling variability available in several pockets of the country needs to be exploited through intensive research efforts. At the same time production technology with regard to plant propagation, tree architecture management, integrated orchard, insect-pest and disease management need to be intensified.

Multitier Cropping System for Crop Risk Minimization and Nutritional Security

Under fragile ecosystem of semiarid regions uncertainty in production is high thus mono-crop culture may not be advisable. Crop failures due drought are common and often leading to great economic losses. Therefore, multi-crop combinations are suggested for these regions. Adoption of multitier cropping system can help to improve sustainability of degraded lands by addition of large biomass to the soil resulting in improved soil health. Fruit trees also help in reducing runoff and soil loss. They also help in improving the environment by sequestering CO_2 from the atmosphere. During young age of the orchard short statured crop like phalsa, guava, acid lime etc as filler crop or leguminous vegetables (cow pea, cluster bean)/ pigeon pea, maize/ seed spices (cumin, fennel, methi)/ medicinal & aromatic plants can be raised as inter crop.

Genetic Resources of Minor Fruits

Jamun: The Jamun (*Syzygium cuminii* Skeels), a member of family Myrtaceae is one of the important minor fruits.

Singh and Singh, 2005 made a survey in Gujarat to identify the elite genotypes of jamun. The study revealed that there was a wide variation among the genotypes. Total soluble solids varied from 10.30 to 12.34 %, total sugar 8.58 to 9.13% and vitamin C 32.12 to 46.37 mg/100g. On the basis of overall performance GJ-18, GJ-19, GJ-23, GJ-24 and GJ-25 were found to be promising among the genotypes studied based on its yield and physico-chemical characters.

At CHES, Godhra, jamun genotypes were evaluated for their growth, flowering and fruiting attributes. After evaluation, Goma Priyanka and Thar Kranti were released as variety. Salient characters are as under:

Goma Priyanka

Goma Priyanka was released at CHES, Godhra in the year 2009. The selection out performed in respect of flowering pattern, fruiting and fruit quality attributes. The tree height and girth was recorded 5.25 m and 64.66 cm respectively. North- South and East- West spread was found to be 5.00m and 4.76 m respectively. It is semi-dwarf, spreading growth habit, dense foliage and drooping branches, early, precocious bearer (starts flowering in 4th year) and suitable for high density planting. It starts flowering in the month of March, ripens in the fourth week of May and recorded 19.86 g average fruit weight, 85.06 per cent pulp, 16.80^{o}Brix TSS, 0.38% titratable acidity, 12.10% total sugar, 6.11% reducing sugar, 45.44 mg/100 g vitamin C, 14.38 mg/100g. It recorded 43.80 kg fruit yield during 8th year of orchard life under rainfed conditions of hot semi-arid ecosystem of western India.

Thar Kranti

It was identified in 2016 at C.H.E.S., Godhra. The selection out performed in respect of flowering pattern, fruiting and fruit quality attributes. It is spreading type, starts flowering in 3rd year, regular bearer, ripens in the fourth week of May and recorded 20.10 g average fruit weight 85.57 per cent pulp and 17.10^{o}

Brix TSS. Mean fruit yield (2015 and 2016) per plant was recorded 65.00 kg under rainfed conditions of hot semi-arid ecosystem.

Tamarind: Tamarind (*Tamarindus indica* L.) a member of sub family caesalpiniaceae of family leguminoceae, is a minor fruit and widely distributed throughout tropic and sub-tropics as stray plantation or avenue. Owing to its varied use in household and export market, it has emerged as a potential fruit crop.

In tamarind, Singh and Singh (2005) reported maximum panicle length (15.12 cm) and number of fruits per panicle (15.00) in T13. Early ripening (4th week March) was recorded in T1, T2, T6, T7 and T12, while T4 and T15 ripened at the last (2nd week May). The highest pulp percent (53.87) was recorded in T10. Total soluble solids and sugars were found maximum in T13. Singh *et al*, 2008 also reported variability in respect of flowering and fruiting pattern of tamarind under semi arid ecosystem of western India.

At CHES, Godhra, total 24 genotypes were evaluated (2001-2009) for their growth, flowering and fruiting attributes and Tamarind-13 (Goma Prateek) was released as variety. Salient characters of Goma Prateek are as under:

Goma Prateek

Goma Prateek was released at CHES, Godhra in the year 2009. It is early, spreading growth habit, regular bearer, semi dwarf and starts flowering in 4th year. Peak period of flowering is first week of June and ripens in the first week of March and recorded 26.70 g average pod weight, 16.70 cm pod length, 50.50 per cent pulp, 14.06% acidity, 71.00°Brix TSS, 55.81% total sugar, 27.27% reducing sugar, 17.53 mg/100g vitamin C, 3.29g/100g protein, 177.70 mg/100g calcium, 44.17 mg/100g magnesium, 70.83 mg/100g phosphorus. Fruit shape is slightly curved with reddish pulp, suitable for processing. Fruit yield per plant is 58.50kg during 9th year of orchard life under rainfed conditions of hot semi-arid ecosystem. These fruits have great potential under semi arid conditions of western India.

Varietal wealth: Some promising tamarind selections as a result of evaluation at different institutions are listed below.

PKM 1: It is a selection made at Horticulture College and Research Institute, Periyakulam. Grafts come to flowering three years after planting. The pods have 35% pulp, 17.1% acidity and 3.90 mg/100 g ascorbic acid.

Urigam: It is progeny of more than 200-year–old tree identified near Urigam by Department of Horticulture, Tamil Nadu. The average length of fruits is 30 cm.

Pratisthan : It was released from Fruit Research Station, Aurangabad. It has 61% pulp, 12% seed, 27% shell and 7-9% acidity. It is sour –sweet tamarind. Its pulp can be stored for long period.

Tamarind - 263: It was also released from Fruit Research Station, Aurangabad. The fruits have pinkish and light yellow pulp and 18-19% acidity.

Yogeshwari: The variety has been released from Taluk Seed farm, Ambajogai (Beed) in Maharashtra .The fruits are large and have red and sour – sweet pulp with 6-7% acidity.

DTS-1: It is a selection identified at University of agricultural Sciences, Dharwad. The pods are straight having semi-curved shape, 23.6 cm length, 3cm width, 19.5 gm weight, 51.00 % pulp and 13.60 % acidity. It is a late variety and takes 310 days from fruit set to maturity.

DTS-2: This is a selection made at University of Agricultural Sciences, Dharwad. The pods are straight having semi- curved shape, 17.60 cm length, 2.60 cm width, 18.00 gm weight, 53.00% pulp and 12.20 % acidity. It is an early variety and the pods mature in 280 days after fruit set.

Bael: After evaluation, 3 verities were identified in Bael.

1. **Goma Yashi:** Released in 2010. It is semi dwarf, precocious, spineless, regular bearer, ripens in April, Pulp 72%, TSS 37° Brix, fruit yield 51 kg / plant during 6th year.
2. **Thar Divya**: Released in 2015. It is very early, vigorous and good for juice making. Fruit ripens in last week of February; fruit yield 50-60 kg / plant during 7th year.
3. **Thar Neelkanth:** Released in 2016. It starts fruiting in 3rd year. The fruit yield per plant is 75.67 kg under rainfed conditions.

Chiraunji

Chiraunji (*Buchanania lanzan* Spreng.) is excellent tree of agro forestry and social forestry. In the wasteland development and dry land horticulture, it assumes great significance due to its multifarious uses and capacity to withstand adverse climatic conditions. It gives monitory reward to the tribal community of the country and seems to be boon for them. The flesh of ripe fruit is very palatable and is largely eaten raw or roasted and the oily kernels are the most important part and are used in preparation of puddings. Very good juice may be prepared from the pulp of chironji fruits. The kernel is highly nutritious and rich in protein (25.0-30.0%) and yields sweet oil, which can be used to substitute olive and almond oil. Kernel contains 33.50% oil, 1.90% of which is unsaponifiable. The saponifiable part contained 20.00 % of linoleic acid. An ointment made out of the kernels is used to cure itch of the skin and to remove blemishes from the face. Chironji provides quality timber wood for various uses.

Singh et al,, 2006 observed earliest flowering (first week of February) in 'CPT 1', 'CPT 5', 'CPT 10', 'CPT 15', 'CPT 17', 'CPT 19' and 'CPT 22', while the latest (4th week of February) occurred in 'CPT 13'. Maximum panicle length (35.11 cm) was found in 'CPT 1', while 'CPT 18' recorded least panicle length (14.20 cm). Highest percentage of perfect flowers (13.10) and fruit per panicle (37.50) was recorded in 'CPT 12' while, it was found least in 'CPT 18'. Ripening time varied from third week of April to second week of May indifferent genotypes. The highest fruit yield per plant was found in 'CPT 7' (28.00 kg/ plant).

At CHES, Godhra, fifteen genotypes of chiraunji were evaluated for flowering and fruit quality attributes. Based on the horticultural traits studied, Thar Priya was identified as variety.

Salient characters of CHESC-7 is as under

Chiraunji -Thar Priya: It was collected from Kada Dam of Panchmahal district, Gujarat. Peak period of flowering and fruit set was in the first week of February and 3rd week of February respectively. It recorded 26 cm panicle length. Peak period of ripening time was third week of April and recorded 1.20 g fruit weight, 22% TSS, 13.00% total sugar and 50 mg/100g vitamin C, 0.12g kernel weight and 30.50 % kernel protein.

Mahua: Survey of Mahua was made in Gujarat. It is one of the important crops for tribal farmers of Gujarat. Singh and Singh, 2005 recorded early flowering i.e. 1st week of March in MH-1, MH-4, MH-5, while it was recorded late (2nd week of April) in MH-6. The number of flowers (45.00) and fruits (8.00) per fascicle were noted highest in MH-2. MH-8 recorded maximum leaf area (128.00 cm^2). Range of ripening span in different genotypes was observed from 2nd week of May to first week of June. Dry flower yield varied from 30.00 to 46 kg / plant being highest in MH-2. Variation in fruit yield among the genotypes ranged from 35 kg to 82 kg / plant and it was maximum in MH-2

After evaluation, Thar Madhu was identified as variety in mahua at CHES, Godhra.

Thar Madhu: This variety was developed by the CHES Godhra. It is spreading type, regular bearer and starts flowering in 5th year. Peak period of flowering is 3rd week of March and it ripens in the month of May. Fresh flowers recorded 26.63^{o} Brix T.S.S., 1.03 % acidity, 23.11 % total sugar and 58.61 mg/100g vitamin C. Dry flower yield was recorded 6.32 kg/plant. The variety recorded 29.00 g average fruit weight, 41.74 % pulp, 14.26° Brix TSS, 0.10% acidity, 10.80 % total sugar and 48.38 mg/100g vitamin C. Fruit yield kg per plant was recorded 20.14 during 10th year of orchard life under rainfed conditions of hot semi-arid ecosystem.

Khirni

Khirni or Rayan botanically known as (*Manilkara hexandra*) is an economically multipurpose tree of the family sapotaceae. Thirty genotypes were selected & evaluated their fruiting and fruit quality attributes. Thar Rituraj variety was identified at CHES Godhra.

Thar Rituraj: It is early, semi-spreading type, regular bearer, semi-dwarf and starts flowering in 4th year. Peak period of flowering is first week of December and ripening is taken place in third week of May. It recorded 5.20 g average fruit weight, 87.49 per cent pulp, 24.73° Brix TSS, 0.32% titratable acidity, 17.80% total sugar, 9.23% reducing sugar and 28.33 mg/100g vitamin C. Fruit yield kg per plant was recorded 13.20 kg during 9th year of orchard life under rainfed conditions of hot semi-arid ecosystem.

Karonda

The karonda (*Carissa carandas*), belongs to the family Apocynaceae. It thrives well as a rainfed crop, the plant hardly needs any care and gives yield with minimum management. It is a sprawling semi vine shrub native to India. It is used as live fencing around the orchards besides providing fruits. The fruit is a berry, ellipsoid and pointed or globose or oval or round, 1.8-2.5cm long, green in colour when unripe and turn to red and finally black or dark purple on ripening. It is highly heterozygous, cross-pollinated fruit crop and as such seedlings exhibit a wide range of variations, which aids in the selection of the superior desirable genotypes. Forty genotypes were selected and evaluated at CHES, Godhra. The average fruit weight ranged from 1.87-6.15 g, fruit diameter from 1.41 cm – 2.06 cm, fruit length from 1.71-2.39 cm, pomace weight from 0.36-0.92 g, seed weight from 0.04-0.47 g, pulp weight from 1.29-4.79 g, total soluble solids from 9.15-10.95 per cent, acidity 0.45-1.22 per cent and vitamin C 22.65-38.32 mg/ 100g in different genotypes. With respect to all horticultural traits studied the genotypes, Karonda-1 and Karonda -2 were found to be promising. Konkan Bold recorded maximum fruit weight (14.90g) and TSS (10.10 ⁰ Brix) but fruit yield was 2.5 kg per plant only.

Thar Kamal was Developed by CHES Godhra

Thar Kamal: It is spreading type, regular bearer, dwarf and starts flowering in 3rd year. Peak period of flowering is 3rd week of March and ripens in the last week of June. Fruit yield per plant was recorded 13.00 kg during 9th year of orchard life under rainfed conditions of hot semi-arid ecosystem. It recorded 4.97 g average fruit weight, 93.66 per cent pulp, 9.54°Brix TSS, 0.64 % acidity, 30.41 mg/100g vitamin C, 6.12 % total sugar, 3.87 % reducing sugar.

Cape Gooseberry (Makoi)

Capegooseberry (*Physalis peruviana* L.) is a short duration underutilized fruit crop of the family Solanaceae. The fruit is rich sources of Vitamin C, carotene, phosphorus and iron. It is excellent for making jam and sweet pickles. The plant is partially erect attaining a height of 60 cm to 1 m. Flowers are large open bell shaped, long or broader, spreading and light yellow. The berry is dusty yellow. The fruit is tasty, quick growing and can be grown as either pure crop or as inter crop. It is famous for its flavour and having good blending of acid-sugar. The fruits are attractive in colour at maturity and if properly packed can easily be sent to distant markets. Ten promising genotypes were grown under field conditions. Highest plant height was observed in CHESC-1, followed by CHESC-3, CHESC-4. Peak period of flowering and fruit set was recorded in November- December. Peak period of anthesis was noted during morning hours in all the genotypes. Maximum number of fruit per plant was noted in CHESC-1, closely followed by CHESC-4 and CHESC-6. Peak period of ripening was observed in the month of February in different genotypes. TSS was found to maximum in CHESC-1 (18.00%), it was closely followed by CHESC-2 (16.00%). Based on the horticultural traits studied, the genotypes, CHESC-1, CHESC-2, and CHESC-4 were found to be promising. Detailed descriptions of two promising genotypes are as under:

CHESC-1: It was collected from Bihar. It ripens in the month of January-February and recorded 11.50 g fruit weight, 12.00 % TSS, 1.64 % acidity, 7.50 % total sugar and 220.10 mg/100g vitamin C.

CHESC-2: It was also collected from Bihar. It ripens in the month of Januar-February and recorded 10.75 g fruit weight, 11.20 % TSS, 1.70 % acidity, 7.25 % total sugar and 190.22 mg/100g vitamin C.

Therapeutic Properties and Value Addition

In view of growing cost health care, people consider that the food should comprise the major part of the healthy diet. In this regard underutilized indigenous fruits cannot only supply adequate amount of vitamins and minerals, but a wide variety of other elements that can contribute therapeutically to human health (Table 2). Consumer awareness on the health promoting capacity of fruits in managing life-style diseases has been increased. There is also renewed interest in therapeutic knowledge of natural plant products. Value added products from such fruits are high in demand in national and international markets.

Table 2: Therapeutic properties and value added products from future fruit crops.

Future fruit crops	Therapeutic properties	Value added
Jamun	• Fruit's syrup is useful for curing diarrhea. • Seeds contain jambosin alkaloid, which reduces conversion of starch into sugars. • Jamun fruits are used as an effective medicine against diabetes and heart and liver trouble.	Squash, RTS, nectar, syrup, vinegar, cider
Phalsa	• Fruits have astringent properties & used for several stomach ailments. • Fresh or dried fruits are suggested to cure heart and blood disorders, fever and diarrhea. • The root bark is used for rheumatism.	RTS, squash, syrup
Custard apple	• The ripe fruits rich in tannins, which is dried, pulverized and employed against diarrhoea and dysentery.	Jam, beverages, ice cream
Manila tamarind	• Fruits rich in phenolics including ûavonoids and showed potent free radical scavenging activity.	RTS, squash, syrup
Wood apple	• Fruit is much used in India as a liver and cardiac tonic. • Unripe fruit pulp, as a means to halt persisting diarrhea and dysentery and effective treatment for hiccup, sore throat, and diseases of the gums.	Jelly, chutney, sauce
Karonda	• Fruits are rich in ascorbic acid, iron and anthocyanins thus antiscorbutic and used to cure anemia.	Candy, jelly, pickle
Aonla	• Treat urinary & cough problems, supports liver. • Acts as body coolant & improve digestion. • Aonla an antioxidant for cholesterol and blood sugar control, delaying development of diabetes.	RTS, squash, syrup, jam, shred, candy
Bael	• Used for the cure of diarrhea, dysentery & other stomach ailments.	Sharbat, squash, murabba, jam
Ber	• Fruits are effective in relieving chest pain & vomiting sensation. • Widely used in traditional Chinese medicine and avurveda • Catechin, caffeic acid, epicatechin, ferulic acid, rutin, p-hydroxybenzoic acid and chlorogenic acid present in fruit.	Murabba, candy, dehydrated ber

Contd.

Future fruit crops	Therapeutic properties	Value added
Tamarind	• Contains tannins, saponins, sesquiterpenes and alkaloids • In Indian ayurveda used for gastric or digestion problems, and in cardio protective activity • Decoction used for treatment of stomach disorders, general pain, jaundice, yellow fever and as blood tonic. • Used as skin cleanser and in malarial fever.	Juice concentrate, pulp powder, jam syrup, candy, tamarind kernel powder
Jackfruit	• Fruit is rich in Vit A, ascorbic acid and isoflavonoids. • Strengthen immune system & protect against cancer. • Aids in healthy digestion. • Maintain a healthy eye, skin & control Asthma.	Canned pieces, nectar, pickle & chips
Mulberry	• Resveratol an antioxidant found in mulberries helps to promote heart health and overall vitality.	Juice, squash, syrup
Khirni	• Seeds contain 25% oil, considered as demulcent and emollient. • Bark is useful in treating gum problems & dental disorders. • Fruits are milky, sweet, aphrodisiac, appetizer, emollient & tonic. • Fruits are used to cure arthritis, jaundice, heat burning deworming, & blood purification.	Dehydrated fruits, fruit bar, RTS, jam
Mahua	• Useful as antipyretic, hematoprotective, anti-inûammatory, analgesic & antitumor • Treats rheumatism, ulcer, bleeding, and tonsillitis	Biscuits, cakes, dried ûowers, seed oil, wine
Lasora	• It has antihelmenthic, diuretic, demulcent expectorant properties.	Pickle, vegetable
Palmyra palm	• Palm sap is stimulant, ant phlegmatic & cures dropsy. • Used in digestive and other stomach ailments.	Toddy, syrup, jaggery, candy, toffee

References

Singh, Sanjay and Singh, A.K. 2005. Studies on variability in *Jamun (Syzygium cuminii)* in Gujarat. *Progressive Horticulture*, 37(1): 144-148.

Singh, Sanjay and Singh, A.K. 2005. Studies on variability in tamarind (*Tamarindus indica* L) from Gujarat. *Progressive Horticulture*. 37: 280-284.

Singh, Sanjay and Singh, A K 2005. Studies on genetic diversity in Mahua (*Bassia latifolia*) under semi arid ecosystem of Gujarat. *Indian Journal of Agricultural Sciences*, Vol 75: 519-523.

Singh, Sanjay, Singh, A K and V V Apparao 2006. Genetic diversity in *Chironji* (*Buchanania lanzan* Spreng) under semi arid ecosystem of Gujarat. *Indian Journal of Agricultural Sciences*, 76: 695-698.

Singh Sanjay and Singh, A. K. 2012. Enjoy new jamun variety. *Indian Horticulture*. May-June, Pp 9-11.

Singh Sanjay and Singh, A. K. 2013. Goma Prateek enriching tamarind wealth. *Indian Horticulture*. September- October, pp9-11.

25

Crop Regulation in Fruit Crops for Improving Quality and Income of Farmers

D.S. Mishra, Sanjay Singh, A.K. Singh, V.V. Appa Rao, Vikas Yadav and P. L. Saroj

Introduction

India is bestowed with diverse agro-climatic conditions which favour the production of a variety of fruit crops. At present India is second largest producer of fruits in the world after China. Fruit culture can be highly profitable by increasing employment opportunities and commercialization of the rural sector. During early sixties, priority was to achieving self sufficiency in food grain production in our country, however, over the years, fruit culture has emerged as an indispensable part of agriculture, offering a wide range of choices to the farmers for crop diversification. It also provides ample opportunities for sustaining large number of agro-industries to generate substantial employment opportunities (Bardhan, 2016). Amongst various tropical and sub-tropical fruit crops grown in India, guava, pomegranate, acid lime, sweet orange and mandarins, if left on its own, give several light crops of the variable quantities and qualities from the various flowering flushes throughout the year. Under natural conditions, these crops produce flowers thrice in a year *i.e.* February-March (*Ambe Bahar*), June-July (*Mrig Bahar*) and October-November (*Hasth Bahar*) with the corresponding harvest during rainy, winter and spring season, respectively (Boora *et al.*, 2016, Lal *et al.*, 2017). Owing to this, potential yield is not obtained during the desired period. Therefore, to obtain higher fruit yield during a particular period, these fruit crops are given a resting period with artificial means so that the natural flowering tendency of the trees is altered. It can be achieved through root exposure and its pruning (Boora *et al.*, 2016), shoot pruning (Tiwari *et al.*, 1992, Thakre *et al.*, 2016), withholding irrigation water (Stern *et al.*, 1993,

Krajewski and Rabe, 1995), deblossoming (Tiwari and Lal, 2007), chemicals/ PGRs application (Babu and Rajput, 1982, More *et al*., 2016, Maji *et al*., 2015, Mahalle *et al*., 2010, Ahire *et al*., 1993), change in fertilizer schedule (Boora *et al*., 2016) and shoot bending (Sarkar *et al*., 2005). However, the results of these experiments were not similar and the responses differed according to cultivars, tree conditions, soil types and agro-climatic conditions (Maji *et al*., 2015). Regulated crops are desired to avoid glut in the market and also ensure the regular supply of fruits (Devi *et al*., 2011). The choice of *bahar* at a particular location is determined by prevailing production constraints like availability of irrigation water, quality of produce, market demand and extent of damage by insect-pests and diseases (Lal *et al*., 2017). To avoid glut in the market and to extend fruiting period, several attempts on flowering induction show significant results in regulating off season flowering in mango, mangosteen, durian and rambutan (Poerwanto *et al*. 2008). Any practice or combination of practices that will produce a favorable carbohydrate: nitrogen ratio (i.e. a moderate stress) is generally beneficial, although the endogenous growth regulator balance in the leaves and stems on a tree actually controls the flowering process (Wilkie *et al*., 2008). Lack of floral initiation in an "on" year of a biennial bearing cultivars in several fruit crops has been attributed to gibberellins in developing fruits and seed; therefore, chemical deblossoming and fruit thinning or equivalent cultural practices will be effective to produce a greater return bloom only if they result in a shift in hormonal balance to overcome gibberellins inhibitory effects (Bardhan, 2016).The basic principle of crop regulation is to induce flowering and fruiting in desired season of the year that contribute to increased fruit yield, quality, profitability and sustainability of the environment by reducing the use of the frequency of the pesticides (Lal *et al*., 2017). Finally, crop regulation in different fruit crops is discussed.

Present Status in India and Global Scenario

Crop regulation in different fruit crops is about identifying, selecting, implementing and monitoring methods to control the yield and quality. It is planned and executed in such a way that it should not have negative impaction people, the environment or the financial bottom line. Different methods of crop regulation have been successfully demonstrated for improving the yield and fruit quality of guava in India (Boora *et al*., 2016, Lal *et al*., 2017) and other parts of the world (Shigeura *et al*., 1975, Chapman *et al*., 1979, Hojo *et al*., 2007, Khan *et al*., 2011, Mamun *et al*., 2012, Atawia *et al*., 2017).Water stress seems to be the major flower-inducing signal for pomegranate and citrus under tropical conditions (Cassin *et al*., 1969, Patil *et al*., 2002). Flowering in citrus invariably occurred following the renewal of rain, subsequent to a shorter or longer period of drought. Water stress has been, and is still being, used in Italy, Israel and California for the production of summer lemons (Spiegel-Roy and Goldschmidt, 1996). Successful off season production of tropical fruits like mango, mangosteen, rambutan etc. was demonstrated by Poerwanto *et al.* (2008) with the use of girdling technique and plant growth regulators. Regulation of flowering in other tropical fruit crops in relation to the effect of the environment, interactions with vegetative growth, the roles of PGRs and carbohydrates and recent advances in molecular biology are discussed by Wilkie *et al.* (2008).

Gaps in Prevailing System

Fruit production is seasonal activity and during the peak season price drops sharply owing to the glut in the market. At same time in multiple flushing species like citrus, guava and pomegranate desired yield and quality is not obtained during the peak demand period in the market. This condition is not economically sustainable. Therefore, to obtain higher fruit yield during a particular period, these fruit crops are given a resting period with artificial means so that the natural flowering tendency of the trees is altered. In Japan, technology for off-season production of Satsuma oranges has been developed and the oranges can be produced year-round (Poerwanto *et al.*, 2008). Various crop regulation methods in different agro-climatic conditions have been standardized for guava, pomegranate and citrus in India (Lal *et al.*, 2017).However, manipulation for flower forcing and fruit setting has not yet been done commercially in other tropical fruit crops in India for its economical benefits as has been standardized in Indonesia (Poerwanto *et al.*, 2008) and other parts of the world (Wilkie *et al.*, 2008).

Approaches for Crop Regulation in Different Fruit Crops

Some of the tropical fruit crops like guava, pomegranate and citrus bloom throughout the year without any resting period and produces two or three crops of variable quantity and quality (Lal *et al.*, 2017) while some of the tropical fruits like mango, litchi did not flower regularly, if flowering regulation treatments are not imposed cautiously (Wilkie *et al.*, 2008). Therefore, it is essential to understand the flowering and fruiting behavior of crops and which bahar will give economically good crop with full consideration to all the factors associated with particular *bahar*. Considering above facts in mind crop regulation is discussed for most of the fruit crops separately.

Guava

Guava (*Psidium guajava* L.) is the fifth most important fruit crop which is successfully grown over a wide range of soil and climatic conditions. It is available at reasonable price and known for its rich nutraceutical values (Kumar and

Mishra, 2012) thus, named as apple of tropics and super fruit (Maji *et al.*, 2015). In general, guava flowers twice in a year, i.e., in March-April (*Ambe bahar*) and June-July (*Mrig bahar*), of which fruits ripen in rainy and winter season, respectively (Table 1). However, in central and south part of India, there is a third crop with flowers appearing in October (*Hastha bahar*), of which fruits ripen in the month of March. This pattern of flowering and fruiting is not desirable for commercial exploitation. Moreover, the fruits of Ambe bahar which are harvested during the months of July-September are insipid, watery, and poor in quality and heavily infested with fruit fly resulting in significant loss to most of the guava growers (Mishra and Tiwari, 2000, Singh, 1995). The winter season fruits are superior in quality, free from pests which fetch high monetary returns (Singh *et al*, 2000). Obviously, regulation of crop will help to overcome such problems to a great extent by giving various flower and fruit thinning treatments during summer months.

Table 1: Flowering period of guava in different parts of India

Region / state	Flowering period	Reference
West Bengal, Eastern India	April-May and Sept.-October	Mitra, 1983, Singh, 1995
Punjab, North India	April-May and August-Sept.	Gupta and Nijjar, 1978, Singh, 1995
Maharashtra, Tamil Nadu, Southern and Western India	April-May, August-Sept. and October	Singh, 1995

Withholding of Irrigation Water

A general practice is to withhold watering of trees from February to middle of May. This results in the shedding of flowers and trees go to a rest period during which accumulation of food materials takes place in branches (Singh, 1995, Sachin *et al*., 2015). This method is suitable for light sandy soils; however, in heavy soils or soils of *Tarai* region, this method was not found useful (Tiwari and Lal, 2000). Cheema *et al*. (1954) advocated withholding of water and removal of soil from the upper roots by June 10 followed by covering tree basin with soil and manure mixture followed by irrigation to get winter crop in Maharashtra. The trees thus treated produce profuse shoot growth followed by heavy flowering leading to enhanced yield in winter season. Effect of soil-moisture stress on relative water content, proline accumulation in leaf, fruiting behaviour and fruit quality was studied by Singh *et al.* (1997) in 'Allahabad Safeda' and 'Sardar' guava. Stress was exerted on trees by withholding water from January to June, and the non-stressed control trees were watered at an interval of 10-15 days, depending on the evaporation. The abscission of flowers was noticed when relative water content reached 60.20% in 'Allahabad Safeda' and 53.40% in 'Sardar' guava. A trend was also noticed in the quantity of leaf proline according to the degree of water stress. The metabolic activity in the plants was maintained better even at low RWC in 'Sardar' than 'Allahabad Safeda'. 'Sardar' appeared more tolerant than 'Allahabad Safeda'. Water-deficient trees had significantly lower yield during rainy season in both the varieties. However, 'Allahabad Safeda' responded more to soil moisture stress for regulation of flower to harvest and better fruit yield during winter season than 'Sardar'. In 'Sardar' in addition to soil-moisture stress a chemical that can inhibit the growth may also be essential to regulate cropping.

Root Exposure and Root Pruning

Carefully, 7-10 cm upper soil around the tree trunk in a radius of 40-60 cm removed so that roots are exposed to the sun which results in reduced moisture supply to the top, therefore, the leaves begin to shed the leaves and tree goes to a rest period. After about 3-4 weeks the exposed roots again covered with soil and manure mixture followed by watering (Singh, 1995, Lal *et al.*, 2017). Consequently, trees put forth new growth, profuse flowering and fruiting. However, in light sandy and shallow soils, exposure of root is not required as withholding of irrigation water for 2-3 weeks is sufficient for leaf wilting and shedding. Water stress can be induced through root exposure and pruning of minute roots to suppress the rainy season crop and to get a good winter crop (Cheema *et al.*, 1954, Sachin *et al.*, 2015, Suresh *et al.*, 2016).

Shoot Pruning

Shoot pruning have been reported to be successful in of guava as guava flowers are always borne on newly emerging vegetative shoots, irrespective of the time of year (Rathore and Singh, 1974). This feature makes guava unique, that it can be pruned (Lotter, 1990) for regulating the crop. Shoot pruning is helpful in reducing the tree size and improving the fruit quality (Singh and Bal, 2006).The beneficial effect of pruning on yield and fruit quality of guava is reported by several workers (Bajpai *et al.*, 1973, Lal, 1983, Tiwari and Lal, 1984, Jadhav *et al.*, 1998, Lal *et al.*, 2000, Singh and Singh, 2001, Dhaliwal and Kaur, 2003, Dhaliwal and Singh, 2004, Kumar and Mishra, 2010). One leaf pair shoot pruning of current season's growth of spring flush in the first week of May recorded the maximum yield (Tiwari and Lal, 2007) and quality (Sharma *et al.*, 2013) during winter season. Similar results of shoot pruning were obtained in high density planting system of guava (Chandra and Govind, 1995, Thakre *et al.*, 2013, Pratibha and Lal, 2013, Thakre *et al.*, 2016, Joshi, *et al.*, 2016). Pruning was found to modify the production curve of guava (Quijada *et al.*, 1999). Also, Salah (2005) recorded the highest bud emergence in guava trees by using severe and moderate pruning. The time and intensity of pruning affected guava cv.

Paluma tree sprout and yield (Serrano *et al*., 2008a). Moreover, (Serrano *et al*. 2008b) reported that the light pruning increased the number of productive branches and number of fruits per branch in guava cv. Paluma. Pruning guava plants at a 20 cm pruning level in early May was the most effective management to reduce yield in the rainy season and to enhance yield and quality in the winter season in Nepal (Adhikari and Kandel, 2015). However, 10 cm apical pruning in May month resulted in higher yield and fruit quality in Cairo, Egypt (Sahar and Hameed, 2014). Prakash *et al.* (2012) also conducted an experiment to improve the fruit yield and quality of monsoon flushed guava crop by regulating flowering of spring-flushed crop using varying shoot pruning intensities (25 to 75%) under subtropical condition of Western Uttar Pradesh. Summer pruning of spring flushed bearing shoot of current season was found significantly effective in increasing the production of superior quality fruits in winter guava. With the increase in summer pruning severity, yield and quality of winter guava was improved significantly up to 33% pruning intensity. Hence, moderate pruning of bearing shoot of current season's growth just before bud opening in summer was significantly most effective in improving the production of monsoon flushed crop and fruit quality of spring and monsoon flushed crops in guava. This might be due to the fact that the plants accumulated food during rainy season which might have been got utilized for the development of winter season fruits (Chandra and Govind, 1995). Pruning results in greater penetration of photosynthetic photon flux in canopies of pruned trees than in unpruned trees during May and June which significantly increased the quantum of fruit yield harvestable during December and January by May pruning. Therefore, it can be concluded that shifting of rainy season crop to winter months in guava cultivars with the help of pruning in May is found economically efficient (Singh *et al*., 2001).

Deblossoming

Different chemicals and PGRs have been used to do deblossoming of rainy season crop which subsequently increased the winter season crop (Singh *et al*., 2016, Lal *et al*., 2017). Manual deblossoming on a commercial scale is economically not viable (Singh *et al*., 2002), however, in contrast to this Das *et al.* (2007) found it economically profitable when 50% of rainy season crop is removed manually. Rathore (1975) noted 96 per centdeblossoming with 100 ppm NAA in guava while 200 ppm NAA was effective for guva cv. L-49 in rainfed plateau conditions of eastern India (Das *et al*., 2007). While Chundawat *et al*, (1975) reported 82 per cent deblossoming with 400 ppm NAA spray. However, complete deblossoming was recorded by Pandey *et al*. (1980) with 400 ppm of NAA while More *et al*. (2016) recorded maximum flower drop (80.40%) and fruit let drop (73.54%) during rainy season crop and higher yield in winter season crop at 800 ppm NAA. Application of NAA (800ppm) has been found useful to get a good winter crop by other workers (Tiwari and Lal, 1984, Singh and Reddy, 1997). Flower thinning by using NAD (Kumar and Hoda, 1977, Mitra *et al*., 1982, Maji *et al*., 2015), 2,4-D (Kumar and Hoda, 1977, Das *et al*., 2007), potassium iodide (Narayan *et al*., 1999, Singh *et al*., 1996, Sachin *et al*., 2015) and ethephon (Singh *et al*., 1996, Singh *et al*., 2000) have been tried with varying degree of success.

The maximum deblossoming has been observed with 20 % urea spray in guava (Dwivedi *et al*., 1990; Singh *et al*, 1996 and Singh *et al*, 2002).Whereas, Singh *et al*, (1994) and Choudhary *et al*. (1997) found 15 % urea most effective for deblossoming of rainy season guava crop. However, Singh and Singh (1994) reported complete deblossoming with double spary of 15 % urea in guava to regulate the crop. However, single application of 25-30% urea virtually eliminated all the summer flower and flower buds within 12 days and drastically reduced the rainy season yield. Among the treatments, trees which received single treatment with 25 % urea produced higher yield in cv. Sardar, while in cv. Allahabad Safeda, two applications of 10 5 urea had the highest fruit yield closely followed by single treatment with 20% urea (Singh *et al*., 2002). Shigeura *et al*. (1975) in Hawaii and Chapman *et al*. (1979) in Australia used urea sprays (25%) as a defoliant on guava to change the yield pattern from scattered bearing to a concentrated peak harvest during the desired period. Addicot (1964) reported that most defoliants have no direct effect on the abscission zone, however, induces abscission by way of leaf injury which sets in motion the chain of events leading to leaf fall. This abscission may be caused by the direct effects of urea or by NH_4^+ released by urease (Singh *et al*., 2002).

Branch Bending

Guava trees in which shoots grow erect bear poorly (Singh, 1995). In this context, branch bending and pruning could be of great use. The cultural practices are known to induce the lateral shoot production by breaking the apical dominance and activating the latent buds present on the branch (Samant *et al*., 2016) which bears flowers and fruits. Branch bending treatment which was imposed in the month of May contains the growth (56.91cm) of newly emerged shoots as compare to control (68.10cm). The same treatment was found very effective in inducing flowering (46.27%) and recorded the highest yield with improved fruit quality parameters (Samant *et al*., 2016). Shoot bending is one of the ways to produce better quality fruits in the offseason (Sarker *et al*., 2005, Mamun *et al*., 2012). In case of bending of branch wood tension of branch is increased and phloem formation decreased. As a result photosynthetic product pass slowly from the shoots of bent branch as to the other parts, maintaining increased C: N ratio and induce more flowering and fruit set (Mamun *et al*., 2012).

Pomegranate

Pomegranate is an important fruit crop of arid and semiarid regions of the world. India is one of the leading producers of pomegranate in the world. Pomegranate fruit has a good consumer preference for its attractive, juicy, sweet acidic and refreshing arils. There is a growing demand for good quality fruits both for fresh use and processed products (Pal *et al.*, 2014). Under the tropical climate of South, Central and western India, with a mild winter, the growth and flowering continues throughout the year. There are three distinct flowering seasons, *i.e.* Ambe bahar (Jan.-Feb.), mrig bahar (June-July) and hasth bahar (Sept.–Oct.) and fruit ripen in the months of July-Aug, Dec-Jan and March-April respectively. Profuse flowering occurs in the monsoon season/ mrig bahar.In the subtropical climate of North India, the trees remain dormant during the cold winter and thus the main flowering season is in the following spring (Singh, 1969), although another flush of flowers is also observed during monsoon (Table 2). However, pomegranate may be induced to bear fruit in any one of the seasons which is determined by the demand, price, pest/disease incidence and the availability of the water in a given locality (Pal *et al.*, 2014).

Table 2: Flowering period of pomegranate in different parts of India (Pal *et al.*, 2014)

Region / state	Flowering period
Central India & Western India	Ambe bahar (Jan – Feb); Mrig bahar (June- July) Hasth bahar (Sep-Oct)
Karnataka	June –August (80-87 days); March (22-30 days) Sept. (20-30 days)
Punjab	April-June
Bihar	February – March; July–Aug.
Rajasthan	Mrig bahar (June – July). In Ambe & hasth bahar, fruit quality is poor
Himachal Pradesh	April - May

Mrig bahar is taken in the North-West arid region (Mann and Pareek, 1974) and dry landsof South India (Singh *et al.*, 1967) where water is scarce during the hot and dry summer. Ambe bahar is taken in South India and Dholka area of Gujarat (Singh *et al.*, 1967) in places where assured irrigation facility is available (Patil *et al.*, 2002). It is preferred in Maharashtra since the yield and quality of this bahar is better as the fruit development takes place during dry months, they develop an attractive colour and quality thus suitable for exports. Whereas mrig bahar, crop is severely damaged by pests and diseases and mature during winter, the colour and sweetness of the fruit is affected. Generally hasth bahar is not taken because, during the preceding monsoon season in July-August, it is not possible to withhold water to induce flowering in September leads to poor flowering and thus affects the yield. However, flowering can be induced in pomegranate during hasth bahar by foliar spray of suitable growth hormones at 2-4 weeks after defoliation (Babu and Pal, 2014).Raturi and Hiwale (1991) also found it highly profitable to take hasth bahar crop under rainfed conditions of south Gujarat. The fruits from hasta bahar are harvested during the month of March to April. They have very attractive rind with dark coloured arils. Since the availability of the fruits during this season is limited, they fetch high prices (Sonawane and Desai, 1989, Lal *et al.*, 2017).To take a bahar of choice, trees are given bahar treatment. In this treatment, the irrigation is withheld 50 days prior to the bahar (Sonawane and Desai, 1989) followed by shallow digging up to a depth of 10 cm in the basin. This facilitates the shedding of leaves. The trees are then medium pruned 40-45 days after withholding irrigation. The recommended doses of fertilizers are applied immediately after pruning and irrigation is resumed. This leads to profuse flowering and fruiting. Ahire *et al.* (1993) suggested use of PGRs for regulating flowering in Maharashtra. Among the PGRs tried, ethrel (500 ppm) was found most effective flower thinning agent which also induced more number of hermaphrodite flowers while Ethrel 200 ppm was found effective for fruit setting, yield and early ripening in Gujarat (Goswami *et al.*, 2013). Similarly, application of ethrel 250 ppm was found to be better for increasing the hermaphrodite flowers and also reducing the days taken for flowering in pomegranate var. Bhagwa in Karnataka (Anwal *et al.*, 2015). However, to get fruits of good size and quality, the flower buds in pomegranate cultivars 'Ganesh-I' and 'Kandhari' Singh *et al.* (2011) recommends manual removal of unopened flower buds from 15 April and 22 April onwards, respectively and to get higher net market returns (Singh *et al.* (2006) under north Indian conditions.

Citrus spp.

Citrus is the most economically important fruit crop in the world, is grown in developed and developing countries and certainly constitutes one of the main sources of vitamin C (Iglesias *et al.*, 2007). Acid lime (*C. aurantifo* Swingle) is the third important citrus crop in India next to mandarins and sweet oranges (Devi *et al.*, 2011). In subtropical regions, citrus major bloom occurs during the spring flush along with the vegetative sprouting. Under these environmental conditions, flowering takes place after a period of bud quiescence and the exposure to the low temperatures and short days of winter (Iglesias *et al.*, 2007). The importance of temperature as a major factor of flower induction is well established and has been recognized for a long time (Moss, 1969; Altman and Goren, 1974, Valiente and Albrigo, 2004, Nebauer *et al.*, 2006). Several workers have proposed that low temperatures may have a dual effect releasing bud dormancy and inducing flowering (Southwick and Davenport, 1986,García-Luis *et al.*, 1992). Moreover, temperatures below 20°C have been demonstrated to contribute to flower bud induction in a time dependent manner (Moss, 1969; Southwik and Davenport, 1986; García-Luis *et al.*, 1992). However, in tropics, bud sprouting and flowering does occur uninterrupted throughout the year (Table 3) although the main bloom still occurs during the spring (Monselise, 1985; Spiegel-Roy and Goldschmidt, 1996). In contrast to the sub-tropical stimuli, in tropical conditions citrus apparently flower in response to drought periods after the first effective rains of the rainy season (Chica and Albrigo, 2013, Cassin *et al.*, 1969). Moreover, water deficit has been proved to increase the ratio of floral shoots and the total number of flowers (Southwick and Davenport, 1986).

Table 3: Flowering period of acid lime in different parts of India

Region / state	Flowering period	Fruit maturity period	Reference
Tamil Nadu	Ambe bahar (Jan – Feb); Mrig bahar (June- July)	July – Aug. and Dec. – Jan.	Thirugnanavel *et al.* (2007)
Karnataka	Dec.-Jan. and July-Aug.	June-July and Jan.-Feb.	Athani *et al.* (1998)
Maharashtra	Dec.-Feb. and June-July	July – Aug. and Dec. – Jan.	Ghawede *et al.*(2002)
North India	Only once in spring season	Aug.-Sept.	Motial (1964) and Singh (1995)
Southern and Western India	Ambe bahar (Jan – Feb); Mrig bahar (June- July) and Hastha bahar (Sept.-Oct.)	June-July, Nov.-Dec. and April-May	Lal *et al.* (2017)

It is also important to note that many citrus species and varieties show a wide range of behaviors regarding both flowering time and response to the inductive conditions. Nir *et al.* (1972) obtained increased intensity of flowering due to stress which showed that flower differentiation occurred during moisture stress and that generative buds formed did not undergo flower development till water was supplied. This effect has been commercially used to induce off-season flowering (Davies and Albrigo, 1994). Goell *et al.* (1981) reported that the moisture stress followed by alleviation was effective in initiating and promoting vegetative flushing. The flowering was delayed under longer period of stress owing to conditions like high and low temperature and low humidity condition. Singh and Chadha (1988) advocated that imposition of stress caused uniformity in flushing and intensity depended on amount of stress as measured by relative water content (RWC) in acid lime plant before alleviation. Stress level of92.26 ± 0.89 or 90.97 ± 1.12 RWC causes high intensity of flowering. Flower promoting genes expressed in sweet orange (Chica and Albrigo, 2013) and lemon (Li *et al.*, 2017). A total of 1638 differentially expressed genes (DEGs) were identified in lemon by RNA sequencing. DEGs were related to flowering, hormone biosynthesis, or metabolism (Chica and Albrigo, 2013). Results Chica and Albrigo (2013) suggest that water deficit induces flowering through the upregulation of

CsFT and that CsFTis the leaf integrator off lower-inducing signals generated by the exposure to water deficit and cool temperatures in sweet orange. Therefore, endogenous expression of CiFT is closely correlated with flowering under various conditions, suggesting that endogenous CiFT may regulate floral induction in citrus (Nishikawa Nishikawa *et al.*, 2007, 2009, 2011, 2013, Nishikawa, 2013).

Girdling and ringing treatments have been used in commercial citrus production to enhance flowering for many years which suppress the growth of vegetative shoots and roots, and promote floral induction in citrus (Goldschmidt *et al.*, 1985; Inoue *et al.*, 1991; Iwahori *et al.*, 1990). Fruit removal also promotes floral induction (Nishikawa *et al.*, 2012). In satsuma mandarin, the presence of fruit on the tree results in a gradual inhibition of floral induction. However, early thinning of fruit increases flower number in the following spring (Garcia-Luis *et al.*, 1986) and the number of fruit on the tree is negatively correlated with flower number in the following year (Nishikawa *et al.*, 2012). The effects of factors such as defoliation, girdling, and fruit removal seem to be associated with nutritional status (Nishikawa, 2013) as defoliation diminishes supply of carbohydrate and fruit removal involves the loss of a major carbohydrate sink and accumulation of carbohydrate in roots (Okuda *et al.*, 2002). Girdling causes accumulation of carbohydrate within the girdled branch (Goldschmidt *et al.*, 1985). Thus, these phenomena indirectly demonstrate the involvement of carbohydrate in flowering.

The citrus productivity depends on various factors, among these the plant growth regulators holds a prime position. The use of plant growth regulators is one of the important management practices in the field of citriculture because of the wide range of potential roles they play in increasing the productivity of cropper unit area (Bons *et al.*, 2015). Desai *et al.* (1982) sprayed cycocel (1000 ppm) on 16th August and repeated on 16th Sept. followed by spray of2,4,5-T (10 ppm) on 30thSept. Resulted in increased percent flowers (58.2%) as against control (16.3%) in acid lime. Whereas, Prasad *et al.* (1980) reported that foliar application of 1000 g N/tree with GA at 150 ppm resulted in the greatest number and size of leaves, fruit set, retention and yields in acid lime trees. Babu and Rajput (1982) recorded earliest flowering with 2, 4-D (10-20 ppm) which was delayed with GA_3 (25-50 ppm). However, duration of flowering was shortest (22-24 days) with GA_3 (50 ppm) and longest in the controls (30-35 days) whereas Davenport (1983) reported that GA_3 markedly inhibited flowering in Tahiti lime (*Citrus latifolia*), producing morphologically typical vegetative growth. Monselise and Halevy (1964) reported that gibberellic acid could inhibit flowering in 'Shamouti' orange. However, the triazole compounds that inhibit gibberellic acid biosynthesis have been reported to promote inflorescence production

(Delgado *et al.*, 1986) as Devi *et al.*, obtained higher yield by application of PBZ (5ml m^{-1} canopy) as compared to control in acid lime. Tripathi and Dhakal (2005) reported that paclobutrazol reported that the earliest (July) application of paclobutrazol was superior among the treatments under Chitwan, Nepal condition to induce and advance early flowering in lime for off season market. Thirugnanavel *et al.* (2007) revealed that application of GA 3 50 ppm in June+cycocel 1000 ppm in September + KNO3 2% in October showed better performance in delaying flowering by nearly two months. Similarly, Mahalle *et al.* (2010) reported that two sprays of cycocel (1000 ppm) at an interval of one month before initiation of flowering that is in August and September resulted in maximum yield in terms of number of fruits and weight of fruits $tree^{-1}$ and this treatment also improved the fruit quality in respect to juice %, TSS, acidity, and ascorbic acid content. For regulating hasta bahar lime crop, the foliar spray with GA_3 50 ppm June + Cycocel 1000 ppm September is the most promising and economically viable technology for maximizing the fruit yield, quality and benefit cost ratio (2.43) for the acid lime growers during summer in Andhra Pradesh (Mukunda *et al.*, 2014). Out of the different bio regulators *viz.*, bromouracil, 2, 4-D and paclobutrazol in variable doses in different agro climatic condition are effective for inductionand regulation of the acid lime (Devi *et al.*, 2011). The concentration of 600 ml^{-1} ethephon applied for two consecutive years was the most suitable to regulate the Ponkan mandarin production (Moriera *et al.*, 2011). Among the PGRs used to promote the thinning, Ethephon has been considered more efficient in comparison to other plant growth regulators (Domingues *et al.*, 2001). However, research findings on PGRs can vary due to cultivars, concentrations and climate changes, making it difficult to extrapolate the results for specific situations (Moriera *et al.*, 2011).

Other Tropical Fruits

Most mango cultivars bear irregularly having distinct 'on' and 'off' years (Pandey, 1989). The timing of flush development is importantfor successful flowering because bud release, for vegetative or reproductive growth, can only occur from mature flush (Nunez-Elisea and Davenport, 1995). Treatments such as pruning that manipulate the timing of flush development and synchronize canopy flushing have been successful in increasing flowering intensity (Yeshitela *et al.*, 2005). Potassium nitrate is thought to induceflowering (Bondad and Apostol, 1979), but seems to be ineffective in many environments (Davenport and Nunez-Elisea, 1997).GA biosynthesis inhibitors such aspaclobutrazol (Rademacher, 1995) hasten and increase the flowering intensity of mango (Kulkarni, 1988, Blaikie*et al.*, 2004, Abdel Rahim, *et al.*, 2011, Gopu and Balamohan, 2015) and also reduce vegetative vigour (Winston, 1992). The application of paclobutrazol followed by application of a dormancy-breaking substance has forced mango, citrus and mangosteen trees to produce off-season flowers. However, the same treatment given to durian and rambutan trees has not shown any satisfactory results. It is possible to induce flowering in durian by the application of paclobutrazol in combination with the application of plastic mulching to prevent infiltration of rainfall into the soil. In rambutan, ringing treatments induced the trees to flower and produce fruit off-season. The ringed trees started to flower 4 months earlier than the control trees without any effect on the size and number of inflorescences, and the number of flowers per inflorescence (Poerwanto *et al.*, 2008). Ethylene has long been used to promote flowering commercially in pineapple (Turnbull*et al.*, 1999). Studies of Potchanasin *et al.* (2009) suggest that cytokinins play a significant role in $KClO_3$induced"off season" floral induction in longan. Jia *et al.* (2009) identified regulatory genes implicated in continuous flowering of longan. Out of different genes they observed ELF4may be a keygene for this trait. The identification of candidate genes related to continuous flowering will provide new insight into the molecular process of regulating flowering time in woody plants.

Potential Benefits of Crop Regulation in Fruit Crops

Cultural and hormonal manipulations for flower induction and fruit setting have been adopted commercially in many fruit crops for various reasons. These are to regulate a uniform and good quality fruits and induce regular flowering and fruiting in alternate bearing cultivars of many fruit crops (Wilkie *et al.*, 2008, Devi *et al.*, 2011, Boora *et al.*, 2016, Lal *et al.*, 2017). In addition to this to get harvesting season extended (Poerwanto *et al.*, 2008), reduce the attack of insect-pest and diseases (Khan *et al.*, 2011, Lal *et al.*, 2017) and increase the nutritional quality, profitability and environmental sustainability (Singh *et al.*, 2000, Das *et al.*, Devi *et al.*, 2011, Maji *et al.*, 2015, Lal *et al.*, 2017).

Future Scope and Research Need

Fruit production is seasonal which leads glut in the market therefore, fetches poor prices and sometime also leads higher fruit spoilage too. This seasonal supply is also limited to few days in litchi to few months in other fruit crops. Therefore, through adoption of various flower forcing techniques at commercial scale, harvesting season can be extended. Thus focus of research should be given in this direction for most of the fruit crops so that fruit availability could be increased during the lean periods in India. Physiology and molecular biology of crop regulation is extensively studied in many fruit crops like mango, apple, citrus, longan, litchi etc. however, attention in this direction needs to be given for guava and pomegranate for better management and understanding in these crops.

Conclusion

The investigations on crop regulation showed that shoot pruning and application of various chemicals were effective to regulate flowering in guava by deblossoming of summer flowers and to produce fruits in winter. Among various chemicals used NAA (600-800 ppm) and urea (10-25%) could be widely used in different agro-climatic conditions for regulating cropping in guava. In case citrus and pomegranate water stress followed by watering was found key to crop regulation in tropics, however, spray of ethephon was found equally good as flower and fruit thinning agent. Similarly in mango, paclobutrazol can be commercially used for regulating the crop.

References

Abdel Rahim, A.O.S., Elamin,O.M. and Bangerth, F.K. 2011. Effects of paclobutrazol (pbz) on floral induction and associated hormonal and metabolic changes of biennially bearing mango (*Mangifera indica* L.) cultivars during off year. *J. Agric. Biol. Sci*., 6(2):55-67.

Adhikari, S. and Kandel, T.K. 2015. Effect of time and level of pruning on vegetative growth, flowering, yield, and quality of guava.*Int. J. Fruit Sci*., 15(3):290-301.

Ahire, G. Z., Desai, U.T. Choudhari, S.M. and Masalkar, S.D. 1993. Crop Regulation in Pomegranate: I. Effect of Growth Regulators on flower induction, sex expression and flower drop. *Ann. Arid*, 32 (2):97-98.

Altman, A. and Goren, R 1974. Growth and dormancy cycles in Citrus bud cultures and their hormonal control. *Physiol. Plant*., 30:240-245.

Anawal, V.V., Narayanaswamy, P. and Yallesh Kumar H.S. 2015. Effects of plant growth regulators on induction of flowering in pomegranate (*Punica granatum* L.) cv. Bhagwa *Int. J. Scient. Res*., 4(12):7-9.

Atawia, A.A.R., El-Gendy, F.M.A., Bakry, Kh. A.I., Abd El-Ghany, N.A. and Singer, M.A.A. 2017. Physiological studies on flowering and fruiting of guava trees. *Middle East J. Agric*., 6(1): 143-151.

Athani, S.I., Hulamani, N.C. and Patil, M.P. 1998. Flowering and fruitset behaviour of kagzi lime strains. *Adv. Agric. Res*., 9: 151-53.

Bardhan, K. 2016. Crop Regulation: Concepts and its Commercial Uses. In: *Commercial Horticulture*. Patel, N.L., Chawla, S.L. and Ahlawat, T.R (Eds.), New India Publishing Agency, New Delhi, pp. 257-275.

Babu, R.S.H. and Rajput, C.B.S. 1982.Effect of Zinc, 2, 4-D and GA3 on flowering in Kagzilime. *Punjab Hort*. J., 22: 140-44.

Bajpai, P.N., Shukla, H.S. and Chaturvedi A.M. 1973. Effect of pruning on growth, yield and quality of guava (*Psidium guajava* L.) var. Allahabad Safeda. *Prog. Hort*., 5: 73- 79.

Blaikie, S.J., Kulkarni, V.J. and Muller, W.J. (2004). Effects of morphactin and paclobutrazol flowering treatments on shoot and root phenology in mango cv. Kensington Pride. *Sci. Hortic*., 101:51–68.

Boora, R.S., Dhaliwal, H.S. and Arora , N.K. 2016. Crop regulation in guava-A review.*Agric. Rev*., 37 (1):1-9.

Bondad, N.D. and Apostol, C.J. 1979. Induction of ûowering and fruiting in immature mango shoots with KNO3. *Curr. Sci*., 48: 591– 593.

Bons, H.K. Kaur, N. and and Rattanpal, H.S. 2015. Quality and quantity improvement of citrus: role of plant growth regulators. *Int. J. Agric. Env. Biotechnol*., 8(2): 433-447.

Cassin, J., Bourdeaut, J., Fougue, A., Furon, V., Gaillard, J.P., LeBourdelles J., Montagut, G. and Moreuil C. 1969. The influence of climate upon blooming of citrus in tropical areas. *Proc. 1st Int. Citrus Symp*. 1:315-323.

Chica, E.J., and Albrigo, L.G. 2013. Expression of flower promoting genes insweet orange during floral inductive water deficits.*J. Am. Soc. Hortic. Sci.*,138:88–94.

Chandra, R. and Govind, S. 1995. Influence of time and intensity of shoot pruning on growth, yield and fruit quality of guava under high density planting. *Trop. Agric*., 72: 110-13.

Chapman, K.R., Saranah, J. and Paxton, B. 1979. Induction of early cropping of guava seedlings in closely planted orchard using urea as a defoliant. *Aust. J. Agric. Anim. Husb*., 19:382-384.

Chundawat, B.S., Gupta, O.P. and Godara, N.R. 1975. Crop regulation in Banarsi Surkha guavacultivar. *Haryana J. Hort. Sci*.,4:23-25.

Choudhary, R., Singh, U.P. and Sharma, R.K. 1997. Deblossoming of rainy season crop of guava. *Hort. J*., 10:93-95.

Das, B., Nath, V, Jana, B.R., Kumar, S. and Dey, P. 2007. Evaluation of crop regulation in guava grown under rainfed plateau conditions of eastern India. *Indian J. Hort*., 64(3): 294-299.

Davis, F.S. and Albrigo, L.G. 1994. Citrus. In: Crop ProductionScience in Horticulture Series, pp.134-135. CABInternational, Wallington.

Delgdado, R., Casamayor, R., Rodriguez, J.L., Cruz, P. and Fajardo, R 1986. Paclobutrazol effects on oranges under tropical conditions. *Acta Hortic*., 179: 545-48.

Desai, U., Choudhari, K.G., Rane, S.D. and Patil, Y.S. 1982. Alteration of flowering period in kagzi lime with plant growth substances. *J. Maharashtra Agric. Univ*., 7: 161-62.

Dhaliwal, G.S. and Singh, R. 2004. Effect of different pruning levels on vegetative growth, flowering and fruiting in Sardar guava. *Haryana* J. *Hort. Sci*., 33 (3-4): 175-177.

Davenport, T.L. and Nunez-Elisea, R. 1997. Reproductive physiology. In: *The mango: botany, production and uses. Wallingford.* Litz RE (ed.), CAB International, 69–146.

Devi, H.L., Srkar, S.K., Dhanbati, L. and Majhi, D. 2011. Flushing – flowering behavior and regulation in acid lime – a critical review and research interventions. *J. Crop and Weed*, 7(2):87-90.

Domingues, M.C.S., Ono, E.O. and Rodrigues, J.D. 2001. Reguladores vegetais e o desbaste químico de frutos de tangor Murcote. *Sci. Agri*., 58(3):487-490. 2001

Dwivedi, R., Pathak, R.K. and Pandey, S.D. 1990. Effect of various concentrations of urea on crop regulation in guava(*Psidium guajava*L.) cultivar Sardar. *Prog. Hort*., 22:134-139.

Garcia-Luis, A., Almela, V., Monerri, C., Agustí, M.and Guardiola, J. 1986. Inhibition of flowering in vivo by existing fruits and applied growth regulators in Citrus unshiu. *Physiol. Plant*., 66: 515–520.

Garcia-Luis, A., Kanduser, M., Santamarina, P and Guardiola, J.L. 1992. Low temperature influence on flowering in Citrus. The separation of inductive and bud dormancy releasing effects. *Physiol. Plant*., 86:648-682.

Ghawade, S.M., Panchbhai, D.M., Jadhao, B.J., Katole, S.R. and Dadmal, S.M. 2002. Flowering behaviour of kagzi lime (*Citrus aurantifolia*Swingle) under sub-humid tropicalclimate.*Agric. Sci. Digest*., 22: 252-254.

Goldschmidt, E.E., Aschkenazi, N., Herzano, Y.,. Schaffer, A.A and Monselise, S.P. 1985. A role for carbohydrate levels in the control of flowering in citrus. *Sci. Hortic*. 26: 159–166.

Gopu, B. and Balamohan, T.N. 2015. Crop regulation using chemicals for flowering and fruit set in mango *(Mangifera indica* L.) cv. Alphonso under ultra high density planting. *Trends in Biosciences*, 8(2):436-441.

Goswami, J.D., Patel, N.M., Bhadauria, H.S. and Wankhade, V.R. 2013. Effect of plant growth substances on growth, fruit setting and yield of pomegranate cv. Sinduri. *Int. J. Agric. Sci*., 9(1): 332-334.

Gupta, M.R. and Nijjar, G.S. 1978. Crop regulation in guava.*Indian J. Hort*., 35: 23-27.

Hojo, R.H., N.N.J. Chalfun, E.T.D. Hojo, R.D. Veiga, C.M. Paglis and L.C. deO-Lima. (2007). Production and quality of guava fruits 'Pedro Sato' submitted to different pruning times. *Pesq. Agropec. Brasil*., 42: 357-362.

Iglesias, D.J., Cercos, M., Comenero-Flores, J.M., Naranjo, M.A., Rio, G., Carrea, E., Rivero, O.R., Lliso, I., Morillon, R., Tadeo, F.R. and Talon, M. 2007. Physiology of citrus fruiting. *Braz. J. Plant Physiol.*, 19(4):333-362.

Inoue, H., Ikoma, Y. and Kataoka, I. 1991. Effects of ringing and temperature on flower bud differentiation of Satsuma mandarin. *J. Japan. Soc. Hort. Sci*., 60: 275–284.

Iwahori, S., Garcia-Luis, A., Santamarina, P., Monerri, C.and Guardiola, J.(1990). The influence of ringing on bud development and flowering in satsuma mandarin. *J. Exp. Bot*. 41: 1341–1346.

Jadhav, B.J., Damke, M.M., Mahorkar, V.K., Dod, V.R. and Wagh, A.P. 1998. Studies on effect of time and severity of pruning on growth and yield of guava (Psidium guajava L.) cv. Sardar. *J. Soils Crops*, 8 (2): 139-141.

Jia, T., Wei, D., Meng, S., Allan, A.C. and Zeng, L. 2014. Identification of regulatory genesimplicated in continuous flowering of longan(*Dimocarpus longan* L.). *PLOS ONE*, 9(12):e114568. doi:10.1371/journal.pone.0114568.

Joshi, P., Lal, S., Nautiyal, P. and Pal, M. 2016. Effect of Plant spacing and pruning intensity on flowering, fruiting, and yield of guava (*Psidium Guajava* L.) cv. Pant Prabhat. *Int. J. Agric. Sci*., 8(7):1064-1068.

Khan, A.S., Khan, M.R.G., Malik, A.U., Saleem, B.A., Rajwana, I.A. and Ahmad, I. 2011. Influence of defoliation and deblossoming on the vegetative and reproductive growth of guava (*Psidium guajava* L.) cv. Gola. *Pak. J. Bot* ., 43(6): 2891-2896.

Krajewski,A. J. and Rabe, E. 1995. Citrus flowering: a critical evaluation. *J. Hort. Sci*.,70: 374.

Kulkarni, V.J. 1988. Chemical control of tree vigour and the promotion of flowering and fruiting in mango (Mangifera indica L.) using paclobutrazol. *J. Hortic. Sci*., 63:557–566.

Kumar, R. and Hoda, M. N. 1977. Crop regulation studies in Allahabad Safeda guava. *Indian J. Hort*.,34:13-14.

Kumar, R. and Mishra, D.S. 2010. Amrud ki labhkari kheti. *Kisan Bharti*, 41(10):5-6.

Kumar, R. and Mishra, D.S. 2012. Guava: A health capsule. *Indian Farmar's Digest,* 45(8):30-32.

Lal, N., Sahu, N., Marboh, E.S., Gupta, A.K. and Patel, R.K. (2017). A review on crop regulation in fruit crops. *Int. J. Curr.Microbiol. App.Sci.,* 6(7):4032-4043.

Lal, S. 1983. Effect of pruning on crop regulation in guava (*Psidium guajava* L.) cv. Lucknow-49. *Prog. Hort*., 7(3): 60-62.

Lal, S., Tiwari, J.P. and Misra, K.K. 2000. Effect of plant spacing and shoot pruning intensity on fruit yield and quality of guava. *Prog. Hort*., 32: 20-25.

Li, Jin-Xue, Hou, Xiao-Jin, Zhu, J., Zhou, Jing-Jing, Huang, Hua-Bin, Yue, Jian-Qiang, Gao, Jun-Yan, Du, Yu-Xia, Hu, Cheng-Xiao, Hu, Chun-Gen and Zhang, Jin-Zhi. 2017. Identification of genes associated with lemon floral transition and flower development during floral inductive water deficits: A hypothetical model. *Front. Plant Sci*., https://doi.org/10.3389/fpls.2017.01013.

Lotter, J. de V., 1990. Vegetative and reproductive habit of the guava (*Psidium guajava* cv. Fon Retief) in relation to pruning methods. *Acta Hort*., 275: 229-237.

Mahalle, S.S., Ingle, H.V. and Sable, P.B. 2010. Influence of plant growth regulators and chemicals on yield and quality of Hasta bahar in acid lime. *Green Farm*, 1: 285- 87.

Maji, S., Das, B.C. and Sarkar, S.K. 2015. Efficiency of some chemicals on crop regulation of Sardar guava. *Sci. Hortic*., 188:66-70.

Mamun, A.A., Rahman, M.H. and Farooque, A.M. 2012. Effect of bending and fruit thinning for offseason production of guava. *J. Agrof. Environ*., 6(1):111-116.

Mamun, A.A., Rahman, M.H. and Farooque, A.M. 2012. Effect of shoot bending and fruit thinning on productivity of guava. *J. Environ. Sci. Nat. Resour.*, 5(2): 167-172.

Mann, H.S. and Pareek, O.P. 1974. Potentialities of growing fruit in the hot arid zone. *Indian Hort.*, 19: 1-3.

Mishra, D.S.and Tiwari, J.P. 2000. Guava: a potential fruit of subtropical and tropical India. *Swarna Sedyam,* 3(9):47-49.

Mitra,S.K. 1983. Studiesonthe physiologyof growth, floweringand fruit growthin guava(*Psidium guajava* L.), Ph.D Thesis, submittedto Bidhan Chandra Krishi Viswavidyalaya, Nadia, West Bengal.

Mitra, S.K; Sen, S.K; Maity, S.C and Bose, T.K. 1982. Effect of growth substances on deblossoming,regulation of cropping and fruit quality in guava. *Hort. J.*, 1: 81-88.

More, D.B., Raut, U.A. and Bongane, Y.S. 2016. Effect of Different Growth Regulator on Vegetative Growth, Flowering and Yield of Winter Season Guava cv. Sardar. *Adv.Life Sci.*, 5(12): 5064-5066.

Monselise, S.P. 1985. Citrus and related species. In: Halevy AH (ed), CRC Handbook of Flowering, Vol. 2, pp.275-294. CRC Press, Boca Raton.

Monselise, S.P. and Halevy, A.H. 1964. Chemical inhibition and promotion of citrus flower bud induction. *Am. Soc. Hort. Sci.*, 84:141-46.

Moss, G.I. 1969. Influence of temperature and photoperiod on flower induction and inflorescence development in sweet orange (*Citrus sinensis* L. Osbeck). *J. Hort. Sci.*, 44:311-320.

Motial, V.S. 1964. Fruit set and fruit drop studies in some non-commercial citrus fruits. *Punjab Hort. J.*, 4: 27-33.

Mukunda, L.L, Ramana, K.T.V., VNP Sivarama Krishna, Yuvaraj, K.M., Lakshmi, T.N., Sarada, G., Sankar, T.G., Gopi, V. and Gopal, K. 2014. Effect of growth regulators and chemicals on fruit yield and quality of hasta bahar flowering in acid Lime (*Citrus aurantifolia* Swingle) cv. Balaji. *Res. Rev. J. Agric. All. Sci.*, 3(3):11-13.

Narayana, C.K., Singh, Gorakh and Singh, A.K. 1999. Cellulase enzyme activityand abscission in guava (*Psidium guajava* L.) as influenced by different crop regulators. *Ann. Agric. Res.* 20: 51-55.

Nebauer, S.G., Avila, C., García-Luis A. and Guardiola, J.L. 2006. Seasonal variation in the competence of the buds of three cultivars from different Citrus species to flower. *Trees*, 20:507-514.

Nishikawa, F. 2013. Regulation of Floral Induction in Citrus. *J. Japan. Soc. Hort. Sci.*, 82 (4): 283–292.

Nishikawa, F., Endo, T., Shimada, T., Fujii, H., Shimizu, T.and Omura, M. 2009. Differences in seasonal expression of flowering genes between deciduous trifoliate orange and evergreen satsuma mandarin. *Tree Physiol.*, 29: 921–926.

Nishikawa, F., Endo, T., Shimada, T., Fujii, H., Shimizu, T., Omura, M and Ikoma, Y. 2007. Increased CiFT abundance in the stem correlates with floral induction by low temperature in satsuma mandarin (*Citrus unshiu* Marc.). *J. Exp. Bot.*, 58: 3915–3927.

Nishikawa, F., Iwasaki, M., Fukamachi, H.and Endo, T. 2013. Leaf removal suppresses citrus FLOWERING LOCUS T expression in satsuma mandarin. *Bull. Natl. Fruit Tree Sci.*, 15: 1–6.

Nishikawa, F., Iwasaki, M., Fukamachi, H., Nonaka, K., Imai, A.and Endo, T. 2011. Seasonal changes of citrus FLOWERING LOCUS T gene expression in kumquat. *Bull. Natl. Fruit Tree Sci.*, 12: 27–32.

Nishikawa, F., M. Iwasaki, H. Fukamachi, K. Nonaka, A. Imai, Takishita, F.Yano, T.and Endo, T. 2012. Fruit bearing suppresses citrus FLOWERING LOCUS T expression in vegetative shoots of satsuma mandarin (*Citrus unshiu* Marc.). *J. Japan. Soc. Hort. Sci.* 81: 48–53.

Nir, I., Goren, R. and Leshem, B. 1972. Effects of water stress, gibberellic acid, 2-chloroethyltrimethylammonium chloride (CCC) on flower differentiation in 'Eureka' lemon trees. *J. Am. Soc. Hort. Sci.*, 97:774-778.

Nunez-Elisea, R. and Davenport, T.L. 1994. Effect of leaf age, duration of cool temperature treatment, and photoperiod on bud dormancy release and floral initiation in mango. *Sci. Hortic.*, 62:63–73.

Okuda, H., Noda, K.., Kihara, T. and Hirabayashi, T. 2002. Fine root volume, flowering, sprouting, and carbohydrate content in the leaves and roots of bearing and non-bearing satsuma mandarin trees in a systemized alternate-bearing orchard. *J. Japan. Soc. Hort. Sci.*, 71: 588–590.

Pal, R.K. Babu, K.D., Singh, N.V., Maity, A. and Gaikwad, N. 2014. Pomegranate Research in India – Status and future challenges. *Prog. Hort.*, 46(2):184-201.

Pandey, R. M., Lal, S. and Kaul, G. L. 1980. Effect of chemicals and flower thinning on regulationof crop in guava.*Indian J. Hort.*, 37:234-239.

Pandey, RM. (1989). Physiology of flowering in mango. *Acta Hort.*, 231:361–380.

Patil, A.V., Karale, A.R. and Bose, T.K.2002. Pomegranate. In : *Fruits : Tropical and Subtropical*, Vol. II. Bose, T.K., Mitra,S.K. and Sanyal, D. (Eds.), Naya Udyog, Calcutta, pp.125-162.

Poerwanto, R., Efendi, D., Widodo, W.D., Susanto, S. and Purwoko, B.S. (2008). Off season production of tropical fruits. *Acta Hort.*, 772:127-133.

Potchanasin, P., Sringarm, K., Naphrom, D. and Bangerth, K.F. 2009. Floral induction in longan (*Dimocarpus longan*, Lour.) treesIV. The essentiality of mature leaves for potassium chlorate induced floralinduction and associated hormonal changes. *Sci. Hortic.*, 122:312-317.

Prakash, S., Kumar, V., Saroj, P.L. and Sirohi, S.C. 2012. Response of yield and quality of winter guava to severity of summer pruning. *Indian J. Hort.*, 69(2): 173-176.

Pratibha and Lal, S. (2013). Response of shoot pruning and planting systems on guava cv. Sardar under Tarai conditions. *Pantnagar J. Res.*, 11(1):75-82.

Quijada, O, Araujo, F.and Corzo, P. 1999. Effect of pruning and hydrogen cyanamide on bud break, flowering, fruit yield and quality of guava (*Psidium guajava* L.) in the municipality of Mara, state of Zulia. *Revista de la Facultad de Agronomia, Universidad del Zulia.*, 16(3): 276-291.

Rademacher, W. 1995. Growth retardants: biochemical features and applications in horticulture. *Acta Hortic.*, 394:57–73.

Rathore, D. S. 1975. Deblossoming of rainy season crop of guava by NAA. *Prog. Hort.*, 7:63-65.

Rathore DS, Singh RN, 1974. Flowering and fruiting in the three cropping patterns of guava. *Indian J. Hort.* 33: 331-336.

Raturi, G.B. and Hiwale, S.S. 1991. Regulation of cropping in pomegranate under rainfed conditions. *Indian J. Hort.*, 48(1): 13-15.

Rodrigo A. Moreira, R.A., Ramos, J.D., Cruz, M.C.M., Neimar A. de Araújo and Marques, V.B. 2011. Production of 'Ponkan' mandarin trees submitted to chemical thinning. *Rev. Bras. Ciênc. Agrár.*, 6(2):287-291.

Sachin,A.J., Ramteke, V. and Bharath, K.R. 2015. Flower Regulation in Guava and Pomegranate. *Pop. Kheti*, 3(4): 53-54.

Sahar, A.F and Hameed, A.A.A. 2014. Effect of pruning on yield and fruit quality of guava trees. *J. Agric. Vet. Sci.*,7(12): 41-44.

Salah, A. El-D.M. 2005. Effect of pruning on growth, flowering and fruiting of some guava cultivars. M.Sc. Thesis, Fac. Agric. Cairo Univ.

Sarkar, A., Ghosh, B. Kundu, S. and Sukul, P. 2005. Effect of shoot pruning and bending on yield and fruit quality in guava cv. L-49. *Env. and Ecol.*, 23S: Special 3: 621-623.

Samant, D., Kishore, K. and Singh, H.S. 2016. Branch bending for crop regulation in guava under hot and humid climate of eastern India. *J. Indian Soc. Coastal Agric. Res*., 34(1): 92-96.

Serrano, L.A.L, Marinho, C.S., Gabetto, E., Silva, M.and Tardin, F.D. 2008a. Phenological and yield characteristics of "Paluma" guava tree pruned in different times and intensities, in north of Rio de Janeiro State, Brazil. *Revista Ceres*., 55(5): 416-424.

Serrano, L.A.L; Martins, M.V.V.,Lima, I. De M., Marinho, C.S. and Tardin, F.D. 2008b. Effect of pruning time and intensity on Paluma' guava trees, in Pinheiros, ES, Brazil. *Revista Brasileira de Fruticultura*, 30(4): 994-1000.

Sharma, A., Wali, V.K., Sharma, R.M., Jamwal, M. and Sharma, B. 2013. Effectiveness of Various Crop Regulation Treatments in Guava (Psidium guajava) cv. L- 49. *Madras Agric. J.*, 100 (1-3): 747-750.

Shigeura, G.T., Bullock, R.M. and Silva, J.A. 1975. Defoliation and fruit set in guava.*Hort. Sci*., 10:590.

Singh, G. and Singh, A. K. 1994. Urea induced defoliation and subsequent yield in guava.*Fertilizer News*, 39:39- 41.

Singh, G., Singh, A.K. and Verma, A. 2000. Economic evaluation of crop regulation treatments in guava (*Psidium guajava*). *Indian J. Agric. Sci*., 70(4):226-230.

Singh, G., Singh, A.K., Rajan, S. and Bhriguvanshi, S.R. 2002. Strategy of crop regulation in guava (*Psidium guajava* L.) through foliar urea sprays and its effect on different N-forms in leaves. *J. Appl. Hort.*, 4(2):93-98.

Singh, G., Pandey, D., Rajan, S. and Singh, A.K. 1996. Crop regulation in guava through different crop regulating treatments. *Fruit*(Paris), 51:241-246.

Singh, G., Rajan,S., Pandey, D.and Singh,A.K., 1997. Effect of soil moisture stress on water relation by plant and cropping behaviour in guava (Psidium guajava L.). *Indian J. Agric. Sci*., 67:303-306.

Singh, G., Rajan, S. and Singh, A.K., 2001. Influence of pruning date on fruit yield of guava(*Psidium guajava* L.) under subtropics. *J. Appl. Hort*., 3(1): 37-40.

Singh, G., and Reddy, Y.T.N. 1997. Regulation of Cropping in guava. *Indian J. Hort*., 54:44-49.

Singh, H.J. and Bal, J.S. 2006. Effect of pruning and growth regulators on physio-chemical characters of guava during rainy season planted at different spacing. *Intr. J. Agric. Sci.*, 2(2): 533-537.

Singh, H.P. and Chadha, K.L. 1988. Regulation of flushing and flowering in acid lime (*Citrus aurantifolia* Swingle) through stress management. *Prog. Hort*., 20: 1-6.

Singh, N.P., Dhillon, W.S. and Gill, P.P.S. 2011. Quality improvement studies in pomegranate under sub-tropics of India. *Acta Hort*. 890: 363-369.

Singh, N.P., Dhillon, W.S., Sharma, K.K. and Dhatt, A.S. 2006. Effect of mechanical deblossoming on flowering and fruitset in pomegranate cvs. Ganesh-1 and Kandhari.*Indian J. Hort.*, 63(4): 383-385.

Singh, R.N. 1969. Fruits, National Book Trust, New Delhi, pp.135-74.

Singh, S.; Krishnamurthi, S. and Katyal, S.L. (1967). Fruit culture in India, ICAR, New Delhi. pp. 207-14.

Singh, S.P. 1995. Commercial Fruits, Kalyani Publ., New Delhi, pp. 148-162.

Sonawane, P.C. and Desai, U.T. 1989. Performance of staggered cropping in pomegranate. *J. Maharashtra Agric. Univ.*, 14:341-342.

Southwick, S.M. and Davenport, T.L. 1986. Characterization of water stress and low temperature effects on flower induction in Citrus. *Plant Physiol*., 81:26-29.

Spiegel-Roy, P. and Goldschmidt, E.E. 1996. Biology of Citrus. Cambridge University Press, Cambridge

Stern, R.A., Adato, I., Goren, M., Eisenstein, D. and Gazit, S. 1993. Effects of autumnal water stress on litchi flowering and yield in Israel. *Sci. Hortic*., 54: 295–302.

Suresh, M., Kumar, S. and Swaminathan, V. 2016. Crop Regulation In Guava (*Psidium Guajava* L) A Review. *Agric. Sci.*, 5(12):458-459.

Thakre, M. B.; Lal, S.; Uniyal, S.; Goswami, A. K.; Prakash, P. 2016. Pruning for crop regulation in high density guava (*Psidium guajava* L.) plantation. *Spanish J. Agric. Res.*, 14(2)1-8. http://dx.doi.org/10.5424/ sjar/2016142-7846.

Thakre, M. B., Lal, S., Goswami, A. K. and Prakash, P. 2013. Effect of various methods of crop regulation in guava under double-hedge row system of planting. *Indian J. Hort.*, 70(2):211-216.

Thirugnanavel, A. Amutha, R., Rani W.B., Indira, K., Mareeswari, P., Muthulaksmi, S. and Parthiban, S. 2007. Studies on Regulation of Flowering in Acid Lime (*Citrus aurantifolia* Swingle.). *Res. J. Agric. Biol. Sci.*, 3(4): 239-241.

Tiwari, J.P. and Lal, S. 2000. Amrud ki kheti tatha bagh prabandh. Dir. Publ., GBPUAT, Pantnagar, pp. 47.

Tiwari, J.P. and Lal, S. 2007. Effect of NAA, flower bud thinning and pruning on crop regulation in guava (*Psidium guajava* L.) cv. Sardar. *Acta. Hort.*, 735:311-314.

Tiwari, J. P. and Lal, S 1984. Research Report. ICAR Workshop on Fruit Research, Lucknow.

Tiwari, R. B.; Tiwari J.P and Lal, S. (1992). Effect of shoot pruning, NAA and urea on cropping pattern of guava. *Indian J. Hort.*, 49: 305-308.

Tripathi, K.M. and Dhakal, D.D. 2005. Effect of paclobutrazol on off-season flower induction in acid lime (*Citrus aurantifolia*, Swingle) land races under Chitwan condition. *J. Inst. Agric. Anim. Sci.*, 26:87-92.

Turnbull, C.G.N., Sinclair, E.R., Anderson, K.L., Nissen, R.J., Shorter, A.J. and Lanham, T.E. 1999. Routes of ethephon uptake in pineapple (*Ananas comosus*) and reasons for failure of ûower induction. *J. Plant Growth Regu.*, 18: 145–152.

Valiente, J.I., Albrigo, L.G. 2004. Flower bud induction of sweet orange trees [*Citrus sinensis* (L.) Osbeck]: effect of low temperatures, crop load, and bud age. *J. Am. Soc. Hort. Sci.*, 129:158-164.

Wilkie, J.D., Sedgley M. and Olesen, T. Regulation of floral initiation in horticultural trees. *J. Exp. Bot.*, 59(12):3215–3228.

Winston, E.C. 1992. Evaluation of paclobutrazol on growth, ûowering and yield of mango cv. Kensington Pride. *Aust. J. Exp. Agric.*, 32: 97–104.

Yeshitela, T., Robbertse, P.J. and Stassen, P.J.C. 2005. Effects of pruning on ûowering, yield and fruit quality in mango (*Mangifera indica*). *Aust. J. Exp. Agric.*, 45, 1325–1330.

26

Soil Health Card and its Importance for Getting Higher Income

M.L. Soni, V. Subbulakshmi, V.S. Rathore, Birbal, N.S. Nathawat K.R. Sheetaland N.D. Yadava

Introduction

Agriculture is the mainstay of the Indian population and contributes about 30 per cent to the gross domestic product of India. The government has laid great emphasis onimproving productivity, quality of produce and welfare of farmers through awareness campaign and enhanced agri-knowledge for the farming community to double the farmer's income by 2022 and make the farmers realise their significant contribution to the nation building. Doubling the farmer's income can only be possible, with proper diversification of crops, developing the existing irrigation system in proper utilization of available water on plant requirement basis, conservation of soil fertility through adding organic manures to the field, best utilization of human excreta from the storage tanks of individual toilets after proper decomposition through bacterial degradation, fulfilling the nutrient requirement of crops on the basis of soil test values, proper financial support from the Government to the farmers in managing stress due to natural calamities etc. The governmentbelieves that farmers' welfare will improve if there is increase in net income from the farms. With this end in view, the approach is to reduce cost of cultivation, enable higher yield per unit area and realize remunerative prices of farm produce. To achieve this goal, action on three pillars needs to be taken.

1. The cost of production should be reduced and productivity should be increased.
2. The farmers need to diversify their crops and adopt other agricultural activities such as animal husbandry, poultry, goat farming, bee keeping, and timber plantation.

3. The farmers get regulated markets close to their farms to sell their produce and earn profitable returns.

Fertilizer usage contributed to about 50 per cent of increased food grains production in the world (Hegde and Sudhakarbabu 2004; Tanwar and Bisvas 2005). In India, fertilizer consumption is concentrated in about one-third of the cultivated area. It is key element to increase sustainable production of agriculture (Painulyand Dev, 1998). Still, the average yield per hectare of crops in India is low due to exhausted soils which have been over cropped from centuries without adequate replenishment for plant nutrients through fertilizers. Such heavy removal of plant nutrients from soil leads to depletion of soil fertility, which shows up in crop yield decline and lowered factor productivity (Yadav *et at.*,1998). Therefore, application of fertilizers in adequate amount is essential to prevent soil degradation, keeping agriculture land productive and economically viable. To reduce the cost of production and increase the productivity, the Government has launched the ambitious scheme of soil health card across the country to advise the farmers on the dosage of fertilizers and the soil amendments, that he should apply to maintain soil health in the long run and increase the income from his produce.

Soil Health Card (SHC) Scheme is a flagship program of the Government of India promoted by the Department of Agriculture, Co-operation & Farmers Welfare under the Ministry of Agriculture & Farmers Welfare to issue SHC to farmers. It is a printed report about the nutrient status for each of his holdings. The implementation is being done through the Department of Agriculture of all the State and Union Territory Governments. This card contains the status of 12 soil parameters, namely nitrogen (N), phosphorus (P), potassium (K), sulphur (S), zinc (Zn), Iron (Fe), copper (Cu), manganese (Mn), boron (B), soil organic carbon (OC), pH and electrical conductivity (EC). The card provides an advisory based on the soil nutrient status of a farmer's holding. SHC help the farmers to fulfil the nutrients demand of crop as per their requirement, improve productivity, increasing income and maintaining soil health. SHCs also promote the judicious use of the fertilizers thus reducing the cost of production for the farmer. It recommends ways to improve the soil quality, appropriate dosage of nutrients for improving soil health, its fertility and analyse the nutrient level. Every two years, the soil health will be reviewed to examine nutrients deficiencies so that corrective measures can be taken accordingly. Some states are already issuing SHCbut, it was found that, there was no uniform norm for sampling, testing and distribution of SHC across the states. The salient features of SHC include adoption of uniformed approach towards samples collection and testing in laboratories, one coverage for all the land in the country, and issue SHC every 2 years.The scheme has been approved for implementation during 12th Plan with an outlay of Rs.5685.4 million. In order to improve quality of soil and ultimately for better nutrient values and higher yields, experts say while at present, general fertilizer recommendations are followed by farmers for primary nutrients, the secondary and micro nutrients are often overlooked. This has become a limiting factor in increasing food productivity.

National Informatics Center (NIC) has developed a web portal (www. soilhealth. dac. gov.in) for generation of uniform SHC and fertilizer recommendation across the country, which has four modules:

a) Registration of soil samples.

b) Testing of samples in soil testing laboratory.

c) Fertilizer recommendation based on Soil Test Crop Response (STCR) equations.

d) Final Reports.

Importance of Soil Health Card (SHC)

Fertilizers are one of the costly inputs but continue to exert significant contribution to produce additional food grains for the ever-increasing population. The green revolution demanded use of chemical fertilizers for increasing production. However, mindless use of fertilizers led the soil in many regions useless for farming. This is mainly because of unwise distortion of the ideal NPK (nitrogen, phosphorous, potassium) proportion. The existing N: P: K consumption ratio in the country is skewed at 6.7:2.4:1 as against the preferred ratio of 4:2:1 (Table 1). A great variability is observed in fertiliser consumption among states from 250 kg ha $^{-1}$ in Punjab, 212 kg ha^{-1} in Bihar, 207 kg ha^{-1} in Haryana to 4.8 kg ha $^{-1}$ in Nagaland and 2 kg ha^{-1} in Arunachal Pradesh, during 2012-13. India is spending nearly Rs. 7,00,000 million on fertilizer subsidy every year. According to the estimates, subsidy amount is about Rs.5000/ha of net cropped area and about Rs.5100/farmer resulting in excessive use of fertilizers, especially NPK at the cost of micro-nutrients and manure. The imbalanced application of fertilisers have caused deficiency of primary nutrients (*i.e.* NPK), secondary nutrients (such as sulphur), and micronutrients (boron, zinc, iron, manganese, copper etc.), in most parts of country. For example, in 2011, the ratio of N:P:K consumption was 19.2:5.5:1 and 20.6:6:1 respectively in Punjab and Haryana. The over-use of Urea is mainly because it comes with a government subsidy. Excessive use of N reduces the health of crop and productivity goes down year after year. With the intensive cropping of high yielding varieties of rice and wheat, deficiency of Zn initially, and subsequently deficiencies of Fe in rice, and Mn in wheat, emerged as threats to sustaining high levels of food crop production. Micronutrient deficiencies are now frequently observed in intensively grown cereals, oilseeds, pulses and vegetable crops. With widespread and regular application of Zn fertilizers, the occurrence of Zn deficiency has declined in recent years, but multi micronutrient deficiencies are now becoming an increasing problem. Studies conducted in different parts of India showed that the scenario of micronutrient deficiency in north India in early eighties was different than now after four decades. Zinc deficiency has increased from 44% to 48% and expected to further increase up to 63% by 2025 as most of the marginal soils are showing higher response to added zinc (Singh 2006). Data of 5823 trials indicated a shift in crop response from <0.2 t ha^{-1} in 44-46% trials during 1980's to response of 0.2-0.5 t ha^{-1} recorded in 40 - 65% trials after nineties. Some of the state government has promoted regular use of 15-25 kg zinc sulphate ha^{-1} to paddy, wheat and other crops instead of multi micronutrient mixtures. This decision overcame the deficiency of zinc. As a result of this, the consumption of zinc sulphate during the years 1988 to 2008 has increased from 16292 to 48,000 t yr^{-1} in Punjab, 1692 to 11470 t yr^{-1} in Haryana and 1752 to 22600 t yr^{-1} in Uttar Pradesh. Zinc deficiency which was a dominant problem in 77, 69 and 53% soils in Haryana, Punjab and Uttar Pradesh respectively before 1980 has declined to 19, 22.6 and 39, which in turn improved the zinc fertility of soils, crop yields and its content in seeds. In middle India prevalence of Zn deficiency was higher 63 and 54% in Madhya Pradesh (M.P.) and Bihar compared to 26 - 36% in Gujarat, and West Bengal. Zinc deficiency remained similar over the four decades despite micronutrient fertilization. Even after four decades, little change in zinc deficiency 63.7, 57 and 36.5% for M.P, Bihar and Gujarat soils is mainly attributed to the inherent low organic matter, high $CaCO_3$, and cultivation of soybean - gram , wheat crops. Deficiencies of Cu, Fe and B in north India zone are not widely observed. However, in calcareous soils of western Rajasthan occurrence of Fe deficiency is prevalent (Chattopadhyay *et al.*, 1997; Soni *et al.*, 2006) and the crops like groundnut and mungbean suffers from Fe deficiency. In calcareous soils of Bihar, management of multi nutrient deficiencies (Zn, B, S) is of utmost important for sustaining higher crop yields. These soils have marginal status of Fe, Cu, Mn .Similarily, manganese (Mn) deficiency emerged as a serious problem in wheat (,Nayyar 1983; Nayyar *et al.*, 1985; Soni *et al.*, 1996) and other winter crops in coarse textured alkaline soils. In rice-wheat cropping system, submergence of rice solubilizes more soil Mn which gets leached down to lower layers (Soni *et al.*, 1997) so next wheat crop badly suffered with severe Mn deficiency as wheat seedling roots are not able to absorb Mn from lower soil horizons (Singh, 2006). Application of manganese sulphate @ 25 kg ha^{-1} as soil application or 0.5% foliar spray correct the deficiency of Mn and improve the crop yield (Soni *et al.*, 1996). Some crop rotations also help in combating Mn deficiency. Deficiency of Mn did not appear in maize - wheat system in similar soils so diversification of crops or cultivars is very much helpful in reducing Mn deficiency in wheat grown in these soils. Hence, there is a need for balanced use of fertilizers, keeping this government of India introduced SHC scheme across India.

Table 1: Consumption of Fertilisers in million tonnes in India

	Million tonnes				NPK ratio		
Year	N	P	K	Total	N	P	K
1956	0.1	0.0	0.0	0.1	10.4	1.3	1
1981	3.7	1.2	0.6	5.5	5.9	1.9	1
1991	8.0	3.2	1.3	12.5	6.0	2.4	1
2001	10.9	4.2	1.6	16.7	7.0	2.7	1
2011	16.6	8.0	3.5	28.1	4.7	2.3	1
2015	16.9	6.1	2.5	25.6	6.7	2.4	1

Aim and Objectives

- To issue SHCs every 3 years, to all farmers of the country, so as to provide a basis to address nutrient deficiencies in fertilization practices, improve soil quality and profitability of farmers.
- To strengthen functioning of Soil Testing Laboratories (STLs) through capacity building, involvement of agriculture students and effective linkage with Indian Council of Agricultural Research (ICAR) / State Agricultural Universities (SAUs).
- To diagnose soil fertility related constraints with standardized procedures for sampling uniformly across states and analysis and design taluka / block level fertilizer recommendations in various district.
- To develop and promote soil test based nutrient management in the districts for enhancing nutrient use efficiency.
- To build capacities of district and state level staff and of progressive farmers for promotion of nutrient management practices.
- Provide soil testing facilities to farmers at their doorstep.
- Online delivery of SHC to the farmers using SHC portal.

History of Soil Testing in India

India's soil testing program began in 1955–1956 with the establishment of 16 soil testing laboratories under the Indo-US operational agreement for the determination of soil fertility and fertilizer use. The program benefited from a substantial revival in the 11th five-year plan (2007–2012), when the "National

Project on Management of Soil Health and Fertility" was launched with an outlay of Rs 4298.5 million (approximately US$65 million at the prevailing exchange rate) to set up new laboratories and strengthen existing laboratories with micronutrient testing facilities. The program was expanded further in the 12th five-year plan, when all states adopted the system of preparing and issuing soil analysis–based SHCs to farmers along with associated fertilizer use recommendations. The recently launched flagship program of SHC will give a major fillip to on-going soil testing efforts. This program aims to provide SHCs to all farmers across the country every three years. States like Gujarat have already implemented a similar program whereby farmers received crop-specific fertilizer application recommendations for all plots of land in the state. Essentially, the SHC scheme is modeled on a successful programme launched by state government of Gujarat.

At present, about 1454 labs exist in India, of which only 700 are equipped with micro-nutrient testing facilities. Very few of the labs could take up micro nutrient analysis. They are neither equipped with skilled personnel or chemicals nor functional equipment. This infrastructure is grossly inadequate by any standard, given that 110 million farmers need to be covered. Under the current PPP model, investments in labs to be done by private companies with an element of subsidy. The quality of such reports should be checked at random by authentic agencies.

Gaps and Need of SHC

Past strategy for development of the agriculture sector in India has focused primarily on raising agricultural output and improving food security. The net result has been a 45 per cent increase in food production per capita, which has made India not only food self-sufficient at aggregate level, but also a net food exporting country. The strategy did not explicitly recognise the need to raise farmers' income and did not mention any direct measure to promote farmers welfare. The net result has been that farmers income remained low, which is evident from the incidence of poverty among farm households. Doubling real income of farmers till 2022-23 over the base year of 2015-16, requires annual growth of 10.41 per cent in farmer's income. This implies that the on-going and previously achieved growth rate of farm income has to be sharply accelerated. Therefore, strong measures will be needed to harness all possible sources of growth in farmers' income within as well as outside agriculture sector. One of such measures is balance use of soil nutrients on soil test basis. The promotion of balanced chemical fertilizers has become a necessity to meet the ever-increasing demands of food grains, pulses, oil crops, sugarcane production, fruits, vegetables, fibres, forage and grasses.

The farmers give top priority to urea, without knowing that the soil may require more phosphorous, sulphur or micronutrients like zinc, iron, manganese, copper or boron. It has been proved that imbalance use of fertilizer not only causes deterioration in soil quality but also affected nutrient use efficiency (NUE). Soil forms the basis for any crop production activity and is the most precious natural resources. Declining soil fertility is one of the important factors that directly affect the productivity. Fertilizer recommendations based on average nutrient concentration of the area have widely been adopted over three to four decades. Therefore, soil fertility management is crucial to ensure productivity and nutritional security while maintaining soil health and sustainability. To get a maximum benefit, enhanced NUE and reduced nutrient losses from fertilizers, they must be applied in the right quantity, sources and combination at the right time using the right methods. The package of practices recommended should cater to the need of soil variability. The response to fertilizers is greatly influenced by the soil type and spatial soil variability that has resulted from complex geological and pedological processes. Spatial variation of soil properties decreases the use efficiency of fertilizers applied uniformly at the field scale. Therefore, application of variable rates of fertilizers has been proposed. Applying fertilizer, best quality seeds and ample water is not enough. Farmers should nurture their soil and know what inputs to use and in what quantities. Nutrients available in soil are rarely present in adequate amounts and in balanced proportion to meet the nutrient requirement of the crops. In order to attain this, it is essential that the amount of nutrients removed from the soil should be replenished through judicious use of fertilizers and manures. This needs a more comprehensive approach for fertilizer use, incorporating components like soil test, field research and economic evaluation of the results. Soil test provides the requisite information about the amounts of nutrients available in the soil and their imbalances, while fertilizer recommendations aim at correcting the imbalances in nutrients according to crop requirements.

Norms of Sampling

The criteria of soil sampling are in a grid of 2.5 ha in irrigated area and 10 ha in rain- fed area with the help of GPS tools and revenue maps. The State Government may collect samples through the staff of their Department of Agriculture or through the staff of an outsourced agency. Based on the grid system, the total 141 million hectares (m ha) of the net cropped area requires a total of 34.6 million grid samples [with 52% of the area under rainfed (73 mha) with 7.3 million grid samples and with 48% area under irrigated (68 m ha) with 27.2 million grid samples]. The target is to cover 17.3 million grid samples per year. This comes to on average 25000 grid samples per district/year and 29 grid samples per village/year. With this, all 103.9 million farmers will be covered in two years. Every year 52 million farmers need to be covered.

The State Government may also involve the students of local Agriculture / Science Colleges. Soil samples are taken generally two times in a year, after harvesting of rabi and kharifcrop respectively or when there is no standing crop in the field. Soil samples will be collected by a trained person from a depth of 15-20 cm by cutting the soil in a "V" shape. It will be collected from four corners and the centre of the field and mixed thoroughly and a part of this picked up as a sample. Areas with shade will be avoided. The sample chosen will be bagged and coded. It will then be transferred to soil test laboratory for analysis. The soil sample will be tested as per the approved standards for all the agreed 12 parameters.

Cross Checking: Mechanisms are put in place for random checking of 1 % of the total samples will be analyzed by external agencies for ensuring the quality of soil analysis. A team of Deputy Director of Agriculture (Ext.), District Agriculture Officer, Assistant Director of Agriculture (STL) and a Soil Scientist from SAU/ ICAR would jointly monitor, inspect and evaluate the functioning of STLs/other concerned agencies pertaining to the issue of SHCs to the respective districts. The joint certificate will be issued from District Agriculture Officer, Sarpanch and Gram Sevak regarding satisfactory sample collection work.

Constraints/Challenges in Effectively Implementing the Scheme

For effective implementation of the SHC project however there are certain hurdles putting bottlenecks on the way of the progress of this scheme. A few of such handicaps are given below:

1. Shortage of trained/skilled/technical manpower for analysis work and proper maintenance of soil testing lab (STL).
2. Many of the equipements in state laboratories are old and non-functional which need immediate replacement and proper care by technical specialists.
3. There are0.1 m ha of saline soils in the Haryana State only which need reclamation. There is similar type of cases where soil is not worth agriculture productivity.
4. Delay in transporting samples from collection centres to labs.
5. Power shortage and interruption in power supply
6. Lack of availability of Funds for this purpose
7. Need of IT trained manpower for interpretation of analysis reports and preparation of recommendations and their online communication.
8. Short term hiring and out sourcing is one of the ways to carry out work. Such persons would need intensive training before undertaking the task.

9. Soil experts and farmers feel that the present scale (2.5 ha grid for irrigated areas and 10 ha grid for the rainfed area) is not representative of all the soil types. Given the high fragmentation in India, each farmer may have number of fragments that too not at one place, farmers want more soil samples to be collected and tested for more reliable results. Number of soil samples per unit area should be based on soil variability. In most of the developed countries, soils are tested as per the requirement by crop.

10. A comparison of indicators included in SHC in India with other countries like the USA (across its states) indicates that only nutrient availability along with some other soil characters like pH and EC are included in India while number of states in USA use many other indicators like physical, topography, water, soil indicators, biological indicators and nutrient holding capacity. Some of these indicators like water quality, soil erosion, slope, crop residue, *etc.*, are very important for proper soil management. Experts feel that soil texture, direction of slope, water quality, irrigation source, cracks on soil surface, depth of the soil, crop residues can be included in the SHC.

How does SHC help the farmers

- With the issue of the card, the farmers will get a well-monitored report of the soil which is chosen for cultivation of crops.
- The monitoring will be done on a regular basis.
- The farmers will be guided by experts to come up with solutions to improve the quality of the soil.
- Regular monitoring will help the farmers to get a long-term soil health record and accordingly can study and evaluate the results of different soil management practices.
- This card can become most helpful and effective when filled out regularly by the same person over a period of time.
- The idea is not to compare the varied soil types but to find out methods to improve soil fertility, to access the different types of soil and their ability to support crop production in spite of their limitations and as per their abilities.

SHC will ensure that farmers do not spend money unnecessarily on purchase of fertilizers by adding more than required. Once there is economy on the use of chemical fertilizers, the cost of production is expected to decrease. Promotion of integrated nutrient system is expected to reduce the consumption of chemical fertilizers by 20%thereby easing the fiscal strain on Government. Fertiliser sector accounts for a significant percentage of the total subsidies and power consumption in the country. India imports large quantity of various fertilizers to meet the demand. About 25-30 % of the requirement of Urea, 90% of the requirement of Diammonium phosphate (DAP) and 100% of the requirement of Muriate of Potash (MOP) are met by imports. The soil test based fertilizer usage will reduce import bill and will also ensure higher yields per unit. Over a period of time SHC can determine changes in soil health that are affected by land management.

Basis of Recommendations by SHC

The recommendation of SHC is based on either Soil Test-Crop Response (STCR) formulae developed by ICAR or general fertilizer recommendations provided by state governments. The soil testing services is closely related to fertilizer use efficiency. It identifies soil specific requirements of different nutrients and suggests the recommendations of nutrients and fertilizers required for the individual crop in the form of SHC.

Appropriate and efficient use of fertilizers for getting the maximum crop yield is possible through target oriented fertilizer scheduling which is based on the principles of balanced fertilization of crop (Ramamoorthy *et al.*, 1967; Kanwar, 1971). Besides, in this approach it is also possible to make fertilizer recommendations to the farmers considering their financial conditions (Sekhon*et al.*, 1976). Khosa *et al.* (2012) also reported the superiority of the target yield concept over other practices for different crops as it gave higher yields and optimal economic returns. The specific yield equation based on soil health besides ensuring sustainable crop production also steers the farmers towards economic use of costly fertilizer inputs depending on their financial status and prevailing market price of the crop under consideration (Bera *et al.*, 2006).

The fertilizer prescription equations are developed for different soils, different varieties and for different regions by AICRP on soil test crop response correlations. They also include IPNS based adjustment equations using different type of local available organic sources. Substitution of soil test values in the prescription equations gives us that particular fertilizer nutrient recommendation which need to be converted in to fertilizer quantity to be applied (using conversion factor). In case of IPNS based equations, FYM composition and the rate are to be substituted in the place of FYM given in the fertilizer prescriptions. These fertilizer recommendations are grouped in two groups mainly food grain crops and horticultural crops. Ready reckoners for some of the crops are given in Table 2-4. STCR report can be consulted for a number of crops including, medicinal, forage, vegetables and horticulture crops.

Table 2: Ready Reckoner of fertilizer doses at varying soil test values for specific yield target for clusterbean

Soil available nutrient (kg ha^{-1})			Fertilizer nutrient required (kg ha^{-1}) for yield target of clusterbean					
$KMnO_4$-N	Olsens'P	Amm.Ac.-K	15 q ha^{-1}			18 q ha^{-1}		
			N	P_2O_5	K_2O	N	P_2O_5	K_2O
80	15	170	44	39	15	60	54	30
90	20	190	39	27	15	55	42	23
100	25	210	35	20	15	51	30	16
110	30	230	30	20	15	46	20	15
120	35	250	26	20	15	42	20	15
130	40	270	21	20	15	37	20	15
140	45	290	16	20	15	32	20	15
150	50	310	12	20	15	28	20	15
160	55	330	7	20	15	23	20	15

Soil testing laboratory : ARS, Bikaner
Soil : Sobhasar, Khiran, Gajnar, Bhamatsar series
Crop and variety : Guar (RGC-986)
Target range : 15-18 q ha^{-1}
Soil nitrogen range : 80-160 kg ha^{-1}
Soil phosphorus range : 15-55 kg ha^{-1}
Soil potassium range : 170-330 kg ha^{-1}

Fertilizer adjustment equation
$FN = 5.38T - 0.46\ SN$
$F\ P_2O_5 = 5.07\ T - 2.46\ S\ P_2O_5$
$F\ K_2O = 4.86\ T - 0.34\ S\ K_2O$

In the equation of STCR, integrated plant nutrient supply (IPNS) system has also been taken into consideration. The nutrient supplied through FYM /compost was adjusted while supplying the inputs and thus farmers can save the money by saving of costly fertilizer inputs and increase his income. Table 3 shows the same targeted yield in IPNS system with less fertilizer input as compared to sole fertilizer application. The experiment was conducted at Rahuri on tomato crop.

Table 3: Fertilizer prescription for targeted yields of tomato for varying soil test values

Soil test values (kg ha^{-1})			Without FYM			With FYM (20 t FYM ha^{-1})		
			30 t ha^{-1}target			30 t ha^{-1} target		
N	P	K	N	P_2O_5	K_2O	N	P_2O_5	K_2O
100	6	200	114	91	105	70	53	52
120	8	300	105	84	80	61	47	30
140	10	400	96	75	55	52	42	25*
160	12	500	86	66	30	44	36	25*
180	14	600	77	55	25*	35	30	25*
200	16	700	68	50	25*	27	25	25*
220	18	800	59	42	25*	18	19	25*
Targeted Yield Equations			FN = 5.33 T – 0.46 SN FP_2O_5 = 3.88 T– 4.16 SP FK_2O = 5.16 T – 0.25 SK			FN = 4.13 T – 0.43 SN – 1.13 FYM FP_2O_5=2.5 T – 2.78 SP – 0.57 FYM FK_2O = 3.44 T – 0.22 SK – 0.75 FYM		

* Minimum dose of K_2O

Table 4 shows the varying nutrient levels for different yield targets in Isabgol after adjusting the nutrients supplied by FYM and soil. Farmers can save their money when nutrients are applied on soil test basis.

Table 4: Ready Reckoner of fertilizer doses for Isabgol at varying soil test values for specific yield target

Soil test values (kg ha^{-1})			Fertilizer nutrient required (kg ha^{-1}) with 5 t ha^{-1} FYMfor yield target of					
KMn O4N	Olsens' P2O5	Amm. Ac.- K2O	10 q ha^{-1} target			12 q ha^{-1} target		
			N	P_2O_5	K_2O	N	P_2O_5	K_2O
70	20	170	49	34	36	68	47	50
90	25	190	42	30	34	61	43	48
100	30	210	39	26	31	58	39	45
110	35	230	36	22	29	54	35	42
120	40	250	32	19	26	51	32	40
130	45	270	29	15	23	48	28	37
140	50	290	26	11	21	45	24	35
150	55	310	23	7	18	41	20	32
160	60	330	19	3	16	38	16	29
Fertilizer adjustment equation			FN = 9.35T – 0.33 SN – 0.65 ON FP2O5 = 6.53 T – 0.76 S P2O5 – 0.86 OP2O5 FK2O = 6.95 T – 0.13 SK2O – 0.47 OK2O Where T= Targeted yield F= Nitrogen required through chemical fertilizer SN, S P_2O_5, SK_2O= respective nutrient present in soil ON, O P_2O_5, OK_2O= respective nutrient present in organic manure					

The above fertilizer adjustment equations were tried on the farmers' fields in Bikaner district with varying yield targets during Rabi 2007-08 and all the yield targets could be achieved at the place tried

Conclusions

Soil Health Card assists in judicious use of fertilizers and improves yield and income of the farmers. However, farmers must be taken into confidence while collecting the soil samples from their crop field and proper monitoring of the work of the agency associated with collection of soil samples and distribution of card is essential. Adequate focus should be given on infrastructure development of soil labs in every state. The high fertilizer subsidy for NPK should be reduced and subsidy on use of recommended dose of micro-nutrients, bio-fertilizers and organic inputs should be encouraged to improve soil health on long term basis.

References

Bera, R., Seal, A. , Bhattacharyya, P. , Das, T.H. , Sarkar, D. andKangjoo, K. 2006. Targeted yield concept and a framework of fertilizer recommendation in irrigated rice domains of subtropical India. *Journal of Zhejiang University-Science,* 7(12): 963-968.

Chattopadhyay, T. , Singh, R. S., Sahoo A. K . and Shyampura, R. L. 1997. Available micronutrient status of Rajasthan soils. *Agropedology* 7:40-43.

Hedge, D.M. and Sudhakarbabu S.N. 2004. Role of balanced fertilization in improving crop yield and quality.*Fertilizer News,* 49(12): 113-114.

Kanwar, J.S. 1971. Soil testing in India- Prospects and Retrospect. Proceedings of International Symposium on Soil fertility Evaluation held at New Delhi. 1:1103-1113.

Khosa, M.K., Sekhon, B.S., Ravi, M.S., Benipal, D.S. and Benbi, D.K. 2012. Performance of target yield based fertilizer prescription equations in rice-wheat cropping system in Punjab, *Indian Journal of Fertilizer* 8(2): 14-18.

Nayyar, V.K. 1983. Manganese deficiency in Punjab soils.*Progressive Farming*19, 18–20.

Nayyar, V.K., Sadana, U.S. and Takkar, T.N. 1985. Methods and rates of application of Mn and its critical levels for wheat following rice on coarse textured soils. *Fertilizer Research*8 (2): 173–178.

Painuly, J.P. and Dev, S.M. 1998. Environmental dimensions of fertilizer and pesticide use: Relevance to Indian agriculture. *International Journal of Environmental and Policy*10(2): 273-288.

Ramamoorthy, B., Narshaman, R.L. and Dinesh, R.S. 1967. Fertilizer application for specific yield targets for Sonara-64. *Indian Farming* 17(5):43-45.

Sekhon,G.S., Singh, B. and Deol, P.S. 1976. Experience of targeting crop yield in Punjab.*Fertilizer News*21 (11):41-45.

Singh, M.V. 2006. Micronutrients in crops and in soils of India.In. Micronutrients for global crop production. (Ed. BJ Alloway), Springer.Business

Soni, M.L. Swarup, A. and Singh, M. 1996. Influence of rates and methods of manganese application on yield and nutrition of wheat in a reclaimed sodic soil . *Journal of Agriculture Science.Camb.* 127: 433-439.

Soni, M.L. Swarup, A. and Singh, M. 1997. Behaviour of native and applied manganese during reclamation of a reclaimed sodic soil. *Journal of Indian Society of Soil Science.* 45(2) 349-359.

Soni, M.L., Beniwal, R.K., Mondal,B.C. and Yadava, N.D. 2006. Micronutrient status of arid soils under three predominant land use systems in arid zone. *Current Agriculture* 30: 53-56.

Tanwar, S.P.S. and Biswas, B.C. 2005. Strategies for food security and agricultural sustainability in changing scenario. *Fertilizer News,* 50(2): 25-28.

Yadav, R.L., Yadav, D.S., Singh, R.M. and Keme, A. 1998. Nutrition recycle. *Agroecosystems* 51: 193–200.

27

IPM Strategies in Arid Horticultural Crops for Getting Higher Income Returns

S.M. Haldhar

Introduction

It is a well known fact that agricultural pests cause substantial crop losses throughout the world. In the past farmers had to manage this problem to secure their basic subsistence needs, and as a response, the farmers practiced and developed cultural and mechanical pest control based on trial and error. Over a period of time, these practices have become a part of their production management system. The first known chemical control dates back to 2,500 years. From the late 1800s to 1940s, the main insecticidal compounds used were oils, soaps and resins, plant-derived poisons and inorganic compounds. The chemical age began in the 1940s with the discovery of dichloro diphenyl trichloroethane (insecticide), ferbam (fungicide), and 2, 4-D (herbicide) (Arneson and Losey, 1997). In 1940s and 1950s, pesticides were thought to be the final word in pest control and their introduction contributed substantially to raising agricultural productivity in many regions of the world. Since then, pesticides have become an integral component of many intensive agricultural systems. Repeated applications of pesticides lead to resistance in the pest (Beaument, 1993). This resistance results in the increased application of pesticides and to the collapse of the agricultural systems characterized by highly resistant pests, with no natural enemies left to control them. Pesticides have been found to causeacute and chronic human health problems, contamination of groundwater, surface water, atmospheric contamination, and negative effects on non-target organisms (Mullen, 1995). Concerns about the negative effects of pesticides led to research and promotion of alternative pest control practices – Integrated Pest Control or simply IPC. This new concept called Integrated Pest Control

(IPC) and later Integrated Pest Management (IPM) was stimulated by symposia organized by the Food and Agriculture Organization (FAO) of the United Nations in 1966. IPM made a paradigm shift in the philosophy of pest control, from pest eradication to pest management. Instead of single tactic control, emphasis was placed on the use of a combination of available tactics in a compatible manner aimed at providing cheap, long term sustainability with minimum of harmful side effect and controls the pest at ETL level.

India is bestowed with a wide variety of agro-climatic conditions and has unique comparative advantage for growing almost all horticultural crops (fruits and vegetables) in the country. Horticulture has remained as a major sub-sector of agriculture and forms an integral part of food and nutritional security, employment generation and enhancing the farm income. The growing economic importance of these crops could be attributed to increasing demands in domestic as well as in overseas market. Remarkable increased in fruit and vegetable crops cultivation under arid region of western India lead to drastic changes in the pest population dynamics. The fruits and vegetables such as ber (*Ziziphus mauritiana*), date palm (*Phoenix dactylifera*), aonla (*Emblica orficinalis*), lasora (*Cordia myxa*), phalsa (*Grewia sp.*), ker (*Cappari deciduas*), kinnow (*Citrus reticulata*), bael (*Aegle marmelos*), watermelon/*Mateera* (*Citrullus lanatus*), *Kachri* (*Cucumis callosus*), snap melon/*Phoot/ kakadi* (*Cucumis melo* var. *momordica*), long melon/*Kakri* (*Cucumis melo* var. *utilissimus*), round melon/*Tinda* (*Praecitrullus fistulosus*), cluster bean (*Cyamopsis tetragonoloba*), bottle gourd (*Lagenaria siceraria*) and ridge gourd (*Luffa acutangula*) are an important fruit crops grown in this area (Haldhar *et al.*, 2013a). The productivity is concern, it is low because of harsh environment (high temperature, low precipitation, high PET, high wind velocity), poor soil fertility and salinity. Beside these, the biotic pressure like pests also plays a critical role, which cause significant loss and adds cost of production. Hence, here we discussed the present status of insect-pests of these arid horticultural crops and their IPM strategies and this information would be highly useful to the farmers of this region for successful management of different insect-pests of arid horticulture crops.

Present Status of Integrated Pest Management (IPM)

A comprehensive IPM program has been worked out in many countries to test the knowledge of the grower to identify the important pests and natural enemies in the agri-horti-eco-system. Although, several advanced technologies have been developed but the extent of their diffusion and rate of adoption among farming communities are very slow. Technology such as high density planting, meadow plantings, micro-propagation technique of planting materials, drip irrigation to enhance the water use efficiency and special technologies for problems areas have not been adopted by the masses. Thus, there is a need to strengthen the means of technology transfer and improve the lab to land programme. ETL for all major pests should be standardized and this scientific information should be disseminated among farmers. But, due to more complexity in determining the ET level, it has been replaced with Agro-Ecosystem Analysis in modern agriculture. Preventive methods comprises cultural practices like summer and pre monsoon ploughing, suitable crop rotation, incorporation of intercrops, use of resistant /tolerant varieties, proper water management by judicious irrigation while intervention methods mainly focused on mechanical, biological and chemical control.

Changing Pest Scenario in Arid Horticulture

In recent past, many new cultivars have been introduced to increase the production growth rate of fruits. These intensive, high yielding hybrids with good agronomic traits and yield attributes have gradually changed the pest situation. Consequently, a shift in pest status has been realized in time and space. Many pests have adapted new hosts, developed resistance to pesticide and often there are secondary out breaks. The stone weevil, *Abeus himalayanus* is recorded as a new pest of ber for the first time from Andhra Pradesh state of India and recently in Bikaner district of Rajasthan during 2010 (Karuppaiah *et al*., 2010; Haldahr *et al*., 2013). Earlier shoot gall maker and *Deudorix isocrates* of aonla known to be minor pest is gradually becoming a regular problem in aonla, whereas, fruit fly severity of occurrence and infestations have been intensified due to climatic change.

A. IPM of Ber Crop

The ber (*Ziziphus mauritiana* Lamark) also called as desert apple tropical fruit tree species, belonging to the family *Rhamnaceae* (Balikai *et al*., 2013). Balikai (1999) reported total of 22 insect and non-insect species in Karnataka. However, the pests such as fruit fly, *Carpomyia vesuviana* Costa, chafer beetle, *Holotrichia consanguinea* Blanch and bark eating caterpillars, *Indarbela tetraonis* Moore; *Indarbela quadrinotata* Walker; ber butterfly, *Tarucus theophrastus* Fabricius and stone weevil, *Aubeus himalayanus* Voss are the major pests of ber in India which causing the significant yield loss (Karuppaiah *et al*., 2010; Haldhar *et al*., 2012; Haldhar *et al*., 2016).

Fruitfly, *Carpomyia vesuviana* Costa (Tephritidae: Lepidoptera)

Fruitfly, *Carpomyia vesuviana* Costa (Diptera: Tephritidae) is the most destructive pest of ber in India. It is a monopahagous pest, infests only on

Zizyphus species and contributes towards low yield and poor quality of fruits (Balikai *et al.*, 2013). The pest causes the yield loss up to 80% under severe infestation (Batra, 1953). The infestation found to be occurs in all the wild and cultivated species *viz.*, *Z. zizyphus, Z. mauritiana, Z. spina-cheresti, Z. numularia, Z. lotus, Z. jujube, Z. sativa* and *Z. rotuntifolia.*

The adult fly lays their eggs singly in the young developing fruits after 2-5 days after emergence. Upon hatching, the maggot starts to feed on the pulp and make galleries with accumulated excreta and results into fruit rot. Infested fruits develop into deformed one and their growth becomes arrested and severe cases a large number of such fruits drop off (Batra, 1953). Adult fly also found to be damages the fruits through ovi-punctures. They lay their eggs inside the epidermis and young maggots feed on fleshy and juicy pulp of fruits (Gupta and Sharma 2006; Singh, 2008). In Northern India infestation occurs during November to April, the fly activity was very high at the time of fruit maturity and there may be a 2 to 3 generation during the active period (Batra, 1953). The incidence found to be abundant in December and least in March and it has 6 to 9 overlapping generation per year (Lakra and Singh, 1989). In the Central India (Gujarat), the fruit fly attack starts around mid October and increased abruptly in mid November and continues till December (Bagle, 1992). However, the physical factors like soil moisture, soil temperature and soil depth also play crucial role for adult fly emergence from the site of pupation. The incidence found to be high when the relative humidity is 62 to 85% and temperature 17 to 25.5° C and low at 2.3 to 4.8° C (Dashad *et al.*, 1999). The pest incidence and temperature reveals positive relation whereas it shows negative relation with relative humidity, wind speed and cloud cover.

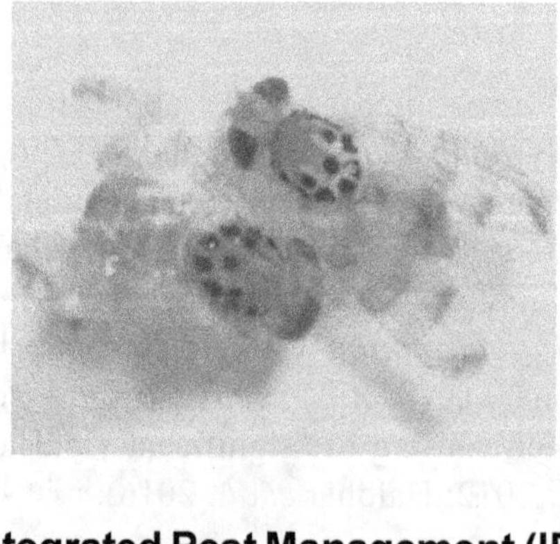

Integrated Pest Management (IPM)

The fruit fly *C. vesuviana* cause internal damage and it is very difficult to manage this pest without insecticides. However, integrated approaches would give better control over single control approach with low cost and environmentally safe (Singh, 2008).

Clean cultivation: Field sanitation surrounding the orchard through destruction of pruned parts of cultivated as well as the wild bushes. Collection of all fallen, bird damaged and infested fruits at periodical interval and proper destruction and feed such fruits to sheep goats, camels or other farm animals or bury them at least on one meter deep in compact soil can avoid the fly's emergence. Bird attack on unripe and semi ripped fruits results in a built up of initial population of tephritids causing heavy losses at later stage of crop (Grewal and Kapoor, 1986). Regular harvesting of matured fruits, before to colour change (green to yellow). Deep and through raking up of soil/ploughing the orchards during hot summer to expose the residual pupa to hot summer it also been destroy the over wintering pupae through mechanical injury during the operations.

Growing resistant varieties: Thin skinned, soft and sweet cultivars are found to be more susceptible to fruitfly. Cultivar with moderate sugar content and hard texture of fruit coupled with resistant to fruit fly (Pramanick *et al*., 1995). The adult flies avoids the egg laying in the fruits which are less than 9 x 4.5 mm size and lays more than 50 % of the eggs in 20 x 9 mm size fruits. The infestation of fruit fly is correlated positively with fruit weight, pulp stone ratio, total soluble solids and total sugars and negatively correlated to acidity, vitamin C and total phenol. Fruit size, fruit weight and pulp ratio also reveals positive correlation with fly infestation and cultivar with round fruits and early varieties contributes higher infestation (Singh and Vashishtha, 2002). The early maturing cultivar with moderate to bigger fruit size, sweet soft and thin skinned, more juicy and attractive flavored pulp are more susceptible to fruit fly (Saxena and Rawat, 1968). The varieties such as Meharun, Tikadi, Dodhia, Kakara, Umran, Mundia, Banarasi and Chhuhara found to be less susceptible to *C. vesuviana* (Table 2).

Table 2: Resistance cultivars to *C. vesuviana*

Cultivar	Remarks	Reference
Tikidi	Resistant	Singh and Vashistha (1984)
Kakro, Gola, Illaichi	Moderately resistant	
Tikidi, Illaichi	Resistant	—
Umran, Tas bataso, Deshi Alwar, Kismis	Moderately resistant	
Cv. Illaichi, Chhuhara	Resistant	Nandihalli *et al*. (1996)
Tikidi, Mundia	Moderately susceptible	—
Tikidi	Highly resistant	Pramanick *et al*. (2005)
Tikadi, Katha, Illaichi	Resistant	Haldahr *et al*. 2017
BS-75-1, Safeda, Dandan, Gola, Goma Kirti, Jogia, Narma, Mundia, Reshmi, Seb, ZG-3, Umran, Akharota	Moderately resistant	Haldahr *et al*. 2017

Biological control: There is no successful record of parasitoids, predators and pathogens against the *C. vesuviana.* Singh (1989) reported that the braconid, *Biostres vandenboschi* Fullaway as a parasitoid of *Carpomyia vesuviana* from India but the proportion observed to be negligible to employ against fruit fly. Likewise, parasitoids *Bracon fletcheri, Opius carpomyia* and *Omphalia* sp also been noticed on ber fruitfly however, there was no significant reduction in the pest population. The wasp parasitoid, *Fopius carpomyia* was found at larval stage of fruit fly and the ovipositor is very suitable to parasitize the hidden host in fruits. The rate of parasitization was 21-27% (Farrar *et al.*, 2004). The successful suppression of *C. vesuviana* with parasitoid can made through the augmentative release. Lack of mass culturing and efficacy testing techniques should be developed to overcome this bottleneck by generating new ideas for the practical application in the integrated management programme.

Chemical control: The pest can be managed with proper insecticide schedule. Spraying 0.05% malathion during January showed better control of fruit fly. Spray of dimethoate and deltamethrin would be most effective to manage *C. vesuviana* on ber (Patel *et al.*, 1990). The syntheytic pyrethroids fenvelarate 0.005% and decamethrin 0.0015 found to be most effective and consistent in action against the fruit fly activity (Bagle, 1992). Spraying of dimethoate followed by eco-neem shows good reduction in infestation (Singh *et al.*, 2000) and lowest incidence of *C. vesuviana* was observed with lambda cyhalothrin 0.0025% followed by 0.0018% beta cyfluthrin (Gyi *et al.*, 2003). The integrated management with dipterex 100g/ acre+ 5% molasses baiting and hoeing and collection of fallen fruits throughout the season proved better than single treatment applied trees. For the control of fruit flies, use traps containing 1 ml methyl eugenol and 2 ml malathion/ litre of water and placing 100 ml of this mixture per trap at 10 traps per hectare would be effective against fruitfly (Ahmed *et al.*, 2005). Application of neem powder and tobacco extracts significantly reduced the infestation of *C. vesuviana* and they are the potential candidates for organic cultivation of ber (Mari *et al.*, 2013).

Fruit borer, *Meridarchis scyrodes* Meyr (Carposinidae: Lepidoptera)

The fruit borer, *Meridarchis scyrodes* (Lepidoptera: Carposinidae) is a serious pest in Southern and western India (Balikai, 2013). The moths lay eggs on fruits at pea stage and upon hatching the newly emerged caterpillars bore into fruits and feed on the pulp near seed and accumulate fecal. The borer causes up to 70% yield loss under severe infestation. The first and second instar larvae feeds on the fruit superficially but third to fifth instar larvae feeds internally and damages the pulp around the seed. At initial stages of fruit development, the full grown larvae found to feed on soft immature seed. The occurrence of fruit borer found be positively correlated with the temperature and negatively correlated with the relative humidity and wind speed. The infestation starts during month of November and the peak incidence occurs during end of December (Gopali *et al.*, 2003).

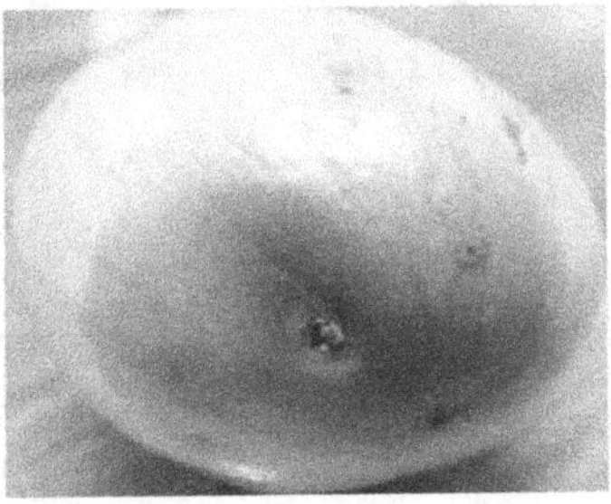 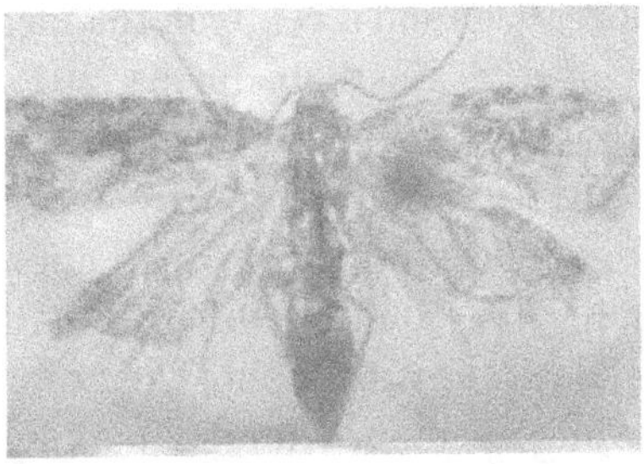

Integrated Pest Management (IPM)

Growing of resistant cultivars like Gola Gurgaon, Jhajjar Special, Kadaka, Banarsi Pewandi and Derakhi. Cultivars *viz.*, Danda, Seb, Elachi, Jogia and Manuki had infestation levels of <10% has been reported by Patil and Patil (1997). Removal of wild ber bushes present around or near to the ber orchard. Soil under the tree or near the trees should be raked to destroy the maggots and residual pupae present in the soil. Collection and destruction of all infested fallen fruits and harvest of fruits at early stage when they are still green but fully mature. Sprays of dimethoate 30 EC (0.051%) + jaggery solution (1.0%) at the marble stage and one spray at the maturation stage of ber resulted in minimum infestation and found to be more economical than at other growth stages (Patil and Patil, 1997). The Neem Seed Kernel Extract at 5% found as effective as like synthetic chemicals fenvalerate, decamethrin and cypermethrin (Yelshetty *et al.*, 1996). Methanol extracts of *Annona reticulata*, *Azadirachta indica* and *Ocimum sanctum* recorded 60-70 % mortality after 48h with 1.0% (Dhembare, 1998). Application of fenthion at 0.1% showed the lowest mean fruit damage of 36.46% (Rajaram and Siddeswaran, 2006). Spray schedule comprises of malathion at 2 ml or fenvalerate at 0.5 ml per liter of water or neem seed kernal extract (NSKE) 5%; first spray at marble stage, second spray at 15 days later and third spray at fruit ripening stage, by alternate use of insecticides would be effective against the fruit borer.

Stone weevil, *Abeus himalayanus* Voss (Curculionidae: Coleoptera)

The stone weevil, *Abeus himalayanus* Voss is recorded as a new pest of ber for the first time from Andhra Pradesh state of India (Gour and Sriramulu, 1994). Later, noticed as a new pest of ber for the first time from Karnataka state of India (Balikai *et al.*, 1998). Likewise, severe damage of this weevil

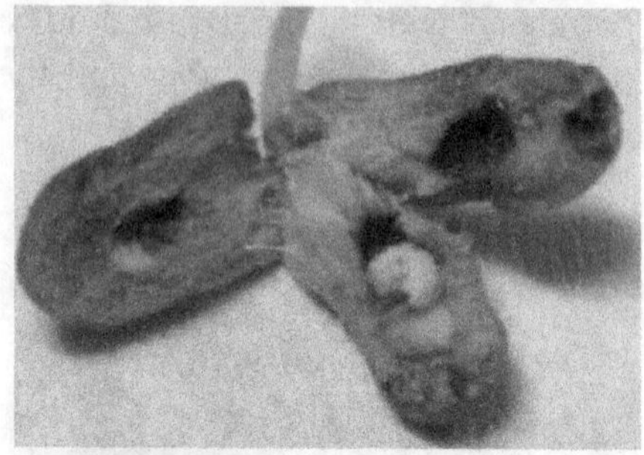

also noticed in the ber orchards of Rahuri, Maharastra and Jobner, Rajasthan in 1996 (Srivastava and Nanda, 1983) and recently in Bikaner district of Rajasthan during 2010 (Karuppaiah *et al.*, 2010; Haldahr *et al.*, 2013). So it seems to be an emerging pest in ber growing region of India. The adult weevils are small dark in colour with snout. The grubs are white in colour with red marking on the bogy, which are present inside the seeds (Balikai *et al.*, 2009; Karuppaiah *et al.*, 2010). The activity of adult beetle could be observed in the field during morning and evening hours and the biology of this pest is not studied so far. The adult female weevil lays their eggs on the stylar end of fruits and newly hatched grubs enter into seed by making puncture in endocarp at immature stage and starts feeding on soft seed coat and later it enters into endosperm moving downward. After entering into the seed, it starts feeding on inner content of the seed, and pupates within the seed by making hallow galleries. The weevil completes its life within a single fruit (Karuppaiah *et al.*, 2010). The developing seed was completely eaten away by the pest. The infested fruits were round in shape and varied in size ranging from pea to pebble. The infested fruits fails attain complete maturity and never increased in size more than pebble. This is because of the fact that the pest fed only on the seed portion of developing fruit and arrested further development of attacked fruit. The entry hole was healed up and closed while the exit hole was clearly seen. Looking to the nature of damage and its incidence pattern, it could be concluded that the pest enters into the fruits in the early stage of its development and the entry hole gets healed up (Balikai *et al.*, 2009).

Integrated Pest Management (IPM)

Periodical collection and destruction of infested fruits around the tree. Manual collection and destruction of adult weevil during morning and evening hours will reduce the population. Spray schedule must be started from flowering to fruit set stage only. Spraying of Spinosad 45 SC or Malathion 50 EC or Dimethoate 30 EC just before the fruit setting and repeat the sprays at three weeks interval was found as effective to control the stone weevil.

IPM of Date palm Crop

Date palm (*Phoenix dactylifera* L.) cultivation is traditional to the arid zones and it is prone to biotic stresses that have detrimental effects on its growth, development, production and productivity. In India, Gujarat, Rajasthan and Punjab are major states for datepalm cultivation. Carpenter and Elmer (1978) reported on more than 50 species of insects and mites as pests of date palms in worldwide. In Israel, approximately 25% of these species of insects and mites are considered serious pests. In India, the main pests are *Parlatoria blanchardii*, *Batrachedra amydraula*, *Rhynchophorus ferrugineus* and *Oryctes rhinoceros* and they are well adapted to the oasis environment, have been reported to cause serious losses to date palm.

White scale, *Parlatoria blanchardii* Targioni-Tozzetti (Hemiptera: Diaspididae)

The white scale, *Parlatoria blanchardii* Targ is widely distributed throughout most of the date growing regions of the world (Howard *et al.*, 2001). It is generally believed to be native to Arabian Gulf countries (as is the date palm). Commerce in date offshoots over the centuries paved the way to spread of this pest into India, Central Asia, the Middle East, North Africa and Turkey, and later to Australia and America. Date palm is one of the preferred host of scale, but it also been recorded on other hosts belonging to four plant families: Arecaceae (Palmae), Apocynaceae, Oleaceae and Rhamnaceae (Ben-Dov, 2006). It mainly infests the leaves and sucking the sap, if full leaf is covered with white scales finally the leaves dry out. In heavy outbreaks, fruits also attacked and falloff before maturity. Nymphs and adults suck the sap from the leaflet, midribs and fruits. Under each scale insect, a discoloured area appears on the leaflet. Heavy infestation causes leaflets to turn yellow and contributes to premature death of the fronds. Respiration and photosynthetic activities are almost stopped resulting in early death of the infested leaf. Damage on fruits is easily noticeable and the production is not marketable, especially due to poor acceptability.

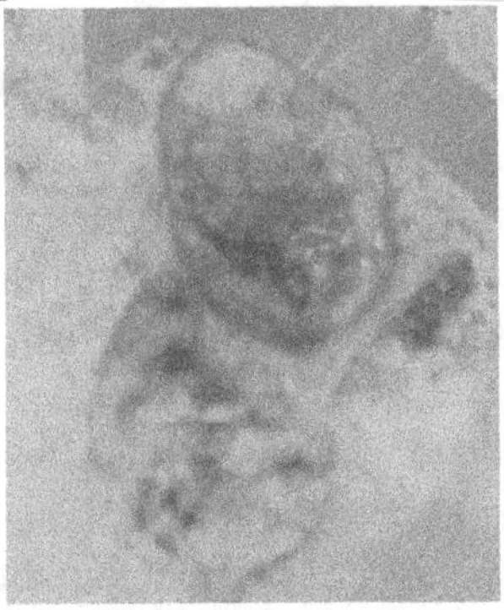

Integrated Pest Management (IPM)

Varietal Resistance: Varieties Mishrig and Gondaila found as most susceptible varieties and Brakawi and Jaw were moderate, and Tamoda found as tolerant (Mohamed, 1991). Natural Enemies: The populations of *Parlatoria blancharidii* (Targ) are normally kept at tolerable levels by indigenous natural enemies such as larvae and adults of a brownish lady bird *Pharosscymnus sp*, (Coccinellidae), *Cybocephalus dudiclri* (Nititulidae), *Crysoperla carnea* (Chrysopidae), *Archenomus arabicus* (Aphelmidae), and some unidentified mites. An introduced predator, lady bird, *Chilocorves bipusrulatus* var. *irpnsis* author was applied also to control the pest but with little success due to unsuitability of climatically conditions or some other ecological reasons. The predacious mite *Hemisarcoptes coccophagus* Meyer (Hemisarcoptidae) was also found to control white scale (Gerson and Smiley, 1990). Re-establishment of local natural enemies: Establishment of parasitic wasps and cybocephalid beetles by human intervantion mainly transferring these natural enemies from unsprayed plantations, where they were abundant, to plantations from which they had been eliminated. However, the re-establishment of coccinellids relied mainly on supply from laboratory mass-rearing cultures. Siddig (1975) reported that removal of infested lower leaves of datepalm was found to be an effective means of decreasing infestation of the insect. Spraying with petroleum oil plus Dimethoate (Roger) also resulted in significant control. Chemical control appears to be conducted occasionally in young plantations. Mineral oils are used for controlling the scales (Djerbi, 1994).

Lesser date moth, *Batrachedra amydraula* Meyerick (Lepidoptera: Batrachedridae)

The lesser date moth (*Batrachedra amydraula* Myer) is a pest that damages fruits in both fields and stores. The pest distribution is in Egypt, Iraq, Kuwait, Libya, Qatar, Oman, UAE, Yemen, Bahrain and Saudi Arabia. It is one of the major pests of date palm cause the yield loss upto 50-70% (Elwan, 2000). Larvae found to damage the large number of young fruits from April to June or July (Zaid, 1999). Since1980s, it has gradually spread throughout most of the date growing plantations of the country, where it has become a major pest of newly set and young green date fruits. The moth is highly speciûc to the date palm and no reports are known of infestation of other host plants by this moth. The larvae attack the newly formed inûorescences, but the main injury is to the young green fruits. Approximately 80% of the fruits are attacked when the size lie between 0.6 and 1.0 cm in diameter. The larva chews a hole near the calyx, through which it penetrates the fruit and feeds on the pulp and the soft immature seed. A damaged fruit is easily recognizable by the black feces attached to the penetration site. The larva also moves from fruit to fruit within the bunch, thus increasing the damage. In severe infestation most of the infested fruits drop to the ground; the bunch ceases to grow, and then dries. Thus lesser date moth injuries to the palms cause considerable fruit drop and losses of upto 75% of the yield (Blumberg, 1975).

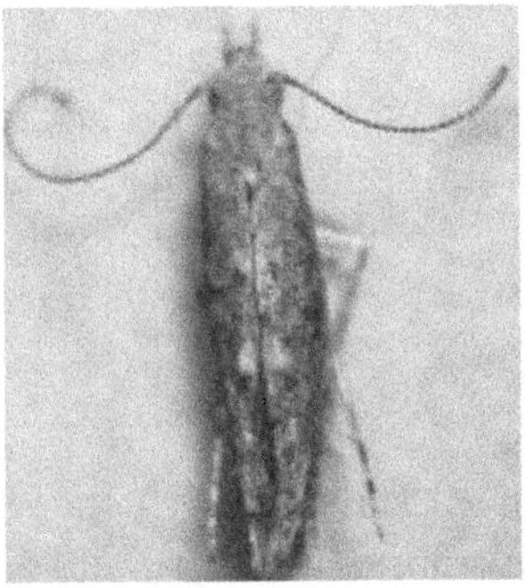

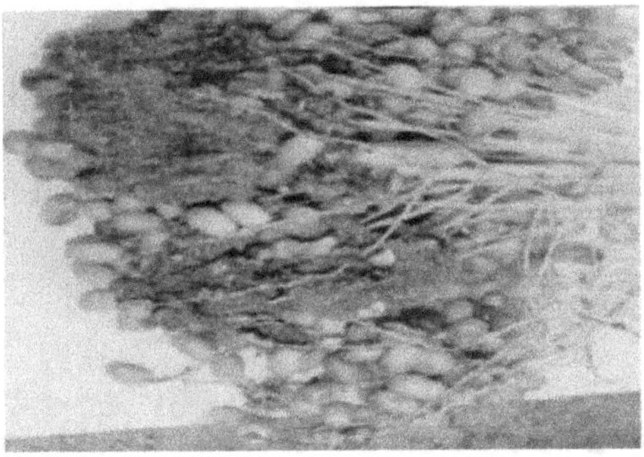

Integrated Pest Management (IPM)

The use of pheromone traps is a best way to control of the dates moth and it is more accurate and reliable as well as saves much of the annual costs of chemical control. Preparation of the pheromone itself is not a big problem and any sticky trap will do the job properly.

Natural enemies: The parasitic wasp, *Parasierola swirskiana* Argaman (Hymenoptera: Bethylidae) attacks the lesser date moth. This parasitoid is common in date plantations in the Arava Valley, and is known also from Jordan and Afghanistan (Argaman, 1991-1992). *Bracon* sp. (Hymenoptera: Braconidae) attacks larvae of lesser date moth as well as those of other fruit moths, but its contribution to the control of date moth has not yet been studied (Blumberg *et al*., 2001).

Microbial control: Several commercial products based on the bacterium *Bacillus thuringiensis* Berliner (Bt) were found to be effective against the lesser date moth and they are used for control, mainly in organic plantations (Navon, 2000). These products are used either in spraying or in dusting formulations; the latter offers better capability of penetrating into the center of the date bunches (Blumberg *et al*., 2001). The mixture of Bitayon and date pollen protects the young setting fruits against ûrst generation moth larvae immediately at the time of pollination. The efûciency of Bt compounds is higher at the beginning of the season, when a larva consumes more than one fruit to complete its development. Later in the season, when the fruits are bigger, the larvae usually complete their development in a single date fruit and are not exposed to the Bt compounds (Blumberg *et al*., 2001).

Chemical control: The growers replaced endosulfan with organophosphates, mainly chlorpyriphos. Later, additional compounds, mostly IGRs such as triûumuron (Alsystin) and teûubenzuron (Molit) were also added to the arsenal (Ucko and Bitton, 2000). In Egypt, in 1999 and 2000, spinosad (Tracer) at 10 ml 100 l^{-1} gave 94% and 86% control of *B. amydraula*, respectively (Sayed *et al*., 2001).

Biological control: Inundating release of the egg parasitoid, *T. cacoeciae* at approximately 1000 adults per palm, together with an application of a commercial product of the fungus *B. bassiana*, found to be potential agents against this moth (Navon *et al*., 1999).

IPM of Aonla Crop

Aonla was supposed to be free from major insect and disease menace. With lapse of time and intensification of its cultivation, number of insects and diseases has started feeding and infecting at various stages and causing considerable damage. Jhala *et al*. (2003) reported occurrence of fruit borer, midge and leaf cutting weevil for the first time in northern and western India. These and other pests, which have been noticed damaging aonla, are described below.

Shoot gall maker, *Hypolamprus stylophora* (Swinhoe) (=*Betousa stylophora*) (Thyrididae: Lepidoptera)

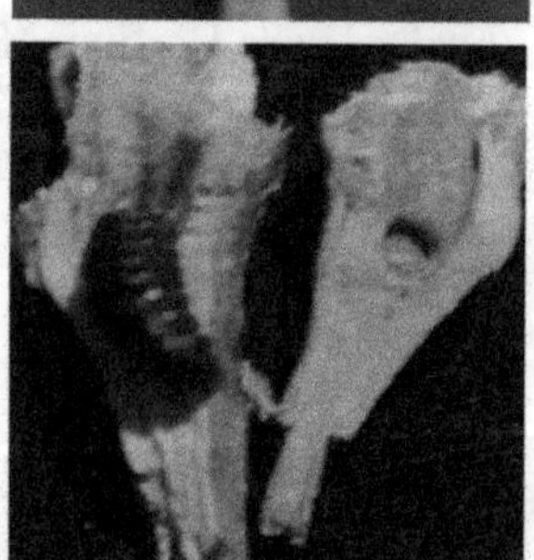

The gall disease of 'aonla' occurs throughout India, Sri Lanka, Myanmar, Bangladesh, China (Hong Kong) and Java. Though known to occur widely in India, but it is minor sporadic pest in most of the areas, its occurrence, excepting a few places where it causes considerable economic losses. It is a specific pest of 'aonla' and it is not known to attack any other host so far. Inside the gall, the larva feeds on the succulent woody tissues and pushes out reddish frass and excreta through a small hole at one end, kept guarded by a mesh work of silken threads. There is only one larva in each gall which remains confined in an ellipsoidal cavity carved out in gall. The gall is a hollow swelling/ localized tumescence of the tender shoot, irregular, roughly, spindles shaped- resembling 'Snake charmer's flute'. In the beginning of the infestation terminal shoots swell, which increases in size with the passage of time. Full size galls can be seen in the month of October-November.

Integrated Pest Management (IPM)

Over-crowding of branches should be discouraged. Galled shoots should be pruned and destroyed along with the pest after harvest. In case of regular occurrence of this pest, spray chlorpyriphos 20 EC or malathion 50 EC or dimethoate 30 EC at the rate of 1.5 t 2.0 ml per liter of water in the beginning of the monsoon season. It may be repeated at fortnightly intervals, if needed.

Bark Eating Caterpillar, *Indarbela quadrinotata* (Walker) (Metarbelidae: Lepidoptera)

The pest is widely distributed throughout the country, infesting a number of fruits, ornamentals and forest trees. This pest is very common in aonla orchards (Lal *et al.* 1996). Generally, incidence of this pest is more common in neglected and poorly managed orchards. The infestation of this pest starts in April with the emergence of moths. The larvae of this moth are recorded feeding and boring on the bark of main trunk, stem and branches. In case of severe infestation, the growth of the tree, flowering and fruiting is adversely affected. Host plants of bark eating caterpillar are citrus, mango, guava, *jamun*, loquat, mulberry, pomegranate, *ber*, drumstick, litchi, rose and a number of forest and ornamental trees. Thick, ribbon like, silken webs are seen running on the bark of the main stem be seen on a tree, one caterpillar or pupa occupying each hole. A severe infestation may result in the death of the attacked stem but not of the main trunk. There may be interference with the translocation of cell sap and thus arrestation of growth of the tree is noticed with the resultant reduction in its fruiting capacity.

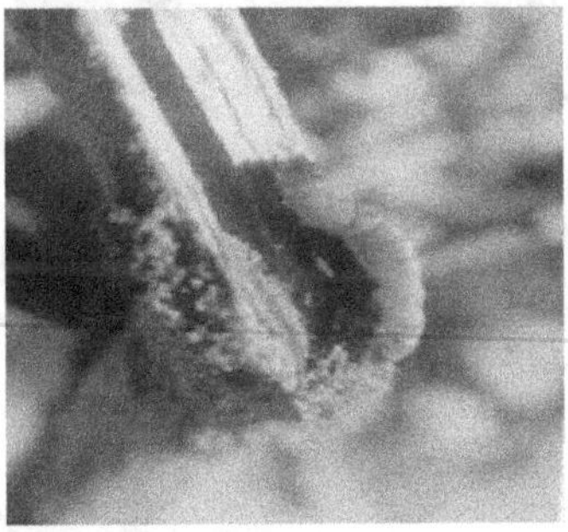

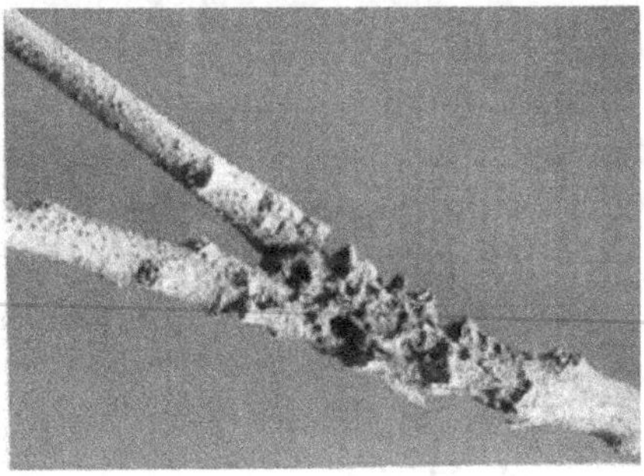

Integrated Pest Management (IPM)

The control of bark eating caterpillars has remained a problem because its attack becomes noticeable when the conspicuous webbings appear on the tree trunks i.e. the time by which the pest has already inflicted considerable damage. The management of this pest involves the following steps: Do not allow overcrowding of branches; keep the orchards clean and healthy to prevent infestation of this pest; detect early infestation by periodically looking out for drying young shoots; kill the caterpillars mechanically by inserting iron spike in shelter holes made by these borers at early stage of infestation. In case of severe infestation, remove the webs and insert the swab of cotton wool soaked in 0.025% Dichlorvos 100 EC or inject water emulsion of 0.05% chlorphyriphos 20 EC and plug the holes. The larvae of this pest are parasitized by entomogenous fungus, *Beauveria bassiana* in nature. This can be used as a potential bio-control agent.

Aonla aphids, *Schoutedonia* (=*Cerciaphis*) *emblica* (Patel & Kulkarny) (Aphididae: Hemiptera)

The incidence of this pest is mainly seen from July to October with the peak period in September (Lal *et al*., 1996, Devi and Rajasekaran, 2011). The new shoots are infested at growing points. The nymphs and adult females suck the sap. Heavy attack of this pest affects the growth and vigour of the tree and ultimately affecting the flowering and fruiting. The population of aphids initiated from second fortnight of April, then increased gradually during May and June and showed peak activity during second fortnight of June. Its population declined during July and then disappeared from end of October. Lal *et al*. (1996) also reported the peak incidence of this pest during first fortnight of September, where its population was maximum. The infested leaves turn yellow and dry up. Infested shoots appear bended and twisted at the growing points. Presence of ants also indicates the infestation of aphids. The new shoots are infested at growing points. The nymphs and adult females suck the sap. Heavy attack affects the growth and vigour of the tree, ultimately affecting the flowering and fruiting. Aphids were observed to congregate on the new flush at the growing points of aonla. The infested leaflets turned yellow and dried up.

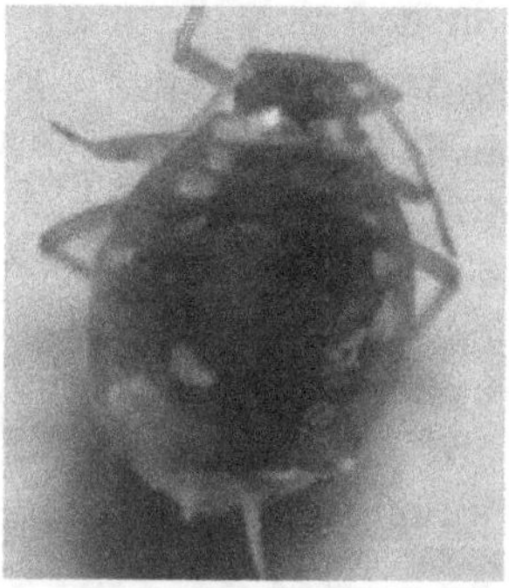

Integrated Pest Management (IPM)

Clipping off and destruction of affected leaf and shoot. Spray Dimethoate 30 EC at the rate of 0.06% or Imidacloprid at 0.005% (3 ml of imidacloprid 17.8 SL in 10 litres of water) at the appearance of aphid. Crop spray with Neem oil at the rate of 0.5% (50 ml neem oil+ 10 gm detergent in 10 litres of water). Two species of Coccinellid beetles, *Menochilus sexmaculatus* Fab. and *Coccinella septempunctata* Linn. (Coccinellidae: Coleoptera) were fond feeding on the aphids.

IPM of Bael crop

Bael (*Aegle marmelos* Correa) belongs to family Rutaceae is an important indigenous fruit crop to dry forest in hills, plains of central and southern parts as well as arid and semi arid regions of India. It also called as Bengal quince or Indian quince. Fruits are famous for its medicinal values as well as edible fruit quality. One of the reasons for it is infestation of insect pests on the vegetative as well as developing fruits which ultimately leads to significant yield loss and quality attributes of the fruits. About 30 species of insect and mites have been reported feeding on *bael* in India (Lakra, 2004).

Lemon butter fly, *Papilio demoleus* Linnaeus (Lepidoptera: Papilionidae)

The lemon butterfly is one of the economically important pests whose larval forms cause serious damage by devouring large quantity of foliage of Rutaceae family with special preference towards both wild and cultivated species of citrus during the later stages of their development. It was reported to be a pest of citrus throughout India. Various types of insect-pests are found to be infesting bael in which *P. demoleus* is major one which defoliate the plant (Haldhar *et al*., 2010). Pathak and Rizvi (2002) reported that the *P. demoleus* is a serious citrus pest in India. The larvae are a serious pest of citrus nursery stock (trees 1-2 ft. in height) and other young citrus trees and are capable of defoliating entire nursery groves. Larvae may utilize young foliage on more mature trees (Narayanamma, 2001). Haldhar *et al.* (2010) observed in bael that population and infestation of *P. demoleus* appeared from August to February. The highest egg, larval population and plant infestation were observed 12.75, 6.15/ 10 branches and 65.0 %, respectively. Similarly, the pest occurred in epidemic form during eighties in Southern zone of Andhra Pradesh (Ramasubbareddy, 1986). Its may cause complete (100%) defoliation of infested young plants. Similarly 83% defoliation of Sweet orange plants in AP was reported by Narayanamma *et al.* (2001). From Rayalaseema region and >40% defoliation reported in acid lime gardens in Tamil Nadu region during the October to January (Usharani *et al*., 2012).

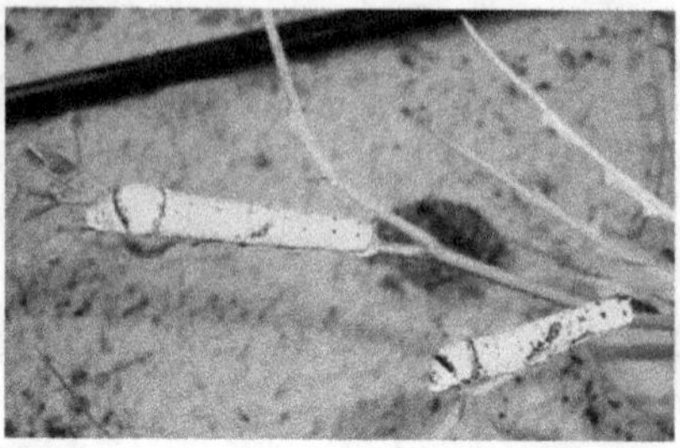

Integrated Pest Management (IPM)

In the management of lemon butterfly through various integrated pest management techniques can be adopted. Many pesticides were tested against this pest and found effective also. Under present globalised situation of the world scenario, keeping in view the side effects of these pesticides on non targeted organisms including humans, environmental pollution, and chemical control is no more a valuable technique of pest management. Hence, pest control with integrated pest management (IPM) system like cultural, physical, biological, botanicals and bio-rational pesticides are effective for control of lemon butterfly. Larvae are easily hand-picked from nursery plants, but this practice is labour intensive if infestation is high.

Biological control: The young caterpillars of lemon butter fly were attacked by the yellow wasp (*Polistes hebreus* F.), preying mantis (*Creobrator gemmatus*) and spiders. The pupae were parasitised by *Pteromalus sp*. and eggs by another unknown hymenopteran parasite. *Trichogramma chilonis* and *Telenomus sp.* are egg parasitoids on lemon butterfly and as high as 75.9% egg parasitism was recorded on both *P. demoleus* and *P. polytes* species (Krishnamurthy and Singh, 1986). Similarly, *Apanteles lunatus* and *Pteromalus puparum* are natural enemies of larvae and pupae on *Papilio demoleus* as observed in citrus orchards of Nepal (Resham *et al*., 1986). Other natural enemies of larvae found in Thailand were a predatory pentatomid bug, *Cantheconidea furcellata* (Wolff); reduviid bugs; birds; spiders; sphecids; and chameleons (Lewis, 1976). Spiders were the important mortality agents in the eggs and young larvae, meanwhile in the older larvae were *S. dichotomus* and *Podisus* sp., *Ooencyrtus papilioni* and *Pteromalus puparum* were the important parasitoids for the eggs and pupae of *P. demoleus*, respectively (Suwarno, 2012).

Biopesticide: Shivankar (1999) that Dipel (*Bacillus thuringiensis* Berl.) spray at 0.05% give good control of the pest. Entomopathoens like bacterium *Serratia marcesscens* and fungus *Fusarium* sp. also killed the pest population substantially. Some plant extracts like Pipal (*Ficus religiosa*), Beshram (*Manchoria hastaefolia*), *Parthenium histerophorus*, Neem (*Azadirachta indica*) and Datura (*Datura stramoneum*) significantly (52.8%) reduced *P. demoleus* larval population (Solunke and Deshpande, 1991). Similarly the plant products allitin, replin, margosol and neem guard @ 1% aqueous extract gave >75% pest control in citrus. The leaf extracts of *Eucalyptus globulus* and *Ageratum conyzoides*, clove extracts of *Allium sativum* cause morphological abnormalities in 5th instar larvae and thus are useful botanicals in the pest management (Suyog pande, 2012). The juvenoid diofenolan severely hampers the normal growth, development and metamorphosis of *P. demoleus* and causes several deformities like delay in larval-larval and larval-pupal ecdysis, ecdysial failure, mortality, severe reduction in pupation, deformed pupae and complete inhibition of adult emergence (Singh and Kumar, 2011).

Chemical control: Similarly, Vyavasaya Panchangam (2007-08) advocated dichlorovos @1.0 ml/l of water or *Bacillus thuringiensis* @ 1 gram/l of water for managing citrus butterfly. Murthy *et al.* (2009) stated that novaluron indicated 94.8 per cent mortality followed by emamectin benzoate (84.5%), chlorpyriphos (59.4%) and neem (59.4%). Horticulture dairy (2004) recommended and suggested that spraying of 0.05% fenitrothion would bring down the incidence and control the citrus butterfly caterpillar.

IPM of Ker Crop

Capparis decidua (Forsk.) commonly known as *ker* is a spiny bushy shrub in dense tufts, 4-5 m high, commonly used in landscape gardening, afforestation and reforestation in semi desert and desert areas and it provides assistance against soil erosion. It is well distributed in the arid regions of north-west India covering parts of western Rajasthan and drier parts of Gujarat, Haryana and Punjab. It is extremely drought-resistant and tolerates some frost. The bitter roots, root bark and fruits are used in preparation of medicines. Root bark and stem contains a spermidine alkaloid, isocodonocarpine, effective in treatment of asthma, inflammation and cough (Ahmad *et al.*, 1989). It is mainly used for food, fodder, fuel and timber (Orwa *et al.*, 2009) as it produces hard, heavy and termite resistant timber and fruits which are rich in proteins, carbohydrates and minerals.

Ker butter fly, *Anaphaeis aurota* Fabricius (Lepidoptera: Pieridae)

A. aurota is found throughout tropical Africa, Arabia and Indian subcontinent. It is migratory and regularly penetrates the Middle East and Egypt. As is the case in most migratory species there is little or no geographical variation. Actual migrations recorded from Arabia have been of modest

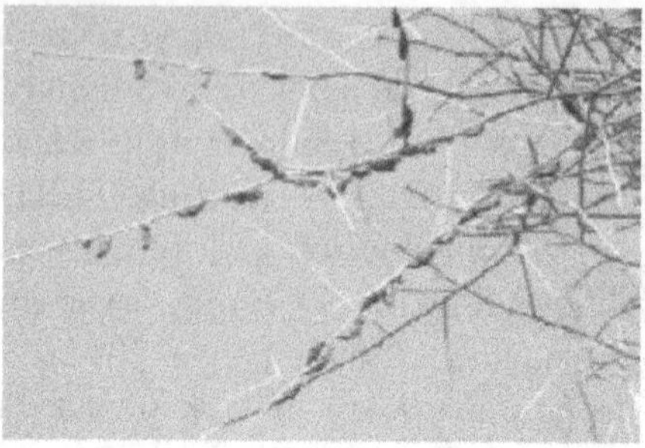

proportions, but in East Africa several million butterflies may cross a narrow front in a single day. *Capparis decidua* was found to be heavily infested with *Anaphaeis aurota* commonly known as the Pioneer or Caper white, in various parts of Rajasthan. Larsen and Nakamura (1983) referred to several occasions citing the migratory behaviour of this butterfly in Lebanon and Palestine. It prefers the warm Jordan Valley; however, it was collected from two localities within the eastern mountains. The species is so abundant that the caterpillars completely strip trees and bushes; bare twigs may literally be festooned with the pupae of those caterpillars which did not starve to death. The regular occurrence of dwarf specimens is almost certainly linked to this type of overcrowding. During the investigation it is found that the grown caterpillars of *A. aurota* easily strip off the branches, devouring leaf after leaf causing great damage in Rajasthan (Ankita and Haldhar, 2012).

Integrated Pest Management (IPM)

Systematic collection of egg masses of butter fly and their destruction. Pest can be checked by handpicking and mechanical destruction of caterpillars during early stage of attack when the caterpillars feed gregariously. Natural control: The pupae of *A. aurota* are parasitized with *Brachymeria albicrus*. The mean per cent parasitization of pioneer butterfly by *B. albicrus* at CIAH farm and at Desnok, Bikaner was 49.5 and 47.5 respectively and the mean per cent emergence of the mature adult parasitoids from the parasitized pupae was 15.5 and 14.0 respectively (Ankita and Haldhar, 2012). Spray application of cypermethrin 1.5 ml, spinosad 0.5 ml and dimethoate 2.0 ml per liter of water affords protection.

IPM of Kinnow Crop

Indian is the major producer of kinnow mandarin in the world and Punjab, Rajasthan, Haryana, Himachal Pradesh, Jammu and Kashmir and Uttar Pradesh are the major kinnow growing states. It thrives well under arid-ecosystem and gives good return. However, pests and diseases are the major constrains in kinnow cultivation in current situation. Among the various insect pests reported on kinnow mandarin the pest such as leaf miner, psyllid, black fly, citrus butterfly and fruit flies are the important pests in this region.

Citrus black fly, *Aleurocanthus woglumi Ashby (Alerodidae: Hemiptera)*

It is considered as sporadic pests of kinnow mandarin causing serious damage to the foliage. The nymphs as well as adults colonize under lower surface of old leaves and suck the plant sap resulting, curling, immature fall of leaves and loss of plant vigour. Honey secretion favour the shooty-mould growth on leaves and it reduce the photosynthetic activity. In severe attack, the trees produce few blossoms which develop into unhealthy fruits. The population build up is observed during May-June and September-October. Adult psylla is minute, shiny black in colour with grey dusting on the body. The nymphs are flattened oval in shap and it gives appearance like scale when the infestation occurred on the leaf. Total lifespan is depends upon the surrounding temperature and it is ranging from 45 to 133 days; usually more than one generation occurred during the season (Nguyen *et al.*, 1983). Adult female lays their eggs in a spiral pattern on the lower surface of the leaves, egg period lasts in 10 days; nymphal stages is six to 50 days and pupal period is about 16 to 50 days (Dowell *et al.*, 1981).

Integrated Pest Management (IPM)

Removal and destruction of infested leaves, followed by application of chlorpyriphos 2 ml or imidacloprid 0.5 ml/lit would reduce the population build up. Release of *Mallada bonenensis* eggs @ 6 eggs and 4 eggs/ shoot were promising (Ingole *et al.*, 2005).

IPM of Lasora Crop

Cordia myxa L. commonly called Indian cherry (synonyms: clammy chery, franrant manjack, lasora, shelu, cinnanakkeru, bahubara, chokri, lahsoda, lahsua, gonda, gondi) belongs to Boraginaceae family, this tree is a multipurpose species distributed in hot arid and semi-arid regions of India, originated in North-Western part of the country (Stewart and Brandis, 1992). According to the National Tree Seed Project (1999), insect damage on *Cordia africana* seeds was 45%, 30% and 20% from Sekoru, Arjo and Wondo Genet places, respectively. Tibebu (2002) studied pre-dispersal insect seed predators on seeds of these two tree species, 10 and found 20% of *Cordia africana* seeds collected from Sekoru, and 8-10% of seeds collected from Denbi, Jimma and Arjo places.

Tingid bug, *Dictyla cheriani* Drake (Hemiptera: Tingidae)

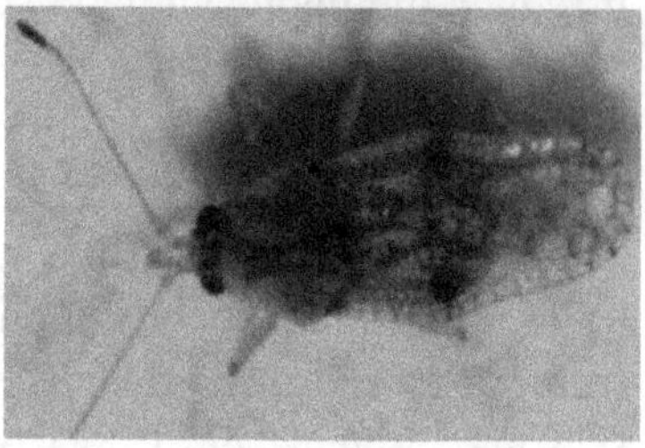

Few studies were made on the bug, *D. cheriani* infestation on *Cordia sp.* (India) and bug, *D. monotropidia* infestation on *Cordia verbenacea* (Brazil) by Daniel *et al.* (2008). Haldhar and Singh (2014) reported that infestation of *Dictyla cheriani* on Indian cherry (*Cordia myxa*), was first noticed in 2010 at Bikaner, Rajasthan. The maximum incidence was observed in October (51.67% on bold & 76.67% on small seeded plants) and minimum in January (11.67% on bold & 21.67% on small seeded plants). The number of this lace bug ranged between (0.5 to 8.8 on bold & 4.5 to 25.97 on small seeded plants) nymphs and adults per leaves. The female was distinctly larger than the male in respect to all body parts. This lace bug was typically characterized as having body oblong, pale testaceous with brownish or fuscous markings, with collar and hood yellowish brown, body beneath reddish dark with thoracic sterna darker. The lace bugs were found to be aggregated on the leaves of the plants. During a survey of the natural vegetation, the population of this pest was found to be higher on small seeded plants than the bold seeded plants. The lace bugs sucked the sap from newly emerging leaves and young branches, which led to the leaves turned yellow and suppression of growth of the tree through drying of leaves and young branches (Haldhar and Singh, 2014).

Integrated Pest Management (IPM)

Provide proper cultural care so plants are vigorous. If damage has previously been intolerable, monitor plants early during subsequent seasons. Take action when populations begin to increase and before damage become extensive. Provide adequate irrigation and other care to improve plant vigor. Prune out damaged foliage if the discoloring is intolerable and relatively localized. Do not remove more than a small percent of a plant's branches during one season and use good techniques so that pruning does not injure plants, such as by exposing inner branches to sunburn. Almost any insecticide will control lace bugs if it is sprayed directly onto the insects. Azadirachtin (BioNeem), insecticidal soap (Safer), narrow-range oil (Green Light, Volck) and neem oil sprays temporarily control lace bugs if insecticide thoroughly covers the underside of leaves where adults and nymphs occur. Imidacloprid 17.5 SL can be applied as a foliar spray for controlling of this pest (Dreistadt *et al.*, 2004).

IPM of cucurbit crops

The cucurbit crops like watermelon/*Mateera* (*Citrullus lanatus*), *Kachri* (*Cucumis callosus*), snap melon/*Phoot/ kakadi* (*Cucumis melo* var. *momordica*), long melon/*Kakri* (*Cucumis melo* var. *utilissimus*), round melon/ *Tinda* (*Praecitrullus fistulosus*) bottle gourd (*Lagenaria siceraria*), ridge gourd (*Luffa acutangula*), *etc.* have great potential in arid pockets area of country because they are well adapted to existing adverse climatic conditions and possess unprecedented ability to withstand against stresses like moisture stress, high soil pH and load of pests and diseases. Insect-pests are the major biotic constraints to cucurbits production. Apart from causing direct damage, many of them also act as vectors for several viral diseases which have aggravated the problem still further. Among major biotic constrains like fruit fly (*Bactrocera cucurbitae*), leaf eating caterpillar, *Diaphania indica*, hadda beetle, *Epilachna viginctioctopunctata* (Fabricius) in cucurbits production in India.

Melon fruit fly, *Bactrocera cucurbitae* Coquillett

It is commonly called melon fruit fly. Originally from the Indo-Malayan region, the fly was introduced into Hawaii from Japan in 1895. It is widely distributed and has been recorded from East Africa, some parts of USA, northern Australia, Taiwan, Okinowa in Japan, South China, South-east Asia and the Indian sub-continent. Melon flies have more than 80 hosts. They are major pests of beans, bittergourd, cucumbers, edible gourds, eggplant, green beans, melons, peppers, pumpkins, squashes, tomatoes and watermelon. The female flies puncture the soft and tender fruits with their stout and hard ovipositor and lay eggs below the epidermis. On hatching the maggots feed inside on the pulp of fruits and the infested fruits can be identified by the presence of brown resinous juice which oozes out of the punctures made by the flies for oviposition. These punctures also serve as an entry for various bacteria and fungi; as a result, the infested fruits start rotting, get distorted and malformed in shape and fall off from the plants pre-maturely (Haldhar *et al.* 2013b, Haldhar *et al.*, 2017).

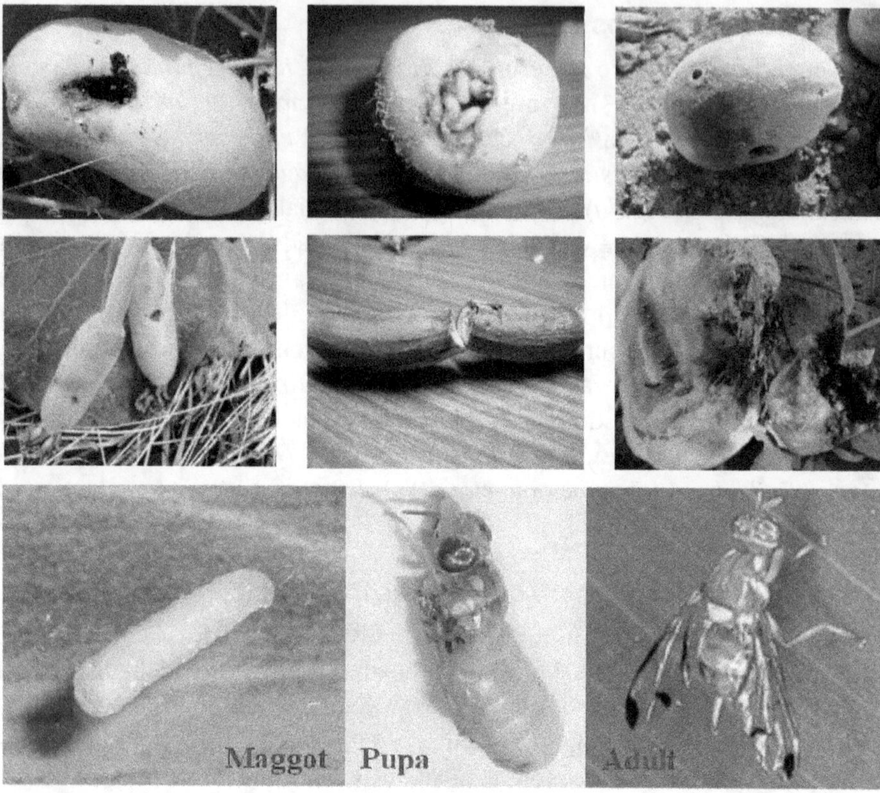

Integrated Pest management (IPM)

The most effective cultural management technique is the destruction of all infested and unmarketable fruit and the disposal of crop residues immediately after harvest. To break the reproduction cycle and population increase, growers need to remove all unharvested fruits or vegetables from a field by completely burying them deep into the soil. Burying damaged fruits 0.46 m deep in the soil prevents adult fly eclosion and reduces population increase. Bagging of fruits on the tree (3 to 4 cm long) with 2 layers of paper bags at 2 to 3 day intervals minimizes fruit fly infestation and increases the net returns by 40 to 58%. Methyl eugenol and cue-lure traps have been reported to attract *B. cucurbitae* males from mid-July to mid-November. A leaf extract of *Ocimum sanctum*, which contain eugenol (53.4%), beta-caryophyllene (31.7%) and beta-elemene (6.2%) as the major volatiles, when placed on cotton pads (0.3 mg) attract flies from a distance of 0.8 km. Although, the protein baits, parapheromone lures, cue-lures, and baited traps have been successful for the monitoring and control of melon fruit fly, the risk is the immigration of protein-satiated females.

The organic IPM system proved to be the most effective and economical approach (B: C ratio 8.80:1) against cucurbit fruit fly (*Bactrocera cucurbitae*) in which the lowest incidence was recorded as compared to other modules. The organic IPM module-III comprised of growing resistant genotype (RM-50), spray of neem oil at 20 DAS, installation of pheromone trap (10/ hectare) at 42 DAS, spray of tumba fruit extract (TFE 5%) at 50 DAS and spray of spinosad 46 SC at 60 DAS was the most effective (Haldhar *et al.* 2014). It is reported *Opius fletcheri* Silv. to be a dominant parasitoid of *B. cucurbitae*, but the efficacy of this parasitoid has not been tested under field conditions in India. More recently, a new parasitoid, *Fopius arisanus* has also been included in the IPM program of *B. cucurbitae* at Hawaii. Application of either 0.05% fenthion or 0.1% carbaryl at 50% appearance of male flowers, and again at 3 days after fertilization is helpful in reducing the melon fly damage. It is also reported that triazophos to be the most effective insecticide against fruit fly on bitter gourd. An extract of *Acorus calamus* (0.15%) reduced the adult longevity from 119.2 days to 26.6 days when fed continuously with sugar mixed with extract (at 1 ml/g sugar). Neem oil (1.2 %) and neem cake (4.0 %) have also been reported to be as effective as dichlorvos (0.2 %).

Hadda Beetle, *Epilachna vigintioctopunctata* (Fabricius, 1775)

This species originated in the far east of Russia and has been expanding its range in the second half of the 20th century and is now found over most of Russia, North-East China, Northern Korea, Australia and Japan. It is found in different parts of South East Asia. The insect is considered to be one of the important pests of brinjal, sometimes becoming serious on cucurbits like snap melon, water melon, musk melon, ridge gourd *etc.* The grubs and adults scrape the leaves in a characteristic manner and feed. They feed on the epidermal layers of leaves which get skeletonized and gradually dry away. They affect the crop in all the stages.

Integrated Pest management (IPM)

Plough the fields just after harvesting destroy the adults. In the initial stage, collection and destruction of affected leaves along with the eggs, grubs and adults. The various natural insect control agents and two disease-causing organisms (*Aspergillus flavus* and *Bacillus thuringiensis*) which were sometimes able to control populations. The aqueous extract sprays of the castor oil plant *Ricinus communis* (Family: Euphorbiaceae) reduced *Epilachna* attack on foliage and capitulum of sunflower *Helianthus annulus* and consequently increased the oilseed crop. Spray application of carbaryl 0.1% or cypermethrin 0.025% or profenofos 0.05%. The organic module comprised of deep summer plouging, growing resistant genotype, spray of neem oil at 20 DAS, spray of tumba fruit extract (TFE 5%) at 50 DAS and spray of spinosad 46 SC at 60 DAS are the most effective for control of hadda beetle in this crop.

Leaf Eating Caterpillar, *Diaphania indica* (Saunders, 1851)

This moth occurs in many tropical and subtropical regions, though it is native to southern Asia; it is occasionally a significant pest of cucurbits and some other plants. The moth attacks cucurbits and eats their leaves. Cucurbits are watermelon, cucumber, melon, pumpkin and snake gourd. The caterpillars do the damage after hatching; they roll the leaves with silken threads and eat the leaves between the veins. They also attack the flowers and reduce the number of fruits set. Young fruits are also attacked by this pest. The caterpillars damage the skin and cause the fruits to rot.

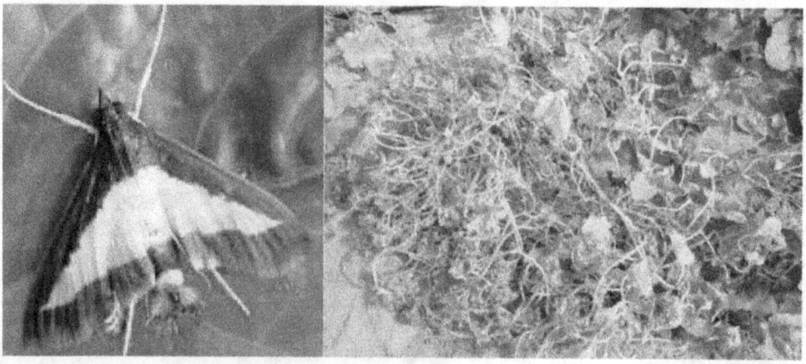

Integrated Pest Management (IPM)

Make regular inspections of the crop, at least once a week. Check the young leaves, looking for those that are stuck together and those where only the veins are left. Look for faeces on the leaves. They are the signs that the caterpillars are present. Hand picking should be the first method of control. Look for rolled leaves: remove them, or squeeze the caterpillars inside. Check the flowers and the fruits for damage. If it is severe, consider using chemicals. But consider carefully which ones to use. The survey for the natural enemies associated with the caterpillar, *Diaphania indica* revealed the presence of 20 species of parasitoids, predators and pathogens. Of these, 16 were parasitoids belonging to the families Braconidae, Ichneumonidae, ethylidae, Elasmidae and Chalcididae. Except for 3 species the remaining parasitoids were new records for *D. indica*. The predators recorded were ants and spiders. *A microsporidia* also was recorded for the first time on *D. indica*. Spray any following insecticides like malathion 50 EC @ 500 ml, dimethoate 30 EC 500 ml, methyl demeton 25 EC @ 500 ml/ ha *etc*.

Future Strategies and Research Need in IPM

Enhancing the knowledge base of farmers with modern science, improving their skills and internalizing the inputs at various levels can bring significant changes to the way they carry out arid horticulture. New solutions could include novel resistant cultivars with multiple resistance genes, suitable epigenetic imprints and improved defence responses that are induced by attack. Plant activator agrochemicals could be used to switch on natural plant defences. Habitat manipulations such as push–pull strategies can improve pest management and yields in less intensive systems. Genomic and transcriptomic information will facilitate the development of new resistant crop cultivars once annotation and availability of data on multiple cultivars is improved. Knowledge of the chemical ecology of pest–plant interactions will be better exploited once the genes for biosynthesis of key plant metabolites are discovered. Genetic modifi ation of crops has the potential for speeding up the development of crops with novel resistance. The presence of private sector in biopesticide production and marketing is marginal, and needs to be improved through economic incentives.

Conclusions

India has successfully reduced pesticide consumption without adversely affecting the agricultural productivity. This was facilitated by appropriate policies that discouraged pesticide use, and favored IPM application. Despite it, adoption of IPM is low owing to a number of socio-economic, institutional and policy constraints. On the supply side, lack of commercial availability of biopesticides and inappropriate institutional technology transfer mechanisms are the critical impediments to increased application of IPM. IPM is a complex process and farmers lack understanding of biological processes of pests and their predators and methods of application of new technology components. The socio-economic environment of farming is also an important factor in adoption of IPM. There are a number of IPM practices that work best when applied by the entire community and in a synchronized mode.

References

Ahmad, B., Anjum, R., Ahmad, A., Yousaf, M., Hussain, M. and Muhammad, W. 2005. Comparison of different methods to control fruit fly (*Carpomyia vesuviana*) on ber (*Zizyphus mauritiana*), *Pakistan Entomology*, 27(2): 1-2

Ahmad, V. U., Ismail, N., Arif, S. and Amber, A.R. 1989. Isocodonocarpine from *Capparis decidua*. *Phytochemistry*, 28: 2493-2495.

Ankita Gupta and Haldhar S. M. 2012. Pupal parasitization of *Anaphaeis aurota* Fabricius (Lepidoptera: Pieridae) infesting *Capparis decidua* (Forsk.) by *Brachymeria albicrus* (Klug) (Hymenoptera: Chalcididae). *Journal of Biological Control*, 26 (2): 155-156.

Argaman, Q. 1991-1992. *Parasierola swirskiana* n. sp. (Hymenoptera: Bethylidae) from the lesser date moth, *Batrachedra amydraula* Meyrick (Lepidoptera: Batrachedridae). *Israel Journal of Entomology*, 25-26: 195-198.

Arneson, P.A. and Losey, J.E. 1997. *Integrated Pest Mangement-Lecture Notes*. Cornell University, Ithaca.

Bagle, B. G. 1992. Incidence and control of fruitfly (*Carpomyia vesuviana* Costa) of ber (*Ziziphus mauritiana* Lamk). *Indian Journal of Plant Protection*, 20(2): 205-207.

Balikai, R. A. 1999. Pest scenariao of ber (*Zizyphus mauritiana* Lamarck) in Karnataka. *Pest Management in Horticultural Ecosystem*, 5(1): 67-69.

Balikai, R. A. 2009. Seasonal occurrence of ber fruit weevil, *Aubeus himalayanus* Voss in India. *Acta Horticulture*, 840: 461-474.

Balikai, R. A., Kotikal, Y. K. and Prasanna, P. M. 2013. Global scenario of insect and non-insect pests of jujube and their management options. *Acta Horticulture*, 993: 253-277.

Balikai, R. A., Naik, L. K. and Yelshetty, S. 1998. A record of ber fruit weevil, *Aubeus himalayanus* Voss (Curculionidae: Coleoptera) from Karnataka. *Karnataka Journal of Agricultural Sciences*, 11(2): 508-508.

Batra, H. N. 1953. Biology and control of *Dacus diversus* Conquillet and *Carpomyia vesuviana* Costa and important notes on other fruit flies in India. *Indian Journal of Agricultural Science*, 23: 87-112

Beaumont, P. 1993. *Pesticides, Policy and People*. Pesticide Trust, London.

Ben-Dov, Y. 2006. A systematic catalogue of eight scale insect families (Hemiptera: Coccoidea) of the World: Aclerdidae, Asterolecaniidae, Beesoniidae, Carayonemidae, Conchaspididae, Dactylopiidae, Kerriidae and Lecanodiaspididae. *Elsevier*, Amsterdam, the Netherlands.

Blumberg, D. 2004. Date palm pests and their control management in: Bernstein, Z. (Ed.) Date Growing in Israel. The Plant Production and Marketing Board, Tel Aviv, Israel (in Hebrew). pp. 398-433.

Blumberg, D., Navon, A., Kehat, M., Eliahu, M., Levski, S. And Nakash, J. 2001. Datepalm pests in Israel at the beginning of the third millennium. *Alon Hanotea,* 55: 42-48.

Daniel, D. D. D. R., Marco, A. M. A. B. and Edson, L. L. E. L. B. 2008. Occurance of *Dictyla monotropidia* Stal (Hemiptera: Tingidae) on *Cordia verbenacea* Al. DC in Brazil. Neotrop Entomol 37: 236-238.

Dashad, S. S., Chaudhary, O. P. and Rakesh. 1999. Studies on the incidence of ber fruitfly (*Carpomyia vesuviana* Costa) in south-western Haryana. *Crop Research*, 18(1): 115-118.

Devi, S. J. and Rajasekaran, B. 2011. Biology and Morphology of Aonla Aphid, *Schoutedonia emblica. Madras Agriculture Journal,* 98 (7-9): 277-278.

Dhembare, A. J., Pondhe, G. M., Mandakini, P. and Shobha, G. 1998. Efficacy of some plant extracts against ber fruit borer *Meridarchis scyrodes* (Mayre). *Advances in Plant Sciences*, 11(1):299-301

Djerbi, M. 1994. Diseases and pests of date palm (Chapter XII), In FAO Corporate Document Repository, www.fao.org/docrep/006/y4360e/y4360e0g.htm.

Dowell, R.V., Cherry, R.H., Fitzpatrick, G.E., Reinert, J.A and Knapp, J.L. 1981. Biology, plant-insect relations, and control of the citrus blackfly. *Florida Agricultural Experiment Station Bulletin,* 818: 1-48.

Dreistadt, S. H., Clark, J. K. and Flint, M. L. 2004. Pests of Landscape Trees and Shrubs: An Integrated Pest Management Guide. Oakland: Univ. Calif. Agric. Nat. Res. Publ. 3359.

Elwan, A. A. 2000. Survey of the insect and mite pests associated with date palm trees in Al-Dakhliya region. Sultanate of Oman. *Egypt Agriculture Research,* 78 (2): 653-664.

Farrar, N., Asadi, G. H. and Golestaneh, S. R. 2004. Damage and host ranges of ber fruitfly *Carpomya vesuviana* Costa (Diptera: Tephritidae) and its rate of parasitism. *Journal of Agric Sciences,* 1(5): 120-130.

Gerson, U. and Smiley, R. L. 1990. Acarine biocontrol agents -an illustrated key and manual. Chapman and Hall, London, UK.

Gopali, J. B., Sharanabasappa and Yelshetty, S. 2003. Incidence of ber fruit borer, *Meridarchi scyrodes* Meyrick (Lepidoptera: Carposinidae) in relation to weather parameters. *Insect Environment,* 9(4): 165-166.

Gour, T. B. and Sriramulu, M. 1994. *Aubeus himalayanus* Voss (Curculionidae: Coleoptera), ber fruit weevil, a new pest of *Ziziphus* from India. *Indian Journal of Plant Protection,* 22(2): 228-229.

Grewal, J. S. and Kapoor, V. C. 1986. Bird damage and its effect on infestation by fruit-ßies in various orchards in Ludhiana. *Indian Journal of Agricultural Sciences,* 56: 370-373.

Gyi, M. M., Lal, O. P., Dikshit, A.K. and Sharma, V. P. 2003. Efficacy of insecticides for controlling ber fruit fly. *Annals of Plant Protect Sciences,* 11(1): 152-153.

Haldhar S M, Choudhary B R, Bhargava R and Sharma S K. 2014. Development of an organic integrated pest management (IPM) module against insect-pests of muskmelon in arid region of Rajasthan, India. *Journal of Experimental Biology and Agricultural Sciences* 2: 19-24.

Haldhar SM, Bhargava R, Krishna Hare, Berwal MK and Saroj PL. 2017. Does bottom-up effects of different host plant resistance cultivars on ber (*Ziziphus mauritiana*)-fruit fly (*Carpomyia vesuviana*) interactions?. *Crop Protection* (Under review).

Haldhar S. M., Sharma S. K., Bhargava R., Singh R. S., Sharma B. D. and Singh Dhurendra 2013a. Insect pests of arid fruit crops: practical manual for field identification and damage symptoms. CIAH/Tech./Pub. No. 42, pp. 1-53.

Haldhar S.M., Samadia D.K., Bhargava R., Singh D. 2017. Host plant genotypes determine bottom-up effect of *Cucumis melo var. callosus* against melon fruit fly. Crop Protection, 98, 157-165.

Haldhar, S. M. and Singh, R. S. 2014. Report of *Dictyla cheriani* (Hemiptera: Tingidae) on Indian cherry (*Cordia myxa*) in Rajasthan, India: Incidence and morphometric analysis. *Indian Journal of Agricultural Sciences,* 84 (1): 128–130.

Haldhar, S. M., Karuppaiah, V., Sharma, S. K. and Singh, R. S. 2010. Population dynamics of lemon butterfly (*Papilio demoleus*) in *bael* (*Aegle marmelos*) as influenced by abiotic factors in arid region of Rajasthan. *Indian Journal of Arid Horticulture,* 5 (1-2): 50-52.

Haldhar, S.M., Bhargava, R., Choudhary, B.R., Pal, G., Kumar, S., 2013b. Allelochemical resistance traits of muskmelon (*Cucumis melo*) against the fruit fly (*Bactrocera cucurbitae*) in a hot arid region of India. Phytoparas. 41, 473-481.

Haldhar, S.M., Deshwal, H.L., Jat, G.C., Berwal, M.K., Singh, D., 2016. Pest scenario of ber (*Ziziphus mauritiana*Lam.) in arid regions of Rajasthan, a review. J. Agricul. Ecol. 1, 10-21.

Haldhar, S.M., Karuppaiah, V., Sharma, S.K., Singh, D. 2012. Insect-pests of ber (*Ziziphus mauritiana* Lam.) as influenced by abiotic factors in arid region of Rajasthan. Global conference on "*Horticulture for food, nutrition and livelihood options*" organised by ASM foundation, New Delhi and OUAT, Bhubaneswar. Odisha during 27-31, May, 2012.

Horticulture Dairy, 2004. Published by Andhra Pradesh Horticultural Officers Association, Public Gardens, Hyderabad. 43 pp.

Howard, F. W., Moore, D., Giblin-Davis, R. M. and Abad, R. G. 2001. Insects on palms. CAB International, Wallingford, UK.

Ingole, D. A., Deotale, V. Y., Deotale, R. O and Ramkumar, J.. 2005. Preliminary study on the performance of *Mallada boninensis* (Okamoto) against *Aleurocanthus woglumi* Ashby on citrus. *Journal of Biological Control,* 19(1): 77-80.

Jhala, R. C., Bharpoda, T. M., Patel, M. G., Patel, V. B. and Patel, J. R. 2003. Survey on insect pest and their natural enemies occurring in aonla orchard under middle Gujarat condition. Paper presented in Nat. Seminar on production and utilization of aonla, organized by *All India Aonla Growers Association,* Salem, Tamil Nadu, India.

Karuppaiah, V., Haldhar S. M., Sharma, S. K. and Singh, R. S. 2010. Incidence of major pests in Bael, *Aegle marmelos* as influenced by abiotic factors in arid region of Rajasthan. National Seminar on '*Precision Farming in Horticulture*' organized by College of Horticulture & Forestry, Jhalarapatan, Jhalawar– 326 023 (Rajasthan), pp 272.

Krishnamurthy, A. and Singh, S. P. 1986. Record of the egg parasite, *Trichogramma chilonis* on *Papilio* spp. in citrus. *Current Science*, 55 (9): 461.

Lakra, R. A. 2004. Insect pests of fruits of arid and semi arid regions of India. In *pest management in horticultural crops* (ed. Verma, L. R., Verma, A.K. and Gaudam, D. C.). Vedams Books Pvt. Ltd., Delhi. Pp 544.

Lakra, R. K. and Singh, Z. 1989. Bionomics *of Zizyphus* fruitfly, *Carpomyia vesuviana* Costa (Diptera: Tephritidae) in Haryana. *Bulletin of Entomology*, 27: 13-27.

Lal, M. N., Joshi, G. C. and Singh, S. S. 1996. Studies on incidence of *Inderbala* spp. *Cerciaphis indica* on aonla in eastern Uttar Pradesh. *Progressive Horticulture,* 28(1-2) 66-75.

Larsen, T. B. and Nakamura, I. 1983. The butterflies of East Jordan. *Entomologist's Gazette*, 34: 135-208.

Lewis, C. B. 1976. New swallowtail butterfly. The appearance of *Papilio andraemon* in Jamaica. Natural History Notes of the Natural History Society, 25: 15.

Mari, J. M., Chachar, Q. I. and Chachar, S. D. 2013. Organic management of fruit fly in jujube ecosystem. *International Journal of Agriculture Technology*, 9(1): 125-133.

Mohamed, G. S. 1991. Studies on the date scale insect, *Paralatoria blanchardii* (Targ), Homoptera: Diaspididae. M. Sc. Thesis, Faculty of Agriculture, University of Khartoum, Sudan.

Mullen, J.D. 1995. "Estimating Environmental and Human Health Benefits of Reducing Pesticide Use through Integrated Pest Management Programs." Master Thesis. Virginia Polytechnic Institute and State University, USA.

Nandihalli, B. S., Patil, D. R., Jagginavar, S. B., Biradar, A. P., Guled, M. B. and Surkod, V. S. 1996. Incidence of fruit borer (*Meridarchis scyrodes* Meyr.), and fruit fly (*Carpomyia vesuviana* Costa) on different varieties of ber. *Advances in Agriculture Research in India,* 6:13-18.

Narayanamma, V. L., Savithri, P. and Rao, R. 2001. Influence of citrus butterfly *Papilio demoleus* L. damage on growth parameters of the sweet orange host plant. *Indian Journal Plant Protection,* 29: 140–141.

National Tree Seed Project 1999. Annual progress report of national tree seed project. Addis Ababa, Ethiopia.

Navon, A. 2000. *Bacillus thuringiensis* application in agriculture in: Charles, J.F., Del′ecluse, A. and Nielsen-LeRoux, C. [Eds.] Entomopathogenic Bacteria: From Laboratory to Field Application. Kluwer Academic Publishers, Dordrecht, the Netherlands. pp. 355-369.

Navon, A., Nakash, J., Blumberg, D., Levski, S., Gan-Mor, S., Bechar, A. 1999. The effectiveness of *Bacillus thuringiensis* in granular feeding baits for use against the lesser date moth, *Batrachedra amydraula* in date palms, in comparison with chemical and biological control agents (1995-1999). *Alon Hanotea*, 53: 484-489.

Orwa, C., Mutua, A., Kindt, R., Jamnadass, R., Simons, A. 2009. Agro forest tree Database: a tree reference and selection guide version 4.0 (http://www.worldagroforestry.org/af/treedb/)

Patel, I. S, Upadhyay, V. R., Vyas, H. N. and Patel, A. T. 1990. Field bioefficacy of different insecticides for the control of ber fruit fly, *Carpomyia vesuviana* Costa (Tephritidae: Diptera).*Gujarat Agricultural University Research Journal*, 16 (1): 57-59.

Pathak, K. N. and Rizvi, P. Q. 2002. Age specific life table of *Papilio demoleus* on different hosts. *Annals of Plant Protection Sciences*. 10: 375-376.

Patil, P. and Patil, B.V. 1997. Management of ber fruit borer, *Meridarchis scyrodes* Meyric (Carposinidae: Lepidoptera). *Karnataka Journal Agricultural Science*, 10 (1): 52-55.

Pramanick, P. K., Sharma V. P. and Singh, S. K. 1995. Genetic behaviour of some physical and biochemical parameters of fruit fly (*Carpomyia vesuviana* Costa.) resistance in ber. *Indian Journal of Horticulture*, 62(4): 389-390.

Rajaram, V. and Siddeswaran, K. 2006. Efficacy of insecticides and plant products against the fruit borer and fruit fly in ber under rainfed condition. *International Journal of Agricultural Science*, 2(2): 538-540.

Ramasubbareddy, K. 1975. Surveillance of insect pests and natural enemies. All India Coordinated Citrus Improvement Project, Tirupati. Annual Report. pp.16.

Resham, B. T., Fanindra, P. N. and Butani, D. K. 1986. Insect pests of citrus in Nepal and their control. *Pestology*, 10 (4): 24-27.

Sayed, A. A., Temerak, S. A. and Vergoulas, P. 2001. Comparative performance of *Bacillus thuringiensis* subsp. *kurstaki*, and the natural product, spinosad for the control of the lesser date moth, *Batrachedra amydraula* Meyer infesting date palm trees in New Valley, Egypt. *Journal of Agriculture Science,* 32: 183-189.

Shivankar, V. J. 1999. Recent trends in insect pest management of citrus. Hi-tech citrus management: Proceedings of International Symposium on citriculture (Nov. 23-27) held at NRC for citrus, Nagpur, pp.773-774.

Siddig, S. A. 1975. Field control of the scale insect, *Parlatoria blanchardii* (Targ) infesting date Palm in Sudan. *Journal of Horticulture Science* 1: 13-19.

Singh, M. P. 1989. Biosteres vandenboschi Fullaway- a new braconid parasite of Carpomyia vesuviana Costa from the Indian desert. Entomon, 14: 169.

Singh, M. P. 2008. Managing menace of insect pests on ber. *Indian Horticulture,* 53 (1): 31-32.

Singh, M. P. and Vashistha, B. B. 2002. Factors influencing infestation of fruit fly *Carpomyia vesuviana* in Indian jujube. *Indian Journal of Agricultural Sciences*, 72 (9): 543-547.

Singh, S. and Kumar, K. 2011. Diofenolan: a novel insect growth regulator in common citrus butterfly, *Papilio demoleus*. *Phytoparasitica*, 39 (3): 205-213.

Singh, S., Gupta, R. N., Awasthi, B. K. and Verma, R. A. 2000. Effective control of ber fruit fly *Carpomya vesuviana* by insecticidal schedule. *Indian Journal of Entomology*, 62(2): 171-174.

Solunke, B. R. and Deshpande, S. V. 1991. Studies on use of plant products for the control of lemon butterfly larvae. *Journal of Maharastra Agric University*, 16 (2): 302-303.

Srivastava, A. K., Nanda, U. S., 1983, Studies on the control of insect pest at Udaipur. Tech.Doc. No.13 Second National Workshop on Arid Zone Fruit Research, Sukhada University,Udaipur, p.288.

Stewart, J. L. and Brandis, D. 1992. The forest flora of North-West and Central India. Reprinted by Bisen Singh and Mahendra Pal Singh, New Connaught Place, Dehradun, pp. 602.

Suwarno, P. 2012. Age-specific life table of swallowtail butterfly *Papilio demoleus* (Lepidoptera: Papilionidae) in dry and wet seasons. *Biodiversitas*, 13(1): 28-33.

Suyog Pande, Pandey, J. P. and Tiwari, R. P. 2012. Effect of some botanicals on haemocytes and molting of *Papilio demoleus* larvae. *Journal of Entomology*, 9 (1): 23-31.

Tibebu H. 2002. Terminal report on forest entomology research activity. -Forestry Research Center, Ethiopian Agricultural Research organization, Addis Ababa, Ethiopia.

Ucko, O. and Bitton, S. 2000. Recommendations for the control of pests and diseases of the date palm. Israel Ministry of Agriculture, Extension Service and Diklaim Organization (in Hebrew).

Usharani, B., Vijayasamundeswari, A., Saraswathy, S., Indra, N., Parthiban, S., Suresh, K. and Selvarajan, M. 2012. Diversity of insect pests of acid lime in Tamilnadu. National Dialogue on Citrus Improvement, Production and Utilization (27-29, Feb) held at held at NRC for citrus, Nagpur. pp.246.

Vyavasaya Panchangam, 2007-08. Published by Acharya N.G.Ranga Agricultural University, Rajendranagar, Hyderabad. pp 156-157.

Yelshetty, S., Balikai, R. A., Biradar, B. M. and Mokashi, A.N. 1996. Management of ber fruit borer, *Meridarchis scyrodes* Meyrick (Carposinidae: Lepidoptera). *Karnataka Journal of Agricultural Sciences*, 9 (3): 529 532.

Zaid, A. 1999. African palm weevil, *Rhynchophorus phoenicis* F. attack on date palm in the Republic of South Africa and Zimbabwe. Middle East Red Palm Weevil Workshop, 26-29 January, 1999; Cairo-Egypt.

28

Increasing Farmer's Income from *Ber* in Arid Regions

Hare Krishna, P.L. Saroj, S.K. Maheshwari, S.M. Haldhar and Lokesh Kumar

ICAR-Central Institute for Arid Horticulture, Beechwal, Bikaner-334 006

Introduction

Arid regions are characterized by stressful edaphic and environmental conditions such as low and erratic rainfall coupled with limited irrigation resources, high evapo-transpiration and extreme temperature regimes. Under such environment, the productivity of agriculture crops is far behind the satisfactory level, rendering cultivation of field crops an unprofitable venture. The recent aware-ness about potential of ecologically fragile lands for cultivation of fruit trees has not only opened up scope for providing economic sustainability for the natives, especially small farmers, but also for bringing new areas under fruit production. *Ber* (*Ziziphus mauritiana* Lamk.) or Indian jujube, which belongs to the family Rhamnaceae and said to be indigenous fruit crop. *Ber* is grown in India traditionally from ancient times where it has been in use for almost 4000 years (Mittre, 1961). It is one of the most suitable fruit trees for arid and semi-arid regions. It is also known as 'King of Arid Fruits', Chinese date, Chinese fig or poor man's fruit. *Ber* can do well even under marginal growing conditions and provides quality yields at low cost (Shukla, 1996). Cultivation of *ber* under the harsh conditions of the Thar desert of India (west Rajasthan) demonstrates its adaptation to desert conditions. Flowering and fruiting of *ber* coincide with the availability of maximum rain water (monsoon rains), which is during July-September. It has a long tap-rootand can withstand high temperatures during the summer. During the dry hot summer (April-June), it undergoes dormancy by shedding its leaves, thus evading the injury of drought. It can produce fruits even under rainfed situation where average annual rainfall is 150-200 mm. It is truly a desert apple for the dwellers of Thar Desert. The tree propagates freely and greatly resists stress conditions in regions experiencing recurrent droughts. It is thus an important tree suitable for integration into the agro-forestry systems in the warm desert eco-regions. The tree can provide economic sustenance to the region and insurance against ecological degradation.

The ripe fruits are rich in nutritive value, having 13-24 per cent total soluble solids and up to 160 mg/ 100g vitamin C (Morton, 1987; Sharma and Kore, 1990). The popular *ber* cultivar Gola contains 80-82% moisture, 12.5 % carbohydrates, 17-19 °Brix TSS and 0.40-0.70 % acidity. The dehydrated fruits are kept for a long time and are consumed in the off season. The ripe fruits have cooling effect, laxative; remove burning sensation, thirst and blood impurities, whereas dried fruits are laxative and an appetizer. Among processed products *ber* preserve, candy, squash, jam, nectar etc. can be prepared. Stem bark, root and leaves also have some medicinal values. Leaves of *ber* are used as fodder in dry regions. *Ziziphus xylopyrus* is a host plant for rearing the *Tachardia lacca,* a lac insect (Sharma and Kore, 1990). The pruned wood is used as fuel wood. *Ber* has been recognized as an underutilized species worthy of further research by government and research organizations under the Indian National Genetic Resources Program and International Centre for Underutilized Crops, University of Southampton, UK (Azam-Ali *et al*. 2006, Krishna and Parashar, 2013) because of the benefits already mentioned. This presents great opportunities for exploitation of *ber* fruits for its contribution toward food security, and value addition from their diversified uses in arid regions.

Present Status in India and Global Scenario

Indian jujube or *ber* originated from Central Asia and then spread to North Africa and India through Afghanistan, South China, Malaysia and Australia. It was in use around 1000 BCE and is now widely naturalized in the tropics (Africa, Central and South America, and the West Indies), in the Pacific Islands and in the Mediterranean region (Orwa *et al*., 2009). It is only commercially important in India and China (Orwa *et al*., 2009). In South-East Asia, it is mostly found in Thailand. It is becoming increasingly important for its wide adaptation, easy management, early fruit bearing, value as a food and feed, and multiple uses (Liu *et al*., 2014). Chinese jujube and Indian jujube are widely cultivated with a total growing area of over 2 million hectares with an annual production of over 6 million tons of fruit. Approximately 30 million farmers rely on jujube production for their livelihood (Liu *et al*., 2014).The *ber* is distributed worldwide including Indian sub-continent, South-east Asia, Australia, China, Africa, Mediterranean region and American centre but its cultivation is confined over drier part of the globe and main cultivation occurs in India. India ranks second among jujube growing countries in the world after China (Heuzé *et al*., 2017).

The important provinces for jujube cultivation in India are Maharashtra, Gujarat, Uttar Pradesh, Tamil Nadu, Jammu & Kashmir, West Bengal, Haryana, Rajasthan, Madhya Pradesh, Punjab, Karnataka, Bihar, Chhattisgarh and Andhra Pradesh. In Punjab, the districts of Sangrur, Bathinda, Ferozepur, Ludhiana, Patiala, and Mukatsar are the most famous for grafted jujube cultivars. In Haryana, the district of Hisar, Rohtak, Jind, Panipat, Mohindergarh and Gurgaon are famous for commercial cultivation in jujube. In Uttar Pradesh jujube orchards are found around Varanasi, Aligarh, Faizabad and Agra. In Rajasthan its orchards are spread around Bharatpur, Jaipur and Jodhpur. In Gujarat, jujube orchards are found around Banaskantha, Sabarmati, Bhavnagar, Surendranagar, Patna, Ahmedabad, Bharuch, Vadodra, Sabarkantha and Mehsana. In South India, jujube cultivation is found around Tirunelveli, Ramanathapuram, Dharmapuri and Salem districts in Tamil Nadu and Bijapur, Bellary, Gulbarga, Belgaum, Raichur and Bidar in Karnataka. In West Bengal, this fruit is quite popular in the districts of Murshidabad, Malda, Bankura and Birbhum (Bal, 2014). The annual production of jujube in India is estimated to be 5, 26,000 metric tons from an area of 50,000 ha during 2016-17 (Anon., 2017).

Gaps in Prevailing System

Certain gaps such as insufficient knowledge about suitable cultivars, propagation technique, insufficient nutrient & water budgeting, non-adoption of moisture conservation practices, lack of knowledge about creation of micro-climate around plants etc. has rendered realization of full cropping potential of ber a herculean task. This could be the reason that ber remained a minor fruit crop and remained confined to specific areas, particularly in rural and tribal belts; though, its fruits are unique in their its own taste, palatability and nutritive value and are incomparable with any other fruit commodity.

Advance Approaches for *Ber* Cultivation in the Arid Region

Cultivars

Very wide variability with respect to plant and fruit characters exists in *ber* genotypes. In India, more than 300 varieties have been listed but only a few, such as Umran, Banarsi Kadaka, Mundia, Gola, Seb and Kaithali, are commercially important (Krishna *et al.*, 2014).

Banarasi Karaka: Tree tall, vigorous, 8-12m in length. Leaf green, elliptic to ovate, fruit globose, oblong in shape, fruit size 5.2x3.0 cm.

Banarasi Pebandi: It is an excellent variety of Uttar Pradesh and is grown widely. Fruit weight 25.2g, pulp 95.2 per cent and yield 110 kg per tree.

Seb: It is a late maturing cultivar with upright growth habit. The fruits are apple- shaped of about 25 g in weight and the flesh has 17.5 ^{0}Brix TSS. Its slightly rough fruit surface remains greenish yellow even after maturity. The cultivar produces little lower yield (40-50 kg/plant) under rainfed while under irrigated conditions the yield is equal to that of Gola.

Chhuhara: Tree semi-tall with spreading branches. Leaf ovate oblong with short petiole and serrated margins from base to apex. Fruit ovate oblong 2.9cm long and 2.1 cm thick, weight 6.8g, stone sub globose and sharply pointed, 1.6 cm long and 0.55cm wide.

Gola: Tree spreading, leaves cordate, broad, fruit almost round, styler end flat, early cultivar, fruit size 3.8-3.5 cm.

Goma Kriti: It is high yielding with 25.8-38.2% increase in yield. Its fruit have good keeping quality. The physiological loss in weight is just 12%, where as it is 27% in early maturing cultivars. The shelf life is also higher (6 days). It can fetch good price in distant market. Owing to earliness, it is less infected by fruit fly and fruit borer. The incidence of powdery mildew is also low.

Illaichi: Tree spreading, leaves obovate. Fruit oblate to roundish in shape, mature fruit 2.3 cm in length and 2.3 cm in breadth.

Jogia: Tree spreading in habit, leaves oval, dark green, apex acute, fruit ovate and fruit size 3.5x2.66 cm.

Kaithli: Tree is medium, bushy, thorns straight. Leaf is ovate, margin serrated, apex obtuse. Fruit ovate oblong, 3.37 cm long and 1.9 cm wide, weight (6.22 g)stone elliptic oblong and pointed, 2.2 cm long and 0.65 cm wide.

Meharun: Meharun is a popular cultivar if Gujarat and Maharashtra. It bears fruits profusely. The fruit size is small (2.44 -1.98cm), average weight (5.2g), oval in shape, base and apex is round, surface smooth, yellow to light brown in colour. Skin is medium thick and firm. The taste of fruits is sweet. The average pulp percentage is 87.30, pulp stone ratio is low (6.87). The seed is large in size, it form 12.69% of the fruit weight. This is supposed to be resistant from fruit fly.

Sanaur 5: This is a seedling selection. It has large oblong smooth fruit with golden yellow skin, TSS 20-21%. The fruit ripens in the second half of March and yield 150 to 200 kg fruit per tree.

Umran: It is known for heavy production, fruit big, oblong is shape weighing 32.5g, skin thick and hard, pulp 96.45%, average production 150-200 kg/ tree.

ZG 3: Tree spreading, leaves oblong, dark green apex, obtuse. The fruit are obovate, roundish or ovate in shape, fruit size 4.0x3.0 cm.

Mundia: It is an early to mid season cultivar. Its tree has semi-erect growth habit. It bears bell-shaped fruits of nearly 40 g weight having soft flesh having about 20 ^{0}Brix TSS. Average fruit yield under irrigated conditions is about 100-120 kg/tree.

Thar Bhubharaj: It is selection from the local material of Bhusawar area (near Bharatpur district) of Rajasthan. Average fruit weight is 27g, TSS 23%, total sugar is 4.29% and ascorbic acid content is 60mg/100 g fruit. Average stone weight is 1.3g, dry matter content is about19.87%. Fruit yield from five year old plant is about 30-35 Kg/ tree.

Thar Sevika: It is developed by the hybridization of Seb x Katha. The average fruit weight is about 25g, TSS 24.3%, total sugar is 5.02% and ascorbic acid content is 88 mg/100 g fruit. It is early maturing variety. Fruit yield from five year old plant is 28-30 Kg/tree.

Thar Malti: Recently a chance seedling selection, which is precocious bearer and high yielding (65-70 kg/tree) has been identified at ICAR-CIAH, Bikaner.It is semi-erect and less thorny. These two characteristics are desirable traits as they facilitate easy harvesting of fruits. It is a late maturing variety; however unlike commercial variety Umran, it has soft texture pulp that is, organoleptically, a desirable trait. Further, even unripe fruits have soft pulp texture.

Apple Ber: Apple Ber is also known as 'Thai Ber' or 'Taiwan Ber' as it has been in cultivation since the period of Japanese occupation in Taiwan and introduced to other parts of world, perhaps, from Thailand. In India, it might have been introduced by nurserymen of the country from Taiwan, Thailand and Bangladesh. Besides, it is popular as 'Telangana apple' among the growers of Telengana. The use of the term 'Apple' refers to its resemblance to green apples. The good reputation of 'Apple Ber' is, primarily, because of its big fruit size. The plant has no thorns; thereby, facilitate intercultural operations.

Apple ber is gradually gaining popularity among medium farmers. Apple Ber cultivation is catching up very fast, especially, in ber growing areas of Gujarat, Maharastra, Karnataka, Telangana, Andhra Pradesh, W.B., Chhattisgarh, Madhya Pradesh and Rajasthan. It can be cultivated from tropical and subtropical and semi arid to arid climates all over India and can thrive well up to an elevation of 1000 m above MSL. Optimum soil pH for its cultivation should vary between 6.0-7.5.

It can be successfully propagated by Patch budding with 70-80% success in the month of June. Ideally it may be planted during monsoon season. Four hundred plants can be accommodated per hectare. It starts flowering six-nine months after planting. Flowering begins from September onwards and fruit ripens during January. From second year, yield of 8-12 kg/tree can be harvested. Full bearing tree can yield 50-90 kg/tree. Single fruit weigh around 60-120 g under rainfed conditions. Because of bigger fruit size harvesting becomes easy. Farmers are reported to earn up to Rs 3 to 4 lakh per annum from over 400 ber trees. Shelf life is comparatively better in comparison to the prevalent ber varieties of country, therefore, fetches higher price in the market.

Both at ICAR-CIAH, Bikaner as well as its regional station CHES, Godhra, Gujarat the performance of Apple Ber is being evaluated for standardization of its cultivation practices and final recommendation to the growers of arid and semi-arid regions.

Propagation

Seeds of *bordi* (*Ziziphus rotundifolia* Lamk) are generally used for raising rootstock, which are easily available and possess the qualities of a good rootstock (Vashishtha, 1982). Seeds of *jharber* (*Zizyphus nummularia*) can also be used as rootstock (Pareek, 2001; Bal, 2013). The seedlings of *jharber* are slow growing and become buddable after longer period than the seedlings of *bordi* (Sharma and Kore, 1990). The *ber* is usually propagated by budding. The most common methods are modified ring, patch and I or T (shield) budding (Kaundal *et al.*, 1984). Rootstock seedlings are raised by sowing seed kernels extracted by breaking the stones which germinate in about one week. Seed stones collected from dropped fruits contain 50-70 percent non-viable seeds. Germination of seed stones can be improved by soaking them for 48 hours in water or by treatment with sulphuric acid. The ber seedlings are raised from *bordi* stones, which are sown during March-April, after fresh extraction, in well-prepared nursery field at a distance of 15 cm in rows 30 cm apart. Nearly one-fourth of the seedlings attain buddable size of a lead pencil by August, while the rest are ready for budding by April next. The stock seedlings should be healthy and vigorous and it is allowed to grow as a single stem only. The budding is done when the stock stem has attained the thickness of a lead pencil. Budding operation should be done when there is proper flow of sap in the stock to be budded.

To minimize root damage during transplanting, seeds can be sown in 300 guage polythene tubes of 25 cm length and 10 cm diameter, filled with a mixture of FYM, sand and clay in 1: 1: 1 ratio. When raised in polytubes, the budlings become ready for transplanting 1-2 months after budding during August-

September. Plants raised by polytube technique have been observed to retain deep rooting tendency and thus prove the most suitable under the low rainfall areas (Pareek, 1978).

Planting

Transplanting of budlings or rootstock seedlings is best done with onset of monsoon. For this, pits of 60 x 60 x 60 cm are dug during summer at a spacing of 6 x 6 m in low rainfall areas and at 8 x 8 m in irrigated and higher rainfall regions. While filling the pits in sandy areas, a layer of bentonite clay can be placed at the bottom and sides of the pit to reduce moisture losses by infiltration. Each pit is refilled with top soil mixed with 15-20 kg FYM alongwith 50 g of chlorpyriphos dust to protect damage by termite. In rainfed areas, the interspaces between the tree rows can be shaped to provide 5 per cent slope towards the plant. This would help to accumulate runoff water during monsoon near the tree roots thereby resulting in higher establishment success. In irrigated areas, *ber* plants can be transplanted during January-March also.

Training and Pruning

The pruning of *ber* trees is highly desirable to maintain their vigour and productivity as well as to improve fruit size and quality. Pruning also saves the fruit from being affected by the powdery mildew disease and strong winds. The *ber* tree remains young upto 30 years, if proper pruning is done regularly. During the first 2-3 years after planting, *ber* trees are trained to develop a strong frame. After planting during July, the main shoot is headed back during March keeping 1-2 nodes above the bud union to induce vigorous new growth from which one upright vigorous shoot is retained to develop into the main trunk. This is kept clean of side shoots up to 30 cm height from ground level and thereafter 3-4 well spaced shoots are allowed to grow. The main trunk is then headed back. During the spring of the second year, the secondaries are pruned to retain their basal buds as these basal buds produce vigorous shoots. More than one shoot may emerge from the basal bud, but only one of these is retained, while rest is removed. These form the main branches of the main tree. On these main branches also, 3-4 upward growing well spaced shoots are allowed to grow. The process is continued to develop tertiary branches. Care is taken to allow only one upright growing shoot at each node to avoid narrow crotches and thereby develop a strong frame. This basic frame of the tree is maintained by removal of water sprouts from time to time.

In unpruned ber tree, the canopies of the trees get un-necessarily enlarged, the growth and branchlets become weak and both fruit size and quality gets impaired. Ultimately, such trees become economically unproductive besides occupying large orchard space. Some thinning out of the branches of *ber* trees is also necessary to avoid too much crowding so as to admit adequate sunlight and facilitate proper aeration. Moreover, Fruit bearing in *ber* on current season's growth and the fruit quality depends on the vigour of shoots. Therefore, annual pruning is done to induce maximum number of healthy new shoots (Singh *et al.*, 2004). The best time for pruning is during the hot and dry season when the tree sheds leaves and enters into dormancy.

Nutrient Management

Ber orchards are not adequately manured. However, productivity of trees can be improved by nutrient application. The doses of manures and fertilizers depend on the fertility status of the soil. Proper nutrition of *ber* tree is necessary to get good crop over the years (Anon., 2014). The fruit becomes large and attractive and get decent price in the market. In sandy soils at Jobner (Rajasthan), application of 750 g N per tree gave the highest yield while at Rahuri (Maharashtra), 250 g N and 250 g P_2O_5 increased fruit yield. Potash application does not give any response. Besides, an yearly dose of 40-50 kg well rotten FYM per tree is also applied. However, when no recommendation is available, 20 kg farmyard manure and 100 g nitrogen (400 g CAN) can be applied for one year old *ber* tree. Similar amount of farmyard manure and nitrogen should be increased every year up to the age of five years. The quantity of farmyard manure and nitrogen should be stabilized at 100 kg and 500 g (2 kg CAN), respectively, after the age of five years. Half of the phosphatic fertilizers may be applied during rainy season (July-August) and the other half at the time of fruit-set (October-November). The fertilizer should be evenly spread in the basins of trees upto the periphery. After adding the fertilizer, light hoeing with spade or khurpa should be given to the basins to mix it thoroughly with the soil. In sodic soils, having pH over 8.5 and ESP over 21 per cent, *ber* can be successfully planted if amendments such as gypsum and pond soil are added to each pit at the time of filling.

Water Management

For establishment of young plants, 10 irrigations are considered essential during the first year. Irrigation is also essential during the development of fruit, *i.e.* from October to February at intervals of 3 or 4 weeks depending upon the weather. Trees will continue to bear even if no irrigation is applied during this period but the yield is substantially reduced because of heavy fruit drops and smaller size of the remaining fruit. Irrigation should be stopped in March as fruits on the branches lying on the ground get damaged and their ripening is delayed. During dormancy, *ber* need little or no irrigation; therefore, if irrigation is applied during this period, the trees would continue to put fourth growth haphazardly which is not desirable. In rainfed orchards, arrangement for *in situ* water harvesting by giving 5 per cent slope to the inter-row spaces towards the trees has been found be useful. For establishment of young budlings in the field, a double walled pot known as *Jaltripti* has been found to save 75 per cent irrigation water (Gupta *et al*. 1991). The plant is planted in the inner pot (hollow on both ends) and the water is filled in the outer pot (sealed at the bottom). Since the pot is made of clay, water seeps through the wall of inner pot and thus available to the plants. The growth of plants grown in *Jaltripti* was better than those planted directly in the pit.

Mulching by use of black polythene conserves soil moisture and improves growth of the *ber* trees. Sprays of antitranspirants like 0.1 per cent power oil and 7.5 per cent Kaolin have also been found effective to conserve water in the her trees.

Fruiting and Yield

The *ber* tree grows quickly and and the first crop can be harvested within 2-3 years of planting. *Ber* is a non-climacteric fruit and all the fruits on the tree do not ripen at one time. Therefore, the fruit should always be picked at the right stage of maturity, i.e. when it is neither under-ripe nor over-ripe. It should be picked when it has acquired normal size and characteristic colour of the variety, e.g. golden yellow colour in Umran. Picking should be done in the forenoon. Time of harvesting depends on cultivar and agroclimatic conditions. In south India, the fruits are harvested during October-November, in Gujarat during December-March, in Rajasthan during January-March and in Haryana, Punjab and Uttar Pradesh during February-April. Early maturing cultivars ripen during middle of February, mid season cultivars during March and the late cultivars ripen by the end of March to mid of April. The average yield in different varieties during the prime bearing period (10-20 years) ranges between 80-200 kg per tree. Under rainfed conditions, 50-80 kg fruit per tree can be obtained.

Some special agro-techniques have to be adopted for successful *ber* production under arid environment:

Shelter Belts

Productivity of fruit plants under arid agro-climate is often poor. This can be improved by creating microclimate conducive for good productivity by plantation of shelter belts around the orchard especially on the western and northern sides. For temporary shelter, fast growing short duration bushes can be planted around individual blocks. To ensure protection from damage by wild animals common in the arid region, thorny bushes (*boradi, karonda, jharber,* etc.) are planted between the shelterbelt tree rows.

Multiple Cropping

Owing to great uncertainty in production monocrop culture in arid and semi arid region is highly risk prone. Crop failures are common and lead to great economic losses. Therefore, planting of more than one fruit species in combination can ensure some production in case a particular crop fails. During young age of the orchard, short duration fruits or crops can be grown as filler or as intercrop. While selecting a filler or intercrop, its suitability in the region with respect to growth, flowering and fruiting seasons and market demand must be considered. Under hot arid regions of Rajasthan (at ICAR-CIAH, Bikaner), ber cv. Gola was planted at three spacing (6 x 6m, 8 x 8m, 16 x 4m) with the aim to standardize the best spacing for ber cultivation in combination with intercrops. Different crop combinations such as Ber + Wheat – Groundnut, Ber + Clusterbean – Mustard, Ber + Indian Aloe and Sole ber were evaluated. Among these combinations, Ber + Clusterbean – Mustardwas found the best one as it gave ber fruits (15.6 q/ha), green fodder (12.5 q/ha), fuel wood (8.5 q/ha), cluster bean (6.78 q/ ha) and mustard (1.22 q/ha); thereby, realization of the net return of Rs 68,264/ha/yr for the growers. Further, results showed that the wider spacing can be followed for taking intercrops on long term basis, else standard spacing can be followed to take intercrops only during the pre-bearing phase of ber orachards (Saroj *et al.*, 2003). Likewisein the semi-arid regions e.g. at Godhra, combination of *soobabool* and *Eucalyptus* with *ber* was found to be good while in arid sandy plains at Jodhpur horti-pastoral management growing buffer grass was found better. Leguminous crop can be preferred as they also enrich the nitrogen in the soil. In irrigated areas, pea, gram, cumin, chilies, fenugreek can be taken whereas, under rainfed conditions *moong*, moth cowpea and *guar* can be taken during *Kharif* season.

Planting Techniques

i. Rain water can be concentrated near the root zone of young plants by sloping the land surrounding it or by diverting runoff from the nearby catchments. While determining the size of microcatchment, factors such as natural rainfall of the region, water requirement of the crop, characters of soil and soil surface, nature of root growth and spread, etc. are kept in view. In any case, the water requirement of the tree species for its optimum production must be met by rainfall and the runoff from the microcatchment. Thus, runoff from the microcatchment at one point of time should not exceed the moisture storage capacity of this soil profile.

ii. Trickling units can be put around each plant which consists of a tube carrying three or four drippers connected to a tin or pot full of water placed near the plant.

iii. To check evaporational loss of water from soil, black polythene or locally available weeds, crop residues and other organic wastes can be used as mulch material. *Bui* (*Arva* sp.), *Sarkanda* (*Saccharum* sp.), paddy straw and paddy husk and groundnut shells can be effectively used for this purpose. Organic mulches also add organic matter to the soil (Bal and Singh, 2012).

iv. Water loss from the leaf surface as a result of intense radiation and high wind velocity can be reduced by spray of radiation reflectants (Kaolin, phenyle mercuric acetate, power oil, waxol, etc.).

v. Shallow tilling can create dirt mulch 2 to 3 inches deep which dries out easily but is discontinuous from the subsurface water, preventing further loss. Tillage must be repeated after each rain to restore the discontinuity.

Disease Management

Many fungal diseases such as powdery mildew, black leaf spot, Fruit rot, *Cercospora* leaf spot and *Alternaria* leaf spot in ber fruit crop have been identified in arid and semi- arid regions.

1. Powdery Mildew

Powdery mildew of ber is caused by fungal pathogen, *Oidium erysiphoides* f. sp. *ziziphi Yen and Wang.*This disease is one of the most serious diseases of ber and causes fruit yield losses in semi-arid and sub-tropical region but it is not serious in Bikaner and nearby regions due to prevailing unfavourable climatic conditions. Initial symptoms appear on young fruits at pea stage of development. White specks may appear on immature fruit and later may cover the entire fruit. The infected fruits become misshapen and corky. Fruits finally drop. In severe conditions, floral parts, whole fruits, tender branches and leaves would appear with powdery mass of fungal conidia. Infected fruits fail to develop, show discolouration and become mummified. Young shoots and spurs are also attacked.

Integrated Disease Management

Successful management is achieved through integrated strategies alone. Following combined applications of management can be useful for disease management.

i) Cultural Control

a) Removal and destruction of infected leaves, twigs and collateral hosts from orchard.

b) Ber orchards should be free from wild species of ber (*Ziziphus nummularia*).

c) Summer ploughing in the orchard.

ii) Chemical Control

a) Karathane (0.1%) proved the best control of powdery mildew, followed by calixin @ 0.1%.

b) Maximum disease control has been found by combined application of *Psendomonas fluorescens* (1%) and Karathane @ 0.05%.

c) Three fungicides namely, bayleton 25 WP (0.05%), karathane 48 EC (0.01%) and sulfex 80 WP (0.25%) were applied during flowering, after fruit set and first week of November against powdery mildew. The studies showed the superiority of bayleton in managing the disease effectively. Bayleton significantly checked the disease, reduced the fruit drop and increased the fruit yield as compared to karathane and sulfex.

d) Three sprays of karathane (0.1%) followed by difenoconazole (0.05%) at 15 days interval proved 51.0% and 40.3% disease control against powdery mildew of ber.

iii) Biological Control

a) Spraying of *Trichoderma* spp. (CIAH- 240) @ 5% also reduced the disease.

b) *Pseudomonas fluorescens* strain CIAH- 196 (1%) was also moderately effective against this disease.

iv) Host Resistance

Plantation of resistant varieties is the best method for reducing the disease. Resistant/tolerant varieties such as Sanur-2, Chinese, Jogia and Vikas must be grown.

2. Black Leaf Spot

Disease is caused by *Isariopsis indica* var. *ziziphi,* wherein sooty tuft like circular to irregular black spots develop on lower surface of leaves. At advance stage, the larger part of lower surface of leaves is also covered with spots, while the upper leaf surface shows brownish discolouration. Leaves and twigs are dried rapidly under severe infection. Cloudy weather with moderate temperature during October-November is favourable for disease initiation.

Integrated Disease Management

i) Cultural

a) Removal and destruction of infected leaves and twigs from orchard.

b) To keep the ber orchard free from all weeds.

c) Pruning should be done properly in the orchard.

ii) Chemical

a) Fungicidal sprays at 15 days interval (mancozeb or copper oxychloride @ 0.2%).

b) Two sprays of propiconazole (tilt), followed by difenoconazole (score) @ 0.1% was found to be effective against this disease and also increased fruit yield.

iii) Host resistance: Use of resistant varieties such as ZG-3, Seo, Bahadurgarhi and Safeda Rohtak.

3) *Alternaria* Fruit Rot

It is caused by *Alternaria alternata*, which manifestson immature as well as young and mature fruits. It starts from pedicel and in bottom portion of fruits. Ripened fruits are nearly free from this rot. Light brown to dark brown spots are produced on *ber* fruits. Mature fruits rot under severe conditions by brown blotching. Such fruits ultimately fall down resulting huge losses. Some fruits have concentric rings with a number of fungal spores. All the young fruits also fall down in severe case and *ber* plants appear with moderate fruiting.

Integrated Disease Management

i) Cultural

a) Removal and destruction of infected leaves and twigs from orchard.

b) Keeping the *ber* orchard free from all weeds.

c) Pruning should be done properly in the orchard.

ii) Chemical

a) Fungicidal sprays at 15 days interval (mancozeb @ 0.2%) was given to minimize the disease.

b) Mancozeb and copper oxychloride (0.2%) are also useful for retarding the fungal growth.

c) Neem oil significantly reduced severity of fruit rot of ber caused by *Alternaria alternata*.

iii) Biological

This disease can be controlled by Trichoderma harzianum and Bacillus spp. and Fluorescent *Pseudomonads*.

4. *Cercospora* Leaf Spot

It is caused by *Cercospora ziziphi*, which results in development of circular to oval discoloured spots (up to 4 mm diameter) on leaves which later turn to brown colour and surrounded by dark brown margins.

Disease management strategies

i) Cultural

a) Removal and destruction of infected leaves, twigs and weeds from orchard.

b) Proper pruning in the orchard.

ii) Chemical

a) Spraying of mancozeb (0.3%) in ber trees was useful for reducing the disease.

b) Systemic fungicides (benlate and bavistin @ 0.1%) are also effective against this disease

5. *Alternaria* Leaf Spot

The causal organism of this disease is *Alternaria alternate*. Infection ismanifested as dark greyish-black colonies of the fungus on leaves. Infection is quite prominent during foggy days on *ber* foliage. At later stage, infected surface turns grey by fungal growth.

Integrated Disease Management

i) Cultural

a) Removal and destruction of infected leaves, twigs and also weeds from orchard.

b) Pruning should be done properly in the orchard.

ii) Chemical

For effective management of disease, maximum per cent disease control (77.33) was recorded with difenconazole (0.1%) followed by propiconazole @ 0.1% (75) and copper oxychloride @ 0.2% (68.71) whereas Neem oil @ 3.0% (53.45) proved least effective.

Insect-Pests Management

Fruit fly

The major insect causing heavy losses in ber is fruit fly. In India, fruit fly is *Carpomyia vesuviana* Costa. The fly puncture the young developing fruits and lays egg. Infection starts in September when the fruits are very small and continue up to harvest. The affectedfruits become misshapen and their growth is retarded. The quality of fruit is reduced and they become unfit for consumption and marketing. The fruits are infested at "pea" stage fruit fly; therefore, it is essential to spray any systemic insecticide like Monocrotophos @ 0.01% at this stage of fruit development. The same should be repeated three weeks after the first spray. However, the decision in this regard will be governed by local conditions. For control, grow fruit fly tolerant cultivars like Umran, Sanaur-1, Safeda, Illaichi, Mirchia, Katha, Tikadi, Dodhia, BS-1 and Meharun.

Bark Eating Caterpillar

The presence of winding galleries of frasing web on the stem/branches joints indicates the infestation. The caterpillar makes holes in the trunk while feeding on bark. The pest becomes active during night and remains hiding in day time in the stem. Infect, it is a major problem of neglected orchards. Affected trees become stunted and their yield potential is reduced. Putting kerosene oil with the help of injection syringe and pasting the hole with mud also helps in reducing the incidence.

Hairy Caterpillar

The caterpillars feed on leaves in initial stage and old caterpillar feeds on fruits and tender shoots. They start eating new foliage emerged after pruning and this is continued by over lapping generation. For control measures, spray of Monocrotophos @ 0.01% is suggested.

Leaf Chafer Beetle

The beetles of light brown colour feeds on leaves during night. The pest is most active in rainy season when new growth started. In severe cases, the whole tree is rendered leaf less. It can be effectively controlled by spraying of Monocrotophos @ 0.01%.

Potential Benefits of Adoption of Advance Approaches for *Ber* in the Arid Region

Adoption of advance approaches in *ber* cultivation, as discussed in previous paragraphs, would fetch remunerative prices to the growers for their produce, which involves judicious uses of inputs for realization of optimum yield without adversely affecting the environment. With the introduction of Apple *ber* in India, the avenues of exports are also opened. Presently, fresh fruits of apple *ber* are being exported to the countries like Maldives and gulf countries like Bahrain, Saudi Arabia and Kuwait. Therefore, in addition to social and nutritive values, ber fruits also contribute in strengthening National economy by fetching foreign exchange.

Future Scope and Research Need

Bal (2016) suggested following areas to work upon keeping in view the dynamism of current horticultural scenario.

Long-term

- High density and canopy management with new training systems on trellis for easy harvesting and effective implementation of different orchard cultural practices.
- Hybridization should be done on large scale for early, mid season, better quality cultivars tolerant to biotic stresses especially powdery mildew disease.
- Hybridization for processing purpose needs to be tapped.
- Evaluation of dwarf and high yielding rootstocks.
- Varietal evaluation and introduction from different sources.

Short-term

- Managing fruit drop with chemicals or other means.
- Insect pest management more efficiently and effectively.
- Technology device for enhancing seed germination.
- Biological weed control in orchards.
- Enhancing shelf life at low temperature for extended market period.

Conclusion

Ber is not so extensively cultivated and its consumption and trade tend to be more limited, geographically and quantitatively, than those of the commercial fruits, despite having high nutritive and medicinal attributes and also their ability to grow under adverse soil and climatic conditions. Further, if exploited properly, it has the potential of transforming the economy of rural or disadvantaged areas because the produce are of explicit quality with great nutritional, medicinal, organoleptic, economic and traditional importance. Therefore, it is important to include *ber* in health promotion campaigns. This is the need of the hour to focus on this important and economically rewarding area of horticulture.

References

Anonymous. 2014. Package of Practices for Fruit Cultivation (Ludhiana: Punjab Agricultural University).

Anonymous. 2017. The Third Estimates of Area and Production of Horticultural Crops 2-16-17. Ministry of AAgriculture & Farmers Welfare, GoI.

Azam-Ali S, Bonkoungou E, Bowe C, deKock C, Godara A and Williams JT. 2006. *Fruits for the Future 2*, p 289. International Centre for Underutilized Crops, Southampton, UK

Bal JS. 2016. Development and production of Indian jujube (*ber*) in India. *Acta Horticulturae,* 1116: 15-22.

Bal, J.S. (2013). The utilization of rootstocks in Indian jujube. Acta Hortic. *993*, 47–50. http://dx.doi.org/10.17660/ActaHortic.2013.993.6.

Bal, J.S. (2014). Fruit Growing, 3rd edn (New Delhi: Kalyani Publishers).

Bal, J.S., and Singh, S. 2012. Mulching in Ber Orchards (Saarbrucken, Germany: Lambert Academic Publishing, GmbH & Co. KG), pp.110.

Gupta, I.C., Singh, P.M., Yadava, N.D. and Sharma, B.D. 1991. Note on performance of ber during establishment under conventional and *Jaltripti* irrigation systems. *Indian Journal of Horticulture*, 48:222-227.

Heuzé V., Tran G., Boval M., Lebas F., 2017. Indian jujube (*Ziziphus mauritiana*). Feedipedia, a programme by INRA, CIRAD, AFZ and FAO. https://feedipedia.org/node/80

Kaundal, G.S., Grewal, S.S., and Bal, J.S. (1984). Influence of meteorological factors on the bud take efficiency of ber (*Zizyphus mauritiana* Lamk.) cv. Umran. Journal of Research Punjab Agricultural University *21*, 372–374.

Krishna H and Parashar A. 2013. Phytochemical constituents and antioxidant activities of some Indian jujube (*Ziziphus mauritiana* Lamk.) cultivars. *Journal of Food Biochemistry,* 37: 571–577.

Krishna H, Parashar A, Awasthi OP and Singh K. 2014. *Ber*. pp 137-155. (*in*) *Tropical and Sub Tropical Fruit Crops: Crop Improvement and Varietal Wealth Part-I.* SN Ghosh (Ed). Jaya Publishing House, Delhi.

Liu, M; Yan, G., 2014. Third International Jujube Symposium. 3rd Int. Jujube Symp., Brisbane, Australia

Liu, M; Yan, G., 2014. Third International Jujube Symposium. 3rd Int. Jujube Symp., Brisbane, Australia.

Mittre, V. 1961. Plant economy in ancient Navdatoli-Maheshwar. Delhi University, Delhi.

Morton, J. 1987. Roselle. p. 281–286. In: Fruits of warm climates. Julia F. Morton, Miami, FL.

Orwa, C.; Mutua, A.; Kindt, R.; Jamnadass, R.; Anthony, S., 2009. Agroforestree Database: a tree reference and selection guide version 4.0. World Agroforestry Centre, Kenya.

Pareek, O.P. 1978. Quicker way to raise ber (Z. *mauritiana*) orchards. *Indian Horticulture*, 28: 13-15.

Pareek, O.P. 2001. Ber, International Centre for Under utilized crops Southampton, U.K., p-290.

Saroj, P.L., Dhandar, D.G., Sharma, B.D. Bhargava, R. and Purohit, C.K. 2003. Ber (*Z. mauritiana*) based Agri-horti system: A sustainable landuse for arid ecosystem. *Indian J. Agroforestry*, 5: 30-35.

Sharma, V.P. and Kore, V.N. 1990. *Ber*. *In*: Fruits: Tropical and Sub Tropical (T.K. Bose and S.K. Mirta; Eds.), Naya Prokash, Calcutta. pp. 592-605.

Shukla, A.K. 1996. Efficacy of drip and mulching on young plantation of *aonla* + *ber* cropping model in sodic soil. M. Sc (Ag) Thesis. NDUAT, Faizabad.

Singh, H., Bal, J.S., and Singh, G. 2004. Standardization of pruning technique in ber. *Indian J. Hortic*. 61, 259–260.

Vashishtha, B.B. 1982. Horticultural qualities of *Zizyphus nummularia* – Bordi, CAZRI Monograph, Jodhpur.

29

Pomegranate- A Potential Fruit Crop for Ensuring Income Security of Farmers

N. V. Singh, N. N. Gaikwad and R. K. Pal

Introduction

Pomegranate is one of the most remunerative fruit crops that grow well under arid and semi-arid agro eco-system with severe scarcity of water. The popularity of pomegranate has grown tremendously due to its health beneûts, high returns on investment, less water requirement, therapeutic and functional properties. Growing health consciousness among the consumers and acquisition of super fruit status by pomegranate has made marketing of pomegranate fruits and derived value added products quite easy. As the result of all these benefits, India is witnessing a pomegranate revolution with more than 80 % expansion in area and 250 % increase in production during last one decade. However, there are many challenges to be overcome so as to double the income of pomegranate growers, these challenges include market glut due to unprecedented increase in production coupled with limited export resulting into suboptimal price of the harvested produce; intensification of diseases and insects and pest due to huge unplanned expansion; dominance of only one variety; very limited value addition; hampered export due to suboptimal quality, unwanted sprays, issue of pesticide residues and lack of sufficient number pesticide label claims in pomegranate; inadequate supply chain management. So, in order to overcome these challenges and to increase the profitability of pomegranate stakeholders, we need to focus on scientifically sound cultivation practices to reduce the cost of cultivation, breeding for improved varieties, improved packaging, transport, storage, marketing, supply chain management, facilitating and encouraging export, practicing and promoting climate resilient horticulture, formation of farmers' cooperatives and farmers producers organization (FPOs), processing and value addition with the aim of total utilization and use of information and communication technologies. This article deals with opportunities and strategies in pomegranate for doubling farmers' income through nursery management; improved horticultural practices; integrated disease, insets pests and nutrient management; use of information and communication technologies; creation of FPOs; processing and value addition.

Present Status in India and Global Scenario

India is the world leader in pomegranate acreage and production with 2.16 lakh ha area and 25.21 lakh tones of annual production (nhb.gov.in, 3rd Advance estimate of Area and Production of Horticultural Crops).Though, India leads the world in total pomegranate production but the export lies at dismal 0.02 % of the total produce, where as countries like Spain and Israel export more than 50 % of their pomegranate produce. Indian private companies like INI Farms and Sam Agritech Limited are exporting modified atmosphere packaged pomegranate arils to the European Countries.

The rapid rise in production of pomegranate by more than 2.5 folds with only about 0.02 % export forces stakeholders to reorient future marketing and utilization strategy for this high value produce. Scientific post-harvest management to improve shelf life of produce by adopting modern handling, storage, packaging and transportation practices is of high importance for marketing fruits to distant places in India and international destinations. Moreover, about 20-25 % of the harvest generally do not fetch optimum price in the domestic market due to fruit spots or other external deformities, if processing industry/ cooperatives/ SHGs/ Growers Associations/ Farmer-Producer companies target this produce, then low value fruits can be processed into high value premium quality products of international standard there by creating win-win situation for farmers as well as processor.

Impact scenario of Tippehalli village in Solapur district of Maharashtra before and after pomegranate innovation system

Before 1997/98	After 2003
40 ha pomegranate was grown	Above 445 ha pomegranate grown, with a profit of 1.5 lakhs/ household/year
Almost entire families of 350 household migrated	Only males from around 40 families migrate
Only two agri- input dealers	Several agri- input dealers
Had rabi sorghum, few vegetables and meat as food items	Had in addition wheat, rice, many vegetables and poultry as food items.
Living standards poor (only 2 bicycles, no literacy or schooling)	Improved living standards (50 motorcycles, 7 trucks, two tractors, 6 jeeps, bicycles in almost all households, residential schooling)

Source: Raina, Rajeswari S, NISTDS, New Delhi

Approaches for Doubling Income and Potential Benefits

Improved Varieties

Pomegranate variety Solapur Lal-Pomegranate variety Solapur Lal has been identified by ICAR-NRCP, Solapur for table purpose. 'Solapur Lal' has early maturity (160-165 days), higher yield (23-27 t/ha from 6th year onwards) and better quality. As this is rich in nutrients *viz.*, iron and zinc, it is useful for supplementation of nutrient deficiency in human beings. This variety is a high yielder (25-35%) over the Standard Check variety.

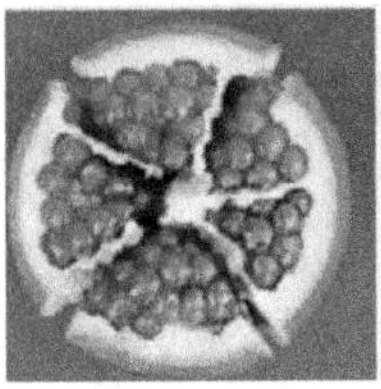
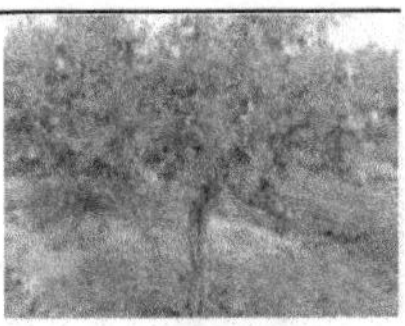

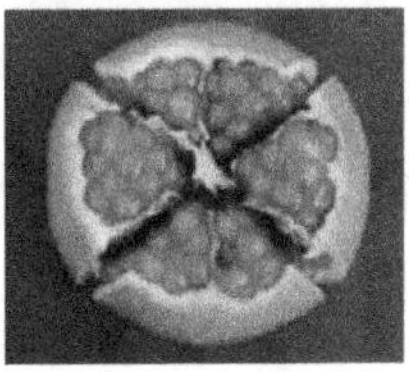

'Solapur Lal'	Characters	'Bhagwa'
Medium (2.0-2.3)	Tree height(m)	Medium (1.8-2.1)
Red	Calyx colour	Red
160-165	Fruit maturity (days)	180-185
23-27	Yield(t/ha)	16-20
280-290	Fruit weight(g)	287-297
Red	Fruit colour	Red
Medium (3.3-3.5)	Rind thickness (mm)	Medium (3.3-3.5)
40-41	100 Aril weight (g)	35-36
Dark Red	Aril colour	Red
Medium	Seed Texture	Soft
45-50	Juice %	43-48
17.5-17.7	TSS(°Brix)	15.7-15.9
19.4-19.8	Vitamin-C (mg/100g)	14.2-14.6
385-395	Anthocyanin (mg/100g)	355-365
5.6 – 6.1	Iron (mg/100g of fresh arils)	2.7-3.2
0.64– 0.69	Zinc (mg/100g of fresh arils)	0.50-0.54

Comparative performance of SolapurLalvs. Bhagwa

Pomegranate variety Solapur Anardana- Pomegranate variety Solapur Anardana has been identified by ICAR-NRCP, Solapur for processing purpose. 'Solapur Anardana' has higher yield (22-24 t/ha), titrable acidity (4.7-4.9%) and anthocyanin (457-467mg/100g) content compared to Amlidana.

Improved Horticultural Practices

Nursery Management and Planting Material Production

The area under pomegranate cultivation has been remarkably increased to about double during the last six years but rapid expansion without the availability of healthy and elite planting material has led to the spread of bacterial blight to new areas through infected saplings. There is an annual demand of more than 10 million healthy saplings of pomegranate annually and this demand can only be fulfilled by supporting traditional methods of propagation (air layering and hard wood cutting) with *in-vitro* propagation. The establishment of certified nursery or secondary hardening units is a highly profitable venture and has capability of increasing farmers' income many folds.

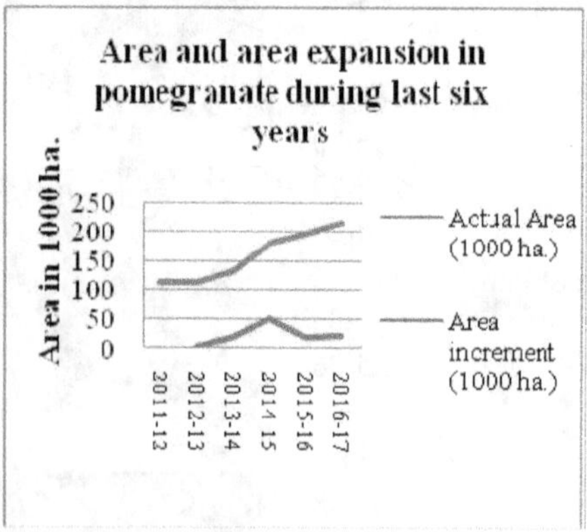

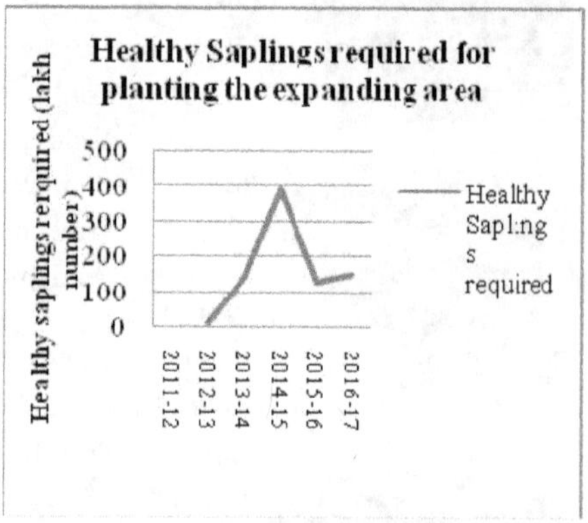

Standardization of Subsurface Drip Irrigation (SDI) System in Pomegranate

Water is a scarce and most precious natural resource in majority of pomegranate growing areas and often acts as an important yield limiting factor. Water productivity needs to be enhanced so as to realize optimum yield from the same quantity of water. The SDI with double in line lateral at 30 x 30cm was found better than other micro-irrigation systems. Water is conveyed in soil through capillary action and gravity thus up to 46 % larger volume is irrigated as compared to drip irrigation. It also eliminates water loss due to evaporation and conserves water upto 70% as compared to surface drip irrigation. Similarly, experiment on other important water saving approaches like partial root zone drying technique, mulching, etc are under progress at ICAR-NRCP.

A novel bio-formulation for potassium fertilizer supplement- A novel carrier based bio-formulation (from *Penicillium pinophilum*) for potassium fertilizer supplement to pomegranate crop has been developed. This bio-formulation has the capacity to supplement 70% requirement of potassic fertilizers in pomegranate. The research data reveal significant saving to the tune of approximately Rs. 40,000/- per ha by the use of this bio-formulation in pomegranate on account of procurement of imported potassic fertilizers. Use of this bio-formulation led to approximately 24% increase in economic yield with better fruit quality of pomegranate owing to improved plant growth, leaf area index and better photosynthetic activity.

Micronutrient management in pomegranate: Foliar application of $ZnSO_4.7H_2O$ @0.3 % four times *viz.* two sprays before flower bud initiation and two after fruit set at 30 days interval resulted into enhancement in fruit yield by 55.42 % and Zn content in edible part by 28.04 % besides improvement in TSS, juice per cent, phenols, ascorbic acid and anthocyanin content

Model pomegranate production practices-comprising of scientific canopy management of plant, sanitation measures, nutrient and water management practices, undertaking spraying operations as per IDIPM schedule were demonstrated in five farmers' orchards in the village Ankoli and Shejbabhulgaon, Mohol taluka, Solapur. The adoption of model pomegranate production practices resulted remarkable reduction in bacterial blight disease incidence by 80-90 % and increased yield by 30-48% compared to farmers' own practices.The fruit yield in adopted orchards ranged from 39.58 to 23.38 t/ha while in non-adopted orchards it ranged from 18.08 to 26.56 t ha^{-1}.The results of Model pomegranate production practices were demonstrated to the farmers of Ankoli and Shejbabhulgaon by organizing farmers' "Field Day" at the project site where from highest fruit yield 39.58 t/ha was obtained. And they were imparted with technological knowledge for enhancing productivity through effectively managing bacterial blight disease.

IDIPM schedule was developed in 2010 and evaluated in Network mode in 2011-12 in 35 orchards in Maharashtra, Karnataka and Andhra Pradesh states. Later demonstrated in 15 acres area in Mohol taluka of Solapur-

- Blight reduction up to 86.04 %
- Yield increase up to 79.5%
- Average Benefit: Cost:: 4.19:1
- Number of sprays/month was reduced to 2 to 4 in IDIPM as against 10 to 15 in traditional system.
- In India, the average pomegranate productivity has increased from 6.94 t/ ha in 2011-12 to11.67 t/ha in 2016-17.

Intercropping: Alternate row intercropping with marigold, rabi sorghum, maize or less shading and fast growing tress like drumstick during initial two years of non bearing period may generate additional income and farm border plantation with wind break/timber trees may generate substantial income later on.

Climate Resilient Horticulture in Pomegranate: Research on suitability of rootstocks is the need of the hour to mitigate the challenges under climate resilient horticulture with special reference to increased soil salinity, drought, insect pests and diseases in commercial cultivars of pomegranate. ICAR-NRCP has already standardized wedge grafting and patch budding in pomegranate to exploit rootstocks. Screening of huge germplasm repository of ICAR-NRCP field gene bank against various biotic and biotic stresses is under progress to identify and utilize potential rootstocks.

Value Addition

In India major part of production is utilized for table purpose fresh fruit consumption. The pomegranate arils, peel, seeds and flowers can be processed in to value added products. The pomegranate can be processed in to traditional and non-traditional value added products. The traditional products like anardana, juice and ready to serve beverage can be prepared from pomegranate. Non-traditional products like minimally processed arils, wine, seed oil, bio colours, peel extracts can be developed.

Minimally Processed Pomegranate Arils

Pomegranate arils contain 85% water, 10% total sugars, mainly fructose and glucose, and 1.5% pectin, organic acid, such as ascorbic acid, citric acid, and malic acid, and bioactive compounds such as phenolics and flavonoids, principally anthocyanins (Viuda-Martos *et al.*, 2010).

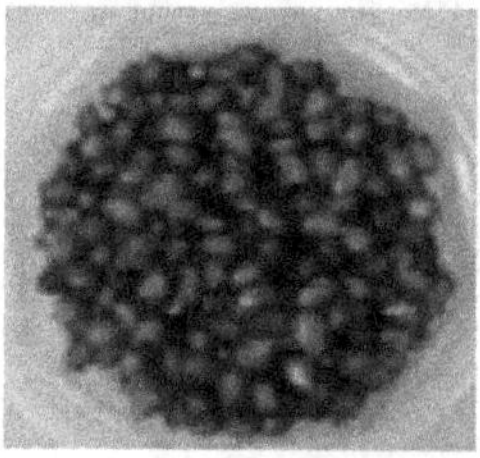

Minimally Processed Arils

Pomegranate arils provide 12% of the Daily Value (DV) for vitamin C and 16% DV for vitamin-K per 100g serving. The pomegranate is consumed mainly fresh, but the difficulty encountered in separating the edible arils from the fruit has several limitations for its direct consumption unlike the other fruits e.g. Oranges, banana, grapes etc. Hence, minimal processing of pomegranate is of great importance for convenience of the consumers. Commercialization of minimally processed and "ready-to-eat" fresh arils is the good alternative. In the minimal processing generally various hurdles are created for spoilage of freshly extracted arils using low temperature, pH regulation and edible GRAS chemicals as anti-microbial agents. The minimal processing consists the washing with sanitizing agents to reduce the primary inoculum load, pH modification, use of antioxidant agents, temperature control and others, to control partially the high perishability of the fruits. On the other hand, the selectively permeable polymeric films for packaging of minimally processed pomegranate arils are used for generation of Modified Atmospheric Packaging (MAP) system in order to develop a micro controlled environment that reduces the respiratory activity and maintains unfavorable conditions for the action of many contaminating microorganisms. There are some commercial units of minimal processing operating in country. The units are processing the fruits to the tune of 5.0 tons/day. These units had unit operations such as washing, manual aril extraction, weighing, gas mix (O_2, CO_2& N_2) flushing, and sealing. The units are packaging and marketing the arils in retail pack of 100 g and commercial pack of 10 or 20 kg. ICAR-NRCP has developed technology for pretreatment of arils for extension of shelf life up to 14 days (Pal and Gaikwad, 2016).

Pomegranate Juice and Ready to Serve Beverage

Pomegranate juice is nutritionally an important beverage since it is consumed frequently for its phenolic compounds (such as anthocyanins, ellagic acid, phytoestrogenic flavonoids and tannins). The red color of juice can be attributed to anthocyanins, such as delphinidin, cyanidin and pelargonid in glycosides. The screw presses are used prominently for juice extraction from pomegranate arils in small scale processing. The processing of pomegranate juice at commercial level involves squeezing juice from the fruit with the seeds and the peels together. The juice extraction from whole fruits is carried out by subjecting the cut fruits to hydraulic press. The pressure of less than 100 psi is used to avoid undue yield of tannin. The juice from crushed whole fruits contains excess tannin from the rind and which is precipitated out by a gelatin process. The juice is clarified by heating in a flash pasteurizer, cooling, settling, racking up and filtering or decanting. The clear juice can be preserved by heat treatment or by using chemicals. The ICAR-NRCP has standardized process protocol for processing of juice and RTS beverage.

Material and Cost for Preparation of 5 litres Ready to Serve Beverage

- Pomegranate fruits @ 40 Rs/kg : Rs. 80
- Sugar: Rs. 30
- Preservatives and CA: Rs. 10
- Packaging: Rs. 50
- Labour: Rs 100 (half man day)

- 25 bottles of 200 ml each can be prepared from 5 l of RTS and at sale cost of 20 Rupees per bottle
- Net Profit: 500-270= 230
- The profit will increase with increase in quantity

ICAR-NRCP Played Important role in Establishing New Startup Enterprise by

Conceptualization of business plan

Providing information on FSSAI licensing

Training on pomegranate RTS beverage development and packaging

Technology licensing Hands on experience at pomegranate processing pilot plant

Guidance for establishment of pomegranate processing unit

TVK Beverages at a glance

Initial investment: Rs.120 lakh

Production Capacity: 1000 l./day

Manpower Required: 5 Persons

Market Linkages: Self Marketing

Present Turnover (Rs): Rs.15 lakh per month

Profit: Rs 1.8 lakh/month

Pomegranate juice/RTS processing plant at M/S TVK Beverages at Solapur (Maharashtra)

Mr. Vijay Kale young entrepreneur established startup enterprise M/S TVK Beverages at Solapur (Maharashtra) -the heartland of Indian pomegranate production for utilizing readily available pomegranate and ICAR-NRCP technology and guidance for skill development. He left his well paid job at Saudi Arabia to pursue his dream.

Carbonated ready to serve beverage can be prepared by application of CO_2 at 80 PSI pressure. The carbonation is carried out at 4 °C. The packaging is also carried at cool conditions of 4 °C

Anardana

Pomegranate raisins or '*Anardana*' is traditionally prepared by drying arils of sour and acidic fruits of wild pomegranate. *Anardana* has a distinct tart flavour, and is commercially available in many Asian countries, where they are consumed in large quantities. *Anardana* is also used in the ayurvedic medicine as digestive and stomachic. The cracked fruits at matured stage can be economically used traditionally for preparation of dried value added product called *anardana*. It is used as acidulent and condiment in Indian culinary or traditional system of medicine. The cabinet drying at 55^{0}C for 7 hours of the arils is best for getting quality *anardana*(Pal and Gaikwad, 2016) . The variety 'Solapur Anardana' developed by ICAR-NRCP is a high yielder and rich in natural acidity and is highly suitable for preparations of *anardana.*

Anardana

Pomegranate Seed Oil

Pomegranate seeds are approximately 10 per cent of the fruit weight. Pomegranate seeds are excellent sources of dietary fiber. The soft seeded varieties of pomegranate contain seeds oil to the tune of 25-26% (V/W). The pomegranate seed oil contains more than 70 % of conjugated linolenic acid The Pomegranate seed oil contains punicic acid (65.3%), palmitic acid (4.8%), stearic acid (2.3%), oleic acid (6.3%) and linoleic acid (6.6%).

Pomegranate seed oil is a unique natural product with only few plant sources known that contains conjugated fatty acids. Conjugated fatty acids inhibit eicosanoid metabolism,thus acts as natural anti-inflammatory agents. Pomegranate seed oil contains other significant bioactive compounds. Pomegranate seed oil is one of the richest known plant sources of a steroidal estrogen, estrone. Other important compounds found in Pomegranate seed oils include gamma-tocopherol, a rare and potent form of Vitamin E and the phytosterols: beta-sitosterol, stigmasterol and campesterol.It has been linked to improve heart health and also indicates to protect against cancer (Hernandez, *et al.*, 2002; Lansky and Newman, 2006) and artheroschlerosis (Boussetta, *et al.*, 2009).

The studies on determination of pomegranate seed oil content revealedaverage per cent oil in cultivars'Bhagwa', 'Ganesh' and 'Arakta' was found to be 28%, 26.43% and 23.70% (w/w). The fatty acid profile of Bhagwa and Ganesh cultivar revealed both prominent cultivars grown in India have high level of linolenic acid content. The price of virgin pomegranate seed oil in the international market ranges from Rs. 15000-20000/ litre due to its high anti-inflammatory and anti-tumour activities (Pal and Gaikwad, 2016)

Fatty acid profile	Bhagwa Seeds	Ganesh Seeds
Linolenic (%)	68.7	69.8
linoleic (%)	13.9	11.4
Oleic (%)	12.0	9.1
Palmitic (%)	2.1	2.4
Stearic (%)	1.9	2.1

Bio-Colours from Pomegranate

Pomegranate seed oil

The pomegranate peel is an important source of bioactive compounds such as phenolics, flavonoids, ellagitannins, and proanthocyanidin compounds, minerals, mainly potassium, nitrogen, calcium, phosphorus, magnesium, and sodium, and complex polysaccharides. The rind of pomegranate contains a considerable amount of tannin, about 19% (Adeel *et al.,* 2009; Tiwari *et al*., 2010). The main coloring agent in the pomegranate peel is granatonine which is present in the alkaloid form N-methyl granatonine (Goodarzian and Ekrami, 2010). This compound gives colour as dye. Similarly, pomegranate flowers are also not put to any use at present. As pomegranate plant parts are rich in tannins, these materials can be used as a tannin source for mordanting of cotton in natural dyeing and also as a natural dye. Today, textiles dyed with natural dyes are preferred by environment conscious consumers due to their eco-friendly nature and fetch higher prices. The natural dyes from rind of pomegranate are used in coloration of lipsticks and other cosmetics. The dried pomegranate peel powder can also be used to prepare organic mouthwash to tackle bad breath causing microbes.

Pomegranate Wine

Pomegranate wine is the product of anaerobic fermentation by yeast in which the sugars are converted into alcohol and carbon dioxide. Pomegranate wines have higher antioxidant activity and total phenol content than the red wine. Melatonin (N-acetyl-5-methoxytryptamine) is a neurohormone related to a broad array of physiological functions and proven therapeutic properties. Melatonin was observed to be absent in pomegranates juices but it is detected in prominent amounts in pomegranate wine (0.54–5.50 ng/ml). The presence of this biogenic amine makes pomegranate wine use as bioactive phytochemicals supplementation. The protocol for preparation of sparkling pomegranate wine is developed by ICAR-NRCP, Solapur.

Formation of Farmer Cooperatives, Growers' Association and Farmer Producer Organizations

The credit for the success of sugarcane, grapes and pomegranate in Maharashtra also goes to strong Growers' Associations in these crops. These associations play critical role in adoption of latest technologies, export, marketing and storage of the harvested produce and creating awareness among farmers on various crop related issues. The Maharashtra Pomegranate Growers Association was also instrumental in obtaining GI status to pomegranate of Solapur which will certainly boost the export. Similarly, cluster or village level small scale processing, grading and sorting units can be established near the major production areas by forming Farmer Producer Organizations and utilizing the schemes of State and Central Governments.The small scale industries besides generating additional income will also generate additional employment during lean season.

Use of Information and Communication Technology

The information and communication revolution should be harnessed for the benefit of farmers by making multilingual, pictorial and easy to understand mobile based applications like 'Solapur Anar' (mobile application developed by ICAR-NRCP). India has one of the largest mobile phone users in the world and with affordable pricing of smart phones it is possible to make available necessary agricultural information to every nook and corner of the country. Information can be disseminated through mobile applications, weather based disease forecasting and warning systems, farmers' mobile advisory services.

Pomegranate Export

India enjoys a comfortable edge over other major pomegranate producing countries as far as availability of fresh fruits are concerned, pomegranate fresh fruits are available round the year in India whereas our nearest rival Spain affords the availability only upto eight months. The ICAR-NRCP, APEDA, MAHA ANAR, ANARNET (APEDA) are working jointly to come out with a comprehensive list of chemicals with their PHIs and permissible limits to have a spray schedule for export oriented package of practices. There is a huge demand of pomegranate and its products in the international market and India has the potentiality to fulfill the demand and become the world leader in pomegranate export. A five-pronged strategy need to be adopted to promote export of pomegranate and its products, this includes Product Segmentation,

Market Diversification, Market Penetration, Value Addition and Agriculture Infrastructure upgradation.

Future Scope and Research Needs

The processing of pomegranate in to various value added products as well as utilization of by-products of processing industry into high value nutraceutical and pharmaceutical products holds the key for doubling income. Similarly, promotion pomegranate export should be on top priority in order to substantially increase income of pomegranate growers. Though, ICAR-NRCP in collaboration with APEDA, pomegranate exporting companies, progressive farmers and other stakeholders has formulated a package of practices for export but an effective export oriented package of practices in pomegranate based on label claims, PHI of chemicals and other important criteria is essential. Some dry pockets of the country can also be selected and promoted for organic pomegranate production to improve export and farmers' income.

Improved horticultural practices like scientific canopy management; efficient utilization of irrigation water; integrated nutrient, disease, insects and pest management, climate resilient horticultural practices, precision horticulture system, use of remote sensing, etc should certainly be the part of modern pomegranate orchard management to increase farmers' income in a sustainable manner.

Conclusion

Pomegranate has potential to provide economic security to masses particularly in natural resource deficit regions if the consorted efforts of all the players can be translated into functional groups, large scale emergence of small scale on site processing hubs serving few villages of tehsils, scientifically sound package of practices to meet all the requirements of export, developing pockets of organic pomegranate production and percolation of latest technologies to the last pomegranate grower. A more holistic approach involving climate responsive technologies, market driven farming, ICT platforms, precise and location specific package of practices at par with international standards, improved agricultural infrastructure, collaborations and consensus between different stakeholders of pomegranate need to be nurtured to have sustained income.

References

Adeel, S., Ali, S., Bhatti, I. A. and Zsila, F. 2009. Dyeing of cotton fabric using pomegranate (*Punica granatum*) aqueous extract. *Asian J. Chem.* 21: 3493-3499.

Zarfeshany, A. , Asgary, S. and Javanmard, S. H .2014. Potent health effects of pomegranate.*Adv. Biomed Res.* 3: 100.

Boussetta, T., Raad, H., Letteron, P., Gougerot-Pocidalo, M.A., Marie, J.C., and Driss, F., 2009.Punicic acid a conjugated linolenic acid inhibits TNFalpha-induced neutrophil hyperactivation and protects from experimental colon inflammation in rats. *PLoS One*. 4: 7.

Goodarzian, H. and Ekrami, E.2010. Wooldyeing with extracted dye from pomegranate(*Punica granatum*) peel. *World Appl Sci. J.* 8: 1387-1389.

Hernandez, F. P., Melgarejo, J. M., Olias, J.M. and Artes, F. 2002. Fatty acid composition and total lipid content of seed oil from three commercial pomegranate cultivars. *CHEAM-Options Mediteranenes*: 205-209.

Lansky, E. P. and Newman, R. A., 2006. *Punica granatum* (pomegranate) and its potential for prevention and treatment of inflammation and cancer.*J. Ethnopharmacol.* 109:177–206.

Viuda-Martos, M., Fernandez-Loaez, J. and Perez-alvarez, J. A. 2010. Pomegranate and itsmany functional components as related to human health: a review.*Comprehensive Reviews in Food Science and Food Safety*, 9:635–654.

Tiwari, H.C., Singh, P., Mishra, P.K., and Shrivastava P. 2010.Evaluation of various techniques forextraction of natural colorants frompomegranate rind by ultrasonic and enzyme assisted extraction.*IJFTR*, 35: 272-276.

Raina, R.S. 2003.Technological and institutional innovations: A case study of pomegranate production and marketing (a case study of micro-irrigations technologies for rural poor) *IDE (I) Report*, No. 21, International Development Enterprises (India): New Delhi.

Pal, R.K. and Giakwad, N.N. 2016. Exploiting pomegranate potential for double benefit. *Indian Hort.*61(6):15-18.

30

Doubling Horticultural Productivity Under Khejri Based Cropping Models in Arid Environment

D. K. Samadia

Introduction

Dry-land horticulture has immense potential in providing nutrition rich food, social security and eco-restoration for the inhabitants of desert or rainfed and tribal land-areas of the north-western parts of India. Most of the recent past advances in Indian Horticulture are with production technology for favorable agro-climate, however, the vast and diverse land-scape of hot desert, arid, semi-arid and tribal area is yet not getting logical thought for resource based horticultural development. The native crop–plants both annual and perennial are equally playing vital role for food, fodder as well income especially in the hot arid and sem-arid areas of Rajasthan where animal husbandry is also an integral component of the farming system. The monsoon supported arid zone crop-plants are drought tolerant and even the harvest failure in kharif season/ conventional cropping is very high and it is estimated 50–60% in most of the years only because of uncertainty in rain's pattern. Based on my research experience and SWOT analysis over 25 years, it is accomplished that this hot arid zone has excellent horticultural potential with wide and varied opportunities of crop production, utilization of resources and eco-restoration (Samadia, 2016).

In-spite of environmental restrictions and bio-physical constraints, the Rajasthan state can be classified into three distinct zone *i.e.* hot arid, semi-arid and sum-humid for horticultural perspective. With changed scenario, climatic resilience and effective resource management, now it is necessary to have new innovative concepts with most prospective plant species (khejri, *Prosopis cineraria*) for enhancing crop-genotype productivity through agri–horti–silvi–pastoral integration (Samadia, 2016) in the arid region. Here, horticultural exploitation with native crop–plant species is generally found to be most appropriate, viable and stable under the restrictions from environmental situations. Presently, systematic horticulture with the native and potential perennial plant species is very negligible, and productivity and quality of produce from the existing crop production technology is quite marginal and unorganized. This situation is primarily because of unavailability of desirable crop–genotypes suited to the prevailing agro-climatic conditions &/or environmentally stressed production sites of defined sub-zones, unavailability of requisite quality planting material of the recommended crop–genotypes and lack of apposite technology. Therefore, there are essentially two complementary requirements for enhancing productivity *i.e.* improvement in the genetic make-up as provisions for suitable crop-genotypes and development of favourable micro-climate at the production sites of the specified zone to minimize the ill-effects of adversity (Pareek and Samadia, 1999; Samadia, 2014 & Samadia, 2016) as technological tool for the crops under consideration.

For success in horticulture with hot arid to semi-arid sub-tropic agro-climate, land-use plan of production site is the pre-requisite considering the environmental factors, resources availability and big sized (4–10 ha) holding of unit fields. Based on strategic and long-term R&D studies at CIAH, an innovative concept (HBCPSMA) is postulated by me and recommended for development and management of horticultural based crop production sites in the state of Rajasthan. In this concept, major emphasis is on creation of favourable micro-climate and protection of crops in production sites, and it is considered as primary requirement for promotion of horticulture, and it was in addition to provision of appropriate crop-genotypes. Besides, adoption of innovative and improved management practices and agro-technique for specific crop-combinations is the key inputs for efficient resource utilization and minimizing the ill-effects from abiotic and biotic factors, and this crop–genotype–environmental interaction technological advancement is recommended as Horticulture Based Crop Production Site Management Approach (HBCPSMA, Samadia–2004).

Khejri Based Technological Advancements

The traditional farming systems have provided sustenance to the desert dwellers. In vast spread arid farm lands, cultivation of pearl millet, cluster bean, moth bean, sesame and native cucurbits is done as component crops between natural plantations of khejri (*Prosopis cineraria*), and in addition some trees/shrubs/grasses (jharber, bordi, lasora, pilu, ker, phog, sewan, *etc.*) are part of traditional agri–horti–silvi–pastoral system prevalent under the rainfed situations. This system of agricultural commodity harvesting can provide sound farm economy, improves nutrition and health standards of the livelihood and stability when the monsoon rains is good during kharif season. With the changed scenario, now, focus is shifted from sustenance to remunerative farming but the mono-cropping is much risky for development of desert horticulture. Therefore, the conventional systems pre-dominantly mixed cropping needs multi-dimensional interventions through the use of native crop–plants and newer technological advancements, and now it can potentially be exploited with HBCPSMA concept – Khejri based crop production site management approach as innovative tool for doubling farm income (Samadia, 2016).

The successful cultivation of bajra, guar, moth, mung, cowpea and groundnut in kharif and wheat, mustard, gram, cumin, methi during rabi, and excellent growth and yield performance in most of vegetables and fruits (ber, aonla, bael, mandarin, sweet orange, lime), desert trees / shrubs (jharber, ker, lasora, pilu, rohida, kumat, phog, khimp) and grasses (sewan, dhaman, bharut) with the khejri plantations exhibited significance of this versatile tree. Therefore, strategic promotion of khejri based crop production site management approach is required with conventional (khejri + mixed cropping and khejri + native sole crops under rainfed) and non-conventional combinations (khejri + fruits, vegetables, seed spices and agronomic sole crops under irrigated), and this can be an innovative tool for breaking-up the crop yield gaps over the existing production system (Samadia, 2015).

The region and crop specific long-term research findings under HBCPSMA concept from 1994 to 2016 enables me to conclude it with recommendations and based on that some principles have been suggested for development and management of khejri based crop production sites. The production sites should be developed in accordance to the topography of sand-dune landscape and soil conditions. Fencing of production site and development of multi-tier rows of seedling plantation with desert tree species (khejri, lasora, rohida, bordi, ker, kumat) and shrubs (jharber, phog) all-around the field block is the prime consideration for creation of favourable micro-climate and it is must for protection of crop production sites from extremes of hot, cold and high wind conditions. The four side field area of production siteis develop as multi-tier rows of seedling plantation (single or paired, 4mx4m) with the directions for khejri and lasora or rohida,kumat (south), khejri and lasora or ker, rohida (west), khejri and kumat or rohida, bordi (north) and khejri and ker or rohida (east).

The innovative practices / technologies under HBCPSMA includes selection and development of khejri planting model based on resources at the production site, preparation and maintenance of production sites for *in–situ* rain water harvesting and soil moisture conservation, and crop cultivation. Adoption of pre–monsoon deep ploughing in June prior to crop sowing and post–monsoon ploughing after crop harvest in November month as techniques resulted into more *in–situ* rain-water harvesting, moisture conservation and weed free fields. Keeping of crop field fallow for 1–2 months either April–June or October–November, and is an essential practice for soil health security and fertility build-up. Besides, seed/plant selection, sowing/planting time and techniques, maintenance of crop plant population and prophylactic measures are good management practices essentially required for improving yield and productivity of resources.

Based on crop potentialities and critical analysis over the years, so far, three horticultural combinations have been suggested for development of khejri based production sites. The first two are primarily with rainfed or limited irrigations, and namely are organic *Panchkuta* production technology (khejri, ker, lasora, kumat and kachri) and native cropproduction technology (khejri, fruits and vegetables) in which khejri planting model KM–1 (4 m x 4 m) or KM–3 (8 m x 8 m) is practiced, respectivelywith a hectare block lay-out. For diverse cropping and irrigated cultivation, wide spacing planting models i. e. KM–9 or KM–11 is the most appropriate and here 4mx4m paired row khejri plantation is done at 24m or 44m distance, respectively with four hectare block layout and it is intensive crop production technology. In addition, model KM–7 is unique for animal husbandry based multiple crop production system and in which khejri is planted in paired rows (4 m x 4 m) at 16 m spacing and the wider area is for jharber (4mx4m) plantation and mixed cropping of sewan and guar is done under rainfed situations.

For systematic block planting, single or paired rows of khejri (Thar Shobha) should be laid-out in east-west direction and wider spacing area is the developed fields for fruit, vegetable, seed spice or other crop cultivation. For arid horticulture, promotion of KM-1 or KM-9 planting model as khejri based production sites is found to be most successful under resource constraints environment and this should be with the highest remunerative crop-genotype such as kachri (AHK-119), cluster bean (Thar Bhadavi), snap melon (AHS-82), mateera (Thar Manak), bottle gourd (Thar Samridhi) and Indian bean (Thar Kartiki). These khejri based crop combinations resulted into higher family employment and returns @ Rs 75,000–2,25,000 ha/year in comparison to conventional (Rs. 23,000–42,000 ha/year) cropping system, respectively under rainfed or limited irrigations. The crop bio-mass / yield and farm income would be stabilized from 4–5 years, and the doubled income is further in increasing order with time period of 8–10 years of establishment of HBCPSMA production site with khejri.

Khejri Planting Models and Crop Production Sites

Khejri (*Prosopis cineraria*) also known as *shami* (Sanskrit), *janti* and *jand* (Delhi, Haryana, Punjab), *sumi* (Gujarat), *banni* (Karnataka), *kandi* (Sindh) is a leguminous tree species of India. It not only tolerates the extreme edapho-climatic conditions of *Thar* desert with lush green foliage but also bears flowers and fruits and that too during the driest period (Pareek, 2002). It is a multipurpose tree and producing nutritious pods, valuable fodder and fuel, besides its favourable impact on ecology and soil fertility. In nature, it is an important perennial component in various agri-horti-forestry systems and obliging in eco-restoration (Samadia *et al.*, 2002; Samadia, 2007 and Samadia *et al.*, 2009).

It readily perpetuates itself from the dispersed seeds giving rise to a widespread seedling population. It is a sacred tree and widely protected for socio-economic and cultural benefits. Besides, its seedling plantation is also promoted under social and agro-forestry systems. The density of plants varies with the rainfall pattern, range management, farming and utilization systems on farm lands, non-culturable wastelands and barren lands. In general, seedlings ranged from 15 - 165 trees/hectare from arid to semi arid areas, and is scattered.

This multiple use species is the life-line of Indian *Thar* desert and is much potential perennial component compatible to almost any companion crop in traditional farming systems in the hot arid and semi-arid areas of north-western parts of India. It grows luxuriantly under the extremes adverse temperature, hot and dry agro-climate and that too without much cultural care. The tree is a major source of nutritious leaf fodder (*loong*) for animals. Its tender, immature and nutritive pods (*sangri*) are extensively used for fresh and dehydrated vegetable and it is high value, and thus exhibited tremendous scope for horticultural exploitation (Samadia, 2015).

Traditionally, due to edible pod quality, the rural people collect it on large scale for fresh or dehydrated use. There is no quality standard and so far what-ever produce of s*angri* is available and the same is being harvested and used it by the desert dwellers. The ripen pods (*khokha*) with full of seeds are protein rich and is good source of animal feed. The ripen pods can be powdered for preparation of bakery items such as biscuits and cookies. From a mature tree (20 years age), 25–30 kg *loong* and 10–15 kg tender pods can be harvested annually. The *sangri* is sold at high cost both fresh and dehydrated (50–80 and 300–500 Rs/kg, respectively) and is eye-catching vegetable. Leaf fodder is also sold at high price (10–15 Rs/kg) and has excellent market opportunities (Samadia, 2016).

The recent year's technological advancement in khejri (bud grafting as vegetative method of propagation and variety Thar Shobha) at CIAH, Bikaner resulted into wide spread popularization of for commercial exploitation. The newer technological advancement results to availability of standard planting material of khejri and recommended for establishing orchards for uniform *sangri* production as horticultural crop produce, and in varying planting models for diversified cropping systems as well exploitation under resource constraint arid environment. Systematic production of standard product from such plantations would add value to the already existing lucrative trade in dehydrated pods (*sangri*) and provide assured income to farmers and quality produce to the people (Samadia, 2015).

Before the year 2005, there was no standard variety for uniform and quality *sangri* production under traditional system of crop harvesting and hence, there was no systematic cultivation of khejri. With the development of vegetative propagation techniques at CIAH, the high yielding variety Thar Shobha was recommended by the institute in 2007 for quality *sangri* production. It is the first variety developed for establishing uniform plantations with better quality and high bio-mass production on commercial scale. The spreading of new technologies resulted into promotion of systematic khejri planting and also in farming systems for perennial component. For horticultural exploitation, a series of technological advancement have been made at CIAH up to 2010 and recommended are budding for mass multiplication of plants in nursery, *in situ* orchard establishment by bud grafting, induction of thornless in plants, variety Thar Shobha, crop regulation practice for harvesting both *sangri* and *loong* annually, low-input diversified crop combinations like organic *Panchkuta* production (khejri, ker, lasora, kumat and kachri) adopting HBCPSMA and on-farm value addition.

The long-term and strategic objective under the flagship project was to develop khejri based planting and crop production models and to understand impact of cropping systems for sustainable resource utilization and maximize income under hot arid agro-climate. The research work on khejri based crop production site management approach as technological advancement over the traditional systems in the arid region was initiated with desert and native crop-plant species. For ground storey combinations, arid legumes and cucurbits were tested both under high density planting model (KM-1, 4m x 4m) and wide spacing planting models (KM-10 and KM-11) having single and paired rows khejri plantation (4m or 4m x 4m) at 44 m distances and studied over period of nine years for plant growth, bio-mass production and soil fertility build-up.

Simultaneously, to undertake innovative crop production studies, two hectare khejri block with varying planting models namely KM-2 (6 x 6m), KM-3 (8 x 8m), KM-4 (8 x 4m), KM-5 (8x 4x 4m), KM-6 (16 x 4m), KM-7 (16 x 4x 4m), KM-8 (24 x 4m), KM-9 (24 x 4 x 4m) and KM-12 as control (scattered) was developed by planting seedling rootstock in 2007 and *in-situ* budding of Thar Shobha was performed under rainfed situations. As per technological advancement recommendations, the inter-space field area of said models were utilized for inter-cropping of cluster bean and native crop culture for 3-5 years of establishment of production site. After that, inter-space field area was used for cropping of potential native crop-plant species to understand their feasibility, bio-mass production and economic analysis. For perennial fruit crop component, ker and jharber (4m x 4m) were studied with wide spacing khejri planting models and these models were also studied with inter-cropping of grasses and native vegetables. For boundary plantation studies, seedling plants of khejri, kumat, rohida and lasora were taken into consideration all-around block area and in between desert shrubs (phog, khimp) were incorporated as crop-plants for development of micro-climate at production site and additional source of income.

In this way, to study khejri planting and cropping models adopting HBCPSMA concept, about five hectare virgin field area of sandy-sand-dune land-scape was developed through *in situ* orchard establishment technique under rainfed conditions of hot arid agro-climate at CIAH. From 2007 to 2016, annual growth and development observations were recorded periodically in response to plant establishment and training–pruning techniques. No much difference is observed in respect to the growth parameters under investigation in variety Thar Shobha plantations with varying planting models and it also at the economic yield stage of 6th and 7th year tree age-group during 2015-16 and 2016-17. It is also observed that close spacing and paired rows (4m x 4m) khejri planting exhibited maximum canopy at about 6th years age and thus this age-group plant growth from which normal pruning is recommended for economic and annual harvesting of pod, long and fuel-wood for the highest bio-mass/year.

The annual observations were compiled on growth, yield and bio-mass production in khejri with varying planting models (12), and based on three years of mean data of economic yield, variety Thar Shobha recorded tender pod yield of 6.58 kg/plant under KM-1, KM-9 & KM-11. Similarly, marketable fruit yield of kachri var. AHK-119 (56.58 q/ha) and tender pod yield of cluster bean var. Thar Bhadavi (54.34 q/ha) was recorded as monsoon supported inter-crop with khejri and adopting recommended techniques of HBCPSMA concept from 2014-15 to 2016-17.

Plant Multiplication and Planting T7echnique

To raise true-to-type khejri plantations of elite genotype, vegetative propagation technique is required. The vegetative propagated plants bear with in 3-4 years as against the long juvenile phase of 10-12 years observed in the khejri seedling trees. Time to time attempt were made to multiply khejri plants by cuttings, air-layering, grafting and tissue culture but results were not encouraging, and thus it was prime call for developing appropriate propagation techniques. Realizing the potentiality and its horticultural significance, systematic interventions were initiated with the establishment of CIAH at Bikaner in 1994 so that it can be commercially exploited. However, for this mission mode R&D work was taken by Dr. O. P. Pareek and his team from 1999 with main objectives (i) to identify elite types for use as scion material studies, (ii) to develop simple vegetative propagation technique, and (iii) to identify rootstocks and standardize nursery technique for fast, easier and mass multiplication and *in situ* orchard establishment techniques.

i) Natural perpetuation: The wide spread distribution of khejri is solely by seed propagation. The natural dispersion of the seeds resulted into self-sown seedlings and as well nursery raised seedling plantations which is protected and promoted by the inhabitants as life-line tree of desert eco-system. The ripen pods with full of seeds falling down in the June and eaten by small animals (goat and sheep) and through their litter its spreads everywhere. The seeds do not show any dormancy and germinates quickly. The monsoon rains in June-August helps in easy germination of the seeds. The young seedlings grown up very fast and develop strong tap-root system within the monsoon season. The plants under protection grow very well and took shape within 4 - 5 years. Whereas unprotected saplings drastically and regularly bruised by animals and its above ground growth hampered but are with well developed root system for the survival (Samadia, 2015).

ii) Nursery seedlings: Freshly harvested and non-infested seeds should be used for sowing in polyethylene tubes (30 × 10 cm) filled with mixture of FYM and sandy soil in 1:1 ratio in the nursery. Prior to sowing, seeds should be soaked in water containing 2 % common salt for two hours, and after that dry it for few minutes and treated with fungicide like captan or bavistin. The seedlings raised under open conditions were with better height, collar and stem diameter than those under the shade net, and ready in 7–9 months for field planting as rootstock or budding in nursery. Irrigation by rose-cane was found better than mini-sprinklers under nursery conditions. The seedlings of February and June month sown nursery are ready for field planting during July and October, respectively and gave very high (80–90 %) field establishment and budding success in nursery. Since, the success in budding and subsequent rate of sprouting of scion depends on vigour of rootstock and the desirable thickness of rootstock, therefore, proper care is required in saplings for nursery budding, and seedling developed *in situ* or raised in nursery followed by field planting and budding (Samadia, 2015).

iii) Bud grafting: Khejri can be successfully multiplied by patch budding or bud grafting technique (Pareek, 2001) for uniform planting material, and thus, systematic studies on vegetative propagation was done at CIAH,

Bikaner (Pareek, 2002; Pareek and Purohit, 2002; Samadia, 2007)). For this, the selected elite trees having a record of producing good quality pods should be pruned during November-December month to induce new growth and bearing juvenile scion buds. Fully grown scion buds can be patch budded in the month of March–April and July–October on the nursery raised seedling rootstocks and *in situ* planted seedlings, and an average 76 per cent success was obtained with variety Thar Shobha under Bikaner conditions. The extent of scion bud success, its growth in terms of length, diameter and linear growth depended on the stem diameter of the seedling rootstock, environmental situations and management practices after bud intake.

The patch budding technique is also found to be the most practicable, simple and result oriented on 12–24 months old rootstocks for *in situ* establishment of khejri plantations as well top-working the wild seedling trees of 3–4 year age-group. The identified mother trees of the variety should be pruned every year during the month of November–December to induce juvenile buds suitable for budding. For best results, the stem diameter of rootstocks should be more than one cm, and in such case it is possible to achieve cent per cent budding success. The intake bud may took 18 to 40 days for sprouting and it is weather dependent too and all the sprouted scion buds were survived (Samadia, 2007; Samadia *et al*., 2009).

iv) Field planting and orchard establishment: For promotion of planting, there are two approaches for developing systematic block plantations of khejri and it can be with multiple objectives. The first approach is for horticultural uniformity and high bio-mass production and for this vegetative propagated and varietal plantation is the primary requirement. The second approach plantation is for forestry and eco-restoration and there seedling should be planted. For developing block plantation, systematic field lay-out is done as per objectives and planting models and therefore, there are four approaches of its establishment such as i) field planting of nursery raised bud grafted plants, ii) field planting of nursery raised seedlings and *in situ* bud grafting on the rootstocks, iii) *in situ* seed sowing in field for rootstocks followed by bud grafting, and iv) field planting of nursery raised seedlings.

v) Training and pruning: Training and pruning are two essential practices for plant structure, canopy management and harvesting of crop produce in khejri. After successful intake of scion buds on the rootstocks either in nursery or *in situ* established seedlings, the upper portion of the stocks should be removed or pruned so that growth of sprouted scion bud may faster. The wild sprouting below the graft should be removed regularly through pruning operations using sharp cutter. For better frame-working of grafted plants, training-pruning operations should be done twice in a year *i.e.* November and June months and it should be continued to four years from field establishment. The regular training and pruning operation should be started after attaining the age-group of 4-5 years for better plant canopy management and harvesting of economic produce, and it should be done after assessing the objectives of crop plantation for maximum yield / bio-mass.

vi) Growth regulation for annual harvesting of both *sangri* and *loong*: In general, khejri trees are lopped (pruned) during November–December to harvest the leaf fodder (*loong*). These pruned trees do not bear flower as the new shoot sprouting after the pruning is still immature during the flowering time of khejri in February–March and thus do not produce *sangri*. When *sangri* is desired to be harvested from a particular tree and it is not lopped during November–December but then the tree would not yield *loong*. With technological advancements, systematic plantations would be developed for *sangri* production and with traditional technique it will have to be left unlopped, and consequently trees would not yield *loong*.

To over-come the situations, studies on plant growth, flowering and fruiting regulation were conducted at CIAH and results revealed that plantations developed primarily for *sangri* production should not followed the traditional practice of lopping during November– December. But regular pruning / tree lopping can be done during June month to obtain *loong* and it can be done after the harvesting of ripen pods. The pruning of plants at this time synchronizes with the start of the second growth flush in khejri during June–July. This growth flush got sufficient time (6–7 months) to mature and differentiate floral buds which develop into flowers and produce good crop of pods. The tender pods are ready for harvesting from last week of April to first week of May, and marketable quality *sangri* should be harvested at 2–3 days intervals for obtaining the maximum pickings and yield from budded plants. Thus, June pruning is recommended to harvest both *loong* and *sangri* annually and in practical it should be done when dried pods (khokha) starts falling down. The harvested *loong* dried quickly and excellent in quality so that it can be stored / used as concentrated animal fodder for the prolong period of time.

Table 1: Potential commodity components of the agri-horti-animal husbandry for developing khejri based production sites (HBCPSMA - Innovative concept) under hot arid agro-climate.

Commodity	Rainfed crop culture	Crop needs life-saving (1-2) or limited (3-4) irrigation water management
(1) Cereals/millets	Bajra	Bajra (zonal genotypes)
(2) Legume/pulses	Cluster bean (RGC-936) Moth bean (RMO-40)	Moong bean, Cow pea, Gram
(3) Oil-seeds	Til, Tumba	Til, Rape-seed& mustard
(4) Spices	Sajii	Methi
(5) Vegetables	Kachri (AHK-119), Snap melon (AHS-82), Mateera (AHW-19), Guarphali (Thar Bhadavi) Khejri (Thar Shobha)	Mateera (Thar Manak, AHW-19& 65), Round melon/tinda (AHRM-1), Ivy gourd/kundru (AHIG-1), Kakri (AHC-2, AHC-13), Indian bean (Thar Kartiki), Sword bean(Thar Mahi), Cowpea (CIAH-1), Brinjal (Thar Rachit), Palak (AHLP-1), Bathua (AHLB-1)
(6) Perennial vegetables	Phog, Khimp	Moringa/sehjna (AHMO-1), Meetha neem (curry leaf), Guarpatha (CIAH Selection-1)
(7) Fruits	Jharber (CIAH Sel.-1&2) Ker (CIAH Thornless-1), Lasora (CIAH Selection-1), Bordi, Pilu, Gundi	Ber (Thar Bhubharaj, Gola), Aonla (NA-7, Goma Ashwarya) Bael (Goma Yashi)
(8) Medicinal & Aromatic	Sonamukhi, Tumba, Hingot	Isabgol, Ashwagandha, Henna
(9) Agro-forestry	Khejri, Kumat, Rohida	Neem, Shisham
(10) Grasses	Sewan, Dhaman, Bhurhat	—
(11) Animal husbandry	Cow (Rathi, Naguri, Tharparkar, Kankrej, Gir), Goat (Marwari, Kachhi, Parbatsari) Sheep (Marwari, Jaisalmeri, Chokhala, Nali, Magra, Poogal, Sonadi) Camel (Bikaneri, Jaisalmeri)	

Table 2: Row-plants accommodation lay-out for developing khejri based crop production sites with model KM-1 (Native Crop Production Technology) under hot arid agro-climate.

Production model	Crop	Number of rows and plants based on four hectare production site area			Number of rows and plants based on one hectare production site area		
		Rows	Plants/row	Total plants	Rows	Plants/row	Total plants
KM-1, (4m x 4m), as main crop single rows	Khejri	44	46	2024	21	21	441
Crop-plants of four side boundary plantation in pair rows of 4m x 4m	Ker, East & West	3 x 2 (6)	50	300	2 x 2 (4)	25	100
	Lasora, South	1 x 2 (2)	44	88	1 x 2 (2)	21	42
	Kumat, North	1 x 2 (2)	44	88	1 x 2 (2)	21	42
	Total	—	—	2500	—	—	625
Specific note	Organic *Panchkutta* crop production technology for promotion of desert horticulture in which khejri is in main plot and ker, lasora and kumat are on four sides as boundary plantation, and kachri / cluster bean as monsoon supported inter-crops.						

Table 3: Row-plants accommodation lay-out for developing khejri based crop production sites with model KM-3 (Native Crop Production Technology) under hot arid agro-climate.

Production model	Crop	Number of rows and plants based on four hectare production site area			Number of rows and plants based on one hectare production site area		
		Rows	Plants/row	Total plants	Rows	Plants/row	Total plant
		Rows	Plants / row	Total plants	Rows	Plants / row	Total plants
KM-3, (8 m x 8 m), as main crop single rows	Khejri	19	21	399	9	9	81
Crop-plants of four side boundary plantation in pair rows of 8 m x 8 m	Ker, East & West	3 x 2 (6)	25	150	2 x 2 (4)	13	52
	Lasora, South	1 x 2 (2)	19	38	1 x 2 (2)	9	18
	Kumat, North	1 x 2 (2)	19	38	1 x 2 (2)	9	18
	Total	—	—	625	—	—	169
Specific note	Conventional crop production under rainfed or with limited irrigations for diversified seasonal crops such as cluster bean, moth bean, cowpea, bajra, methi, gram, wheat, groundnut and potential vegetables under resource constraints arid environment						

Table 4: Row-plants accommodation lay-out for developing khejri based crop production sites with model KM-7 (Native Crop Production Technology) under hot arid agro-climate.

Production model	Crop	Number of rows and plants based on four hectare production site area			Number of rows and plants based on one hectare production site area		
		Rows	Plants/row	Total plants	Rows	Plants/row	Total plant
KM-7, 16 (4 m x 4 m) main crop & wide-space	Khejri, pair rows 4 m x 4 m	10 x 2 (20)	46	920	5 x 2 (10)	21	210
Wider area for intercrop	Jharber,rows at 4 m x 4 m	10 x 3 (30)	46	1380	5 x 3 (15)	21	315
Crop-plants for two side boundary plantation in pair rows of 4m x 4m	Lasora, South	1 x 2 (2)	50	100	1 x 2 (2)	25	50
	Kumat, North	1 x 2 (2)	50	100	1 x 2 (2)	25	50
	Total	—	—	**2500**	—	—	**625**
Specific note	Intensive system with native crop-plants as monsoon supported agri-horti-silvi-pastoral model (a) khejri + jharber + legume or grasses (guar/sewan/dhaman/bhurhat), and animal husbandry as primary source of livelihood.						

Table 5: Row-plants accommodation lay-out for developing khejri based crop production sites with model KM-9 (Intensive Crop Production Technology) under hot arid agro-climate.

Production model	Crop	Number of rows and plants based on four hectare production site area			Number of rows and plants based on one hectare production site area		
		Rows	Plants/row	Total plants	Rows	Plants/row	Total plant
KM-9, 24 (4 m x 4 m) main crop & wide-space	Khejri,pair rows 4 m x 4 m	7.5 x 2 (15)	46	690	4 x 2 (8)	21	168
Wider space also for main crop	Jharber, rows at 4 m x 4 m	35	46	1610	17	21	357
Crop-plants of two side boundary plantation in pair rows, of 4m x 4m	Lasora, South	1 x 2 (2)	50	100	1 x 2 (2)	25	50
	Kumat, North	1 x 2 (2)	50	100	1 x 2 (2)	25	50
	Total	—	—	**2500**	—	—	**625**
Specific note	Intensive production model for diverse agri-horti and high value crops with irrigation and maximizing resource productivity.						

Table 6: Row-plants accommodation lay-out for developing khejri based crop production sites with model KM-11 (Intensive Crop Production Technology) under hot arid agro-climate.

Production model	Crop	Number of rows and plants based on four hectare production site area			Number of rows and plants based on one hectare production site area		
		Rows	Plants/row	Total plants	Rows	Plants/row	Total plant
KM-11, 44 (4 m x 4 m) main crops in wide area	Khejri, pair rows 4mx4m	5 x 2 (10)	46	460	2.5 x 2 (5)	21	105
Crop-plants of two side boundary plantation in pair rows at 4m x 4m	Lasora, South	1 x 2 (2)	50	100	1 x 2 (2)	25	50
	Kumat, North	1 x 2 (2)	50	100	1 x 2 (2)	25	50
	Total	—	—	**660**	—	—	**205**
Specific note	Intensive and diverse production model for agri-horti and high value crops with irrigation and maximizing resource productivity						

Table 7: Marketable horticultural harvest in native crop-plants from established plantations adopting khejri based crop production site management approach under hot arid agro-climate.

Crop-plant (genotype)	Start of harvest (plant age, years)	Stabilized harvest (plant age, years)	Yield / plant (kg)	Marketable produce
Khejri, budded (Thar Shobha)	4 - 5	6 - 7	6.58 – 7.86(7.22)	Tender pods (sangri)
Lasora, seedling (CIAH Selection-1)	4 – 5	6 - 7	24.5 – 38.7(31.6)	Tender fruits
Ker, seedling (CIAH Thornless-1)	7 - 8	9 - 10	1.78 – 2.54(2.16)	Tender fruits
Kumat, seedling (CIAH Selection-1)	7 - 8	9 - 10	2.26 - 4.15(3.21)	Mature seeds from tender pods
Jharber, seedling (CIAH Selection-1)	4 - 5	6 - 7	1.85 – 2.56(2.35)	Ripen fruits
Phog, seedling (CIAH Selection-1)	6 - 7	8 - 9	0.556–0.895(0.725)	Mature flower buds (phogla)

Table 8: Marketable yield in native horticulture crop-plants with the varying planting models of the khejri based production site management approach under hot arid agro-climate at CIAH, Bikaner.

Crop-plant	Potential khejri planting models (marketable fruit yield, q/ha from accommodated plants under the model)				
	KM - 1	KM - 3	KM - 7	KM - 9	KM - 11
Khejri (sangri)	31.84	5.84	15.16	12.12	7.58
Lasora (tender fruits)	13.27	5.68	15.82	15.82	15.82
Ker (tender fruits)	2.16	1.12	Not applicable	Not applicable	Not applicable
Kumat (Mature seeds)	1.35	0.58	1.61	1.61	1.61
Jharber (ripen fruits)	Not applicable	Not applicable	7.40	8.38	Not applicable

Table 9: Marketable fodder yield with varying planting models of khejri based crop production site management approach under rainfed conditions at CIAH, Bikaner.

Crop-plant	Potential khejri planting models(marketable fodder yield, q/ha from accommodated plants under model)				
	KM - 1	KM - 3	KM - 7	KM - 9	KM - 11
Khejri (loong)	35.94	6.61	17.11	13.69	8.55
Jharber (dry leaves)	Not applicable	Not applicable	10.01	11.35	Not applicable

Table 10: Marketable fresh yield in native vegetables from varying planting models adopting khejri based crop production site management approach under hot arid agro-climate at CIAH, Bikaner.

Vegetable crop (genotype)	Fresh produce	Yield (q/ha) from available area of crop cultivation in planting model (mean with three best monsoon year)		
		KM-1 & KM-3	KM-9 & KM-11	Average
Kachri (AHK-119)	Fruits	42.48	88.56	56.58
Cluster bean (Thar Bhadavi)	Tender pods	45.32	85.45	54.34

Table 11: Marketable fodder yield in native crop-plants with varying planting models adopting khejri based crop production site management approach under rainfed conditions at CIAH, Bikaner.

Crop-plant	Fodder yield (based on three best monsoon year from 2012–2016)	
Khejri	8.15 kg dry leaf fodder/ plant (loong), yield from stabilized crop	
Jharber	3.18 kg dry leaf fodder / plant, yield from stabilized crop	
	Fresh weight basis (q / ha)	Dry weight basis (q / ha)
Sewan	278.15	140.62
Dhaman	134.37	70.65
Bhurhat	92.53	36.87
Cluster bean	303.12	84.37
	Fresh weight basis (kg/ plant)	Dry weight basis (kg / plant)
Phog	4.28 – 5.25 (4.76)	2.85 – 3.16 (3.01)

References

Anonymous 1994–2017. Annual reports. National Research Centre for Arid Horticulture / Central Institute for Arid Horticulture (ICAR), Bikaner (Rajasthan) India.

Arya, S., R. Tomar and O.P. Toky 1994. Effect of plant age and auxin treatment on rooting response in stem cuttings of *Prosopis cineraria. Journal of Arid Environments*, 27(1): 99-103.

Burkart, A. 1976. A monograph of the genus *Prosopis* (Leguminoceae subfamily Mimosaceae) (Part 1 and 2). Catalogue of the recognized species of *Prosopis. Journal of the Arnold Arboretum*, 57: 219-249 and 450-525.

Dwivedi, N.K., D.C. Bhandari and S.K. Verma 1997. Collecting *Prosopis cineraria* germplasm in western India. IPGRI Newsletter for Asia, The Pacific and Oceania, 22: 19.

Harsh, L.N. and J.C. Tewari 1998. *Prosopis* in the arid regions of India: some important aspects of research and development. In: *Prosopis* species in the arid and semi-arid zones of India (Ed. J.C. Tewari, N.M. Pasiecznik., L.N. Harsh and P.J.C. Harris). Conference Proceedings, Central Arid Zone Research Institute, Jodhpur (India). November 21-23, 1993; 5 -10pp.

Kumar, D., D.K. Samadia, M.K. Jatav, S.R. Meena, B.D. Sharma and D. Singh 2017. Soil fertility status of khejri based cropping system under varying land use pattern in dry land conditions. In: Abstract book of 5th National Seminar - Climate Resilient Saline Agriculture: Sustaining Livelihood Security, Organized by Indian Society of Soil Salinity and Water Quality, Karnal, Haryana from January 21–23, 2017 at SKRAU, Bikaner. p140.

More, T.A., D.K. Samadia, O.P. Awasthi and S.S. Hiwale 2008. Varieties and Hybrids of CIAH. Technical Bulletin - CIAH/TECH/PUB. No., 30: 1-11 pp.

Pareek, O. P. 2001. Vegetative propagation of khejri. ICAR News,7(4): 12.

Pareek, O. P. 2002. Top-working wild khejri trees. Indian Horticulture, 47 (1): 11-12.

Pareek, O. P. and A. K. Purohit (2002). Patch budding in khejri (*Prosopis cineraria*). *Indian Journal of Horticulture*, 59: 89-94.

Pareek, O. P. and V. Nath 1997. Variability in horticultural traits of khejri in Thar Desert. IPGRI Newsletter for Asia, the Pacific and Oceania, 24: 23-25.

Pareek, O. P., A. K. Purohit and D. K. Samadia 2006. Graft compatibility among the Asian and Algaobia *Prosopis* species. *Indian Journal of Arid Horticulture*, 1 (2): 1-5.

Purohit, A.K., D.K. Samadia and O.P. Pareek 2003. Studies on growth enhancement of khejriseedlings and *in-situ* budded plants. *Annals of Arid Zone*,42(1): 1-5.

Rawat, T.S., H.C. Sharma and J.P.S Pundir 1982-83. Air-layering–a success in propagation of khejri(*Prosopis cineraria* L.). *Udyanika*, 5 (1&2): 43-44.

Samadia, D.K. (2004). Horticultural Based Crop Production Site Management Approaches (HBCPSMA – Innovative concept, unpublished). CIAH, Bikaner.

Samadia, D.K. 2007. Techniques for annual harvesting of sangri. ICAR News, 12 (2): 3.

Samadia, D.K. 2014. Khejri – Bagwani Anusandhan anv takaniki vikas. Technical bulletin (Hindi) - CIAH/TECH/Pub. No., 53: 1-44 p.

Samadia, D.K. 2015. Khejri – Thar Shobha: Grow for horticultural exploitation. Technical Folder, CIAH: Bikaner, 6p.

Samadia, D.K. 2016. Thar Shobha: New khejri variety. *Indian Horticulture*, 61 (4): cover page back & 12–13.

Samadia, D.K. 2016. Horticulture based crop production site management approach (HBCPSMA) - an innovative concept for doubling farm income under dry-lands. *Journal of Agriculture and Ecology* (http://saaer.org.in), 2016-1: 1–9.

Samadia, D.K., A.K. Purohit and O.P. Pareek 2002. Genetic diversity in vegetable type khejri.*Indian Journal of Agro-forestry*, 4 (2): 132-134.

Samadia, D.K., O.P. Pareek, A.K. Purohit and T.A. More 2009. Technological advancement for transformation of *Prosopis cineraria* in to a horticultural crop. Abstract of Paper In: International conference on nurturing arid zones for people and the environment: Issue and agenda for the 21st century. November 24-28, 2009; Jodhpur (India), 144-146 pp.

Samadia, D.K., O.P. Pareek and A.K. Purohit 2006. Techniques for annual harvesting of *sangri* and *loong* - An approach to popularize *khejri* as a horticultural crop. Abstract of paper In: National Symposium on Livelihood Security and Diversified Farming Systems in Arid Region. January 14-16, 2005. Arid Zone Research Association of India, CAZRI, Jodhpur. 2.5: 30 p.

Samadia, D. K., Hanif Khan, S. R. Meena, D. Singh and P. N. Sivalingam 2012. Production technology for horticultural exploitation of khejri and moringa. Chapter in compendium of winter school: Exploitation of underutilized horticultural crops for sustainable production. September 05–25, 2012. CIAH, Bikaner. pp. 168 - 177.

Sarolia, D.K., J.S. Gora and D.K. Samadia 2016. Genetic diversity in khejri (*Prosopis cineraria* (L.) Druce) in Thar Desert. Chapter in compendium of winter school: Exploitation of underutilized fruits of arid and semi-arid region, Organized by Directorate of Research, RCA, MPUAT, Udaipur from October 4–24, 2016. pp82–85.

Sivalingam, P.N., D.K. Samadia, D. Singh, H.K. Changal, Hanif Khan and S.K. Sharma 2011. Characterization of *Prosopis cineraria* (L.) Druce germplasm with suitable horticultural traits from the hot arid region of Rajasthan, India. *Genetic Resources and Crop Evolution*, 58: 1095–1103.

Sivalingam, P.N., D.K. Samadia, D. Singh and S. Chauhan 2016. Molecular markers to distinguish 'Thar Shobha' a variety of Khejri (*Prosopis cineraria* (L.) Druce) from trees in natural populations. The Journal of Horticultural Science and Biotechnology (http://dx.doi.org/10.1080/ 14620316. 2016.1160545). pp1–9.

Solanki, K. R., N. L. Kackar and S. K. Jindal (986. Air layering in *Prosopis cineraria* (L.) Mac Bride. *Indian Forester*,112(3): 202-207.

31

Cultivation of Cactus Pear for Higher Income in Arid Zone

Kamlesh Kumar and Dhurendra Singh

Introduction

Climate change has become one of the biggest challenges for the sustainable crop production. Prolonged droughts and desertification are among the issues faced by Indian hot arid zone where the rural poor and smallholders are most heavily affected. If people are to survive in these ever harsher conditions, their crops need to withstand drought, high temperatures and poor soils. Cactus crops are gaining increasing interest across the globe, in particular cactus pear or prickly pear {*Opuntia ficus indica* (L) Mill.} because of its unique characteristics which provide resilience to the harsh ecological conditions. Cactus pear is able to grow on land where no other crops are able to grow; it can be used to restore degraded land. It is the only crop that can be relied on when everything else fails. Cactus pear is grouped as CAM plant (Crassulacean Acid Metabolism), a photosynthesis mechanism evolved in some plants as an adaptation to arid conditions. In a plant using full CAM, the stomata in the leaves remain shut during the day to reduce evapotranspiration, but open at night to collect carbon dioxide (CO_2). At night carbon dioxide is stored as 4 carbon acid malate in cell vacuoles. The malate is transported to chloroplasts where it is converted back to CO_2 during day which is used in the process of photosynthesis (Ting, 1985). The cactus pear plants though shallow rooted but have capacity to absorb and store water in its parenchyma even under unfavourable climatic conditions due to high mucilage production in both cladode and fruits. Besides, peculiar adaptations to water scarcity and high as well as low temperature is because of reduced leaf tissues and cuticular wax covering of cladodes and fruit surfaces. (Saroj *et al*, 2017). The cactus pear fruit is an oval, elongated berry, with a thick pericarp and a juicy pulp and, in general, many hard seeds generally known as tuna. The fruits are of different colours, such as red, purple, orange, yellow and green in colour which show the presence of various antioxidants. The high sugar and low acid blend of the fruit makes it delicious and palatable. The pericarp of ripe fruits accounts for 33% to 55%, while the pulp is 45% to 67%, and seeds 2% to 10%. The pulp is the edible part of the fruit and is composed of water (84% to 90%) and reducing sugars (10% to 15%). The pH value of fruits ranged (5.3 to 7.1) and the very low acidity (0.05% to 0.18% in citric acid) of the pulp, which strongly influences the processing operations. Sugars range from 10°Brix to 17°Brix and are mainly of the reducing type (Russel and Felker, 1987; Stintzing *et al*., 2003). Cactus cladode (nopal) is rich in pectin, mucilage, minerals, polyphenols, nicotiflorin, vitamins, polyunsaturated fatty acids (palmitic acid, oleic acid, linoleic acid, linolenic acid *etc.*) and amino acids (glutamine, leucine, lysine, valine, arginine, phenylalanine and isoleucine). The nopal contains antioxidants and various flavonoids, particularly quercetin 3-methyl ether, a highly efficient radical scavenger. Chemical characterizations of nopal have studied by several workers and reported that it contains vitamins and minerals. Dehydrated nopals contain high levels of potassium content. Considering the lactose intolerance in certain parts of the population, nopals could be an alternative as source of calcium. Cactus nopal pulp has numerous compounds (dietary fibre, vitamin C, phenolic compounds) with the potential to provide important benefits like intestinal, cardiovascular, hepatic health, antioxidant activity and cancer prevention. The fresh young pads so called nopal are excellent source of proteins including essential amino acids and vitamins. The value added products of cactus pear could also supplement to nutritional security and human health. The succulent vegetative parts are called as pads or cladodes which are modified stem (Kumar *et al*., 2017).

The propagation and cultivation of spineless varieties may provide the feed stuffs to domestic animals during summer season in arid ecosystems, when there is lack of forage/grass due to harsh climatic conditions. Even today cactus pear is treated as underutilized crop in India, though it has multiple utility. In general, no special treatments (fertilizer application, pruning and treatment against pests and diseases) are necessary for cactus establishment and cultivation. Supplemental irrigation may be applied during early period of establishment on first year. Cacti are ready for commercially use after 3- 4 years of planting. A well managed cactus orchard can remain productive for about 4-5 decades.

Present Status in India and Global Scenario

Cactus pear is now-a-days part of the agricultural systems and natural environment. The most important economic species in the world is *Opuntia ficus-indica* (L.) Mill. Presently, it is grown in Mexico, Malta, Spain, Sicily,

Italy, Greece, Libya, Tunisia, Morocco, Algeria, Lebanon, Syria, Egypt, Saudi Arabia, Yemen, Israel, Chile, Brazil, Turkey, France, Bulgaria, Portugal, Albania, Cyprus, United States (Kauthale *et al*, 2017). The area covered under cactus cultivation is presented in table 1.

Table 1: Current world scenario of cactus cultivation

S.N.	Country	Cultivated area (000 ha)
1.	Brazil	600
2.	Tunisia	600
3.	Mexico	230
4.	Morocco	150
5.	Algeria	150
6.	Italy	70
7.	Other South American Countries	75
8.	Other North American countries	16
	Total	1891

Source: International Cactus Pear Workshop: Development of a cactus pear agro-industry for the Sub-Sahara Africa Region Bloemfontein, South Africa, 27-28 January 2015.

It can be found growing well from sea level to 5100 meters above sea level in Peru and from Canada to Patagonia and Argentina. The first record of modern cactus pear cultivation techniques using the best varieties were found in Mexico, Zacatecas, San Luis Potosi, Aguascalientes, Jalisco and Guanajuato in the 1940s and 1950s. Cactus pear is now the most reliable and even profitable option for utilizing rainfed semi-arid lands in the semi-arid regions of central Mexico. It is the crop of choice over corn or dry beans in areas exposed to drought, while in slightly improved areas, it supplemented the income of the growers (Josh *et al*, 2017).

Cultivated of cactus pear was started long back in many countries of the world but, in India, its commercial cultivation is yet to start. In India 33 Opuntia clones were introduced by Dr. Peter Felker, Texas, USA at the Nimbkar Agricultural Research Institute at Phalton, India in 1987 as a part of an Indo-US collaborative research programme on Opuntia. All the introduced clones grown well under the semi arid agro-climatic conditions of western Maharashtra and it was reported that some clones produced fruits also (Singh, 2003). Till so for (Kauthale *et al*, 2017), it is at the research stage with limited field trials initiated by Central Arid Zone Research Institute (ICAR- CAZRI), Jodhpur and in Kutch district of Gujarat in few select areas with the support of International Center for Agricultural Research in the Dry Areas (ICARDA). The comprehensive work on Cactus was started by NARI with the collection of good number of imported accessions. However the work could not be pursued to its logical end. During the last two decades the research was conducted by many public sector research Institutes in India especially those are working in arid agricultural crops, but the outcome of this work is yet reach to the farmers.

Recently, in India the research work on Cactus pear has been undertaken by various Research Institutions *viz.*, ICAR-CAZRI, Jodhpur supported by ICARDA, ICAR-CIAH, Bikaner ICAR- IGFRI, Jhansi, ICAR- CSSRI, Karnal. Field adaptation trials on farmer's field are being conducted through ICARDA programme by CAZRI and in Bundelkhand region by IGFRI and CAFRI. NDDB actively involved in spreading cactus among dairy farmers of Gujarat (Kauthale *et al*, 2017). Trials on planting seasons and irrigation management with cactus in various cropping models are in progress at CAZRI. Recently, research work on planting methods, performance of various accessions, exploitation and uses of vegetable type cactus (Kumar *et al.*, 2017) and biomass production, was made at CIAH, Bikaner, CAFRI, Jhansi and CSSRI, Karnal. BAIF Development Research Foundation, Pune, a non-government organization initiated a comprehensive work on Cactus with the financial support of NABARD in 2015. The initiative was to cultivate this new crop on the farmer's field after standardizing the production technology with the aim to cultivate cactus for food and fodder, but later on the research at BAIF was focused mainly on providing an alternate fodder crop to the farmers living in arid and semi-arid regions of India and to address their livelihood issues (Kauthale *et al*, 2017).

Advance Approaches for Cactus Cultivation

Micropropagation technique of spine-less, vegetable type cactus pear (nopal cactus) was standardized by ICAR-CIAH, Bikaner recently in 2009 (Fig. 1&2) and its morphological and nutritional evaluation (Table 1&2) was done under green house conditions (Saroj *et al*, 2017; Kumar *et al.*, 2017).

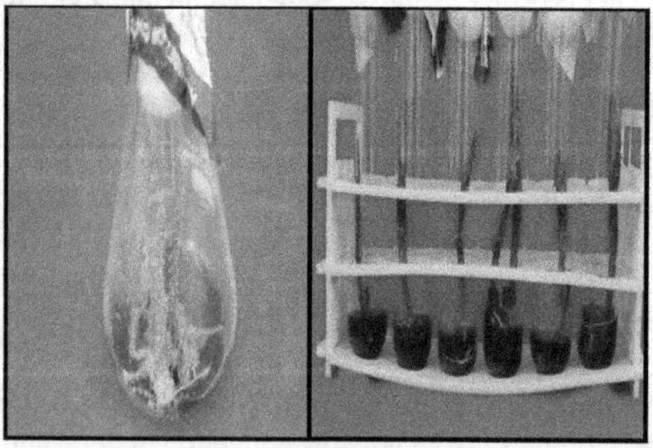

Fig. 1&2: Tissue cultured plants of nopal cactus pear

Table 1: Morphological observations on nopal Cactus pear (vegetable type)

Serial/ Plant Number	Number of cladode/plant (04.02.17)	Number of cladode/ plant (4/3/17)	Days required to reach edible stage	Weight of nopal at edible stage (g)	Length of nopal (cm)	Width of nopal (cm)
1	17	22	23	36.32	16.5	8.0
2	16	22	18	22.52	12.0	6.5
3	16	23	23	39.34	16.7	9.0
4	18	21	18	33.20	13.3	8.2
5	22	29	19	30.73	13.7	7.5
6	25	31	18	22.21	12.0	6.4
7	11	16	24	46.31	18.0	8.7
8	08	11	23	36.94	16.0	7.0
9	19	23	20	27.86	15.4	6.0
10	14	19	19	23.61	13.0	7.0
Average	16.6	21.7	20.5	31.90	14.66	7.63

Table 2: Nutritional evaluation of vegetable type cactus pear

S.N.	Nutrient content	Unit
1.	Moisture content	91-93 %
2.	Dry matter	7-9 %
3.	Ash content	13-15 % (dry weight basis)1 % (wet weight basis)
4.	Mucilage	15-22 % (dry weight basis)
5.	Total phenolics	4.5-7.0 mg/g (dry weight basis)
6.	Flavonoids	200 μg/g (dry weight basis)
7.	Total Antioxidant Activitymg Equi. Vit C/g (dry wt.)	10-16 μg/g (dry weight basis)

Establishment of Greenhouse Unit

After successful development of tissue culture technique at ICAR-CIAH, Bikaner a green house unit of cactus has been established (Fig. 3). Since demand of cladodes of vegetable type cactus pear is increasing day by day. Therefore, we are at CIAH establishing a field repository in open condition.

Fig. 3: Greenhouse unit established for tissue cultured cactus pear genotype

Field Evaluation of Different Cactus Pear Genotypes

The cultivation of vegetable type (nopal cactus pear) for human consumption is depends on selection of the spineless varieties. The most important species for nopales production is *Opuntia ficus-indica* (L.) Mill. Now-a-days, it is parts of kitchen or family gardens in arid and semi-arid regions due to nutritional and medicinal properties. Therefore, field evaluation of this nopal cactus is being carried out with other cactus pear existing germplasm at CIAH farm (Fig. 4 & table 3). Five genotypes of cactus pear were evaluated under field condition for field establishment, sprouting, growth, flowering and fruiting. The data regarding these observations is presented under here in tabular form.

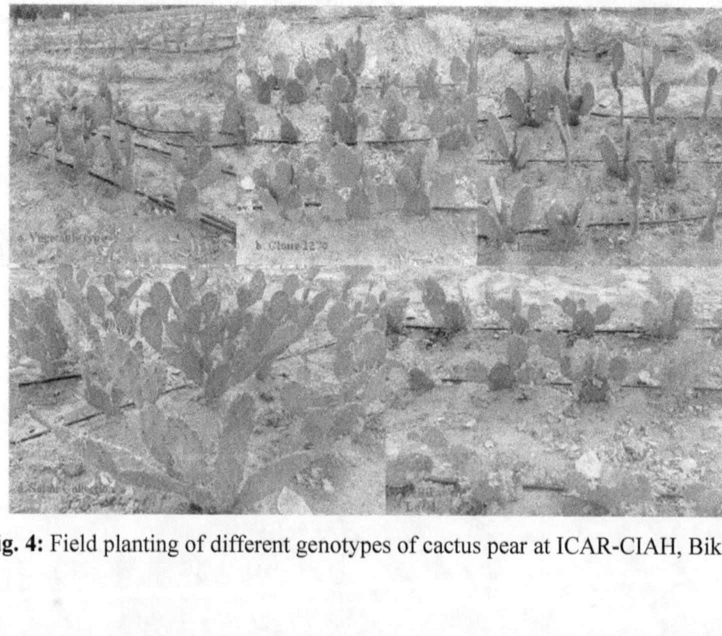

Fig. 4: Field planting of different genotypes of cactus pear at ICAR-CIAH, Bikaner

Table 3: Field evaluation of different cactus pear genotypes

Genotype	Field establishment (%)	Mortality (%)	Sprouting	Avg. plant height (cm)	No. of cladode/ plant	Flowering	Fruit per plant	Use
Vegetable type/nopal cactus	76.93	23.04	III	43.90	6- 23	Occurred	1-9	Vegetable, salad, pickle, squash, animal feed, edible, fruit production
Clone 1270	100	0.00	II	45.20	4- 14	Occurred	1-3	Suited for animal feed
Clone 1271	100	0.00	IV	56.60	4- 7	No flowering	-	Animal feed, biofencing
Solan Collection	100	0.00	I	53.30	16- 32	Occurred	1	Biofencing, fruit production
Bikaner Local Collection	100	0.00	V	41.30	7- 10	Occurred	1-8	Biofencing, fruit production

Cladode, flower and fruit variability in cactus pear genotypes available at CIAH, Bikaner

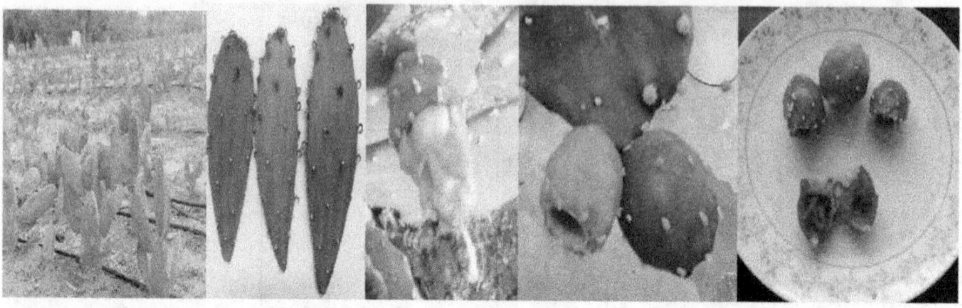

Spineless, vegetable type cactus pear, nopales suitable for human consumption, flower and fruits

A view of clone 1270 and their cladodes, flower and fruits suitable for animal feed

A field view of clone 1271 and their cladodes suitable for animal feed

Solan collection 1 their cladodes, flowers, fruits suited for fruit production and fodder purpose

Fig. 5: Variability of different cactus pear genotypes existing at ICAR-CIAH, Bikaner

Contd.

Solan collection 2, their cladodes, flower and fruits suitable for biofencing

Genotype Bikaner Local, their cladodes, flower, fruits, suitable for fruit production and biofencing

Fig. 5: Variability of different cactus pear genotypes existing at ICAR-CIAH, Bikaner

Earlier Pareek *et al.* (2003) evaluated 51 clones of cactus which were obtained from Texas in pots at CIAH, Bikaner. Clone 1308 was sprouted first (24th day) while clone 1379 took 135 days for sprouting. The average cladode weight was ranged from 66.8 g to 1300 g. Only 3 clones (1269. 1270 and 1271) were flowered after 3 years of planting but fruiting occurred in clone 1270 and 1271. Singh and Singh (2001) conducted a study on growth, cladode yield and quality of vegetable type cactus under Agra (UP) agro-climatic conditions and reported that clone 1308 (vegetable type) was better in all growth, yield and quality attributes than the Nopalea type. Planting of cladodes in East-West direction keeping 1/3 portion under the soil significantly increased the number of roots/pant, cladode length, early sprouting and biomass production in clone 1270 which is a fodder and fruiting type cactus pear (Singh *et al.* 2002).

Very recently at CIAH, Bikaner, 30 plants of vegetable type cactus pear along with Mount Abu Collection, clone 1308 and 1269 had been planted in the field and evaluated for growth and survival. Clone 1308 was sprouted first than other genotypes. Plant survival was observed good under field condition and growth of clone 1269 was found better than other genotypes.

Potential Benefits of Cultivating Cactus Pear in the Arid Region

Cactus pear is cultivated for about hundreds of years as a food crop, a defensive hedge, rearing cochineal insect for production of dye, as a fodder crop and as a standing buffer fodder for animals during drought periods. It plays a key role in erosion control mechanism and land rehabilitation programmes particularly in arid and semi-arid regions of the world, a shelter and refuge. Cactus nopals have been long back used as natual, herbal medicine to treat diseases such as ulcer, allergies, fatigue, rheumatism, antiuric and as diuretic agent. Alleviating effects towards alcohol hangover symptoms cactus pear have been used very recently and associated with reduced inflammatory responses after excessive alcohol consumption. The nutritional and medicinal properties of the fresh cactus cladodes have long been known. The bark is a tonic and diuretic, the fruits have cooling and tonic properties and the young buds and the milky juice are astringent (https://books.google.co.in/books?isbn=0124080642). Some of the epithets used for the plant and the fruit such as "Green gold", "fruit for the poor", "treasure under its spines", "world vegetable dromedary", "future plant", "sacred plant" and "monster tree" (Arias Jimenez, 2013a), these names telling the importance of cactus in the work and lives of people because of their resistance to drought and high temperatures and their adaptability to poor soils.

A number of medicinal applications have been reported in several cactus species. Medicinally, components and extracts have been used in treatments of diabetes, cholesterol and immune system health (Knishinsky, 2004). Nopales products played as preventive and therapeutic effects against alcoholism and alcohol addiction. The flavonoids which are extracted from cactus nopales, quercetin 3-methyl appears to be the most potent neuroprotector. A number of traditional foods are prepared from cactus pear, including fruit-based products: jams, juices and nectars; dried fruit; juice concentrates and syrups; and liquors. Pickles, juices, jams and a number of other minimally processed products can be made from cladodes (FAO, 2013). Cacti have been used as live biofencing material for protecting fields from wild animals.

Different types of *cactus species* are being used for research purposes mainly as fodder in arid, semi- arid, rainfed and drought prone areas. However, few institutes like CIAH, Bikaner, CAZRI, Jodhpur, CAFRI, Jhansi, CSSRI, Karnal the focus has recently shifted to the production of vegetable type, spineless cactus genotypes for human consumption as well as animal fodder. Tender nopales (tender cladodes of vegetable type, spineless cacti) can be consumed during the early stages of growth (till light green in colour). Cactus nopal is suitable to use as fresh. It was observed that nopales can be harvested regularly at an interval of 15-20 days when they reach 10 to 15 cm in length and about average weight of 30 g/nopal which can provide an average yield of 1.5 kg tender nopales/plant/year. Cactus nopales are an important part of the human diet and are also used as feed for livestock.

A number of potentially active nutrients and their multifunctional properties are found in nopal which makes it suitable for the production of health-promoting food and food supplements. Mainly nopales are consumed fresh form in Mexico and canned or pickled nopales exported to particularly Southwestern United States and Texas. Opportunities of several economic products such as fruit, squash of fruit, pickle, ready to serve drinks, colour from fruit and cochineal, cladodes for culinary and salad were prepared at ICAR-CIAH (Fig. 6, 7 & 8), Bikaner and demonstrated to several beneficiaries which can satisfy the product diversification need, increased shelf life and improved human health.

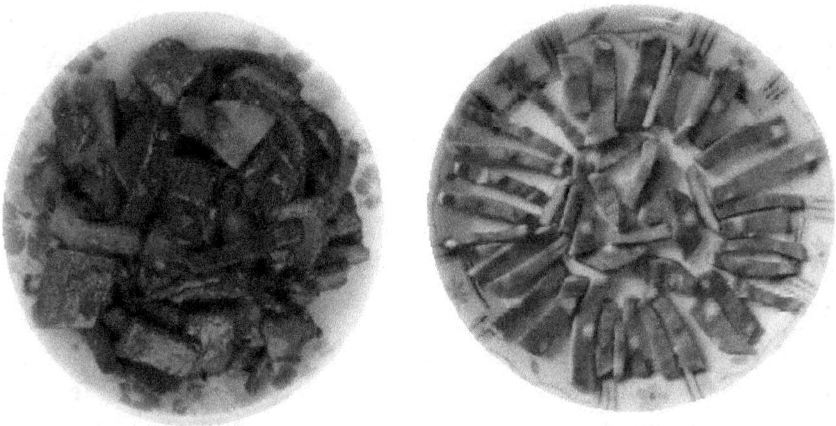

Mix vegetable Salad

Fig. 6: Culinary exploitation of spineless, nopal cactus pear at CIAH, Bikaner

Fig. 7: Syrup, minimal processed nopal, fresh fruits, cochineal insect (*in plate*) for dye preparation and fruit squash made at CIAH, Bikaner

Fig. 8: Cochineal insect reared on nopal cactus, cochineal for dye preparation (*in plate*) and cochineal dye

Economic Returns

It has been proved from the intensive R&D efforts made at ICAR-CIAH, Bikaner that cactus pear can be cultivated successfully in arid region with enhanced nutrition and income from the different products of nopal cactus pear. Further, round the year production can be taken under low cost green house condition. Under greenhouse, nopales can be harvested regularly at an interval of 15-20 days with an average yield of 1.5 kg tender nopales per plant per year. Thus, the farmer can get regular income from this crop. Preliminary studies indicated that Rs 150 to 200 per plant per year may be obtained from different fresh products of vegetable nopales, fruits and processed items (such as squash, RTS, pickle, jam, candied products *etc.*) from a full grown plant after 3-4 years of planting. Thus, a farmer can get an income of Rs 500 to 600 per square metre area under green house condition with an estimated benefit cost ratio of 3:1 (Kumar *et al*, 2017).

Future Scope and Research Need for Cactus Cultivation

Cactus pear traditionally has been used as a valuable health promoting food crop and also has applications in pharmaceutical industries. It has several uses and immense potential to be the food of future. Cactus fruits are popular in several countries and contains important nutritional and health components such betalains, polyphenols and dietary fibre. Although, human interest is increasing with the fruits, recently research also has been carried out to know about its bioactive components and functional properties which have been generated. There are so many future uses of the cactus have been suggested, such as nopal pulp use for the production of shampoos, conditioners, face and body lotions, soaps, hair gels, sun protectors and bio-ethanol production. Cactus juice extracted from cladodes is one of the most commonly used additives in earthen plaster or improving house paint and mucilage found to purify water. Veterinary phytotherapy using *Opuntia* appeared to be a promising field of research and applicability. Cactus plant can be used for rearing of Cochineal insect for production of attractive red and/ or pink colour dye. Human health concerns about artificial food additives have opened the door of popularity of this dyes and the demand is increasing which encourage the rearing of cochineal insect is a golden opportunity for this crop.

Different new processing technologies have been developed but great challenges are still remain. Therefore, it is important to continue research onto this wonderful crop. There are several future opportunities and advantages of growing cactus pear as a crop discussed hereunder in key points.

a. Different food products and alcoholic and non-alcoholic drinks can be made from young nopales (to promote food and beverage industry)

b. Supplements and feed from nopales and waste from fruit processing, including peel and seeds (for livestock feed industry)

c. Gastric mucosal protectants from mucilage extracts, tablets and capsules of nopal powder and flower extracts (for pharmaceutical industry)

d. Creams, shampoos and lotions from nopal (for cosmetic industry)

e. Fibre and flours from nopales (food supplements industry)

f. Gums from nopales and colorants from fruits (natural additives industry)

g. Binding compounds from mucilage/nopales (importance in construction industry)

h. Biogas from digestion of nopales and factory waste streams, alternatively, lignified nopals burned as fuel-wood (role in energy sector)

i. Soils, organic materials and improved drainage from the use of cactus products (as an agricultural inputs)

j. Artisan crafts can be made from lignified nopales (importance for tourism sector)

k. Use of natural colorants, such as carmine from cochineal insects (role in textile industry)

Conclusion

Cactus pear plays an important role in providing food (human), fodder (livestock) and several economic products for human and animals living in arid and semi-arid regions, especially during critical periods of the year. There is a need for research to evaluate the effect of different cactus products on human and animal beings as well as methods to improve the nutritive quality of this valuable crop in these regions. Previously, a number of crops were known of little importance and thus not collected, conserved and researched but later on they have been recognised by international research organisations as necessary for agricultural sustainability and food security. This is also happened with cactus pear. The increased interest in this crop specially from potential contribution to agricultural diversification, application to the exploitation of marginal lands and changing environments, benefits as additional income source for farmers living in arid and semi-arid regions of the world. Cactus pear can tolerate and grow successfully in poor and marginal soils, limited water conditions and high temperatures. Therefore, cactus is being cultivated in arid and semi-arid regions of the world.

References

Arias Jimenez, E. 2013. Importancia de la tuna. *Cactusnet Newsletter*, 13: 9–12.

FAO. 2013. Agro-industrial utilization of cactus pear. Rome.

Jose, C.B., Dubeux, Jr., Hichem B.S. and Nefzaoui, A. 2017. Forage production and supply for animal nutrition. *In*: IX International congress on cactus pear and cochineal- CAM crops for a hotter and drier world (Eds. Inglese, P., Mondragon, C., Nefzaoui, A. and Saenz, C.), Coquimbo, Chile.

Kauthale, V., Aware, M. and Punde, K. 2017. Cactus an emerging fodder crop of arid and semi arid India. Booklet by BAIF Development Research Foundation, Pune

Knishinsky, R. 2004. Prickly Pear Cactus Medicine: Treatments for Diabetes, Cholesterol, and the Immune System. Inner Traditions/Bear and Co, Rochester, VT, USA. pp.705.

Kumar, K., Singh, D. and Berwal, MK. 2017. Cactus pear: A multipurpose climate smart crop for higher income in arid zone. In the compendium entitled "Doubling Income through Advance Approaches for Fruits and Vegetables in the Arid Region" (*Eds*. Saroj, P.L., Sharma, B.D. and Jatav, M.K.), ICAR-CIAH, Bikaner, Rajasthan.

Pareek, O.P., Singh, R.S. and Vashishtha, B.B. 2003. Performance of cactus pear (*Opuntia ficus-indica* L. Mill.) clones in hot arid region of India. *J. prof. Ass. Cactus dev*. 5: 121-130.

Russel, C. and Felker, P. 1987. The prickly-pears (*Opuntia spp*., Cactaceae): A source of human and animal food in semiarid regions. *Econ. Bot.,* 41:433–445.

Saroj, P.L., Singh, D. and Kumar, K. 2017. Culinary exploitation of nopal cactus pear in arid region. *Indian Hort*., 62 (4): 13-16.

Singh, G. 2003. General Review of Opuntias in India. *Journal of the professional association for cactus development*. 5: 30-46.

Singh, R.S. and Singh, V. 2001. Performance of vegetable type cactus under Indian conditions. *Curr. Agri.,* 25 (1-2): 99-101.

Singh, R.S., Bhargava, R., Sharma, B.D. and Pareek, O.P. 2002. Effect of depth and direction of cladodes planting on growth of cactus pear. *Curr. Agri.,* 26 (1-2): 119-121.

Stintzing F.C., Schieber, A and Carle, R. 2003. Evaluation of colour properties and chemical quality parameters of cactus juices. *Europ. Food Res. Tech*., 216 (4):303-311.

Ting, I.P. 1985. Crassulacean Acid Metabolism. *Ann. Rev. Plant. Physiol*., 36: 595-622.

32

Harvesting and Handling of Fruits and Vegetable Crops for Higher Income

M. K. Verma, O.P. Awasthi, Chavlesh Kumar, Tanushree Sahoo Rajesh Jalwania and B. L. Nagar

Introduction

Harvesting and handling are the equally important activities as the entire production system as an individual activity. Therefore, proper harvesting and handling is essential to assure the farm income. There are various factors which directly affect the storage life of fruits and vegetables as below. The natural limits to the post -harvest life of all types of fresh produce are severely affected by other biological and environmental conditions.

***Temperature*:** An increase in temperature causes an increase in the rate of natural respiration of all produce and food reserves and water content become depleted. The cooling of produce will extend shelf-life by slowing the rate of respiration.

***Transpiration and water losses*:** High temperature, low humidity in the store and injuries to produce can greatly increase the loss of water from stored produce beyond that unavoidable lost from natural causes. Maximum storage life can be achieved by storing only undamaged produce at the lowest temperature tolerable by the crop and at the relative humidity appropriate for the produce.

***Mechanical damage*:** Damage caused during harvesting and subsequent handling (i.e. injuries, impact bruising) increase the rate of deterioration of produce and renders it liable to attacks by decay organisms even under refrigerated storage. Mechanical damage to root crops will cause heavy losses owing to bacterial decay and must be remedied by curing the roots or tubers before storage. Curing is a wound –healing process.

***Decay in storage*:** Decay of fresh produce during storage is mostly caused by the infection of mechanical injuries provoked by micro organisms, mostly bacteria and fungi. Furthermore, many fruits and vegetables are attacked by decay organisms which penetrate through natural openings or even through the intact skin. These infections may be established during the growth of the plant in the field but lie dormant until after harvest, often becoming visible only during storage or ripening.

Harvesting

Efficient crop commodity management is generally aimed at maximizing the yield of a crop from the area of land under cultivation while at the same time maximizing the return (doubling the income). The time, labour and capital expended in bringing the crop to maturity are rewarded by the financial return obtained by the grower during marketing. For fruits and vegetables the magnitude of this return mainly depend on -quantity of fresh produce harvested and marketed, which mostly depends on production planning, crop selection, varietal selection, production practices (which include among other irrigation, cultural practices, fertilization and chemical treatments); and quality attributes to satisfy market and consumer requirements. The overall quality of fresh produce cannot be improved after harvest, there are a few exceptions like controlled ripening to improve colour and flavour and refrigerated and controlled cold storage to extend shelf-life. These exceptions are however limited by pre-harvest conditions.

***Varietal selection*:** The desired market characteristics of a commodity can often only be obtained by changing the variety being grown. This will often means that growers have the difficult task of adapting to new varieties which may have very different growing characteristics compared to traditional ones. However a resistance to change may mean a major loss of income as the traditional varieties become less popular on the market.

***Size*:** It is the major quality factor and farmers through varietal selection followed by production practices, should try to control it. The aim is to produce a crop of an average desirable size which is not necessarily the largest possible size.

***Mechanical defects*:** The need to reduce surface defects is an inevitable consequence of market development. These defects can be physical damages which can be controlled pre-harvest by better plant management to protect the produce from wind, sun, hail, chemical residues, etc.

***Pest and diseases*:** A better control with appropriate and controlled application of chemicals and field sanitations practices to control fungal, bacteria and insect attacks and avoid the presence of unsightly marks.

Factors Affecting the Harvesting

The quality and conditions of produce sent to market and their consequent selling price are directly affected by the care taken during harvesting and field handling. Whatever the scale of operations or the resources of labour and equipment available, the planning and carrying out of harvesting operations must observe basic principles. The objective of the grower should be; to harvest a good quality crop in good condition; to keep the harvested produce in good condition (protected from rain, sun or animals) until it is consumed or sold; and to dispose of the crop to a buyer or through a market as soon as possible after harvest.

To meet these objectives, success for the harvesting, field handling and marketing must depend on planning from the earliest stages of production, with particular regard to:

- crop selection and timing to meet expected market requirements;
- contact with buyers so that the crop can be sold at a good price when ready for harvest;
- planning harvest operations in good time, arranging for labour, equipment, material, cover space to protect the inputs and the harvested crop and transport;
- it is economically sound in terms of returns on investment to improve grading, packing and handling of the produce before it leaves the farm;
- considering the above the grower must ensure that all those working on the farm are properly trained;
- provide full supervision at all stages of harvesting and field handling.

Harvest Maturity

Fruits and vegetables are considered to be commercially mature when at the stage of physiological development that consumers consider to be the most desirable. However the stage, at which a commodity reaches commercial maturity, varies greatly. Many leafy vegetables are commercially mature at an early stage of plant development while fruits are often ready for harvest at a fully developed stage. Maturity standards have been determined for many fruits and vegetables (Table 1 and 2). Harvesting crops at the proper maturity allows handlers to begin their work with the best possible quality produce. Produce harvested too early may lack flavour and may not ripen properly, while produce harvested too late may be fibrous or overripe and have a shorter shelf-life. Pickers can be trained in methods of identifying produce that are ready for harvest (Kader, 1992).

Table 1: Maturity indicators for fruits and crops

Indicators	Fruit and vegetable crops
Elapsed days from full bloom to harvest	Apples, pears
Change in the fruit and aril colour	Pomegranate
Doka stage for non astringent and dang for astringent	Datepalm
Mean heat units during development	Peas, apples, sweet corn
Development of abscission layer	Some melons, apples, guava
Surface morphology and structure	Cuticle formation on grapes, tomatoes Netting of some melons Gloss of some fruits (development of wax)
Size	All fruits and many vegetables
Specific gravity	Cherries, watermelons, potatoes
Shape	Angularity of banana fingers Full cheeks of mangos Compactness of broccoli and cauliflower
Solidity	Lettuce, cabbage, Brussels sprouts
Textural properties	
Firmness	Apples, pears, stone fruits
Tenderness of the fruits, softness of the stone	Gonda
Tenderness	Peas
Colour, external	All fruits and most vegetables
Internal colour and structure	Formation of jelly-like material in tomato fruits Flesh colour of some fruits
Compositional factors	
Starch content	Apples, pears
Increase in ascorbic acid, change in fruit colour, change in stone/pulp ratio	Ber, aonla
Red colour of the fruit, better sugar-acid blend	
Sugar content	Apples, pears, stone fruits, grapes
Acid content, sugar/acid ratio	Pomegranates, citrus, papaya, melons, kiwifruit
Juice content	Citrus fruits
Oil content	Avocados
Astringency (tannin content)	Persimmons, dates
Internal ethylene concentration	Apples, pears

Source: Kader, 1983

Table 2: Maturity indicators for vegetable crops.

Vegetable crops	Indicators
Root, bulb and tuber crops	
Radish and carrot	Large enough and crispy (over-mature if pithy)
Potato, onion, and garlic	Tops beginning to dry out and topple down
Yam, bean and ginger	Large enough (over-mature if tough and fibrous)
Green onion	Leaves at their broadest and longest
Fruit vegetables	
Cowpea, yard-long bean, snap bean, sweet pea, and winged bean	Well-filled pods that snap readily
Lima bean and pigeon pea	Well-filled pods that are beginning to lose their greenness
Okra	Desirable size reached and the tips of which can be snapped readily
Snake gourd, and dishrag gourd	Desirable size reached and thumbnail can still penetrate flesh readily (over-mature if thumbnail cannot penetrate flesh readily)
Snap melon	Slip from vine, development of aroma
Kachari	Sugar:acid ratio, change in skin colour, development of aroma
Moringa, kheep, Indian aloe, cactu pear,	Tenderness of the pods
Eggplant, bitter gourd, chistophine or slicing cucumber	Desirable size reached but still tender (over mature if colour dulls or changes and seeds are tough)
Sweet corn	Exudes milky sap when thumbnail penetrates kernel
Tomato	Seeds slipping when fruit is cut, or green colour turning pink
Sweet pepper	Deep green colour turning dull or red
Muskmelon	Easily separated from vine with a slight twist leaving clean cavity
Honeydew melon	Change in fruit colour from a slight greenish white to cream; aroma noticeable
Watermelon	Colour of lower part turning creamy yellow, dull hollow sound when thumped, metallic sound of the fruit, deep red flesh colour
Flower vegetables	
Cauliflower	Curd compact (over mature if flower cluster elongates and become loose)
Broccoli	Bud cluster compact (over mature if loose)
Leafy vegetables	
Lettuce	Big enough before flowering
Cabbage	Head compact (over mature if head cracks)
Celery	Big enough before it becomes pithy

Source: Bautista, O.K. and Mabesa, R.C. (Eds). 1977.

This variation in physiological development means that there is no single factor that can be used to determine the optimum harvest date for all commodities. In addition to eating quality, the effect of maturity on the ability of a commodity to be stored or transported can influence the preferred harvest stage. Climacteric fruits are generally harvested prior to the onset of natural ripening so they can be transported in a hard green stage and then ripened in the market. Some other fruits and vegetables are harvested slightly immature as they are then less susceptible to attack by microorganisms. In contrast, apples required for long cold or controlled atmosphere storage are harvested slightly overripe if we do not want the development of "superficial scald", unless we want to control the disorder with a post-harvest chemical treatment.

Ideal Time to Harvest

When the decision to harvest has been taken the preferred time for harvest varies from crop to crop. However the preferred time is the coolest part of the day, usually in the early morning or late afternoon. This is particularly so for leafy vegetables. The selected time should be that which minimize the time between harvest and transport to a packing shade. Local weather conditions could affect the harvesting time:

- It is not desirable to harvest produce when it is wet from dew or rain as this greatly increases the risk of post-harvest spoilage and the tissue is more prone to physical damage.
- It is also not advisable to harvest during a hot and sunny day if the produce is left in the field and cannot be protected by the sunrays and the heat.

Harvesting Technique

Harvesting is basically a simple operation involving the removal of the produce from the parent plant and placing it in containers for removal from the field to the market, or to the packing shade in the farm itself or to the packing – house. In India most of the produce for internal rural and urban markets is harvested by hand. Larger commercial producers may find a degree of mechanization an advantage, but the use of sophisticated harvesting machinery will be limited for the most part to agro-industrial production of cash crop for processing or export. Hand-harvesting is usually preferred when fruits, such as peaches, and other produce, such as green vegetables, are at different stages of maturity and there is need on repeated visits to harvest the crop over a period of time. When the crop is ready for harvesting, labour and transport available and harvesting operations organized the decision as to when to start harvesting will depend largely on the weather conditions and the state of the market.

***Fruits*:** Many ripe fruits such as apples and some immature seed-bearing structures such as legumes pods have a natural breakpoint of the fruit stalk, which can be easily broken by twisting and lifting the stalk taken between the thumb and index. Fruits and other seed-bearing structure harvested in the immature or unripe green stage are more difficult to pick without causing damage either to the produce or the plant. These are best harvested by cutting them from the plant, using clippers, secateurs or sharp knife. The clippers can be mounted on long poles for tree fruits such as soursop, with a bag attached to the pole to catch the fruit. Plucking methods vary according to the kind of produce being harvested:

- Ripe fruits with a natural break – point, which leave the stalk attached to the fruit are best removed by a "lift, twist and pull" series of movements, e.g. tomato and passion fruit.
- Mature green or ripe fruits with woody stalks which break at the junction of the fruit and the stalk are best clipped from the tree, leaving up to a centimeter of fruit stalk attached. If the stem is broken off at the fruit itself, diseases may enter the stem scar and give rise to stem end rot, e .g. mango, citrus, avocado;
- Immature fruits with fleshy stem can be cut with a sharp and clean knife, e. g. zucchini, okra, papaya, capsicum; these can also be harvested by breaking the stem by hand, with the risk of damaging the plant or the fruit and the rough cut will make the produce and the plant more susceptible of decay.

***Vegetables*:** Either the whole or a part of vegetative growth can be harvested by hand only or sharp knife. Knives may be kept sharp and clean at all times to avoid spreading of virus diseases from plant to plant. Harvesting methods varies in accordance with the plant part harvested:

- Leaves only (spinach, rape, *etc*.) and lateral buds (Brussels sprouts, etc): the stem is snapped off by hand;
- Above–ground part of the plant (cabbage, lettuce, *etc*.): the main stem is cut through with a heavy knife, and trimming (roots and external unsuitable leafs are discarded) is done in the field. Do not forget that the cut stem must not be placed in the soil;
- Immature green onions can usually be pulled from the soil by hand; leek, garlic and mature bulb onions are loosened by using a digging fork as for root crops such as carrots and lifted by hand. Simple tractor implements are available for undermining bulbs and bringing them to the surface.

Flower structure vegetables: Immature flower heads (cauliflowers, broccoli) can be cut with a sharp knife and trimmed in the field. Mature flowers (squash, chayote, pumpkins) are plucked individually by hand or shoot bearing flowers are harvested as a vegetable.

Root and tuber crops: Most roots and tubers that live beneath the soil are likely to suffer mechanical damage ay harvest because of digging tools, which may be wooden sticks, machetes, hoes or forks. Harvesting of those crops is easier if they are grown on raised beds or mounds, or "earthed up" as is common with potato growing. This enables the digging tool to be pushed into the soil under the roots or tubers, which then can be levered upward, loosening the soil and decreasing the possibility of damage to the crop. Other root crops, such as taro, carrots, turnips, radishes, etc. can be loosened from the soil at an angle and leaving the root upward. This method can be used also for celery if it has been earthed up or buried to blanch the stem. Following the harvesting, other field operations, which may vary or have a different sequence (given below) according to the produce, may include: inspection; trimming; cleaning; washing; sorting; grading; sizing; packing in field crates; transport to a cover shade which may protect from rain or sun and may permit packing in the final container for the market; and transport to the packinghouse and packing in the final container for the market.

All operations should be carried out to minimize any damage or loss of quality. Key elements in the maintenance of quality are to keep the handling operations to a minimum, but ensure that each operation is carried out with maximum care; minimize the time between harvest and transport from the harvesting place to the packing place; and the detrimental effects of poor harvesting practices are often not evident.

Packaging

The principal benefit of packaging is to provide protection against physical damages through inadequacies in handling and transport of a fresh produce. There are additional benefits, which can be derived from packaging: Barrier protection, to prevent contamination of produce by undesirable environmental agents such as dust and micro-organisms; No utilization of produce, to create relatively small units of produce, which are easier and quicker to handle than unpacked goods. It also gives better utilization of space during storage and transport. Trading aid, to provide a standard size unit for market trading and thereby eliminating the need to weigh or count all items of produce being traded. Marketing aid, through the display of basic market information and attractive designs on the package.

Post harvest Handling

Post-harvest handling operations encompass those steps required for harvested produce to be selected or modified to meet market quality standards and to be packaged in a form suitable for storage or marketing.

Field operations: The simplest operation is where produce is placed into a container in the field directly after harvest, which is then placed into storage or transported to a market without additional sorting or packing. A common variation is to trim off excess leaves or stems before placing into the field container. These simple packing operations are adequate where there is only a small distance between farm and market and produce is sold directly to consumers. This is commonly the case for vegetables grown close to the towns and/or consuming areas. Field sorting is also recommended for produce, which is highly sensitive to repeated handling. All sorting and grading should be conducted in the field in one operation and the produce packed directly into containers suitable for retailing. For most produce, sorting and trimming of excess plant material in the field is desirable only to minimize the volume of produce transported from the farm.

Handling of Fruits and Vegetables

Fruits and vegetables are living biological organisms, having a respiratory system, similar to that of humans. They continue their living processes after harvest. Respiration is the process by which plants take in oxygen and give out carbon dioxide. On the basis of their respiration rate and ethylene production patterns during maturation and ripening, fruits can be classified in two groups: climacteric fruits (they exhibit a large increase in carbon dioxide and ethylene production rates coincident with their ripening) and non-climacteric fruits (which exhibit no changes in their generally low carbon dioxide and ethylene production rates during ripening). In accordance with the respiration rate most horticultural commodities can be classified as follows:

a) Low respiration rate: Nuts, dates, dried fruits and vegetables, apples, citrus, grape, garlic, onion and sweet potato.

b) Moderate respiration rate: Banana, cherry, plum, cabbage, carrot, lettuce, pepper and tomato.

c) High respiration rate: Cauliflower, avocado, berries and green onion.

d) Extremely high respiration rate: Broccoli, peas, spinach and sweet corn.

Ethylene is a natural product of plant metabolism and is produced by all tissues of higher plants. It is considered the natural aging and ripening hormone and is active even at small traces. Horticultural commodities can be classified as follows, based on the amount of ethylene they produce:

a) Low ethylene production: Cauliflower, cherry, citrus, leafy vegetables, root vegetables, potato, cucumber, eggplant, pepper, pineapple, pumpkins and watermelon.

b) Moderate ethylene production: Banana, guava, honey dew melon, mango, plantain and tomato.

c) High and very high ethylene production: Apples, avocado, cantaloupe, papaya, kiwi, pear, plum, passion fruit, sapote and cherimoya.

Fruits and vegetables contain 65 to 95 per cent of water and their postharvest life depends on the rate at which they use up their stored food reserves and their rate of water losses. When food and water reserves are exhausted the produce dies and decays.

Need of Post harvest Handling

Both quantitative and qualitative losses occur at all stages in the post-harvest handling system of the distribution chain of perishables (from harvesting, through handling, packing, storage and transportation to final delivery of the fresh produce to the consumer). Factors affecting post-harvest losses vary widely from place to place and are more and more difficult. Producing fruits and vegetables for the market requires special attention to retain the best quality and get the best return. Therefore the objective of post-harvest handling is, the creation of an understanding of all the operations concerned from harvesting to distribution so as to enable people to apply the proper technology in each step and in such a way to minimize losses and maintain quality as high as possible during the distribution chain. The grower must give, among others, special and careful attention to the important points of the post-harvest chain like market demand for the produce they are planning to grow; market requirements and buyers; knowledge of the fresh produce; cultivation practices; factors affecting post-harvest deterioration; harvesting and field handling; packing in the field; handling and packing in the packing house; common storage and refrigeration; transport; sale to agents, traders or consumers; market handling, and shelf-life of the produce.

Technological Innovations for Efficient Handling of Harvested Produce

Temperature and humidity are primary factors affecting the shelf life of the harvested produce. Therefore, management of these factors under control results in the enhancement of the fruits and the vegetable quality. There are various techniques have been refined which are as under:

Temperature management practices: Temperature management is the most important tool that we have to extend shelf-life of fresh horticultural commodities after harvesting the produce. Temperature management begins with a rapid removal of the field heat by using one of the following cooling methods includes hydro cooling; in package ice; top icing; evaporative cooling; room cooling; forced air cooling; serpentine forced air cooling; vacuum cooling; and hydro-vacuum cooling. Cold storage facilities should be well constructed and adequately equipped. They should have good construction, and insulation and vapour barrier; strong floor; adequate doors for loading and unloading; effective distribution of refrigerated air; properly located controls; enough refrigerated coil surface; capacity adequate to expected needs; appropriate stacking of the produce.

Relative humidity management practices: Appropriate relative humidity is important to control the water losses; decay development; incidence of some physiological disorders; and uniformity of ripening. Proper relative humidity should be 85-95% for the majority of the fruits, 95-98% for vegetables (except dry onions and pumpkins at 70-75%) and 95-100% for some root vegetables. Relative humidity can be controlled by the following methods viz., addition of moisture to air by humidifiers; regulation of air movement in relation to produce; maintaining coil temperature to 1ÚC difference to air temperature; wetting the floor in the storage room; and addition of crushed ice.

Controlled atmosphere: Controlled atmosphere means the addition or removal of gases resulting in an atmospheric composition surrounding the commodity that is different from that of the air 79% of nitrogen, 21% of oxygen and traces of carbon dioxide). Usually this involves reduction of oxygen and elevation of carbon dioxide, in a perfectly sealed room. The use of controlled atmosphere can be considered only as a supplement to the proper temperature and humidity procedures. Controlled atmosphere is used for a certain number of crops to extend shelf-life, reduce disorders such as chilling injuries, reduction of pathogens and some insect control.

Harvest and Handling for Selected Fruits and Vegetables

Papaya

Harvesting: Papaya is one of the most susceptible fruits to skin damage. Papaya handlers can minimize fruit damage by taking protective measures throughout all handling stages: use the harvesting tools properly, and avoid dropping the fruits into the boxes and overfilling, latex stains should be avoided and the fruits transported carefully.

Harvest maturity: Papayas are ready for harvesting when most of the skin is yellow-green, called the "one yellow stripe" If harvested before this stage papayas will fail to ripe completely. For the local market, particularly in the winter months, papaya should be allowed to colour fairly well, before picking; in this case the fruits are more susceptible to damage and bruising during handling.

Harvesting method: Growers usually twist the fruit to break the stem. In commercial operations it is preferable to use a sharp knife to cut the stem and then trim it to the base of the fruit. Harvesting of high fruits can be done with a harvesting pole with a small hoop and a small mesh bag at the end.

***Field handling*:** The fruit must be handled with extreme care to avoid scratching and licking of latex which provokes stain on the fruit which may burn the skin. After harvest fruits are placed gently in plastic or wooden field crates, preferably with a foam layer as cushioning.

***Post-harvest handling*:** Post-harvest handling of papaya requires care and attention to detail. The fruit is susceptible to several factors (fluctuating temperature, moisture, diseases and mechanical damage) which reduce its commercialization. Fruits could be held at 30 °C and high humidity for 48 hours to enhance coloring before packing. The standard decay control consists of 20 minutes submersion of the fruits in water at 49 °C, followed by a cool rinse. Application of fungicide as thiabendazole (TBZ) or a similar one has proven to be satisfactory also to control the pathological diseases. Heat treatment can help to control insects present on the skin of the fruit.

***Selection/Grading/Packing*:** Grading and packing should be carried out as soon as possible after harvest. Fruits, after separation of broken or heavily diseased ones, are graded according to size and stage of ripeness. Papaya should be packed in single layer cartons. Internally, packaging material could include shredded paper. Papayas are best packed on their sides in rows with the stem end at an angle.

***Storage*:** Fruits harvested at the proper harvesting stage (one yellow stripe) will ripen to 60-70% yellow within 4 to 6 days under ambient tropical conditions (25-28 °C). Fruits harvested at "one strip stage" and stored at low temperature (10 to12 °C) will last from 14 to 21 days, if post-harvest disease incidence can be controlled.

Post-harvest losses: They may include: chilling injuries. They include pitting, uneven ripening, skin scald, hard core and increased susceptibility to diseases; and anthracnose. It is the major cause of post-harvest losses. It can be partially controlled with the heat treatment.

Mango

***Harvest maturity*:** Mangoes normally reach maturity in 4 to 5 months after flowering. When the first large fruits start showing a yellow-pink colour on the tree, all the fruits of the same size or larger are ready and have to be picked. Harvesting must be repeated when the remaining fruits, which in the meantime have grown in size, start coloring. When the fruit is fully grown and ready for picking, the stem will snap easily with a slight pull. If a stronger pull should be needed, it means that the fruit is still immature and should not be harvested. In the red varieties an additional indicator of maturity is the development of a purplish-red blush at the base of the fruit.

***Harvesting method*:** When harvesting by hand from the ground is not possible the picker should use a long pole with a cutting blade and a small bag under the blade which holds only a few fruits. For bigger trees the pickers should climb the branches with cotton or fiber bags that they fill and lower to the ground. Fully mature fruits will detach easily at the natural abscission point, whereas half-mature fruits will not. Optimum picking should be done with shears and cutting the stem 1 to 2 cm away from the fruit. This technique permits the reduction of latex exudation and staining of the skin and the entrance of fungus diseases.

The quantity of latex flow depends on:

- Fruit maturity: younger fruits exude more latex;
- Time of the day of harvest: in early morning harvest more exudes;
- Level of rainfall: more latex flows when harvesting after rain;
- Length of stem at harvest: if fruit is picked with the stem, latex flow is minimal.

***Field assembly*:** Place the field crates with no more than two to three layers of fruits (separating each layer with sponges to avoid bruises and scratches) and baskets with liners in a sheltered or shaded area of the field or under the shade of the tree. Exposure to the sun of fruits with stains provoked by the latex may be corrosive for the skin. After harvest, latex should be allowed to drain away from the fruit by placing the mangoes with the stem downward on grass below the tree, cutting the stem 1 to 2 cm away from the fruit.

***Post-harvest handling*:** Wash/clean the fruits or use a clean soft cloth if latex stains and/or residues of chemical treatments are present on the surface of the fruit. In India, it is a habit at the beginning of the harvesting season to artificially ripened mangoes. In the south of the country mangoes are picked at a quite green stage to avoid bird damage and reduce damage during long transport of badly packed produce, on roads which are in a bad condition. On arrival in the final market dealers layer them with paddy straw, covered with plastic, often burning calcium carbide to increase underneath the quantity of acetylene and induce surface colour and ripening, which will take place within 5 to 7 days at15 to 20° Celsius.

Selection/Grading: Select out all immature, ripe, damaged, scarred or otherwise damaged fruits. In accordance with quality specifications required by the purchaser fruits should be graded in each carton according to variety, size and maturity.

Packaging: Mangoes for export or sophisticated internal markets such as supermarkets should be packed in cardboard boxes in single layer and the stem end facing downward or slightly on one side to avoid the latex contamination and with separators Ventilation holes are recommended to improve air circulation.

Storage: Mangoes ripen within 4 to 6 days, after harvest at the mature green stage. The ripe fruits have a shelf life of 2 to 4 days. Storage at 10 to 12 °C and 85 to 90 % of relative humidity, will maintain fruits inacceptable conditions for up to 4 weeks and fruits will then ripe satisfactorily at higher temperature. These conditions will however depend on the variety, maturity at harvest and tome of harvest.

Post-harvest losses: Chilling injuries. Storage below 10 °C for a few days will result in this damage. Anthracnose. Consists of small, black, circular spots on the skin which increases with the ripening of the fruits. Post-harvest applications of specific fungicides (imazalil or thiabendazole) may assist in disease control. Dipping in hot water (at 50 °C for 5-10 minutes, depending from the fruit size in combination with the fungicide) is also utilized.

Watermelon

Harvest maturity: Ripeness in watermelons is difficult to determine because the fruit remain attached to the wine, rather then 'slipping off". Harvesting generally begins about 30 days after full bloom and continues with periodical cuts at 3 to 5 days intervals. The most recommended method for timing the harvest of watermelons is to cut a few representative fruits in the field. However, also the following indicators of ripeness:

- The tendrils of the leaf closest to the fruit attachment become dry;
- Change in ground spot colour from greenish-white to pale yellow;
- The rind becomes hard to pierce with the fingernails;
- Blossom end "fill out'; and

- The sound of a fruit, when thumped with a finger, is a muffed, dull tone if it is ripe.

***Harvesting method*:** Watermelons are harvested in the afternoon by cutting the stem with a sharp knife rather than pulling the fruit out. Afternoon harvest may reduce the risk of cracking if the field has received abundant water during the previous night. If plants are not too turgid field heat can be minimized by harvesting in the morning. Watermelons do not ripen off the vine. However holding them for 7 days and over at room temperature will improve flavour and colour in seeded varieties. Take care of the wet and dirty soil; if watermelons get dirty clean them with a wet and clean cloth.

***Post-harvest handling*:** Cleaning and packing is done in the field itself, when the field is easily accessible. At 10 to 15°C and 90% relative humidity the fruits can be kept for 2 to 3 weeks after harvest. Prolonged storage will be responsible for reduction in crispness and colour. Chilling injuries (the pit in the ring will be invaded by micro organisms) will take place after storage below 5°C.

Bananas

***Harvest maturity*:** The maturity standards for export dessert bananas do not generally apply to bananas grown for local consumption. When bananas are to be sent to distant urban markets they are best harvested in hard mature but unripe green state, which reduces the risk of deterioration during transport.

***Harvesting method*:** The method of harvesting will depend on the height of the plant. Low-growing varieties can be harvested by cutting through the bunch stalk about 30 to 35 cm above the top hand. With taller varieties, the stem of the plant will be partly cut through to bring the bunch down within the harvester's reach, and then the bunch stalk can be cut through. Harvested bunches are best carried on a foam-padded tray to reduce damage during carrying.

***Field handling*:** It is customary in most banana-growing countries to transport the fruit to market on the bunch. This practice injures the fruit during handling and transport, and it is not recommended. Bananas for urban markets will suffer less damage and look better if they are de-handed and packed in suitable boxes.

***Post-harvest-Selection and grading*:** Bananas which are very immature and small, badly damaged or decaying should be discarded. Size and quality grading will depend on the demands of the market. In the more sophisticated urban markets (e.g. supermarkets), size-graded and good-looking fruit may command a higher price.

***Packing*:** All harvested bananas should be kept dry and in the shade before and after packing. Packing is best done in or as near to the field as possible. As soon as the hands of bananas are cut from the stem, they should be laid, curved side uppermost, across the midribs of fresh banana leaves. This will prevent latex from the cut crown contaminating the fruit. Latex flow should stop in 12-15 minutes, after which the banana may be packed into wooden or, preferably, cardboard boxes, which can be of the slotted or telescopic type.

Storage**:** Bananas have a very short post-harvest life at ambient conditions. This is four to ten days when mature green and two to four days when ripe. Both green and ripe bananas are sensitive to cold and are damaged by temperatures less than 13° Celsius.

Ripening**:** Bananas harvested in the mature green stage will normally ripen under the local ambient conditions in which they are grown, but some types will not develop their full ripe skin color. Where urban high-value markets demand fully colored fruit, ripening under controlled conditions is best carried out on a large scale at the urban distribution point. The operation requires special equipment, good management and technical skills. Where the ripening operation is to be undertaken locally, advice should be sought from specialists.

Citrus (Oranges, grapefruit, mandarins; lemons and limes)

Harvesting maturity**:** To reach their full flavour and sweetness they must be left on the tree to ripen;in the tropics citrus often remain green when they are fully ripe internally and do not develop an orange-yellow colour on the tree. The development of the orange-yellow skin colour can be artificially induced after harvest (de-greening). To assess the readiness of their fruit on several counts, this will vary in different situations, for example:

Skin colour**:** maturity may be indicated by a change in the shade of green shown by the skin; lemons change from dark green to a silvery-green appearance at maturity;

Size: experienced growers may evaluate maturity by considering size as well as with other characteristics, such as slight changes in skin colour; and

Internal condition of fruit**:** if a few typical fruit thought to be mature are cut in two, they can be considered ripe if:

- The juice has developed full flavour and is sweet;
- The fruit pulp has developed to the normal colour; and
- Juice drips from the half-fruit when the cut surface is held vertically.

Harvesting method**:** Clippers or secateurs should be used to remove the fruit from the tree. Fruit may be pulled by hand, but there is danger that the stem may be pulled out of the fruit, damaging the skin, or of damage to the tree providing an entry point for field diseases. Not more than 0.5 cm of stem should be left attached to the fruit. If the fruit is mature or ripe this piece of stem will dry up and fall off, leaving only the flower calyx (button) attached to the fruit. As it is harvested, citrus fruit should be placed in picking bags worn by the harvester or in plastic buckets.

***Field handling*:** Harvested fruit is taken in the harvesting container either directly to the packing facility or to the field assembly point, where it is emptied into field containers. At either point the fruit should be protected from exposure to sunlight and rain while awaiting packing or movement to the packing house.

***Post-harvest –Selection and size grading*:** Before it is packed the fruit should be sorted to eliminate all foreign material, such as leaves and twigs. The fruit is then inspected and pieces which are unripe, immature, undersized, damaged or decaying should be discarded. Where citrus fruits are to be pattern-packed in custom-made cardboard boxes it is usually an advantage to grading it into size categories. Suggested minimum sizes and grade category differences for different commodities are:

Commodity	Minimum (mm)	Grade difference (mm)
Oranges, lemons, mandarins	50	5-10
Grapefruit	70	15-20

Limes are not normally size-graded. Citrus sent to local markets in wooden crates will usually be size-graded by the retailer at the point of sale.

***Packing*:** Baskets, wooden boxes, sacks, bags, factory-made wooden crates and cardboard boxes are all used. Most citrus from large scale commercial production are now packed in telescopic cardboard boxes. The recommended outside dimensions for the boxes are 50x 30x 30 cm. These can be stacked eight boxes per layer on standard 1 x 1.2 m pallets. The capacity of these boxes is about 18-20 kg. Wooden crates can also be used for citrus provided they do not have sharp edges or splinters which will damage the skin of the fruit. Wooden crates should not exceed 25 kg capacity.

***Post-harvest treatments*:** De-greening will only change the colour of the skin of citrus fruits. It will not ripen them internally. De-greening is carried out by exposing the fruit to ethylene gas under controlled environmental conditions. It can only add to the cost of the fruit to the consumer, without any compensation in eating quality.

Storage: Citrus fruits can be held up to three weeks under ambient conditions, depending upon the temperature and moisture content of the air. In dry air they may lose moisture and shrink after a few days. Damaged fruit may become infected and decay quickly after harvest.

- Grapefruits: at 10 to 15 °C, 85-90% Relative Humidity for 6 to 8 weeks.
- Lemons: at 10 to 13 °C, 85-90% Relative Humidity for 1-6 months.
- Limes: at 9 to 10 °C, 90-96% Relative Humidity for 6 to 8 weeks.
- Oranges: at 0 to 9 °C, 90-95% Relative Humidity for 3 to 8 weeks.

Onions and Garlic

Harvest maturity: When the bulbs developing from the leaf bases of both garlic and onions are fully formed, the leafy green tops begin to yellow and eventually collapse at a point a little above the top of the bulb, leaving an upright short neck. When the tops "go down" in this way, the bulbs are ready for harvesting. Because all the onions or garlic in a crop do not mature at the same time, large-scale commercial growers harvest them when about half the tops have gone down. Small-scale growers can, if they wish, harvest their crops progressively as the tops go down, especially so if they intend to store the dry onions for sale or use at a later date.

Harvesting method: Harvesting is done by loosening the bulbs with a fork or hoe before lifting them. In dry, sunny weather the harvested crop is left in windrows in the field for a few days until the tops are dry. Where the harvested bulbs are exposed to high-intensity sunlight (e.g. at high altitudes in the tropics), the windrows should be made so that the green tops cover the bulbs to protect them from sunburn.

Post-harvest-Selection and grading: All damaged or decaying onion and garlic bulbs should be discarded. Onions with thick necks should be put aside for immediate use because they will not store well. If the onions or garlic are to be made up into strings for storage or sale, as described below, it is an advantage to separate them into sizes so that the bulbs will be more or less uniform in size in any string.

Post-harvest treatment: The only post-harvest treatment required for the long storage of bulb onions is a thorough curing of the bulbs. Curing is a drying process intended to dry off the necks and outer scale leaves of the bulbs to prevent the loss of moisture and the attack by decay during storage. It can be carried out in the field under dry conditions by windrowing the bulbs. The essentials for curing are heat and good ventilation, preferably with low humidity. This dries out the neck and the two or three outer layers of the bulb.

Packaging: Both onions and garlic may be made up into strings. These are of 2 kg for garlic or 5 kg to 10 kg for onions. This is, however, a labour-intensive operation suited to small-scale production using family labour. It is not cost-effective on a commercial scale.

Storage: The storage environment must be dry and well-ventilated. Dried onions can be stored in cold storage at 0 °C, 65-70% relative humidity for 1 to 8 months. Garlic may be stored in trays, strings or bunches in the same manner as onions, except that with garlic strings are made by plaiting the dry top leaves of the garlic. Garlic can stored in cold storage at 0 °C, 65-70% Relative Humidity for 6 to7 months.

Leafy vegetables and immature flower heads (cabbage, chinese cabbage, kale, rape, mustard, broccoli, chard, spinach beet, spinach, lettuce, celery, green onions)

Harvest maturity: All are harvested in the immature state before the plant has developed to the point of seed production. The older parts of these commodities become fibrous or woody.

Harvesting method: The parts of the plant harvested vary with the crop:

- Cabbage, Chinese cabbage, lettuce, celery and green onions form more or less compact heads; the entire head is harvested at one time;
- Kale, rape, mustard. young shoots, with or without immature flower heads, are picked by hand-breaking; can usually be harvested over a period of time as long as new shoots continue to develop; and
- Chard, beet and spinach are harvested as individual young leaves; sometimes young shoots of spinach are harvested; harvesting is repeated as new leaves continue to develop.

Those crops forming a head, such as cabbage, are cut with a sharp knife. Young shoots and leaves are broken off by hand. Celery and green onions are either pulled by hand or dug from the soil. They should be harvested under dry conditions when soil can be readily shaken from the roots. The roots are then trimmed with a sharp knife.

Field handling: They should be packed loosely in field containers, which must not be overfilled or the produce will be damaged when the containers are stacked. The harvested produce must be kept free from contamination by soil. Leafy vegetables and immature flower heads deteriorate very quickly after harvest because they lose water fast and produce a great deal of heat. The following care is necessary to keep losses to a minimum:

***Packaging*:** For local rural markets traditional containers are likely to remain in use. Rough handling of heavy packages results in damage to produce. Packaging of leafy vegetables and immature flower heads for urban markets will vary with the type of commodity.

The following are examples:

- Cabbages: woven sacks, net bags or field crates of 20-25 kg capacity are suitable.
- Lettuce: one layer wooden plastic crates or ventilated cardboard boxes each containing 24 heads of lettuce.
- Celery: wooden crates holding 20-30 heads of celery.
- Green onions: normally tied in bunches by the grower; they are best transported in small wooden crates holding 10-15 kg.
- Chard and spinach: crisp, brittle and easily broken by rough handling; they are best packed loosely in cardboard, wooden or plastic field boxes of 5-10 kg capacity; over-packing will cause crushing of leaves and bruising and rapid discoloration of stems.
- Kale, rape and leafy brassicas: may be tied in bunches or packed loose; they can be marketed in nets, cardboard or wooden or plastic field crates of 5-10 kg capacity.

***Storage*:** Leafy vegetables and immature flower heads have a very short post-harvest shelf-life, especially under ambient conditions. Even under refrigeration most remain in good condition only up to two weeks. Ideally, they should reach the consumer within two days of harvest.

Tomatoes

***Harvesting maturity*:** If tomatoes are to be used in the ripe condition, they should be picked at the earliest when they are at least mature green. Immature tomatoes do not ripen after harvest. Tomatoes have reached the mature-green condition when they are fully rounded and have changed from dark to medium or light green, and the skin develops a waxy gloss. As ripening is initiated, the fruit shows a pale pink or yellow tinge, which develops through a definite pink to full red. Most tomatoes are harvested at the early ripening or pink stage, depending on market preference and the time they take to reach the retailer. Tomatoes to be consumed immediately can be harvested when fully ripe.

***Harvest maturity*:** Tomato fruit stalks have a natural break-point. Mature fruit readily breaks away from the cluster when pressure is placed on this point while lifting the fruit upwards. Tomatoes are best harvested into plastic buckets (pails) and transferred if necessary to plastic field crates holding not more than 20 to 25 kg weight.

***Post-harvest - Selection and grading*:** All decaying, damaged, undersized and sunburned tomatoes should be discarded.

***Post-harvest treatments*:** Only those tomatoes which are in good condition are marketed, there should be no need for any post-harvest treatments. Tomatoes produced on a large commercial scale may be subjected to artificial ripening; but in countries where production is mostly on a small scale, this is not necessary since tomatoes are normally harvested at maturity and ripen naturally.

***Packaging*:** For local markets tomatoes can be packed in baskets or other traditional containers assuring careful handling, i.e. rigid enough to protect the contents from being crushed. For urban markets cardboard telescopic boxes or wooden or plastic trays with capacities of not more than 10 kg, should be used. Size-graded tomatoes can be pattern-packed in 2 layers for green produce and one layer for ripe red produce to make best use of the box.

***Storage*:** Tomatoes have a relatively poor storage capability. Green mature fruit can be held for up to two weeks at 18-20 °C and 90-95% Relative Humidity, but for less time under ambient tropical temperatures. Fully ripe tomatoes have only 4 to 7 days' storage life, at 13-15 °C and 90-95% Relative Humidity.

Brinjal

***Harvest maturity*:** The harvest index is based principally on size and colour. The fruits of eggplant should be harvested somewhat before they reach full maturity. The colour should be dark purple, firm and glossy, with green calyx and stem. Browning of the calyx and shrivelled skin are indication of aging and water loss.

***Harvesting method*:** Harvesting should be done once or twice per week, depending on the variety. Secateurs are needed for hand-harvesting, taking care to handle nicely the fruits and leaving 2 to 4 cm long stem above the calyx. The harvesting of wet fruits should be avoided.

***Field handling*:** Ventilated field crates should be placed when full, in a shaded place to avoid over-heating. Sacks and bags should be avoided to reduce development of mechanical damage and heat build-up.

***Post-harvest*:** Clean with a soft cloth to remove soil and residues from the fruits. Fruits showing yellowing of the skin are over-mature and should be separated, along with fruits presenting signs of decay, insect damage, cracks, sunscald and other mechanical damages. After size grading, the fruits of the same size are loose packed in cardboard boxes.

***Storage*:** At optimal storage conditions of 7 to 12 °C and 85-90% Relative Humidity, eggplants can be stored for 7 to 10 days. Stored below 7 °C the fruits will develop chilling injuries such as surface pitting, decay internal breakdown and shrivelling.

Cucumbers

Harvesting maturity and harvesting method. For table or slicing cucumbers quality is primarily based on uniform shape, firmness, and a dark green skin colour. Additional quality indices are size, freedom from growth or handling defects, decay and absence of yellowing. Cucumbers are harvested at a range of developmental stages. Depending on cultivar and temperature, the time from flowering to harvest may be 55 to 60 days. Cucumbers generally are harvested at a slightly immature stage, near full size but before seeds fully enlarge and harden. Firmness and external glossiness are also indicators of a pre-maturity condition. At the proper harvesting maturity a jellylike material has begun to form in the seed cavity. Harvesting is done by hand by cutting free of the vine with a sharp knife, rather than by tearing. Cucumbers are packed in two layer boxes, cardboard, wooden or plastic.

Storage: Storage of cucumbers at 10 to 12 °C and 95% Relative Humidity is generally less than 14 days as visual and sensory quality deteriorate rapidly. Short term storage or transit temperature such as 7 °C is often used but after 2 to 3 days chilling injuries will result. Cucumbers are highly sensitive to ethylene present in the storage room or in the transportvehicle. Do not mix bananas, melons, tomatoes and other ethylene producing products with cucumbers.

References

Bautista, O.K. and Mabesa, R.C. (Eds). 1977. Vegetable production. University of the Philippines at Los Banos.

FAO "Improvement of post-harvest fresh fruits and vegetables handling- a manual" RAPA/, AFMA Bangkok, Thailand, 1986. Available also at :http/www.fao.org/inpho/EN/resources/library/index.asp

Kader, A. A. 1983. Post-harvest quality maintenance of fruits and vegetables in developing countries. *In*: Lieberman, M., *Post-Harvest physiology and crop preservation. Plenum Publishing Corporation.* p.455-469.

Kitinoja, L, and Kader, A.A. 1995. "Small- scale post-harvest handling practices: a manual for horticultural crops.3rd.Edition

Davis, California, 1995. Available also at: http/www.fao.org/inpho/EN/resources/library/index.asp crops export from the Caribbean" Fintrac.

Medlicott, A. "Products specification and post –harvest handling for fruits, vegetables and root

UNCTAD-GATT "Manual on the packing of fresh fruits and vegetable" Geneva, Switzerland, 1998. Wills S, R.B.H. Post-harvest and handling and storage of fruits and vegetables. FAO/GCP/CPR/ 008 /BEL, Beijing, China, 1988.

33

Waste Utilization from Arid Horticultural Crops for Supplementing Farmers' Income

Birbal, K.R. Sheetal, V. Subbulakshmi, P.S. Renjith and M.L.Soni

Introduction

The Indian arid zone covers nearly 31.8 million ha of the country's geographical area. It mainly spreads over parts of Andhra Pradesh, Gujarat, Haryana, Karnataka, Maharashtra, Punjab and Rajasthan states of India. Six of the eleven districts in Western Rajasthan i.e. Barmer, Bikaner, Churu, ShriGanganagar, Jaisalmer and Jodhpur lie wholly in the arid zone. High temperature, low precipitation and high velocity winds are some of the main culprits of Arid zone which limit the scope for higher crop productivity in these areas. Under this situation, arid horticulture based farming practices has ample scope to raise the productivity and living standard of peoples of the hot arid and semi-arid regions. Arid horticulture based farming practices greatly favour the development of high quality produce in many fruit and vegetable crops. Horticulture based farming practices are now considered to be the most ideal strategy to provide food,nutrition and income security to the people especially in arid and semi-arid region of the country (Chundawat 1993; Chadha 2002).

Today India is the world largest producer of many fruits and vegetables. Wastes and by-products generated from fruits and vegetables are organic in nature and contribute a major share in soil and water pollution. Also, accumulated fruit and vegetable wastes (FVWs) cause greenhouse gas emission which is a matter of serious environmental concern. Proper disposal and handling of these wastes can help reduce pollution risks, at the same time be a source of several useful and commercially valuable by-products. Organized processing of wastes can help farmers achieve higher incomes.

Present Status in India and Global Scenario

There exists a huge gap between per capita demand and supply of fruits and vegetables due to enormous waste during post-harvest storage and handling. According to recent estimate around 18% of India's fruit and vegetable production – valued at INR 133 billion – is wasted annually. In developed countries, 40-60% of production is processed, thus reducing wastage, and at the same time providing wastes in bulk for utilization in making other by-products. However in India, around 1.8% of horticultural produce is processed mainly by small-scale units, with no controlled method of waste disposal.

Most of the vegetables and fruits are sources of fibre, oils, and other useful products. Composition of different fruit and vegetable wastes are given in Table 1. In general, seeds contain bioactive lipids and polyphenols while peels are a rich source of dietary fibres.

Table 1: Composition of different fruit and vegetable wastes (per 100g)

Waste	Moisture (g)	Protein (g)	Fat (g)	Minerals (g)	Fibre (g)	Carbohydrate (g)
Banana peel	79.2	0.83	0.78	2.11	1.72	5.00
Sweet orange seeds	4.00	15.8	36.9	4.00	14	-
Watermelon seeds	4.3	34.1	52.6	3.7	0.8	4.5
Muskmelon seeds	6.8	21	33	4	30	-
Pumpkin seeds	6	29.5	35.4	4.55	12	12.53
Apple pomace	-	2.99	1.71	1.65	16.16	17.35
Mango seed kernal	8.2	8.5	8.85	3.66	-	74.49

(*Source*: Maini and Sethi, 2000)

Even though, some value added products are being made from this waste, majority of the waste is currently unutilised and discarded. At present there are few possibilities for the utilization or recycling for most of these wastes, the residues are thus disposed of or fed to animals.There are at present few examples small scale industries which utilize fruit and vegetable wastes and generate income.

- Mandarin essential oil is extracted in small quantity at different processing units at Bangalore, Nagpur, Abohar and Sikkim. Lime oil is also extracted at small quantities.
- Watermelon rind after removing green portion is used to prepare 'Tuti-frutti' in UP.
- Seed kernels of watermelon, muskmelon, cucumber, pumpkin are used in sweets, beverages and bakery.

- Researchers at CIST, Lucknow have developed dietary fibre, Vitamin C and antioxidant enriched biscuits by incorporation of aonla pomace (a by-product generated during aonla juice processing) as one of the biscuit ingredients.

Gaps in Prevailing System

Currently, the fruit and vegetable processing is done in India mainly through small-scale industries which mainly fall under informal sector and, thus, processing waste is considered to be of negligible value compared to the processed fruit. Also the amount of produce being processed is low and mainly consumption is in fresh form. This leads to not much waste being available at a single site for collection and utilization. Data from developing countries indicate that large scale industries process wastes into biogas or compost, while from smaller industries is for the most part disposed of through municipal waste disposal systems or just discarded near processing plants. There is neither organized system nor awareness among people of importance of collection and disposal of this biodegradable and high organic matter rich waste. Also the concept of use of wastes for purposes other than as cattle feed or compost is a novel concept in India.

Advance Approaches for Fruit and Vegetable Waste Utilization in Arid Regions

Various compositional studies of the waste suggest presence of a wide range of bioactive compounds in different residual fractions. The first phase of waste utilization is mainly the substance characterization. Optimal recycling and application areas and possibilities are based on substance characterization. The characteristics of fruit and vegetable wastes, some recoverable valuable by-products (Table 2) and their alternate uses are briefly describedin this section.

Table 2: Possible by-products from fruit and vegetable wastes

Fruit/Vegetable	Type of waste	Possible by-products/usage
Orange	Peels, seed, pulp	Essential oil, pectin, cattle feed, peel candy, phytochemicals
Lime	Peels, seed, pulp	
Mango	PeelPulp waste (fibre)Kernal	Pectin, cattle feed, alcohol, phenolic acids,Wine, vinegar, juiceFat, tannin, starch
Tomato	Core, peel, seeds	Feed, seed oil, flavanols
Potato	Peel, coarse solids	Feed, single cell protein
Apple	Pomace	Juice, wine, vinegar, pectin, feed
Pomegranate	Peel	Flavanols (preservatives)

Contd.

Guava	Seeds	Phenolic acids
Grape	Pomace	Anthocyanidines (Food colour additive, Antioxidant)
Watermelon	Rinds	Dietary fibre, food additives
Carrot	Peel and pomace	Dietary fibres, food additives

(Maini and Sethi, 2000; Banerjee *et al*., 2017)

Process Fruit and Vegetable Culls/Waste to Separate Juice from Pulp

A press is used to separate the fruit and vegetable waste into juice and pulp. Screw presses are typical systems that can effectively separate the juice from the pulp. After separation, each fraction has different utility and can be used for different purposes. If the waste/culls are of good food quality, they can be used as juices in food applications based on available markets. The pulp has also potential to be used as a component of foods. The separated pulp from culls which are not of human food quality can be used as one component of compost or animal food. The juice can be utilized as a feedstock for ethanol production.

Candied Peel Preparation

Citrus peelscan be candied and used either in baked goods or as a snack food.Additionally, shreds of peel are used in marmalades.The process to make marmalade is similar to candying and involves boiling the shreds of peel in 20% sugar syrup for 15-20 minutes and then gradually increasing the sugar concentration to 65-70°Brix (% sugar by refractometer) in the syrup as the food is soaked for 4-5 days.It is then taken out, rinsed and finally dried in the sun or a hot air drier. If big food companies are available and willing to buy the candied peel as an ingredient for their foods, this can therefore form a second product for a fruit juice or jam processor.

Fig.1. Candied orange peels

Source of Oils and Fats

Some of the stone fruits contain considerable quantities of oil or fat, some of which have specialised demand and markets for culinary or perfumery/toiletry applications. Mango kernels contain 12% fat and as per a study on mango stones, 30,000 tonnes of fat can be produced from 0.3 million tonnes of dry mango kernals available every year.

Tomato pomace can be used to produce tomato seed oil. Processing waste constitutes 5-13% of the whole tomato. Tomato seed contains 24-25% of oil. The seeds are the largest component of tomato processing wastes, in food industries manufacturing tomato juice, sauce, ketchup and food colours.

Kernel oil from palm seed is commonly used now both as cooking and industrial oil. Specialized markets are also present catering to other seed oils like passion fruit, papaya, grape *etc*. Steam distillation of citrus peel oils is well established at a small-scale. Citrus wastes (mandarin, lime) are rich source of essential oils and other by-products. It is also possible that the sale of seeds or stones or peels to larger oil processors could generate additional income for small-scale fruit processors.

Source of Pectin

Pectinis a gelling agent used in jams, sweets, pharmaceuticals etc., found to a greater or lesser extent in most fruits. Mainly, pectin is extracted from citrus peel and apple pomace (the residue left after apple juice has been removed). The presence of up to 30% pectin in dried residual matters like sugar beet pulp, carrot pomace, potato pulp or lemon peel and its availability in large quantities have made extraction worthwhile. Some other tropical fruits also contain high levels of pectin, for example the rind of passion fruit is a rich source. The utilisation of the 'shells' remaining after pulp removal offers possibilities for pectin extraction. India imports about 160 tonnes pectin for its processing industry, and hence there would be a good market for supplying local fruit processors with pectin to substitute for imports. Mango peels are also good pectin sources.

Source of Starch and Dietary Fibre

Mango seed kernals and banana pseudostem are sources of starch. Potato peels also contain starch, the content varying with the peeling process. Steam peels contain around 28%, while abrasion peels had 51% as it led to removal of more flesh with peels (Camire *et al.*, 1997). Resistant starch can be obtained from mango kernels, banana peels, etc.

The high crude fibre content of the vegetable residues (like orange peels, mango peels, soybean or oat hulls, fruit pomace, grape pomace) suggests its utilisation as a crude fibre source. Enrichment of different products with crude fibre compounds can thus raise the dietary fibre uptake, which is beneficial for human health. Fruit fibres also have better quality than cereal fibres due to higher total and soluble fibre contents, lower phytic acid contents, colonic fermentability and water and oil holding capacities. Carrot pomace besides being rich in crude fibre, is also rich in provitamins, colour and natural acids. It can thus substitute sourdough in bread, is acidifying agent, preservative or antioxidant in several food products (Laufenberg *et al.*, 2003). A study by Sharoba *et al.* (2013) reported orange waste, carrot pomace, potato peels and green pea peels can serve as a good source of dietary fibres. It was also seen that processing of the vegetable residue by fermentation with lactic acid bacteria improved its shelf life, colour stability and nutritional value.

Preparation of Phytochemicals

Olive pomace is used as a nematode controlling agent for tomatoes (Rodriguez-Kabana *et al.*, 1995), citrus waste streams are used in horticulture (Widmer and Montanari, 1995) and mandarin peel flavonoids are interesting due to their fungistatic activity (Chkhikvishvili and Gogiya, 1995) may be applied naturally to protect vegetables and fruits from molding. The limonoid compounds in citrus peel and seeds have recently been found to have important pharmacological properties as well as potential in the use as an insect antifeedant for agricultural crops (Manthey and Grohmann, 1996). Pomegranate peels are sources of phenolic compounds, gallic acid, catechin, flavones, flavonones and anthocyanidins (Dhineshand Ramasamy, 2016).

Preparation of Wine/Vinegar

Many of the potential products from fruit and vegetable residues have been developed using the SSF technique, and such products include ethanol, methane, lactic acid, citric acid, enzymes and food ingredients (Zheng and Shetty, 1998). It is technically feasible to produce these products from both solid and liquid fruit wastes, though these products should be produced from fresh, high quality fruit juices in order to obtain high quality products.

Sources of Natural Flavours and Food Colors

Vegetable and fruit residues are good substrates for the generation of fruity food flavours and natural colors. Currently used synthetic dyes being unhealthy, are rapidly being replaced by natural colors or plant pigments.

The microbial synthesis of these natural flavours is generally carried out by SMF. In case of bioflavour production, the SSF of residual matter is a fairly new technology of waste utilisation, based on a very old preservation method, which bioconverts secondary raw materials to natural flavours. Almosnino and Belin (1991) described the use of the intrinsic enzyme system of apple pomace for the biotransformation of fatty acids into potential flavors. 20-fold flavourcon centrate can be recovered from mango peel (Nanjundaswamy, 1997). Onion oil flavours can be made from waste onion (Brose, 1993).

Stalk of pumpkin and paprika can be used to produce natural colours. Natural colours of fruits like phalsa,jamunetc has been studied for their suitability as food colours. Tomato skin and peels used as natural colorants (Al-Wandawi *et al.*, 1985). It has beenobserved that in beverages, carrot pomace or citrus waste will stabilize the natural colour, increase the vitamin and fibre content, improve the viscosity (mouth feel) (Laufenberg *et al.*, 1996) and augment or adjust the cloudy appearance (Sreenath *et al.*, 1995). Grape skin extract in powder form is commercially available as a natural food colorant. Besides the blue–red colour, the food will be enriched with "healthy" polyphenols.

Bioadsorbents for Wastewater Treatment

Onion peels (Kumar and Dara, 1981, Bankar and Dara, 1982); lemon peel and pulp, olive, apple and grape pomace (Torre *et al.*, 1995); orange peel for dye removal (Namasivayam *et al.*, 1999), peanut shells for heavy metal removal (Randall *et al.*, 1974), peanut skin (Randall *et al.*, 1975).

Fruit and Vegetable Waste feed to Livestock

Feeding fruit and vegetable waste to livestock may be a good option but before doing so, farmers should consult with veterinarians to confirm the effects of feeding fruit and vegetable waste to livestock (Table 3). The sale of fruit and vegetable waste for feed can produce additional income flow to the farm.

Orange wastes can be fed to cattle. Ripe banana peels and mango seed kernel may be used as poultry feed after drying. Dried mango kernals and peels make high class energy food for animal feed. Egg size and production enhanced with carrot and orange peels. However, not all wastes may be fed to animals. Laufenberg *et al.* (1996) described that a protein concentrate made of potato fruit water could only be fed to cattle due to the high potassium content, while olive cake is not recommended for feeding due to its low digestibility (Clemente *et al.*, 1997).

Table 3: Effects of fruit and vegetable processing co-products (either as bulk material or high value component) on farm animal product quality (Kasapidou *et al.*, 2015)

Waste type	Animal species	Effect
Grape pomace and seed extract	Broilers	Antioxidant
Tomato extract	Broilers, quails	Antioxidant, Yolk colour enhancement
Citrus pulp	Broilers	Improved fatty acid composition
Pomegranate by-products	BroilersGoats, Cows	Antioxidant, Improved fatty acid composition, Hypocholesterolemic Improved milk fatty acid composition

Contd.

Tomato pomace	Sheep	Improved milk fatty acid composition
Olive leaves	Sheep	Improved milk fatty acid composition
Olive cake	Sheep, lamb	Antioxidant, Improved milk fatty acid composition
Tomato waste	Goats	Improved milk fatty acid composition
Citrus (orange) pulp	Goats	Improved appearance, taste and texture (cheese)

Composting Fruit, Vegetable Culls and Biogas

Composting of culled fruits and vegetables is a good optionto lessen the problem of high volumes of culls as well as other waste materials in a community. The culled fruits and vegetables would be mixed with other organic materials to produce compost that can be applied into growing fields to provide stable nutrients and organic matter for the next crop or for be sold for profit. Fruit and vegetable waste (FVW) can also be used for production of biogas by anaerobic digestion (AD) method.

Potential Benefits of Adoption of Advance Approaches for Fruit and Vegetables in the Arid Region

A good utilisation potential for solid vegetable and fruit wastes can thus be seen from above studies. There is the possibility to reduce wastages in the food production cycle and also reduce raw material costs; at the same time provide safe and ecofriendly sources of food additives. The use as bioadsorbents for waste water treatment is a cheaper and more environment friendly substitute for commercial ion-exchange resins. Using food residue as a livestock feed is an efficient way to scale up low quality materials into high quality foods. Utilization of agroindustrial by-/co-products in farm animal nutrition reduces the environmental impact of the food industry and improves profitability of the farm.

Very less work has been done in this area in India. The fraction of discarded materials in the majority of fruit processing industries is typically very high (Laufenberg *et al*., 2003) depending on the location and method of harvest (e.g. mango 30–50%, banana 20%, pomegranate 40–50% and citrus 30–50%). Detailed studies are being carried out on value-addition potential of fruit processing waste. The widespread availability of these wastes and its untapped potential has encouraged researchers to perform detailed studies on value-addition potential of fruit processing waste.

Income Generation from Utilization of Fruit and Vegetable Wastes

It was estimated that around 70% of India's fruit and vegetable production is consumed fresh, leading to approximately 30% being wasted every year. The major contribution to this wastage is accounted to the lack of refrigerated transport and cold storage facilities for both cultivators and sellers, apart from wastage from households (Anon, 2014).

As of now, only about 1.2% produce is processed. Value adding to discoloured or different shaped fruits into chutney, pickle, sauce, jam, and other processed products and phytochemicals, could sometimes be more valuable than sale of fresh, unprocessed fruits and vegetables. Doing so will enable the farmers to generate more income and could even account for up to 25-30% of their income.

Future Scope and Research Need

Fruit and vegetable wastes still remain an underexploited source of bio-compounds and value- added products. Fruit and vegetable processing products as sources of phytochemicals or other by-products is a fairly area and there is limited knowledge on their availability, extractability, bioactivity, applications and functions. Irregularity in waste consistency due to the inherent variations in its biological composition and due to processing conditions also pose a major limitation to the application of fruit and vegetable wastes utilization. The function of the co-product should be specific in order to survive competition in the market. Simultaneously, the bioconversion of residues will become economically attractive only if high value products are produced.

Commercial application of fruit and vegetable wastes as functional food ingredients and other by-products provide challenges and opportunities for researchers. Targeted multidisciplinary research on utilization aspects these residues is needed, if it is to reach farmers. The segregation and study of wastes as a particular type of food waste helps in the development of additional biorefinery processes and ultimately improve the economics of food waste based bio-refinery concept. The composition of different fruit and vegetable wastes, the possible by-products that can be made, mode of extraction, need to be studied and standardized for further expansion in this field. And farmers need to be made aware of the scope in this avenue, in terms of economical benefits.

Conclusion

India is one of the largest producers of fruits and vegetables and hence, wastes and by-products generated from fruits and vegetables are a big source of soil and water pollution due to their organic nature. Besides this, they generate greenhouse gases, which is a matter of serious environmental concern. Keeping in mind these problems, proper disposal and handling of these wastes can help to reduce pollution risks, and can provide useful and commercially valuable by-products. Organized processing of wastes can help farmers to achieve higher incomes by proper handling of waste and post harvest management of fruits and vegetables.

References

Almosnino, A.M. and Belin, J.M. 1991. Apple pomace: an enzyme system for producing aroma compounds from polyunsaturated fatty acids. *Biotechnol.Letters*. 13 (12): 893–898.

Al-Wandawi, H., Abdul-Rahman, M. and Al-Shaikhlyet, K. 1985.Tomato processing wastes as essential raw material source. *Journal of Agriculture and Food Chemistry*. 33: 804–807

Anon. Fruit waste utilization. Accessed on September 25, 2017. Available at: http://www.daenvis.org/technology/n%20fruitwaste.htm

Anon. 2014. India wastes Rs 13,300 cr of fruits, vegetables every year. FirstPost. Available at: https://www.firstpost.com/business/economy/india-wastes-rs-13300-cr-of-fruits-vegetables-every-year-1259881.html

Banerjee, A., Singh, R., Vijayaraghavan, R., MacFarlane, D., Patti, A.F. and Arora, A. 2017.Bioactives from fruit processing wastes: Green approaches to valuable chemicals. *Food Chemistry*. 225: 10–22.

Bankar, D.B. and Dara, S.S. 1982. Binding of calcium and magnesium by modified onion skins.*Journal of Applied Polymer Science*. 27: 1727–1733.

Brose, D.J. 1993. Novel process technology for utilisation of fruit and vegetable waste. SBIR Phase I project USDA ICSRS, Washington, DC.

Camire, M.E., Violette, D., Dougherty, M.P. and McLaughlin, M.A. 1997.Potato peel dietary fibre composition: Effects of peeling and extrusion cooking processes. *Journal of agricultural and food chemistry*. 45: 1404-1408.

Chkhikvishvili, I.D. and Gogiya, N.N. 1995. Flavonoids of mandarin fruit wastes and their fungistatic effect on the fungus *Phomatracheiphila.Applied Biochemistry and Microbiology*. 31 (3): 292– 296.

Clemente, A., Sanchez-Vioque, R., Vioque, J., Bautista, J. and Millan, F. 1997.Chemical composition of extracted dried olive pomaces containing two and three phases.*Food-Biotechnology*. 11 (3): 273– 291.

Dhinesh, K.V. and Ramasamy, D. 2016. Pomegranate Processing and Value Addition: Review. *J Food Process Technol.* 7: 565.

Kasapidou, E., Sossidou, E. and Mitlianga, P. 2015. Fruit and vegetable co-products as functional feed ingredients in farm animal nutrition for improved product quality. *Agriculture*. 5: 1020-1034.

Kumar, P. and Dara, S.S. 1981. Binding heavy metal ions with polymerized onion skin.*Journal of Polymer Science: Polymer Chemistry edition.* 19: 397–402.

Laufenberg, G., Grub, O. and Kunz, B. 1996. New concepts for the utilisation of residual products from food industry—Prospects for the potato starch industry. *Starch*. 48: 315–321.

Laufenberg, G., Kunz, B. and Nystroem, M. 2003. Transformation of vegetable waste into value added products: (A) the upgrading concept; (B) practical implementations. *Bioresource Technology*. 87: 167–198.

Maini, S.B. and Sethi, V. 2000.Utilization of fruits and vegetables processing wastes.Verma, L. R. and Joshi, V.K. (Eds.). Postharvest Technology of Fruits and Vegetables: Handling, processing, fermentation and waste management. Volume 2.Indus Publishing Co. pp. 1006-1018.

Manthey, J.A. and Grohmann, K. 1996. Concentrations of hesperidin and other orange peel flavonoids in citrus processing byproducts. *Journal of Agriculture and Food Chemistry*.44: 811– 814.

Namasivayam, C. and Kadirvelu, K. 1999. Uptake of mercury (II) from wastewater by activated carbon from an unwanted agricultural solid by-product: coirpith. *Carbon*. 37(1):79-84.

Nanjundaswamy, A.M., 1997. Processing. In: R.E. Litz (Ed.), The Mango, Botany, Production and Uses, Cab International, Wallingford, pp. 535–539.

Randall, J.M., Hautala, E., Waiss, A.C., 1974. Removal and recycling of heavy metal ions from mining and industrial waste streams with agricultural by-products. In: SO Proceedings of Minerals Waste Utilization Symposium 4th. pp. 329– 334.

Randall, J.M., Reuter, F.W., Waiss, A.C., 1975. Removal of cupric ion from solution by contact with peanut skins. *Journal of Applied Polymer Science* 19, 1563–1571.

Rodriguez-Kabana, R., Estaun, V., Pinochet, J., Marfa, O., 1995.Mixtures of olive pomace with different nitrogen sources for the control of Meloidogyne spp. on tomato. Journal of Nematology 27 (45), 575–584.

Sharoba, A.M., Farrag, M.A. and Abd El-Salam, A.M. 2013.Utilization of some fruits and vegetables waste as a source ofdietaryfiber and its effect on the cake making and its qualityattributes. *Journal of Agroalimentary Processes and Technologies*.19(4), 429- 444

Sreenath, H.K., Crandall, P.G., Baker, R.A., 1995. Utilization of citrus by-products and wastes as beverage clouding agents. *Journal of Fermentation And Bioengineering* 80 (2), 190–194

Torre, M., Rodriguez, A.R., Saura-Calixto, F., 1995. Interactions of Fe(II), Ca(II) and Fe(III) with high dietary fibre materials: a physicochemical approach. *Food Chemistry* 54 (1), 23–31.

Widmer, W., Montanari, A.M., 1995. Citrus waste streams as a source of phytochemicals. In: 107th Annual Meeting of the Florida State Horticultural Society, Orlando/Florida, USA, vol. 107. pp. 284– 288.

Zheng, Z., Shetty, K., 1998. Cranberry processing waste for solid state fungal inoculant production. *Process Biochemistry* 33 (3), 323–329.

34

Economic Impact Evaluation of Agricultural Management Projects in Arid Region

G.L. Bagdi, V.S. Rathore, N.S. Nathawat, M.L. Soni, N.D. Yadava and R.C. Balai

Introduction

Any project is undertaken and an investment is made with expectation of realizing benefits out of it. This applies to projects both in the private domain as also in the public domain. Hence, it is imperative to assess the benefits derived from the project investment over its entire economically viable. The extent to which project contribute to various aims of development (benefits) and make use of scarce resource (costs) can differ widely. A proper project preparation requires an assessment of the various benefits and costs associated with execution of a project. Methods to determine, analyze and assess the relevant benefits and costs of a project are known as cost-benefit analysis. The cost- benefit analysis ensures that all the relevant factors that contribute to a project's net benefits are systematically taken into account, whether they relate to an individual, a firm, a social group, the public sector or the country as a whole.

There are many tools or criteria for the financial evaluation of projects. However, the important and commonly used evaluation criteria are simple Rate of Return, Payback Period, Net Present Worth, Benefit- Cost Ratio and Internal Rate of Return.

Economic appraisal and evaluation helps the decision makers who are saddled with the responsibility to prioritize measures in a situation with enormous needs and quite limited resources. Government planners as well as donor agencies are all in need of enough data in order to make rational choice or decision in selection of investment alternatives. Most of the economists believe that economic analysis cannot give the final answers. There will always be lack of data, uncertainty about the future, disagreement on the methods, and similar issues. However, economic analysis, if properly done, can indeed be of much help to decision makers.

Cost-benefit Analysis

Cost-benefit analysis (CBA) is a highly structured method to organize information and quantify social advantages (benefits) and disadvantages (costs) in terms of a common monetary unit. Benefits and costs are primarily evaluated on the basis of individual's willingness to pay (WTP) for goods and services, whether they are actually marketed or not. Unquantified effects (intangible) are described and put against quantified values. CBA analysis is required for the following reason:

(i) Market deficiencies, lack of information, externalities, risk, monopolies, etc. may create a difference in the private versus of the social perspectives thus, reliance on the market forces is insufficient.

(ii) Governmental distortions, trade barriers, wage legislations, pricing policies, etc., affect land management and hence need to be analyzed. Thus, there may be conflicts between conservation, promotion and other social goals.

(iii) Resources are limited; somehow the allocation of resources between sectors (industry or agriculture) and within sectors (conservation here or there) has to be decided. CBA can be a guide.

(iv) Attraction of external resources "the aid fatigue" among donor country and public opinion can be fought by demonstrating the efficiency.

To apply CBA, some key concepts are given as under:

Willingness to pay (WTP) is considered to be basis for CBA. This is the amount of other goods or services that an individual is prepared to give up to be able to acquire something. It may be noted that there may be a WTP for non-marketed goods (fresh air, etc.) hence it is a benefit relevant to CBA.

Another crucial concept of CBA is "opportunity cost". This is the benefit foregone by not using a resource (such as labour, land and capital) in its next best alternative (such as the job somebody would have without our project). In other words the opportunity cost is the WTP for a resource's alternative use.

Criteria for Appraisal and Evaluation

Project analyst should determine the worth of a project under three situations:

(i) Determining the project worth in the planning itself – appraisal or ex-ante evaluation.

(ii) Evaluation of on-going projects – concurrent evaluation.

(iii) Evaluation after completion of project – ex-post evaluation.

Ex-ante evaluation is carried out before actual implementation of the project to assess its economic feasibility in advance and to know whether the particular project is capable of achieving the basis objectives of the plan i.e. whether the project is economically sound or not.

There may be one or many on-going evaluations in between ex-ante and ex-post evaluations so that one can know the way things are happening during the project implementation period i.e. whether the project implementation and its outcome is going on the stipulated path or not. Such method of evaluation of the project would help in putting the project implementation process on right direction and hence, help in reducing the gap between what was anticipated and what is actually to be realized from the project concerned.

Nevertheless, the ability to define failure is even more important than ability to document successes. It is possible only by knowing where thing where things have gone wrong in the past that one can hope to achieve success in the future. Thus, ex-post evaluation of project is equally important. Ex-post evaluation is done after the completion of the project so as to know whether the project has actually achieved its objectives or not. If not, up to the desired extent then the valuation will reflect the deficiencies in the present project that may be rectified in future ones.

Technical Data Collection

It is very important to collect and analyze the current information on input use, output produced and their prices. This represents the key components in the evaluation of agricultural project. The information is collected on cost paid for items such as, (a) equipment and power, (b) seeds and fertilizers, (c) labour, (d) irrigation, and (e) construction operation and maintenance of resource management systems. Such information should be maintained at district or state level so that it is easily available. The cost of resources is estimated by systematic listing of physical resources used in the production and it provides value of each resource on hectare basis.

Discounting the Present Value of Future

A future rupee is worthless today than a present rupee. If the rupee represents a benefit, the value of the benefit is less; if it represents a cost, the cost is less in the future than in the present. Of all the concepts of benefit-cost analysis, this is one of the most difficult to comprehend. Yet, it is certainly one of the most indispensable concepts in the analysis. The discount rate is yet another name for the rate of interest society should charge itself for the opportunity cost of time. Time is the most irretrievable loss of all. As time goes on, investment opportunities are inevitably lost. Therefore, one wishes to use time as efficiently as one can. This is the ultimate objective of discounting. The appropriate rate of discount is that rate which represents the rate of return on thus represents the opportunity cost of giving up the total, overtime, return accruing to an alternative project. Discounting is not done for the purpose of real value of money or inflationary or deflationary trend of price adjustment.

Techniques of Benefit-Cost Analysis

An economic problem exists when a person (or persons or society) has limited resources that have alternative uses, seeks to maximize some economic goals and must choose between alternative courses of actions. One of the most important guiding principles which help in selection of project/alternative investment opportunities is economic evaluation of these opportunities. The basis objective of economic evaluation is to measure the project worth by taking all the effects of action taken into account. It should be clear that economic is only one of the tools which help in decision making. The most essential component of economic analysis is listing of all benefits and costs of the project year by year over the period of analysis (called flow table) which is similar to a cash flow chart in financial analysis.

Benefit-cost analysis is a systematic way of comparing the streams of benefits and costs from different projects. Some of the most commonly used economic tools in measuring the project worth (evaluation) are discussed. There are two broad categories of measures of project worth: (i) Undiscounted measures of project worth, and (ii) discounted measures of project worth.

Undiscounted Measures

The following six measures come under this:

(i) Ranking by visual inspection.

(ii) Cut-off period- it specifies the period within which cost/ capital has to be recovered. It does not give the figure of cost/ benefit after this period.

(iii) Payback period or capital recovery criterion – it denotes length of period (years) takes to repay initial outlay

(iv) Undiscounted sum of benefits per unit of initial capital investment.

(v) Average annual benefits per book value of capital investment. It is biased to long term project.

(vi) Net average rate of return – benefits minus costs divided by useful life of project.

It is based against long term projects.

The common weakness in the all the above mentioned six methods is that they do not consider the "time value of money" as commented against each.

Discounted Measures

Appraisal and evaluation criterion falls under this category which has one essential element i.e. accounts for the time value of money. Thus, entire stream of values is brought together at particular point of time by the use of discount rate. The measures under this category are described below:

(i) New present worth (NPW)

This is most widely used measure. It is simply the value of gross benefits minus the discounted value of discounted gross costs.

The general formula for NPW is,

$$NPW = \sum_{t=1}^{N} \frac{B_t - C_t}{(1+r)^t}$$

Wherem B_t = Gross benefits from the a project in year t

C_t = Gross costs from the project in year t

r = Discount rate

t = time, year 1 to n (life of the project)

The decision rule is to select the projects when NPW is greater than zero, otherwise reject. Higher the NPW, better is the project. Select that project which has highest NPW.

The demerit of this criterion is that it does not reflect the return per rupee invested on the project. The project 'A' may lead to more net present worth than project 'B' But 'A' also may have more cost than 'B' and the return to per rupee invested (cost) on project 'A' may be less than project 'B'.

(ii) Internal rate of return (IRR)

Internal rate of return is frequently used in the evaluation of projects by bank and other organizations. It is the rate of discount which makes the present value (value today) of benefits equal to present value of costs, thus, IRR does not use a pre- determined discount rate.

The IRR is the discount rate, r such that,

$$\sum_{t=1}^{n} \frac{B_t}{(1+r)_t} = \sum_{t=1}^{n} \frac{C_t}{(1+r)_t} or \sum_{t=1}^{n} \frac{B_t - C_t}{(1+r)_t} = 0$$

The only procedure to determine IRR is the successive approximation.

In an analysis using IRR measure, the calculated IRR is compared to some prescribed discount rate (generally the interest rate chargeable on money to be invested or the value that invested resources would earn in other opportunities (the opportunity cost of capital) to decide whether the project is worth to take or not? The decision rule applied is to select the project if calculated discount rate IRR is greater than predetermined discount rate, otherwise reject it.

The advantages of this measure are: IRR is well known; oftenly used and easily understood; and ranking of project can be done easily on this basis. However, IRR has some disadvantages too. It may be misleading in a situation where selection has to be made among mutually exclusive projects. The reason is that IRR may be higher for project 'A' that in fact produces less benefits than project 'B' having lower IRR but produces more benefits.

Benefit-cost Ration (BCR)

This ratio simply compares the present value of benefits to present value of costs, obtained by dividing the present value of all benefits (sum of discounted benefits) by the present value of all costs (sum of discounted costs).

$$BCR = \sum_{t=1}^{n} \frac{B_t / (1+r)^t}{C_t / (1+r)^t}$$

The decision rule applied is, if the B: C ratio is greater than unity (1), select the project otherwise reject it. The advantages of this measure are: easily understood; easy to convince the planner or decision maker; easy to compute; and it can easily see how much cost is to be allowed to increase before a project is to be rejected.

Decision Criteria for Project Selection

The decision rule of cost-benefit analysis is to select one or more projects such that discounted net benefits are maximized. Three possible decision situations may exist: (i) select or reject only one project, (ii) select one of many projects, or (iii) select several of many projects.

If we are to simply evaluate one project, the decision criterion is to accept the project as economically feasible if its net present worth (NPW) is greater than zero. Equivalently, the project is accepted if its benefits-cost ratio (B: C) is greater than one its internal rate of return (IRR) exceeds the appropriate (pre-determined) rate of discount.

If one of many projects is to be selected, the decision criterion is to choose that project with the greatest NPW. Neither the B:C nor the IIR will provide unambiguous decision criterion in this situation .

If several projects out of many projects are to be selected, then it is necessary to first determine whether or not the projects are independent and if a budget constraint exists before the appropriate decision criterion can be selected. Projects are independent of one another, if the NPW of each project is unaffected by either the existence or scale of operation of the other projects.

If the projects are independent and a budget constraint exists, then the appropriate decision criterion is to select projects in order of their benefit-cost ratios (B:C) until the budget is exhausted. If the projects are independent and no budget constraint exists, then all projects with net present worth (NPW) greater than zero are to be selected.

References

Ram Babu, Dhyani, B.L.; Agarwal. M.C.; and J.S. Samra. 1997. Economic Evaluation of Watershed Management projects (Concepts, Methodologies and Case studies). Central Soil and Water Conservation Research and Training Institute, Dehradun, pp: 1-12.

Suryaprakash, S. 2005. Economic Impact Assessment of Rural Development Projects- Methodological Issues. In: Impact Assessment studies: Agricultural Extension and Rural Development Programmes. Baldeo Singh *et al.* (Eds.). Indian Society of Extension Education, Division of Agricultural Extension, IARI, New Delhi, pp: 44-49.

35

Interfacing Technologies and the Needs the Farm Women in Horticultural Sector

Kanak Lata, Raj Kumar and Shakti Khajuria

Introduction

Some historians of agriculture believe that it was women who first domesticated crop plants and there by initiated the art and science of farming. While men went out hunting in search of food, women started collecting seeds from the native flora and began cultivating those of interest from the point of view of food, feed, fodder, fiber and fuel. This view is strengthened by the fact that women have been traditionally seed selectors. Even today this tradition has continued in many parts of the developing world. Women have played and continue to play a pivotal role in the conservation of basic life support system such as land, water, flora and fauna. They have protected the health of the soil through organic recycling and promoted crop security through the maintenance of varietal diversity and genetic resistance. In many remote areas agriculture is largely in hands of women. Men tend use to go to towns and cities in search of salaried jobs for fulfillment of family income. Therefore, without total intellectual and physical participation of women, it will not be possible to popularize alternative systems of land management for shifting cultivation, arrest gene and soil erosion, and promote the care of the soil and the health of economic plants and farm animals. Women also tend to look at problems in their totality from the sowing of the crop to its ultimate utilization either as food, feed or raw material for industry. Since they are usually incharge of marketing, they generally possess a clear understanding of market preferences and prejudices. Therefore, for an economically and ecologically sustainable agriculture the involvement of women farmers and farm women (agriculture farmers with and without assets) in the process of modernization of farming practices and village industries is absolutely essential. Horticulture is a major part of Indian agriculture, natural foods of man, fruits and vegetables provide protective nutrients like vitamins and minerals besides, some high quality sugars, proteins, oils and fats. Although belated, their importance has now been recognized in India not only to provide the much needed balanced nutrition and gainful occupation to our populace but also to develop viable internal and export trade. Fruits like banana, mango and minors fruit like mahuva, chironji, rayan etc. vegetables like potato being energy rich are considered prominent food crops e.g. banana can produce as high as 37 million calories per hectare. Unfortunately the component of fruits and vegetables in Indian diet is hardly six percent which should be raised at least five times so that the per capita per day consumption becomes 85 g. fruits, 75 – 125 g. leafy vegetables 85 g. other vegetable and 85 g. roots and tubers. To achieve this present production of fruits and vegetables would need to be pushed up considerably by increasing production levels and the area under fruits and vegetables would need to be pushed up considerably by increasing production levels and the area under fruits and vegetables (Lata and Singh, 2010).

Present Status in India and Global Scenario

There is great scope to improve the yield levels in fruits since the present productivity is far lower than the potentials. This is also evident from the fact that yield levels in some fruits and vegetables in India are one third to one fifth to those obtained in the developed countries. Therefore, the technologies developed as a result of research efforts need to be quickly adopted in our production system. Since women play a prominent role in national horticultural productivity, it is necessary to identify the appropriate technologies for them, the constraints in their adoption and the need of new technology development. Ignorance about the new technologies, lack of skill for their adoption and in case of some operations, need of more physical power, element of drudgery and attention at odd periods come in the way of their adoption by farm women.

The earning capacity of females in agricultural employment is sustaintial all over the world. The female earning varies from 64.9% in UK, 64.9% in India and 110.8% in Denmark (Amtul and Viraktamath, 2013).

Pandit Jawaharlal Nehru at the centenary Celebrations of M. Karve on 18th April, 1955 said, "One of the truest measures of a nation's advancement is the state of its women. For out of the women comes the new generation, and it is from their lops and laps that begins to learn. Social revolution forms the basis of economic revolution which lead to progress. Women play the most important part in social revolution".

The important trades for horticultural productivity consist of nursery technology related to ornamental planets, fruits and vegetables, flower- gardens, homestead gardens fruit and vegetable processing industry and marketing of horticultural produce. At present, in North India, women mostly participate in fruit and vegetable processing, flower gardening ornamental nursery, homestead gardening, kitchen gardening and to some extent in vegetable growing. In south India, besides these operations, women also participate in vegetable and flower marketing and nursery technology (Table-1)

Table 1: Compression of North and South Indian women participation in horticulture activities

Trade	North India	South India
Ornamental nursery	50	60
Fruits and vegetable nursery	40	50
Fruit growing	30	40
Vegetable growing	50	50
Flower gardening	60	80
Homestead gardening	75	80
Fruits and vegetable processing	80	90
Fruit marketing	20	40
Vegetable and fruit marketing	40	60

The extent of participation of women in different operations in horticultural production systems depicts interesting regional differences in India. In northern states like Haryana, women are hardy and work related to heavy implements like spade, machines and those requiring greater skill, e.g. pruning and training, raising nursery plants, lifting and planting of plants, spraying, grading, packing, marketing *etc*. In the southern states although women do not take part in operations requiring heavy manual work, they participate to some extent in nursery technology planting, irrigation, grading *packing etc*. In general, major operations looked after by them are sowing and planting, irrigation, lifting of small vegetables and ornamental seedlings in nursery, application of manures and fertilizers, weeding and hoeing harvesting of fruits, vegetables and flowers, transport of produce to short distances and processing of horticultural produce Goyel, *et al.* (2003).

Training for farmwomen on institutional basis was started in India under the centrally sponsored schemes of farmers training and educational programmes in 1966-67 which now cover all the districts operated by state governments, in these programmes. However, training component for farm women remain restricted to only home science aspects like nutrition health care and some new recipes. For this bilateral assistance is being obtained from the government of Denmark, through DANIDS, is assisting government of India to develop leadership qualities amongst farm women so that they can guide other farm women and to provide follow-up through extension support which will be integrated in the existing T and V system. The government of Netherlands is to provide assistance in many states of India. All these centers are to be provided lady training officer and female demonstrator. Major emphasis in these projects is to be given on food crops, oilseed and pulses and that on horticultural component is negligible. The present state of horticultural education and training in women is rather dismal particularly in the northern states of India. The situations are somewhat better in the southern states. In the northern states, there are hardly few women taking part in the horticultural education institutions. This is of a great significance since women would pick up the improved technologies faster if imparted by women trainers (Waris, 2008).

There are many Mahila Mandals and SHG with a membership of several lakhs rural women. A study of the working of SHG members 300 in 2009 has shown that 20 out of 30 SHG were involved in economic activities like kitchen gardening, community gardening, fruit orchards, compost making, nursery growing, but most of the members reported lack of guidance and proper training facilities. All over the India 578 KVKs working hard to provide the guidance of new horticultural technologies along with agriculture, animal husbandry and allied subjects.

Gaps in Prevailing System

The Gender Gap Index assesses countries on how well they are dividing their resources and opportunities among their male and female populations, regardless of the overall levels of these resources and opportunities. India ranks 123 in terms of economic participation, 121 in educational attainment, 134 in health and survival and 17 in political empowerment. India, with 1.3 million elected women representatives, has the largest number of women participating in local governance among the Asian countries, exceeding its own 33% reservation. A study of Indian local governments by Munshi and Rozenweig (2008) said that women in local government roles came out with better outcomes for communities in budgetary decisions and were more competent in procuring resources despite significantly lower education and labour market experience, so politically empowering women is a good thing.

The analysis of gender pay gap in India based on a voluntary online salary survey conducted by Paycheck India with 16,500 online observations (13,729 males and 2771 females) indicated that a wide gender pay gap exists in India. The average gender pay gap is approximately 54% for years 2006 to 2011. The data also revealed that gender pay gap has narrowed over the years. It was above 70% before 2008 and has come down to almost 40% in 2011. Gender pay gap changes with increase in wages, for wages below Rs1.00 Lakh and above Rs 50 Lakhs, the gender pay gap is negligible. However, for the wage group between Rs 1 Lakh to Rs 50 Lakhs the gender pay gap is above 40%.

The pay gap increases with age, the gender pay gap is highest for the age group 50-60 years at 157% and lowest for the age group 20-30 years at 38.

It was observed that as the education level increases, the gender pay gap increases in most of the cases. The gender pay gap is lowest for Plus 2 or equivalent education level at 11.54% and is highest for Post Doctoral Education level at an alarming 180. Data reveals that men and women with same work experience are not paid equally. Gender pay gap is low at junior level and high at senior levels. The extent of gender pay gap also depends on the occupation data shows that the gender pay gap is highest for health professionals at more than 65%. It is lowest for cleaners and helpers under elementary occupations at just a little above 1%. It is also worthy to note that the gender pay gap is remarkably low for information and communication technicians at only 4.37%.

Marital status is also one of the factors for gender pay gap. For women who were single, the gender pay gap is the lowest at 27%. It increases for married women and divorced women at a little above 40%. It is highest for widowed women at more than 60. The gender wages gap varies across Indian States. The gender wage gap is highest in the state of Assam and Rajasthan at 64% and 59% respectively. The gap is relatively low in the southern part of India. The gender wage gap is lowest for New Delhi at 20%. (Amtul and Viraktamath, 2013)

Advance Approaches for Fruit and Vegetables in the Arid Region

The traditional area in which women already participated to great extent include fruit and vegetable processing, ornamental nursery, light operations in fruit and vegetable growing. Women can effectively participate in several activities and can take part in horticultural production system. As a result of research efforts, many new technologies have been developed and are considered appropriate for farm women Manju, (2002).These are given below:

Nursery Technologies

Raising plants in polythene containers: Traditionally plants were raised in nursery beds which required space and care for longer period resulting in their difficult disposal and higher expenditure. Modern techniques of raising grafted fruit plants in polythene containers using detached vegetative propagation methods are more elegant and conducive for adoption by farm women.

Raising plants in polytubes: Raising plants in poly tubes is a very simple and drudgery effective method, women can take part in all operations right from filling of tubes, sowing, raising seedlings, budding operation until disposal of budding. In case of citrus plants, budded plants are raised in polythene tubes which could easily be adopted. The tubes' raised plants can be easily lifted, packed and transplanted and prove more vigorous growth to their straight uncoiled tap root unlike the pot grown or seed bed grown plants. Moreover, even farm ladies who do not own land. Easily establish small nurseries on any available space like roof top, court yard etc.

Propagation benches with sub-surface irrigation units: One of the most prominent operations done by farmwomen is irrigation of nursery plants by a watering can. The amount of drudgery involved is obvious. This can be easily avoided by provision of propagation benches on which the plant containers having holes at their bottoms can be placed for irrigation by capillarity. The water level in the soil on the benches is maintained by simple water level cabinets as used in every modern house hold, connected to a water source.

Standard growing medium: Potting medium consists of mixtures like sand: clay: FYM in 1:1:1 ratio or loom: Leaf mold : FYM in 1:1:1. Very little care is at present given to sterilization of the medium and addition of the component which would increase its water holding capacity. Now a day, potting mixture is available ready to use mode. There is need to aware the farm women to use this media so as to reduce the drudgery in respective operations to prepare such mixtures, avoid difficulty in obtaining the different components and to avoid risk of infection of the propagules by diseases.

Mist chamber: Traditionally, frequent sprinkling of water has been used to prolong the life of the planted propagules so that they regenerate into complete plants. Now several designs of mist chambers or humid chambers are available to suit different conditions of climate. A simple underground trench with polythene covers or above ground polythene chambers are cheaper structures. In these cases humidity is kept high by sprinkling water and the excessive heat is avoided either by partial shading or by provision of exhaust fans. The other designs with automatic timer devices to put the mist nozzles on and off at calibrated and pre-decided intervals are expensive but more efficient.

Tissue culture: Tissue culture techniques have now been developed in several horticultural crops like banana, grapes, papaya, date palm, fig, orchid and many ornamental plants which can be commercially exploited. The various operations in a tissue culture unit relate to preparation of growing media, aseptic transfer in to the media tubes maintenance, cutting and transfer of differentiated tissue until final separation and transfer of the plantlets into the soil medium. Thus all the operations are conducive for work by women but require a high degree of skill. Now a day, tissue culture technique has been commercially exploited. Women have successfully adapted to the technique and have been considered better technicians than their male counterparts.

Use of bottom heat: In temperate regions and during the severe winters in the sub-tropical region of the country, these elegant techniques can now be used to increase rooting. Women can be involved in this work after skill training.

Fruit production technologies: **Knowledge of quality cultivar:-** Although the country already possesses a rich germplasm resource of high yielding commercial varieties of fruits, several new varieties have now been developed having good productivity and desirable quality characters. Women should make aware about these improved cultivars.

Plantation of orchards: Plantation of new orchard involves operations like layout, digging of pits, filling of pits and planting and care of young plants. Although layout and planting requires some skill but women can easily be trained in these operations. However, digging of pits requires more physical efforts and may not suit women. This operation can be easily mechanized. A post whole digger attached to tractor can dig 1 m deep and 1 m wide pit in 1 – 2 mts depending on soil type. Of the after care operations, frequent watering of the young plantation involves great amount of drudgery and is often performed by women. The modern techniques like pitcher irrigation, micro-tricklets, double wall pots etc. can greatly reduce the drudgery as well as the irrigation cost considerably. These techniques not only eliminate the drudgery for the farm women but are highly economic owing to saving on water and by reduction in mortality of plants.

Dwarfing techniques high density plantation: Orchards operations can now be very easily carried out by women in orchards owing to the small stature of trees of newly bred varieties. Geometry plantation allow high density plantation to obtain higher and quicker returns. Thus 1600 plants of mango and 500 grape vines can be planted per hectare. Similarly 5000 plants of banana and 53000 to 63500 plants of pine apple can be planted per hectare by paired row planting. In such intensive planting system omen can remain engaged profitably even on small holdings.

Training and pruning technique: all though all fruit plant require some amount of training but the fruits like grape can be trained to suit different cultivars and situation. Bower system is the most expensive but head system is much cheaper. The reduction in yields in the later system can be compensated to some extent by the high density plantation. The farm women can be easily trained in the skill of training. Similarly pruning also requires good amount of skin in fruits like grape, ber, peach, fhalsa etc.

Use of growth regulators: Gibbrellic acid for cluster thinning and berry enlargement in grape, ethephon and NAA to induce uniform growing and harvesting in pineapple. NAA for control of mal formation in mango, NAA, 2,

4-D and 2, 4, 5-T for reducing fruit drop in mango, citrus and apple have given commercial success and can be widely adopted. Besides this growth regulators like IBA and NAA are effective in inducing rooting in cuttings for propagating new plants

Orchard management: The different operations involved are application of manure and fertilizer, foliar feeding, irrigation; weeding, intercropping etc. farm women are traditionally very intimately involved in all these operations which incidentally also have a great amount of drudgery. Women often don't prefer operations requiring a lot of spade work like mixing of manure; diverting water for irrigation etc. irrigation is often given at odd times or during night time which does not suit women fortunately modern techniques are now available which can greatly help. Part of the nutritional need can be met by foliar feeding particularly for micro nutrients. The major nutrients which give response in fruit plants are nitrogen, zinc, copper, iron and manganese.

Trickle irrigation: although expensive but is very economic in term of water use efficiently. Moreover, once installation, it can be easily be operated by skilled farmwomen. The cost of installation can be greatly reduced by use of gravity flow in place of a pump. Simple trickles in place of costly nozzles, etc. the other expensive irrigation technique is a bubbler system which cuts the conveyance water losses completely. It involves underground layout of water pipes hydrants are connected at appropriate spots in the orchard and by connecting a hose pipe to these hydrants, required amount of water can be poured in each three basin, once installed this technique can also be easily be operated by farm women. One of the most important orchard operations done by women is weeding. Weed control by the use of herbicides is not only more economic but also eliminates the laborious manual weeding operations. The pre emergence weedicides like gramaxone @ 3Kg/ha. Have proved effective in most fruit orchards. Combinations of a pre-emergence and a post emergence weedicides is often more effective. Weed control can also be done by biological methods. Release of insect feeders has already been found effective to control water hyacinth, Slavonia, parthenium, chromolaena doctorate etc. Intercropping in orchard with perennial or annual legumes vegetables crops and small fruits is a very good way to increase productivity per unit area and for high orchard efficiency.

Vegetable Growing and Ornamental Gardening

Seed production: seed production, particularly hybrid seeds , already become a profitable business both in vegetable growing as well as in ornamental gardening , since quality plays a significant role in the seed production trade are done by women. Since this operation require a great amount of patients and delicate handling.

In recent years, hybrid seed production in vegetable is gaining importance. The operations like sowing rouging controlled pollination, harvesting and seed processing are done by women. In the process of controlled pollination emasculation is very tedious as well as dedicates operation. However the technique has now been improvised development of male sterile lines in tomatoes and onion has avoided the need of emasculation and caging. Block planting of the two parents can be done in case of monocots cucurbitaceous crops and male flowers, flowers from female plant can be removed to produce hybrid seeds. In cruciferous crops like cauliflower having protogyny bud pollinations can be used for this purpose for controlled pollinations after emasculation, bagging or caging is another important operation which can be done efficiently by women. Certain specialized operations like selection and preparation of planting materials in biannual crops can also be done by women. The seed processing's operations like harvesting , thrashing , cleaning , dragging , sieving and packing , sampling , labeling are best done by women . Whereas fermentation method has been developed in vegetables like brinjal and tomatoes physical seed suppression is done in Cole crops, root vegetables, okra chilies etc. In case of ornamental crops, seed production is generally done in large number of annual flower like marigold chrysanthemum, sweet pea, annual dahlia, nasturtium etc. in some cases , hybrid seed production has also been done like sweet pea petunia etc .

Vegetable and ornamental plants nursery: Traditionally women perform most of the operations related to these flourishing horticultural trades. The demand of plants of a large number or annual and perennial flowers, foliage plants besides several vegetables crops has increased over the years particularly near the cities. There is a great demand of seeds, bulbs and corms both in internal as well as export market. The bulbs of amaryllis, tuberose gerbera, lilies, gladiolus, cannas etc can be easily transported. Modernization in operations like pot culture, green house / glass house management, packing etc, will significantly add to the efficiency in operation of these nurseries and would remove a lot of respective difficult operation for women.

Vegetable growing technology: Some of the significant achievement which would provide more income to women growers is as follows:-

Early Varieties all the year round bear varieties processing suitable Varieties. Attractive color, aroma and pungency rich verities would greatly add to the income of vegetable growers. Several new vegetable production technologies like offseason production glass house / green house protected cultivation enable transplanting to catch early market and production of vegetable in adverse climate will help in providing sustain and assured income to the women growers. Multiple cropping sequences for rain fed and irrigated conditions have also been worked out. Drip irrigation is highly profitable in vegetable crops since they are sown in lines in an intensive cropping system. Foliar feeding of nutrients has also given significant response in almost all vegetable crops and thus the technique could be utilized to ensure case in fertilizer application commonly done by women. Inter-culture in vegetable crops is an important operation which require great amount of labor and is commonly done by women. New implements have been developed which can decrease the necessity of the great physical effort. Similarly harvesting by clippers in some vegetable and flowers commonly done by women would be more trouble free and efficient.

Vegetable nursery on raised bed: In high rainfall area during monsoon, it is one of the important technology for production of healthy plating material of vegetable crops.

Mulching: Mulching means covering of soil surface to reduce moisture loss. The various mulching material FYM, straw, polythene etc is widely used in rainfed conditions to conserve soil moisture, reduce weed infestation, increasing micro organism population in soil and yield of crop.

Home stead garden: Since nutrition and health care of family has been the domain of women in both rural and urban area, raising and maintenance of home garden can be undertaken by them which can become important basis for the removal of rampant mal nutrition . More over the production of the nutritive horticultural crops which are often perishable can be produced at the consumption site itself so as to eliminate the great amount of spoilage losses. Some very good models of nutrition gardens have been done by different horticultural institution in the country. At the Indian institute of horticultural research efficient model have been developed for adoption by population group of different sizes considering that 200m plot would be enough for an average family of 5-6 members. Fruit trees like aonla, guava, acid lime, oranges, and papaya banana, Pomegranate, Jamun, Sapota, mango etc being rich in vitamins and minerals, could be important component of such gardens. Efficient management of a home stead garden consists of adoption of crop rotation and continuous and efficient use of all the available space including fencing and ridge space etc.

Technologies for Growing of Ornamental and Medicinal Plants

The important commercial flowers of India are rose jasmine, tuberose, gladiolus, bougainvillea among the perennial flowers and marigold chrysanthemum, nasturtium, petunia etc in annual flowers. Among the other ornamental plants the important ones are hibiscus, crotons, orchids, anthorium, rubber plant, aurocria, palms, Acalypha and several others shrubs and trees. Several new Varieties of these plants has been evolved to suit various uses like garland making, bouquets, wreaths, flower vases, for presentations and decorations etc and the agro-techniques for their cultivation have also been standardized which can be easily adopted. One of the important industry based on flower gardening is perfumery like rose oil, rose water essence and several other perfumery product from other flowers. Women play an important role not only in the growing of these crops but also in the related trades and industries. The medicinal plants of commercial importance are vinea, khus, and safed musali. Ocimuim sanetum, mint etc has gained importance in recent years and can provide a great amount of occupation and employment to women

Mushroom production: In India , the present internal consumption of mushroom is estimated to be about 250 tons per annum which is expected to rise with increasing awareness about its high nutritive value. Commercial mushroom cultivation in our country is very less; so far marchella is collected from coniferous forest of Himachal Pradesh, Jammu Kashmir and Uttarakhand by women workers. Now standard techniques of mushroom cultivation are available. Mushroom cultivation is mostly home bound and requires very little space and thus can be easily done by women. However its cultivation should be handled only after proper training. Training facilities are available at Krishi Vigyan Kendra, Solon, Himachal Pradesh is very famous for mushroom research institute. Of the several varieties of mushroom, the paddy straw mushroom (*pleurotus.sp.*) is very easy to cultivate. It can be grown in polythene bags and complete its life cycle from spawning to harvest in just 30-40 days. A net profit of about Rs. 20,000 can be obtained from medium size mushroom farm taking 20 crops per annum using 800 kg (600bags) straw for each crop. The button mushroom requires more careful attention but fetches higher income. For this various steps like preparation of compost, pasteurizations, cooling, spawning, casing, picking etc have been standardized. A net income of Rs. 71,000 can be obtained from a medium size farm having 600 trays taking 4-5 crops per annum.

Post harvest Technologies: From time immemorial, women have engaged themselves in post harvest aspects like preparation of dehydrated products, preserves pickles, chutneys, murabbas, sarbat etc. from fruits and vegetables. Improved technologies for the preparation of these products from different horticultural crops are now available which can be adopted by women. At present mast of processing units are of cottage scale. Now large industrial units are also developing particularly for the preparation of concentrated juices and beverages. The other important post harvest technology operation like gardening, packing, marketing, storage of produce, so far did not receive much attention. however great advancement have been made on these aspects also and thus women can adopt these technologies with respect to grading of produce

, pre-packing treatments like curing , hot water and fungicidal dips, pre cooling, waxing. Similarly techniques of packaging in corrugated fiber board boxes and storage in zero energy cooling chambers or refrigerated stores can also be adopted by women to minimize post harvest losses. At present, there is scarcity of both zero energy chambers as well as cold storage units in the country which need to be improved.

Apiculture: Although a few women take part in bee keeping perhaps owing to fear complex and ignorance. Since bee keeping requires tender handling and can increase productivity if orchards communities, it can be easily adopted by farm women. The exotic *Apis Mellifera* is much more productive than the indigenous *a. cerana indica.* The operation like hiving of bees and bee swarms , occasional feeding , division of colonies by mass rearing of queens, uniting the colonies , queen introduction , prevention of absconding , swarm control , honey extraction etc can be easily learnt by women.

Sericulture: This is an important trade in which women already participate. Sericulture is largely based on mulberry and caster plantation and involves operations like collection and storage of eggs at low temperature and their hatching to coincide with appearance of foliage in the plantation, feeding tender leaves to the worms. Clearing and changing of the feeding trays, bailing of cocoons, reeling of silk, etc which are easily done by women

Disease and pest management: Chemical control of disease and pest is usually practiced particularly in well managed horticultural gardens. Control measures are available for most of the diseases. Similarly, integrated pest management practices involving insecticides, light traps, trap crops, pheromones, and sterility hormones are also available. Women do not usually participate in spraying operations although with some training they can easily perform these operations. To avoid of pollution by chemical residues in the produce and health hazard by toxic chemical to the operators. Modern techniques of biological control have been developed. The operations like rearing of parasites / predators their release in the field are simple and can be performed by women. Use of proper root stocks can also eliminate the risk of disease and nematodes (table-2).

Table 2: Women participation in IPM practices at tribal area in horticultural crops

S.No.	IPM components	Women participation (N=60)	Women participation (%)
1	Cultural methods		
	Clean cultivation	49	81.67
	Soil manuring and fertilization	25	41.67
	Trap cropping	8	13.33
	Prunning and thinning	19	31.67
	Crop rotation	10	16.67
2	Mechanical methods		
	Hand picking	24	40.00
	Light trapping	6	10.00
	Yellow sticky Traps	8	13.33
	Pheromone trapping	11	18.33
3.	Biological methods		
	Biopesticides sprays	7	11.67
	Preparation of Botanical pesticides	12	20.00

Data based on interviewed with 60 female farmers

Strategies for Women Empowerment

For the quick strengthening women participation in precision farming in horticulture few necessary points are notable.

Training: Exact training needs of different categories of women should be identified. Training programs should incorporate the following aspects. Training in different horticultural trade like nursery management, tissue culture , apiculture, sericulture, mushroom growing, processing of produce besides several other skilled jobs for production of fruits, vegetables and flowers should be organized. The program of Krishi Vigyan Kendras should be expanded and the training program for women be intensified to enhance their productivity, liberate them from drudgeries of farm and home activities and also to improve their skills leading to employability. The training content for farm women in the existing program needs to break new grounds in conformity with the changing times and advancement of technologies. The agricultural extraction organization should try to eliminate the sex bias in offering the training content and skills to farm women. Technologies cannot remain in isolation nor create cultural shock; it needs to be considered in the techno- economic perspective along with socio culture traditional and environmental factors, the exciting adult literacy / education programs must therefore focus on liberating farm women for these sociological impediments. The training program to farm women would be more acceptable and useful if these are imparted by women trainers to enable free participation and discussion, therefore:

(a) More and more women horticulturist should be posted as extension officers.

(b) At least 50% of contact farmer under extension should be farm women.

To ensure group support and mutual consent for adoption of the new technologies and training programs should be organized in groups of farm women from homogeneous social strata. Education tour of such exclusive women groups would be useful to expose them to the development taking place in different aspects of horticulture in different part of country.

Credit and input: - the role of women in horticulture can be enhanced if the access of women to input agencies like back, cooperative societies, self help group, and market yards are increased. This can be done by:

- Suitably, amending the existing rules for credit by bank specially NABARD which place women in disadvantageous position because they insists for immovable properties as collateral securities for credit. Since in rural area the immovable property is commonly held by male head. This rule inhibits credit facilities to women. The cooperation for development of women being set up in different state should take up this matter at appropriate level with the Reserve bank of India, NABARD and state cooperative development.
- SHGs are very good source for farm women to financial support. Organizing SHGs and women cooperatives on functional basis so they can arrange credits, technical inputs, marketing facilities etc. credit lending instruction should support horticultural projects floated by these groups.

Marketing: Efficiency in marketing of perishable horticultural products largely decides the economics of production and avoids gluts and spoilage at production sites. Therefore marketing system incorporating grading, packaging, transport and disposal of produce to market centers should be properly organized. For these Women cooperative should be revitalized with a focus on horticulture related activities. At present there are very few women cooperative in the country and none functioning to the desired level owing to poor membership and lack of leadership. This state of affairs has to be improved.

- Facilities should be created for short term and long term storage of perishable horticultural produce at appropriate locations.

Publicity: Good publicity is required to make the farm women and their organization aware of the facilities available to them for increasing horticultural productivity. For this:

- Direct rote of extension at all level should prepare video cassettes on the cultivation of all horticultural crops which should be made available to all the training organizations in different institutions and departments in the country
- Film division / Doordarshan should prepare documentaries of different aspects where women can participate in horticultural activities and this should be relay through out the country.
- There is dearth of adequate, scientific national data base with regard to farm women role expectations, role performance and implications of transfer of technology. Research program of 'home Science' institution all over the country need to be geared in this direction.
- Recently, social media *viz.* internet, whatsApp, facebook is used for transfer of technology by various KVKs, Institutes etc.

Drudgery, Time and Labor Management

- Strategies should be evolved to discriminate to farm women drudgeries on the farm and in the home. A sound feedback mechanism based on systematic monitoring and evaluation of such dissemination and implantation need to be established.
- For relieving women from the drudgeries of the house hold work related to child care and enchasing the productivity of farm women, the existing child care facilities through crèches and balwadis in the rural area need to be expanded in terms of coverage and these must be made available to women in conformity with their requirement at different farming season.

Although farm women have been participating in horticultural production in the country but that has so far remained poor particularly because of ignorance, lack of skills and owing to different work condition. The available modern technology can eliminate many of these constraints to successful horticultural trades and operate them successfully with great ease. It is however necessary to modernize some of the farm technology above all , training in various skills to farm women would go a long way in increasing horticultural activities in different parts of the country.

Future Scope and Research Need

Women constitute nearly one half of the world's population having enormous potential but being underutilized or unutilized for the economic development of the nation. Indian situation is not an exception for this. The 2001 census of India estimated 0.53 million females against a total population of 1.02 million. In

India, women have remained as a neglected section of work force. They were not considered on par with men. Women are lagging behind men mainly because of the relative absence of economic opportunities and minimal participation in decision making process. Women's access to land ownership is extremely limited. Educational backwardness is another major reason why women are lagging behind men. The latest census reports 2001 revealed that female literacy rate in India was 54 per cent, as compared to the male literacy rate of 76 per cent. However, the life expectancy of female was 65 years in contrast to 64 years of male. Althrough small businesses owned by women have traditionally focused on fashion, fond and other services sector, but recently women entrepreneurs have been moving rapidly into manufacturing, construction and other industrial filed. Women owned business are lightly increasing in the economics of almost all countries. The hidden entrepreneurial potential of women has gradually been changing with the growing sensitivity to the role and economic status in the society. It means women have the potentials, skill, knowledge and adaptability to run a business successfully. The development of women friendly implements are required for carried out activities in production and post harvest management of horticultural crops.

Potential benefits of adoption of advance approaches for Fruit and vegetables in the arid region

- The optimum utilization of available resource in arid and semi arid area.
- Increase productivity of horticultural crops.
- It help in employment generation
- Reduce cost of cultivation and help in increasing income of farmers

Conclusion

One-third of agricultural workers are women and many crops depend on extensive labour provided by agricultural labour, yet when it comes to wages, on an average, their wages are 30% lower than men's wages. Some of the steps needed for imparting a women user's perspective in R&D work in the fields of agriculture and rural development are indicated in Table-3. While action on the above lines is feasible and should be taken, it is important to recognize that science is not a magic wand with which sex inequalities in workload and economic returns can be made to vanish. They should be emphasized clearly as otherwise false hopes will be aroused about the capacity of science and technology to remove deep seated social maladies. In the ultimate analysis, it is only the concern, commitment, and concerted action of agricultural research systems and policy makers that can lead to meaningful results in imparting a women users perspective in research priorities and strategies.

References

Amtul, W. and Viraktamath, B. C. 2013. Gender gaps and women empowerment in India Issues and strategies *International Journal of Scientific and Research Publications*, 3 (9):1-9.

Goyel, G. Randhawa, V. Pannu, K. and Kaur, R. 2003. Empowerment women in agriculture *Rajsthan J. Ext. Edu.*, 11:5-6.

Lata, K. and Singh, S. 2010. Strengthening women participation in precision farming in horticulture in *compendium National seminar on precision farming in Horticulture* pp 658-669.

Manju, S. (2002) Farm women's participation in agricultural activities Maharashtra *J. Ext. Edu.* 21 (1):105-106.

Munshi, K. and Rosenzweig, M. 2008. The Efficacy of Parochial Politics: Caste, Commitment, and Competence in Indian Local Governments. Economic growth Center Yale University P.O. Box 208629 New Haven, CT 06520-8269 http://ssrn.com/abstract=1272500.

Waris, A. 2008. Farm women's participation in agricultural activities in semi arid region of Andhra Pradesh *Green Farming* 2(2):137-138.

36

Adoption of Improved Varieties of *Kachri* and its Economic Impact in Arid Region

S. R. Meena, M. K. Jatav, P. L. Saroj and Anita Meena

Introduction

The hot arid regions of India are spread over 32 million ha (0.32 million Sq.Km.) in the state of Rajasthan (61%), Gujarat (20%), Andhra Pradesh (7%), Punjab (5%), Haryana (4%), Karanataka (3%), and Maharashtra(0.4%) which are characterized by hostile agro-climatic conditions and fragile eco-system. Out of total hot arid area of the country (India), more than 60% area falls under the state of Rajasthan which covers 12 districts viz., Jaisalmer, Barmer, Bikaner, Churu, Sikar, Jhunhjunu, Naguar, Jodhpur, Pali, Jalor, Sriganganagar, and Hanumangarh of the state. In general, the hot arid region of Rajasthan receives very low rainfall, varies from 100 mm annun^{-1}in north-western district of Jaisalmer to 450 mm per annum in the eastern boundaries of arid district of Rajasthan. The potential evapo-transpiration varies from 2063 mm in Jaisalmer to 1503 mm in Sikar districts of Rajasthan. The rainfall is very erratic and often experienced prolong drought. The ground water table is very deep and often brackish in nature. The extremes of temperature having severe winter during December-January with temperature as low as -4 °C and very hot summer during May-June with temperature as high as 48-50 °C is another important characteristic of the hot arid regions. The solar radiation is very high. Atmospheric vapour pressure deficit reaches as high as 30 mb during summer (May-June). The wind velocity in June is very high which leads to heavy wind erosion and formation of sand dunes. Some time strong sand storms of very high speed (60-0 km h^{-1}) with huge amount of sand particles also occurs. The soils are coarse textured with low silt, clay and humus content and water holding capacity is very low and infiltration rate is very high. The soil fertility is very poor having low organic carbon (0.02-0.06%). Soil salinity, calcareousness and gypsiferous nature of soil add another dimension to these constraints in adoption and production any crops in hot arid regions (Dhandhar and Saroj, 2004). Thus, the hot arid region of Rajasthan is full of challenges, threats and constraints which inhabit advancement of the crop production and farming system. However, the hot arid regions of the country have been blessed with some of the strength also which disclose the rays of hope and scope of advancement of horticultural development in the regions. The among such strength of the hot arid regions, the availability of potential bio-diversity/genetic resources, existence of promising germplasm and drought hardy species of ccucurbitaceous vegetables, etc., are the major strengths which can play vital role in horticultural development in these regions. The cucurbitaceous vegetables are the largest group of vegetables among all vegetables grown in hot arid region of Rajasthan. There are various potential and unique cucurbitaceous vegetables which are grown traditionally and used/consumed widely by the dwellers of the region in various ways using their own traditional knowledge and methods. The major traditional cucurbitaceous vegetables of the hot arid region of Rajasthan are kachri (*Cucumis melo var. callosus* Rott. Cong.*)*, snap melon (*Cucumis melo var. momordica*), *mateera loia* (*Citrullus lanatus* Thunb.), round melon (*Citrullus vulgaris var. fistulous*), etc. Among these, kachri is the most favourite and accepted vegetable among the local people of the arid region of Rajasthan. It is highly drought hardy cucurbitaceous traditional vegetable of the region which is grown there from ancient time. *Kachri* belongs to the family- *Cucurbitaceae,* genus-*Cucum*is, species- *melo* and *var. callosus/agrestis*. It is mainly grown during the rainy season under the mixed cropping system at large scale or as sole crop at small scale. Some of the farmers who had irrigation facilities grew the *kachri* as sole crop during the summer season also. *Kachri* is an annual climber. Its vines grow up to 1.5 m and spreads on the ground. The fruits are obovoid/ellipsoid/oval-round shaped with green variegated stripes, 4.0–7.5 cm in size, generally with dark green stripes looks like a miniature of *mateera*/ watermelon. *Kachri* is the most liked traditional vegetable in hot arid regions where it was utilized/ consumed by > 90 % people as fresh for vegetable purpose and in the forms of various traditionally value added products.

Considering immense socio-economic importance and industrial potentiality of kachri, ICAR-Central Institute for Arid Horticulture, Bikaner addressed the improvement and scientific cultivation of *Kachri* (*Cucumis melo var. callosus* Rott. Cong). The Institute initiated the systematic exploration and research work on improvement of kachri. In 1994, intensive crop specific surveys and explorations were undertaken in hot arid and semi-arid regions and a large number of germplasm (land races, local types and semi-cultivated) of kachri were collected, documented, conserved and potential among the same were evaluated. After an extensive evaluation and hard work in research, the Institute developed and released a unique variety of kachri in 1998 and named as “AHK 119”.

Adoption of Improved Variety of Kachri (AHK-119)

A large number of farmers realized the potentiality and beneficial features of this variety (AHK-119) like very early variety, can grown twice in a year (*kharif* & summer season) with limited water/soil moisture, high yielding, best suitable in existing harsh climatic conditions, uniform shape, size and colour of the fruits, very good taste and aroma, highly suitable to prepare the value added products, has long shelf life, high market demand, comparatively high return/has, etc. and started to grow it on their fields at commercial scale. Year to year the demand of seeds of this variety increased so high that the police force had to apply to control the crowd of the farmers while distributing the seeds of the variety at the Institute every year. The graph of high demand of seeds of this variety is hiking year to year and it becomes a challenge for the Institute to meet the high demand of seeds of this variety. The area and production of this variety increased very fast year to year and continue to increase with increasing rate. In past, the farmers of the hot arid region of Rajasthan adopted this variety very speedily and spread among large number of farmers and area within a few years. But, it was not crystal clear picture and actual position of adoption/spread, production, productivity, economic return and other impacts of adoption of improved variety of kachri "AHK-119". Keeping these facts in mind, an intensive impact study was conducted in hot arid regions of Rajasthan "to assess the impact of adoption of improved variety (AHK-119) of kachri in hot arid region of Rajasthan".

Comparative Economic and Yield Potential of Improved Variety (AHK-119) of Kachri

A comparative study was carried out to show the production and yield potential of improved and local variety of kachri. During the study, 100 farmers were selected those who grew improved variety (AHK-119) of kachri and 100 those who grew local variety of kachr; only. The data/information were collected from both categories of the farmers and comparison was made between the gross and net return achieved/ha from improved variety as well as and local variety (local check) of the kachri. It was found that the average production of improved variety of kachri (AHK-119) was 89 q/ha while that of local variety (local check) was 64.0 q/ha. It means, the average production/ha of improved variety of kachri (AHK-119) was 39.06 % higher over the production of local one. The average gross return and net return achieved from the improved variety of kachri (AHK-119) were Rs. 137060 and 100860, respectively while the average gross return and net return achieved from the production of local variety (farmers practices) of kachri were Rs. 77440 and 48640, respectively (Table 1 and Figure 1). It means the net return achieved from the improved variety of kachri (AHK-119) was 107.35 % higher over the net return achieved from the local variety (local check) of kachri. Similar kind of findings was reported by Balai, *et al.*, 2013 in their study.

Table 1: Comparative cost of cultivation and net return (round figures) from production of improved variety (AHK-119) and local variety of kachri.

Particulars/operations	Average cost cultivation in growing the improved variety (AHK-119) of kachri by 100 farmers (Rs./ha)	Average cost of cultivation growing in the local variety of kachri by 100 farmers (Rs./ha)
Land Preparation	6000	4600
Seed	1600	1500
Fertilizer	7200	4000
Irrigation	2000	1500
Agro Chemicals (including pesticides)	2400	1200
Other labour charges (including harvesting and transportation)	17000	16000
Total cost of cultivation	36200	28800
Kachri production (q/ha)	89	64
Selling price of kachari (Rs/q)*	1540	1210
Gross return from kachari/ha*	137060	77440
Net return from kachari/ha*	100860	48640

*The return/income includes both from fresh as well some value added products of kachri.

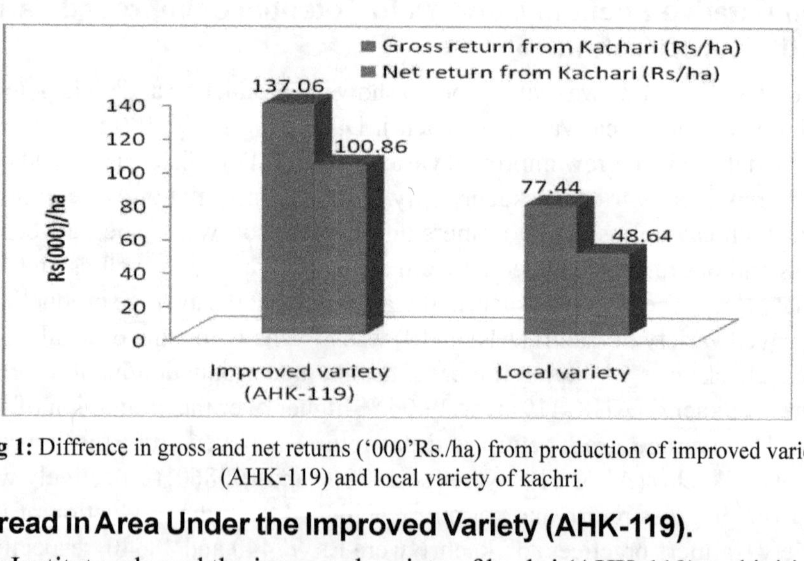

Fig 1: Diffrence in gross and net returns ('000'Rs./ha) from production of improved variety (AHK-119) and local variety of kachri.

Spread in Area Under the Improved Variety (AHK-119).

The Institute released the improved variety of kachri (AHK-119) and initiated the work on dissemination and popularization of the same among the farmers to adopt it. A large number of farmers realized the potentiality and benefits of this variety. Initially, a few farmers adopted it and started its cultivation. The popularity of the variety increased very fast among the farming community and its unexpected demand hiked widely within 3-4 years of after it's releasing and dissemination work of the Institute. The area and production of this variety was increased tremendously year to year and continue to increase with increasing rate. In past, the farmers of the hot arid region of Rajasthan adopted AHK-119 very speedily to large scale and it was calculated that this variety spread on about 4000 ha area within a few years with large production. The yearly calculation/estimation of the status of spread in area and production of improved variety of kachri (AHK-119) and local one is given in Table 2.

The data in table 2 reveals that the total area under cultivation improved variety (AHK-119) of kachri in entire hot arid region of Rajasthan was 1277 ha in 2007 which increased to 3865 ha in the year 2017. In case of different district of the hot arid region of Rajasthan, maximum area of improved variety of kachri was under in Bikaner district followed by Nagaur district which increased from 630 to 1720 ha Bikaner district and from 348 to 1081 in Nagaur districts during the year 2007 to 2017, respectively. The other district where the improved variety of kachri is grown on more than 100 ha of land in 2017 was Jodhpur, Barmer, Sikar and Jhununjhunu. However, the area under this improved variety of kachri was observed fluctuated year to year which might be due to uncertainty in climatic conditions in hot arid region and other factors like erratic and low rainfall, high temperature, frequently occurrence drought and strong sand storms, scarcity of water, low availability of seeds, attack of pests and diseases, low supply of canal water, electricity, *etc*.

Gross and Net Return Both from the Improved Variety (AHK-119) and Local Variety of Kachri in Entire Hot Arid Regions of the Rajasthan

On the basis of information/data collected during the study, year wise gross and net return from the improved variety (AHK-119) of kachri in entire hot arid region of Rajasthan were analyzed and statistically calculated The results revealed (Table 3) that the gross return from the improved variety of kachri (AHK-119) in entire hot arid region of the Rajasthan was 17.50 crores in 2007 which increased to 52.97 crores in 2017. It is also found that gross return from the improved variety of kachri (AHK-119) in entire hot arid region of Rajasthan increased three times in 2017 in comparison to the year 2007. Likewise the net return from the improved variety of kachri (AHK-119) in entire hot arid region of Rajasthan was 12.88 crores in 2007 which increased to 38.98 crores in 2017 which means, the net return from the improved variety of kachri (AHK-119) in entire hot arid region of Rajasthan increased three times (202.63%) in 2017 in

Table 2: Year wise area under improved verities of Kachri (AHK-119) in different districts of the hot arid region of Rajasthan

(Area in ha.*)

Districts	2007	2008	2009	2010	2011	2012	2013	2014	2015	2016	2017
Barmer	32	35	36	105	62	92	78	122	103	122	145
Bikaner	630	707	472	1280	1087	1465	1292	1480	1405	1505	1720
Churu	32	23	29	58	52	62	65	80	72	84	96
Sriganganagar	10	10	9	36	28	33	31	35	28	35	41
Hanumangarh	11	10	11	28	25	31	26	36	30	34	46
Jaisalmer	19	18	18	52	68	78	51	68	52	72	86
Jalore	17	21	24	64	59	71	54	71	62	83	85
Jhunjhunu	40	18	20	55	64	74	61	89	78	89	137
Jodhpur	50	37	36	104	94	131	112	152	141	152	181
Nagaur	348	230	267	688	572	774	695	1035	902	925	1081
Pali	39	16	20	42	68	79	52	89	74	92	110
Sikar	48	25	25	78	74	95	80	108	110	122	137
Total area under AHK-119 in hot arid region of Rajasthan	1277	1152	967	2590	2253	2985	2597	3365	3057	3315	3865
% area of the country (India) under AHK-119 in hot arid region of Rajasthan.	62.11	67.17	64.96	64.29	79.25	87.41	87.11	74.69	79.77	85.48	63.44

* *Kharif + summer* season

comparison to the year 2007. It is also worth to mention here that contribution of the improved variety kachri (AHK-119) in total net return achieved from kachri crops (improved + local) in entire hot arid region of Rajasthan was > 60 % in the year of 2007 which increased > 80 % during the year 2017. Similar kinds of findings with respect to net return per hectare and cost benefit ratio were reported by Jatav, *et al*, 2016a, 2016b and 2016c in their studies.

Table 3: Year wise total area (ha), production ('000'tons), gross and net income (in crorers) from improved variety of kachri (AHK-119) in hot arid regions of Rajasthan.

Years	Area*	Production	Gross return	Net return
2007	1277	11.37	17.50	12.88
2008	1152	10.25	15.79	11.62
2009	967	8.61	13.25	9.75
2010	2590	23.05	35.50	26.12
2011	2253	20.05	30.88	22.72
2012	2985	26.57	40.91	30.11
2013	2597	23.11	35.59	26.19
2014	3365	29.95	46.12	33.94
2015	3057	27.21	41.90	30.83
2016	3315	29.50	45.44	33.44
2017	3865	34.40	52.97	38.98

* *Kharif* + *summer* season both

Development of Seed Chain of Improved Variety of Kachri (AHK-119) and Income Generation

The area under improved variety of kachri (AHK-119) is increasing year to year speedily. The major reason behind fast increasing in area in hot arid region was development of seed production chain of the variety among the farmers. Initially, the farmers purchase the seeds of improved variety of kachri (AHK-119) from the Institute (ICAR-CIAH, Bikaner) and multiply the same at their own fields under the technical guidance of scientists of the Institute and use it to grow on their own fields in next season/year or sell out the surplus seed to their fellow farmers to earn money. There are several progressive farmers who multiply/produce the seed of improved variety of kachri (AHK-119) on their own level each up to extent of 10-30 kg/season under the scientific guidance of the experts and sell it (@ Rs. 2000 – 3000 or more per kg) to other needy farmers/clients to earn money. As the demand of seeds of this improved variety (AHK-119) is very high among the farmers of the hot arid regions, hence, several progressive farmers have made their profession to multiply/produce seeds of this variety to earn in considerable amount of money. The trend of such multiplication of the seeds of improved variety of kachri (AHK-119) is increasing year to year which is encouraging more and more spread of this variety in hot arid region of Rajasthan. These finding are in the line of the findings reported by Meena *et al.* 2016 in study.

Constraints in Adoption and Production of Improved Variety (AHK-119) of Kachri at Large Scale in Hot arid Region of Rajasthan

Despite the apparent need and advantages associated with adoption and production of improved variety (AHK-119) of kachri at large scale in hot arid region of Rajasthan, farmers are not much success in adoption and production/ growing of the improved variety of kachri on their fields in harsh climatic conditions of the region. They have to face various "problems/ constraints" which leads to low/ non adoption of these crops at large scale in hot arid climatic conditions of the hot arid region of the Western Rajasthan. Major general constraints observed during the study which hinder the adoption and production of improved variety of kachri at large scale in hot arid regions of western Rajasthan India were as follows.

(i) Ecological Constraints

The major observed ecological constraints which affect the large adoption and production of improved variety of kachri at large scale in hot arid region of Rajasthan were as follows.

- Acute shortage of water
- Poor and erratic rainfall
- Occurrence of drought and famine very frequently.
- Very deep and salty/brackish ground water
- Extremes of temperature during summer (upto 50° C)
- Perpetuation of high hot winds and sand storms
- Shifting of sand and sand dunes
- High evapo-transpiration and low relative humidity
- Poor soil with low water retention capacity.

(ii) Socio-psychological Constraints

During the study, there were observed various socio-psychological and cognitive constraints also which led to low or non adoption and production/growing the improved variety of kachri at large scale in hot arid regions of Rajasthan which were as follows.

- Lack of awareness, interest and knowledge
- Deep faith in traditional cropping system
- Lack of reciprocal technical interactions between scientists and farmers.
- Lack of technical - social institutions
- Predominance of myths and misunderstandings
- Illiteracy and heterophilic population
- Fear of crop failure
- Dissonance about the success of crops/technologies
- Lackadaisicalness in farmers and lack of inspirations/motivations.

(iii) Techno-economic constraints.

Techno-economic constraints which were responsible for low or non adoption and production of improved variety of kachri at large scale in hot arid region were as below.

- Low/non accessibility of seeds of improved variety of at local level as per desire.
- Non availability technical guidance and reliable inputs at local level.
- Lack of awareness and knowledge improved production technologies of kachri.
- Lack of local markets/*mandies* to sell the produces/products of kachri.
- Poverty, scare resources and low risk bearing capacity of the farmers.
- Perishable nature of the produces / products.
- No standard technologies for value addition of fruits of kachri .
- Lack of subsidy/credit facilities to support to the farmers.
- The commercial production front of kachri is legging behind.
- No security and support against crop failure and low income.
- Monopoly of brokers in selling/purchasing produces/ products.

(iv) Constraints related to essential infrastructural facilities.

There were observed some infrastructural constraints also which were considered responsible for low adoption of the improved variety of kachri in hot arid region were as follows.

- Lack of reliable sources of technical information and inputs.
- Poor extension and communication system
- Lack of the specific facilities for farmers' trainings and skill development.
- Poor networking on modern/improved production technologies of kachri.
- Lack of safe storage / preservation facilities.
- Limited and irregular supply of electricity
- Unfamiliar and long distance of research and extension institutions.
- Absence of regulated local markets and transportation facilities.

Conclusion

It can be concluded that there is wide scope for cultivation of kachri in hot arid region of north-western Rajasthan. The kachri cultivar (AHK-119) is agronomically most suitable and efficient in resource a poor condition which gives high yield, better quality and good return as compared to other vegetable in the arid region. In fact, kachri has great potential in these regions as the availability of vegetables is poor in the arid region of the country throughout the year. If good quality seed, proper marketing system, sufficient cold storage and micro-irrigation facilities are made available, definitely the prospect of kachri production will be bright in hot arid regions. It is therefore necessary to educate the farmers through front line demonstrations as well as imparting training on improved package of practices of kachri production. There is tremendous scope of establishment of processing units based on products and bio-products of kachri. This will increase the demand and will further motivate the farmers to grow kachri in this region.

References

Dhandhar, D. G. and Saroj, P. L., 2004. Changing Scenario in Arid Horticulture. *In* : (Eds) P. L. Saroj, B. B. Vashishtha and D. G. Dhandar. Advances in Arid Horticulture, Vol. 1: Present Status. International Book Distributing Co., Luclnow, UP (India). Pp. 1 - 28.

Meena, S. R. Singh, R. S., Sharma, B. D. and Singh, D., 2016. Most favourite traditional cucurbitaceous vegetables and their utilization pattern in Thar desert of the western Rajasthan, India. *Indian Journal of Traditional Knowledge*, 15(3): 385-394

Balai, C. M., Jalwania, R., Verma, L. N., Bairwa, R. K. and Regar, P. C., 2013. Economic impact of Front Line Demonstrations on vegetables in tribal belt of Rajasthan. *Current Agriculture Research Journal*. 1(2): 69-77.

Jatav, M.K., Sharma, B.D., Samadia, D.K. and Meena, S.R. 2016a. Yield of *kachri* (*Cucumis callosus*) as influenced by organic and inorganic sources of nutrients in arid zone. *Indian Journal of Agricultural Sciences* 86 (7): 961–3, July 2016.

Jatav, M.K., Sharma, B.D., Samadia, D.K. and Meena, S.R. 2016b. Site specific nutrient management of kachri using omission plot technique in hot arid region of Rajasthan. *International Journal of Basic and Applied Biology*, Vol 3, 2016, pp. 173-175.

Jatav, M.K., Sharma, B.D., Samadia, D.K. and Meena, S.R. 2016. 2016c. Effect of different sources of nitrogen on *Kachari* (*Cucumis melo*) performance in the arid region. *Annals of Arid Zone* 55(1&2): 35-39.

37

Socio-Economic Importance of Cucurbitaceous Vegetables in Hot Arid Zones of India

S. R. Meena, P. L. Saroj, M. K. Jatav, R. K. Meena and R. C. Balai

Introduction

The hot arid zones in India are spread over nearly 31.7 million hectare area of which 41.5 per cent is arable and 19 per cent is cultural wasteland. These zones are perpetuated mainly in North –West and Southern part of the country. The major parts of the country under hot arid climatic conditions are Western Rajasthan (19.62 Million ha), North-western Gujarat (6.2 M. ha), South-western Punjab (1.45 M. ha), South-western Haryana (1.28 M. ha), Andhra Pradesh (2.16 M. ha), Karnataka (0.86 M. ha), and Maharashtra (0.13 M. ha). These areas/ zones are characterized by hostile agro-climate and fragile eco-system, an annual rainfall between 100 – 500 mm with a coefficient of variation (CV) varying from 40 – 70 per cent, low and erratic rainfall combined with extremes of temperature (450-500 cal/cm^2/day); low relative humidity; high potential evapo-transpiration value ranging from 1600 mm in eastern part and 1800 mm in western part of the hot arid zones. Other important characteristics of the hot arid regions are hot winds with high velocity, poor soil condition, poor ground water with brackish and saline in reaction, poor vegetation, frequent occurrence of drought and frost, difficult to execute agro-techniques, difficulty in post harvest handling and marketing owning to limited and inefficient transportation and marketing facilities, etc (Yadav and Soni, 2008).

Despite the various bio-physical constraints, there is very good scope of growing the cucurbitaceous vegetables like *mateera*, snapmelon, *kachri*, round melon, bottle gourds, to a considerable extent. The farmers/dwellers grow the above vegetables at large scale for their own use and sale the surplus for earning money and livelihood security. These vegetables are used not only as fresh in current season but also throughout the year by converting them in the form of preservation or value added products. They are the major source of nutrition and daily food stuff. The farmers/dwellers used to prepare various traditional value added products of above vegetables using their own talent/traditional knowledge for their own consumption and selling surplus to local people/markets to earn money too.

Major Cucurbitaceous Vegetables Grown in Hot Arid Zones of India

Cucurbitaceous vegetables are the largest group of vegetables among all vegetables grown in hot arid zones of the country (India). Among these, the favourite and most accepted cucurbitaceous vegetables are: *kachri* (*Cucumis callosus* Rott. Cong.*)*, snap melon (*Cucumis melo var. momordica*), *mateera loiya* (*Citrullus lanatus* Thumb.), round melon (*Citrullus vulgaris var. fistulous*) as given in Table 1.

Table 1: The favourite cucurbitaceous vegetables grown extensively and used widely by the farmers/dwellers in hot arid zones of India.

S.No	Common name of vegetable	Scientific name of vegetable	Growing season	Part utilized/consumed
1	*Kachri*	*Cucumis callosus* (Rott.) Cong.	*Kharif* & summer	Fruits, seeds, roots, leaves
2	Snapmelon	*Cucumis melo var. momordica*	*Kharif* & summer	Fruits, seeds, roots, leaves
3	*Mateera*	*Citrullus lanatus* (Thumb.)	*Kharif* & summer	Immature small fruits (*loia*), seeds, roots, leaves
4	Roundmelon	*Citrullus vulgaris var. fistulous*	*Kharif* & summer	Fruits, seeds, roots, leaves

Most of these vegetables are consumed throughout the year either in fresh or in preserved/different value added forms. The people of the hot arid zones utilized the immature/mature fruits and other parts of these vegetables (Table 1) to prepare the nutritious vegetables and various kinds of value added products to nourish themselves and others. In hot arid zones of India, the cucurbit vegetables are of the great economic and nutritional importance. They are excellent source of vitamins, minerals, carbohydrates and traditional herbal medicines. They are very good sources of staple food stuff and income/entrepreneurship both in fresh as well as preserved/value added forms.

Socio-economic Importance of Cucurbitaceous Vegetables in Hot arid Zones of India.

1. Kachri (*Cucumis melo var. callosus* Rott. Cong)

Kachri the most favourite, suitable and drought hardy cucurbitaceous vegetable in hot arid zones of the country which grown here from ancient time. *Kachri,* belongs to the family- *Cucurbitaceae,* genus-*Cucum*is, species- *melo* and *var. callosus/agrestis*. It is commonly known as small gourd/wild musk melon (English), *Kachri* (Gujarati), *Kachri/Kachriya* (Hindi), *chibdin* (Konkani), *chibbad* (Punjabi), *shinde* (Marathi), *gurmi* (Nepalese), etc. It is mainly grown during the rainy season under the mixed cropping system at large scale or as sole crop at small scale. Some of the respondents who have irrigation facilities grow the *kachri* as sole crop during the summer season also. *Kachri* is an annual climber. Its vines grow up to 1.5 m and spreads on the ground. The fruits are obovoid/ellipsoid/oval-round shaped with green variegated stripes, 4.0–7.5 cm in size, generally with dark green stripes looks like a miniature of *mateera*/watermelon. Initially, the immature fruits of *Kachri* are green in colour with hairy surface and have bitterness. As soon as the fruits reach at maturity, their hairs and bitterness disappear and become smooth and with relish taste having sourness/acidic taste[2].

Importance and use Pattern

(i) *Kachri* as source of nutritious vegetable/food stuff: *Kachri* is the most favourite vegetable in hot arid zones where it is utilized/consumed by 90-100% respondents as fresh vegetable and in value added forms (Table 2).

Table 2: Uutilization/consumption pattern of cucurbitaceous vegetables for the nourishment and sustenance of the farmers/dwellers in hot arid zones of India.

Sr. No.	Vegetable	The modus operandi of utilization and consumption forms of the traditional vegetables in nourishment	Producer & Consumers (%)
1	*Kachri*	Preparation of vegetable (pure and mixed both), dehydrated slice/pieces/ whole fruit (*kachri goite*), pickle, *fried chutney, dry chutney, powder,* juice, hajmola, squash, *jam, papadi, spices* mouth freshener.	90-100
2	Snapmelon	Preparation of vegetable (pure and mixed both), *Khelara (Dehydrated pieces/slices), pickle, Jem, shek, chutney,* cold drinks, sweets, *laddu of kernel of seeds,etc.*	87-100
3	*Mateera*	Preparation of vegetable of *loia*, dehydrated nutritive rousted seeds, highly nutritive *magaj,*	68 – 88

Contd.

		mateera oil, juice, chutney, *rayata,* cold drinks, sweets, *laddu of kernel (magaj) of seeds,etc..*	
4	Roundmelon	Preparation of vegetable, *foflia* (dehydrated pieces/slices) pickle, *rayata, chutney, etc.*	67 – 84

The fresh as well as dehydrated fruits of the *kachri* were mainly used for preparing vegetable (pure/ mixed with other vegetables), pickle and *chutney.* It has high nutritional value and work as mean of food seasoning or relish of the diet of the people of the hot arid zones of the country. The *kachri* is one of the components of the dry mixed vegetable stuff, locally called *Panchkuta* [consist of dehydrated tender pods of *khejri* (*Prosopis cineraria* (L.) Druce), dry fruits of *ker* (*Capparis deciduas Edgew*), dry slice/fruits of *kachri* (*Cucumis melo var. callosus* Rott. Cong), dry fruits of *lasora* (*Cordia myxa* Roxb) and seeds of *kumat* (*Acacia senegal*)]. *Kachri* fruits have very good amount of nutrients like vitamins, mineral, carbohydrates, etc. The 100 gm of fresh edible portion of *kachri* has 88.8% water/moisture, 7.45% carbohydrate, 47-54 Kcal energy, 0.28% protein, 1.28% fat, 1.21% fibre, 1.46% total ash, 0.09 mg calcium, 0.0029 mg phosphorus, 0.182 mg iron, 0.0046 mg copper, 0.052mg zinc and 29.81mg vitamin C. In100gm dry powder of *Kachri,* 63.68% carbohydrate, 2.41% protein, 11.0 % fat, 10.41% fibre, 12.50mg minerals, 17.02mg Vit.c, 0.8 mg calcium, 1.56mg iron, 0.025mg phosphorus, 0.45 mg copper, etc. (Samadia, 2007 and Dahot *et al.,*1999) . The mature fruits of *Kachri* (*Cucumis callosus* (Rottler) Congniaux) are available abundantly in South Western Rajasthan also and it is one of the components of the delicious mixed vegetable popularly known as *panchkutta.* It has 1.28 mg fat, 1.21mg fibre, 43 Kcal energy and 29.81 mg vitamin C in 100 gm^{-1} of fresh fruit of Kachri (Nandal and Bhardwaj, 2015).

(ii) *Kachri* as source of preserved food/value added products: *Kachri* is considered as a source of pure form of vegetable/food stuff in hot arid zones. In addition of vegetable, the respondents of the desert region prepare various traditional value added products of *kachri* using their own experiences/traditional knowledge/methods for their own consumption and selling out the surplus in the markets. The utilization and consumption pattern of traditional value added products of *Kachri* for the nourishment and sustenance of the people of the hot arid zones were recorded during a study. The major value added products of *kachri* prepared and used by the respondents of the study areas in their daily diet and selling out the surplus of the same in local market/*mandies*/localities are depicted in Table 2. The leaves with tender stems of the *kachri* are used as feed of animals too.

(iii) ***Kachri* as mean of income generation and trade specific opportunities:** The *kachri* production is a very good mean of income generation and trade specific activities for the respondents in hot arid zones. The farmers/

people of the regions grow *kachri* in considerable amount for own consumption and sale surplus in the local markets/*mandies*. In addition to preparation vegetable, the growers, prepare several value added products of *kachri* fruits (Table 2). These products are sold in the local markets/*mandies*/localities at high price to earn substantial amount of money from the same. The major value added product prepared by the farmers/rural dwellers of the zones the income generation and business purpose are: dehydrated *kachri* (whole/as pieces/slices), *kachri* pickle, dry chutney, fried wet chutney, spicy powder, mouth freshener, etc. The fresh and dehydrated *kachri* is sold by the grower/producers in the markets/*mandies*/ localities @ Rs. 25-60 and Rs. 180-275 per kg, respectively (Table 3). Likewise, the pickle, readymade dry *chutney* and fried wet *chutney* of *kachri* were sold by respondents in the markets/*mandies*/localities @ Rs. 120-160, 140-200 and 130-150, respectively and a thumping amount of money is earned for their livelihood security and prosperous life. The production of *kachri* seeds are also a very good source of income and business opportunities for the farmers/rural dwellers in hot arid regions. Majority of the farmers produce the seeds of the available potential landraces as well as improved varieties of *Kachri* and sell them in local markets/*mandies*/ localities @ Rs. 800 - 1500 per kg, depending on type, quality, locality, season, demand and production factors of the seeds. Some of the innovative farmers produce the seeds of improved varieties (AHK-119) of *kachri* and sell the same up to Rs. 2000–3000 per kg. Thus, the production and selling of *kachri* seeds at such high rates and earning of a great amount of money are very good source of income generating and business activities among the respondents in hot arid regions of the country which leads to livelihood security and sustenance of the farmers/dwellers.

(vi) ***Kachri* as mean of traditional herbal medicine and therapy:** The *kachri* is not only used as a vegetable in hot arid regions of India but also it works as a good source of herbal medicinal/therapeutic uses in the regions. The people the hot arid regions of the country use the *kachri* fruits, seeds, juice or their combinations for health care and to cure different health problems. The *kachri* fruits are used as a cooling light cleanser or moisturizer for the skin. They were also used as a first aid treatment for burns and abrasions. The flowers of the *kachri* are used as expectorant and emetic while fruits are used stomachic. The seeds are used to cure digestion and urinary problems and also in vermifuge activities. They are considered to possess significant antioxidant, anti-inflammatory and analgesic value. Roots of the *kachri are* used as diuretic and emetic.

Thus, the *kachri* plays very important role as a nutritious food stuff, means of income generation, entrepreneurial activities and source of traditional medicine/ therapeutic value in proper nourishment and sustenance of the socio-economic life of the hot arid zones of the country.

2. Snap Melon (*Cucumis melo var. momordica*)

Snap melon is also considered as another most liked, most suitable and drought hardy vegetable of the hot arid zones of India. It belongs to family- *Cucurbitaceae*, genus- *cucmis*, species-*melo* and *var.-momordica.* It is locally known as '*phoot or photo kakadi, kakadia.* Snapmelon is a monoecious, an annual climber, its stem are generally covered with rough hairs and vines grow up to 1.5 m which spreads on the ground. The flowers of the snapmelon are small, yellow, solitary or rarely in pairs or threes. Initially, the immature fruits of the snap melon are green in colour with hairy surface and have bitterness. As soon as the fruits reach at maturity, their hairs and bitterness disappear and become smooth with relish acidic taste. There is wide variability among the shape, size, appearance and colour of fruits of the snapmelon. Most of the fruits of snapmelon are cylindrical, 10-22 cm length with 18-32 cm central girth, green-yellow/radish-yellow/creamy-yellow or green with yellow/greens strips/pots at maturity stage. After maturity, fruits reach to over maturity stage and at this stage the fruits start to crake automatically which are locally called '*phoot'*.

Importance and use Pattern

(i) Snap melon as source of nutritious vegetable/food stuff: Like *kachri,* the snapmelon was also one of most liked vegetable of the hot arid zones of India where it was consumed by 87 -100 % dwellers of the zones as vegetable and in different value added forms (Table 2). The fresh as well as dehydrated slices of fruits of snap melon were mainly used for preparing vegetable (pure/ mixed with other vegetables). The fruits of *phoot* (over mature fruits) stage have a special pleasant smell and taste like pine apple. These *phoots* are of sweet-sour in taste and eaten directly as fresh fruit very eagerly by the people of the zones. The fresh mature fruits of the snap melon are also used for *salad* purpose. Like *kachri*, snap melon is also considered as a source of pure form of vegetable/food stuff in hot arid zones of the country. It has high nutritional value and work as a mean of food seasoning or relish/palatability of the diet for the respondents of the study areas. The 100 gm of fresh edible portion of snap melon has 79.93gm water/moisture, 15.56gm carbohydrate, 79.29 Kcal energy, 0.36gm protein, 1.12gm fat, 1.34gm fibre, 1.64gm total ash, 0.76mg calcium, 0.0088mg phosphorus, 0.843mg iron, 0.1.04mg copper, 0.202mg zinc and 18.6 mg Vit. C. In 100gm dehydrated slices (*khelra*) of the snap melon has 77.66gm carbohydrate, 395.28 Kcal energy, 1.84gm protein, 5.6gm fat, 6.72gm fibre, 8.18gm ash in which 3.8 mg calcium, 4.3mg iron, 0.44mg phosphorus, 5.2 mg copper,1.1mg zinc, 10.30 vitamin C, *etc* (Samadia, 2007).

(ii) **Snap melon as source of preserved food/value added products:** The snap melon is one of the important source nutritious pure organic food stuff in hot arid zones of the country. The people of the zones prepare various kind of traditional value added products of snap melon using their traditional knowledge and methods for their own consumption and selling out the surplus in the markets. The modus operandi of the exploitation and production of traditional value added products of snap melon and their mode of consumption for the nourishment and sustenance of the dwellers/farmers of the hot arid regions of the country is depicted in Table 2. The major value added products of snap melon as prepared and used by the dwellers/farmers of the hot arid regions in their daily diet and selling out the surplus of the same in local market/localities are dehydrated slice/pieces *(khelra)*, pickle of snapmelon, fried *chutney, jam* of snap melon, *etc*.

(iii) **Snap melon as mean of income generation and trade specific opportunities:** The production of snap melon and its value added products provide very good opportunities for generating the income and trade specific activities for the growers/producers in hot arid regions. The farmers/producers produce the snap melon fruits in large amount and sale them freshly in local vegetable markets/ *mandies*/localities price ranging from Rs. 10-40 per kg., depending on quality of produce, season, selling location, demand and supply/production factors. In addition to preparation of vegetable, the respondents prepared several value added products of fruits of snap melon as shown in Table 2. These products were sold in the local markets/*mandies* and respondents earned thumping amount of money from the same. The major value added products of snap melon prepared and used by the respondents of the study areas in their daily diet and selling out the surplus of the same in local market/localities were: dehydrated slices (*khelra*) and pickle of snap melon, *etc*. The dehydrated slices (locally called *khelra*) of snapmelon were sold by the respondents in the markets/*mandies*/ localities @ Rs. 120-160 per kg (Table 3).

Table 3: Generation of income from cucurbitaceous vegetables for sustenance and livelihood security of farmers/dwellers of the hot arid zones of India.

Sr.No.	Vegetable	Scientific Name	Edible part	Market Price/ Kg (Rs.)
1	*Kachri*	Cucumis callosus (Rott.) Cong.	Fresh Fruits	25 – 60
			Dehydrated fruits/ slice	180-275
2	Snapmelon	*Cucumis melo*	Fresh Fruits	10 – 40
			Dehydrated fruits/ slice	120 – 160
3	*Mateera loiya*	*Citrullus lanatus* (Thumb.)	Immature small fruits	30 – 80
4	Roundmelon	Citrullus vulgaris	Fresh Fruits	20 – 40
			Dehydrated fruits/ slice	110-155

* Farmers' market price of concerned vegetable/value added products depends on season, quality of produce, production, market demand and supply factors of the produce, bargaining of middleman, etc.

The dehydrated slices (*khelra*) of snap melon are used to prepare vegetable (mixed or pure), fried *chutney* or as a food seasoning agent throughout the year. The some people of the hot arid regions prepare pickle of the snap melon and sale it @ Rs. 100-125 per kg and earned substantial amount of money for their livelihood security. The production of seeds of snap melon is also a very good source of income and entrepreneurship/ business opportunities for the farmers/dwellers of the region. More than 70% farmers/clients produce the seeds of the available potential landraces and improved varieties of the snap melons which were sold in localities/local markets/*mandies* @ Rs. 800 - 1500 per kg., depending on type, quality, and locality of sale, demand and production factors of the seeds. The seeds of local varieties/land races were sold at lower rates while the seeds of improved varieties were sold at higher rates. Some of the farmers used to produce the seeds of improved variety (AHS- 82) of snapmelon (as developed by the ICAR-CIAH, Bikaner, Rajasthan) and sale the same @ Rs. 2000-2500 or more per kg. Thus, the production and selling of seeds of the snap melon at such high rates and earning of a great amount of money are very good income generating and entrepreneurial/business activities among the farmers/people of the hot arid zones of India.

(iv) **Snap melon as mean of traditional medicine and therapy:** In hot arid regions of the country, the snap melon is mainly consumed as a vegetable but it has herbal medicinal/therapeutic value also. The rural masses of the hot arid zones of the country, use the snap melon's fruits, seeds, juice and their combinations for health care and to cure different health problems. The fruits of the snap melon are also used as a cooling light cleanser or moisturizer for the skin. They are also used as a first aid treatment for burns and abrasions. The seeds of snap melon are used to cure digestion and urinary problems. Roots of the snap melon are used as diuretic and emetic. Plant seeds were considered to possess significant antioxidant, anti-inflammatory and analgesic value. Thus, it is to say that the snap melon played very important role in nourishment and sustenance of the people/dwellers of the hot arid zones of in India.

3. *Mateera loia* (*Citrullus lanatus* Thumb.)

In hot arid zones of India, another most preferable cucurbit is *mateera.* The tender fruits of *'mateera'* locally known as *'loia'* (immature fruits of 100-150gm) are very good source green vegetable. The scientifically, *mateera* is known as *Citrullus lanatus* (Thumb.) Matsum and Nakai which belongs to the family Cucurbitaceae. It is commonly known as *Kalind* (Sanskrit); *Kalingad* (Marathi); *Tarbooj* (Hindi); *Mateera* (western Rajasthan); *Kalingda* (Gujarat). It an annual vine/creeper and *mateera* has well developed strong tap root system with a highly branching/lateral roots extending up to 1 m deep or more into the soil. The stem of the *mateera* is herbaceous, highly branched vines up to 3 m long (but may be up to 8-10 m). The younger shoots are covered with long, woolly curved hairs protecting the plant from overheating. Flowers are staminate (male), perfect (hermaphroditic), or pistillate (female). However, monoecious types are most common, but there are found andromonoecious (staminate and perfect) types also. The surface of fruits varies from single colour to various striped patterns. The mature fruits of the *mateera* consist of exocarp, mesocarp and endocarp. The endocarp (placenta) is seed containing part that is consumed as food (edible part) and the mesocarp and exocarp are usually referred to as rind having thickness of 10 to 40 mm. The colour of pulp may be white, pink or reddish-pink (Guner and Wehner, 2004: Kumar and Wehner, 2013 and Monika *et al*., 2016)

It is one of the most drought hardy and multipurpose cucurbit. It is preferably grown during rainy season under traditional mixed cropping system extensively. However, the farmers which have irrigation facilities grow it during the summer season as a sole at small scale also. It is grown for vegetable purpose, for fresh fruits, its protein and oil rich seeds, etc., which supports the livelihood of the people in a big way. The *mateera* has a great importance not only as a vegetable and oilseed crop but also plays a vital role in livelihood and nutritional security in hot arid zones of the country.

Importance and Use Pattern

(i) *Mateera* as source of nutritious vegetable/food stuff: The tender fruits of *'mateera'* locally known as *'loia'* (immature fruits of 100-150gm) are traditionally used by the people widely to prepare vegetable or *raita, chutney* and also for sale surplus in the local market/*mandies*/localities. The mature fruits of *mateera* are eaten freshly and much relished. The roasted seeds of *mateera* are generally taken as a common snack in hot arid regions. After removal of seed coat, the kernels of the *mateera* seeds, locally known as '*magaj*' are mainly used in preparing sweets, *laddu* and cold drinks. The kernels (*magaj*) of the seeds are used in restaurants/hotels as source of flavor, taste and thickeners in vegetable preparation. After extraction of oil, the byproducts seeds serve as a quality animal feed which is supposed to increase the milk yield of the milch animals. The *mateera* fruits and seeds are very rich source nutrients and minerals. *Mateera* fruits have 1.75 % nitrogen, 10.93% protein, 0.011 % phosphorus and other valuable nutrients. The seeds of the *mateera* are very rich source of crude protein and oil besides other minerals. The *mateera* seeds oil contains > 80.0% unsaturated fatty acids with linoleic acid being the dominant fatty acid. The seed kernels (*magaj*) are rich in edible oil (40%), crude protein (30%), etc (Monika *et al*., 2019). The another study revealed that the seeds of *mateera*/watermelon has 28% crude fat and 23% crude protein while in kernel corresponding values observed were 49 and 40%, respectively (Das *et al,* 2002) .The further analysis of seed kernel for mineral composition and reported considerable amount of phosphorus (1279 mg/100 gm), potassium (1176 mg/ 100 gm), magnesium (542 mg/100 gm) and calcium (150 mg/100 gm) (Adawy *et al.,* 2001). The leaves with tender stems and fruit rinds of the *mateera* are used as a very good feed of animals.

(ii) *Mateera* as source of nutritious of preserved food stuff/value added products: The dwellers /people of the hot arid zones prepare various preserved food stuff /value added products of the fruits and seeds of *mateera* which play very important role in nourishment and sustenance of the population of the hot arid regions. Such kind of major preserved food stuff/value added products of the *mateera* prepared by the dwellers of the hot arid zones are: dehydrated nutritive rousted seeds, highly nutritive *magaj* (kernel of seeds), *mateera* oil, juice, cold drinks, sweets, *laddu* and powder of *magaj* and some other preparations in different combinations (Table 2).

(iii) *Mateera* as mean of income generation and trade specific opportunities: The immature young fruits of *mateera* locally known as *loia* are used not only as most relished green vegetable but also work as a very good source of income for the farmers/people of the hot arid zones of the country. The '*loia*' are picked up when weighing around 100-150 gm, generally 8-10 days after fertilization. The Table 3 reveals that tender fruits (*loia*) which have enormous demand and sold in local market to fetch very good price (Rs. 30-80 per kg). Hence, the farmers/people can get very good income after 40-50 days of crop sowing, particularly under aberrant weather conditions where crop failures are very common phenomenon. The *mateera* seeds are of great importance in the markets due to its protein and oil rich kernels popularly known as *magaj*. The seeds of the *mateera* offer good remunerative price in local market (5000-9000 per quintal). The green as well as dry rind portion of the fruits locally called as *khuparia* is used as livestock feed which substantially reduce the feeding cost and increase the milk production of the milch animals.

(iv) *Mateera* as a mean of traditional medicine and therapy: Traditionally, fruit juice of *mateera* is used to treat the problems arises due to sun strokes/ high temperature during the summer season. The regular use of the *mateera* fruits is supposed to cure the stone formation in body/kidney. *Mateera* fruits are used to relieve constipation and also supposed useful in cardiac and kidney troubles. The seeds of *mateera* are supposed to regulate blood sugar levels, increase energy, maintain the nervous system and promote healthy skin. The people of the hot arid regions are of the opinion that the fresh eating/consuming of *mateera* (water melon) led to increase the blood, purification of blood, improve blood flow and may helps reducing the risk of heart disease. It helps in reducing the hypertension of the body. The *mateera* juice contain electrolytes, hence it has cooling effect and save the people from harmful effect of high temperature (*Loo*) and dehydration of the body. The people believe that the use/eating of *mateera* and its products work as an anti-inflammatory, anticancerus, diuretic, laxative, antihypertensive, and antidepressant. The *mateera* is used to eradicate the urinary problems, weakness (Singh *et al.*, 2004).

4. Round Melon (*Citrullus vulgaris var. fistulous*)

Citrullus vulgaris var. fistulous, commonly known as squash melon, roundmelon, round gourd (in English) and *tinda* (Hindi) belongs to family- *Cucurbiteceae,* genus- *Citrullus,* species- *vulgaris and var.- fistulous.* It is an annual climbing or trailing herb with tendril, slender, robust, villous hairy stem. The plant is monoecious. The flowers of the round melon are usually solitary in leaf axils, comparatively small, regular, pentamerous, calyx campanulate, petals connate, yellow, and hairy. The male flowers are with three stamens and female flowers with inferior, globose and hairy ovary. The fruits of the roundmelon are globose or depressed-globose berry (hispid when immature), pale to dark green outside and creamy white to pale green inside with many seeds. Seeds are ovate-oblong, compressed, long, with ridged margin, smooth with blackish seed coat (Meena *et al.*, 2016).

Importance and Use Pattern

(i) **Round melon as source of nutritious vegetable/food stuff:** The round melon (*tinda*), especially local type, is one of the important sources of green vegetable in hot arid zones of India. It is consumed by 67-84 % of the dwellers/ people of the zones (Table 2). Round melon is considered as a pure form of vegetable/food stuff in the zones. The fresh as well as dehydrated fruits of the round melon are mainly used for preparing vegetable (pure/ mixed with other vegetables), *rayata* and chutney. It has high nutritional value and work as food seasoning item of the diet. The 100gm of edible portion of round melon has 93.5% moisture, 1.4% protein, 0.2% fat, 3.4% carbohydrates, 21 Kcal energy, 1.0% fibre, 21-25mg calcium, 14mg magnesium, 24mg phosphorus, 0.9mg iron, 35mg sodium, 24mg potassium, 0.12mg copper, 23I.U. vitamin A, 0.04mg thiamine, 44mg chlorine, 0.08mg riboflavin, 3.3mg nicotinic acid, 18mg vitamin C, 2.0mg oxalic acid, 13 µg carotene and 0.3mg niacin (Mukemre *et al.*, 2016: Khulbe *et al.* 2014: Antonious and Kochher. 2009). During the scarcity of fodder, the whole green vine, fruits, rinds of fruits, *etc*., were used as good quality fodder for the animals.

(ii) **Round melon as source of preserved food/value added products:** The people of the hot arid zones of the country prepare/produce various traditional value added products of round melon using their own experiences/traditional knowledge/methods for their own consumption and selling out the surplus. The Table 2 reveals the utilization and production pattern of traditional value added products of round melon and their mode of consumption for the nourishment and sustenance of the masses of the hot arid zones. The major value added products of round melon as prepared and used by the people of the hot arid zones in their daily diet and selling out the surplus of the same in local market/ localities are: *fofalia* (dehydrated pieces/slices), pickle, *chutney, etc*.

(iii) **Round melon as source of income and trade specific opportunities:** The farmers/clients of the hot arid zones produce the local type (green) round melon extensively under mixed cropping system during the rainy season and prepared the various value added products of the same. Thus, the production of round melon in large amount and its value added products (Table 2) create very good opportunities for generating the income and business activities for the farmers/people of the hot arid zones. The farmers produce the round melon fruits in large amount and sale them freshly in local vegetable markets/*mandies*/ localities with price ranging from Rs. 20 - 40 per kg., depending on quality of produce, season, selling location, demand and supply/production factors. The dehydrated slices (locally called *fofalia*) of snapmelon are sold by the farmers/ producers in the markets/*mandies*/ localities @ Rs. 110-155 per kg. These dehydrated slices (*fofalia*) are used to prepare vegetable (mixed or pure), fried *chutney* or as a food seasoning agent throughout the year. Some of the people of the zones prepare the pickle of snapmelon and sell it @ Rs. 80 -120 per kg and earned substantial amount of money for their livelihood. More than 50% farmers/producers produce the seeds of the available potential landraces of round melon and sale it in localities/local markets/*mandies* @ Rs. 350 - 800 per kg. Thus, the production and selling of seeds of the round melon at high rates and earning of a great amount of money are very good income generating and business activities among the farmers/producers in hot arid zones of the country which led to livelihood security of the people/farmers of the zones.

(iv) **Round melon as mean of traditional herbal medicine and therapy:** Traditionally, the fruit of the round melon are used to improve the digestion of the human being. The fruits of the round melon are used to cure the constipation and also considered useful in cardiac and kidney troubles. It helps in reducing the hypertension of the body. Like *mateera*, the juice of round melon is full of good electrolytes, hence it has cooling effect and save the people from harmful effect of high temperature and dehydration of the body. The juice of round melon is useful in reducing the urinary problems, kidney stones, weakness, etc., as reported by the local people. The people/dwellers of the hot arid zones of the country believe that the round melon and its products work as a good source of antioxidant, diuretic, anti-depressant laxative, anti-hypertensive, *etc*.

Conclusion

The hot arid zones of India have unique plant biodiversity. This biodiversity is a great source various drought hardy and nutritionally rich vegetables which are grown traditionally by the farmers/local people or grow naturally in the zones. Among these, major cucurbitaceous vegetables are *Kachri* (*Cucumis callosus* Rott. Cong.), snapmelon (*Cucumis melo var. momordica*), *mateera loia* (*Citrullus lanatus* Thumb.), round melon (*Citrullus vulgaris* var. *fistulous*). These vegetable are the life line for the dwellers of the hot arid zones of the country. They play a great role in socio-economic development, nutritional and livelihood security of the farmers and dwellers as a whole. These vegetable are unique source of potential germplasm/genotype for the vegetable crop improvement and advancement of horticulture in harsh climatic conditions like hot arid zones of the country. There is tremendous scope exploiting the potentiality of the above and such other vegetables with respect of their improvement, production, value addition and commercialization of the same at large scale for the socio-economic up-liftment of dwellers of the zones. Hence, these opportunities should be harnessed at any cost for the welfare and development of the arid and semi-arid zones of the country.

References

Adawy EI, Tarek A and Taha KM. 2001. Characteristic and composition of watermelon, pumpkin and paprika seed oil and flours, *J Agric Food Chem*, 49:1253-1259.

Antonious GF and Kochher TS. 2009. Mobility of heavy metals from soil into hot pepper fruits: A field study, *Bull Environ Contam Toxicol,* 82: 59-63.

Dahot MU, Mangrio SM, Khaskhely MH and Dewani VK. 1999. Nutrient composition of chibber fruit, *Commun Soil Sci Plant Anal,* 30: 75 – 82.

Das M, Das SK and Suthar SH.2002. Composition of seed and characteristics of oil from karingda (Citrullus lanatus (Thumb) Mansf.), *Int J Food Sci Technol,* 37: 893-896.

Guner N and Wehner TC. 2004. The genes of watermelon, *Hort Sci,* 39, 1175-1182.

Khulbe RK, Sood S, Sharma A, Agrawal PK and Bhatt JC. 2014. Value addition and nutritional fortification of finger millet [Eleusine coracana(L.) Gaertn.] using bark of Gethi (Boehimeria regulosa Wedd.) tree, *Indian J Tradit Knowle,* 13(3): 519 - 524.

Kumar R, Dia M and Wehner, TC. 2013. Implications of mating behavior in watermelon breeding, *Hort Sci,* 48, 960-964.

Meena S R, Singh RS, Sharma B D and Singh D. 2016. Economic return from the landrace vegetables and their importance in sustainable production system in hot arid region. *Economic Affairs,*61(1): 1- 9

Monika, Savitri, Kumari A, Angmo K and Bhalla TC. 2016. Traditional pickles of Himachal Pradesh, *Indian J Tradit Knowle,* 15(2) 330 - 336.

Mukemre M, Behcet L and Cakilcioglu U. 2016. Survey of wild food plants for human consuption in villages of Catak (Van- Turkey), *Indian J Tradit Knowle,* 15(2): 183-191.

Nandal U and Bhardwaj RL. 2015. Medicinal, nutritional and economic security of tribals from underutilzed frutis in Aravali region of district Sirohi, Rajasthan, *India, Indian J Tradit Knowle,* 14(3): 423 - 432.

Samadia D. 2007. *Kachri Utpadan avam Prasanskaran (in Hindi),* A technical bulletin, ICAR-CIAH Bikaner, Rajasthan, Publication, 2-3.

Samadia D. 2007. *Phoot kakadi Utpadan avam Prodhyogiki (in Hindi),* A technical bulletin, ICAR-CIAH Bikaner, Rajasthan, Publication, 4.

Singh JP, Beniwal RK and Yadava ND.2004. Ethnobotanical importance of cucurbits in India, In: Advances in Arid Horticulture: Present Status, edited by Saroj PL, Vashishtha BB & Dhandhar DG, Vol. I (International Book Distributing Co., Lucknow, 203-219.

Yadav ND and Soni ML. 2008. Integration of horticultural crops into farming system in hot arid zone of western Rajasthan, In: Hi-tech Production of Arid Horticulture, edited by More TA *et al.*, (Central Institute for Arid Horticulture, Bikaner–334 006, Rajasthan), 367–371.